THE TECHNOLOGICAL EVOLUTION OF INDUSTRIAL DISTRICTS

Economics of Science, Technology and Innovation
VOLUME 29

The titles published in this series are listed at the end of this volume.

Economics of Science, Technology and Innovation

THE TECHNOLOGICAL EVOLUTION OF INDUSTRIAL DISTRICTS

edited by

Fiorenza Belussi
Giorgio Gottardi
Enzo Rullani

KLUWER ACADEMIC PUBLISHERS
Boston / Dordrecht / New York / London

Distributors for North, Central and South America:
Kluwer Academic Publishers
101 Philip Drive
Assinippi Park
Norwell, Massachusetts 02061 USA
Telephone (781) 871-6600
Fax (781) 871-6528
E-Mail <kluwer@wkap.com>

Distributors for all other countries:
Kluwer Academic Publishers Group
Post Office Box 322
3300 AH Dordrecht, THE NETHERLANDS
Telephone 31 78 6576 000
Fax 31 78 6576 474
E-Mail <orderdept@wkap.nl>

Electronic Services <http://www.wkap.nl>

Library of Congress Cataloging-in-Publication

The Technological Evolution of Industrial Districts /
editors, Fiorenza Belussi, Giorgio Gottardi, Enzo Rullani
ISBN: 1-4020-7555-3

Printed on acid-free paper.

Printed in United Kingdom by Biddles/IBT Global

Contents

List of Contributors

Vito Albino, Professor of Management and Engineering, Bari Polytechnic, Bari, Italy.

Giacomo Becattini, Emeritus Professor of Political Economy, University of Florence, Florence, Italy.

Marco Bellandi, Associate Professor of Political Economy at the Department of Economic Science and Faculty of Economics, University of Florence, Florence, Italy.

Fiorenza Belussi, Associate Professor of Economics and Management, Department of Economics, University of Padua, Padua, Italy.

Ronny Bianchi, Research Fellow, CREI, Paris, France.

Lucio Biggiero, Associate Professor of Organisation, University of L'Aquila, L'Aquila, Italy; Institute of Business Studies, LUISS Guido Carli University, Rome, Italy.

Guido Capaldo, Associate Professor of Management and Engineering, DIEG, University of Naples Federico II, Naples, Italy.

Aldo Enrietti, Associate Professor of Industrial Economics, Department of Economics, University of Turin, Turin, Italy.

Giorgio Gottardi, Professor of Management and Engineering, University of Padua, Padua, Italy.

Roberto Grandinetti, Professor of Marketing, University of Udine, Udine, Italy; University of Padua, Padua, Italy.

Luca Iandoli, Research Fellow, DIEG, University of Naples Federico II, Naples, Italy.

Luciana Lazzeretti, Professor of Economics and Marketing, Faculty of Economics, University of Florence, Florence, Italy.

Mauro Lombardi, Associate Professor of Political Economy, Department of Economics, University of Florence, Florence, Italy.

Peter Maskell, Professor, Centre for Economic and Business Research and Department of Industrial Economics and Strategy (IVS), Frederiksberg; Copenhagen Business School (CBS), Copenhagen, Denmark.

Maurizio Mistri, Associate Professor of Political Economy, Department of Economics, University of Padua, Padua, Italy.

Ivana Paniccia, Research Fellow, Italian Agency for the Quality Control of Public Services, Rome, Italy.

Pascal Petit, Director of Research, CEPREMAP/CNRS, Paris 75013, France.

Luciano Pilotti, Professor of Economics and Management, Faculty of Science Policy, Department of Economics and Business, University of Milan, Milan, Italy.

Michel Quéré, Professor of Economics, CNRS-IDEFI, Sophia Antipolis, 06560 Valbonne, France.

Mario Raffa, Professor of Management and Engineering, DIEG, University of Naples Federico II, Naples, Italy.

Enzo Rullani, Professor of Economics and Management, University of Venice, Venice, Italy.

Alessia Sammarra, Research Fellow, Institute of Business Studies, LUISS Guido Carli University, Rome, Italy.

Gianni Schiuma, Associate Professor of Management and Engineering, DAPIT, University of Basilicata, Potenza, Italy; Cranfield School of Management, Cranfield University, Bedford, England.

Fabio Sforzi, Associate Professor of Economic Geography, Faculty of Economics, University of Turin, Turin, Italy.

Stefano Solari, Researcher in Political Economy, Faculty of Law, University of Padua, Italy.

Dimitri Storai, Research Fellow, University of Florence, Florence, Italy.

Margherita Turvani, Associate Professor of Economic Policy, DP-IUAV University of Venice, Venice, Italy.

Giuseppe Zollo, Professor of Management and Engineering, DIEG, University of Naples Federico II, Naples, Italy.

Preface

Fiorenza Belussi, Giorgio Gottardi, and Enzo Rullani

This volume collects some papers presented at the Vicenza conference "The Future of Districts", held in June 1999, organised by the Department of Technology and Management of Industrial Systems of the Faculty of Engineering of Padua University, with the collaboration of several engineers, industrial economists, and experts in the issue of technology management. This was the starting point of a long-lasting and painful collective discussion, the results of which are documented here, during many meetings of this "itinerant" group, including the workshop in Padua, organised by Professor Luciano Pilotti and held in May 2001, "Systems, governance & knowledge within firm networks" at the Department of Economics of the University of Padua, and the recent international research seminar, held in May 2002, in Rome at the Tagliacarne Institute, within the EU sponsored project "Industrial districts' re-location processes: identifying policies of EU enlargement West-East ID". The reason we decided to organise this book was not only to underline the importance of the industrial district (ID) model as a tool of propulsive local growth in a country like Italy. On the contrary, the idea that moved us was the theoretical dissatisfaction with the way in which the phenomenon of local development and industrial clustering of specific industries was treated in the international approach of the various disciplines.

To the roots of the Marshallian concept of clustering

If we disregard the determinism of neoclassical equilibrium, we become aware of the real complexity of the economic processes which generate the many possible patterns of evolution at local level. Our intent was thus to go back to the origin of such a fruitful discussion which started with Marshall, when he tried to highlight the reasons that render the agglomeration of specialised industries a territorial phenomenon worth analysing (Storper, 1997; Porter, 1998).

The term ID has been widely utilised in the economic literature for different purposes and within different theoretical frameworks. During the 1980s, Becattini re-used the Marshallian concept of ID as a new unit of analysis for his theorisation of industry-specific enterprises clustering in his theory - based on a socio-economical approach - of industrial organisation (Becattini, 1987). Subsequently, Porter (1989), starting from his interest in the combination of competitive forces which give impetus, within the economic system, to specific specialised clusters formed by inter-organisational networks (functional and/or territorial), discovered the model of the industrial district that he called territorial cluster (TC). However, the examples of territorial clusters he described in his work were often mentioned by Becattini, the Italian economist who originally studied them, as typical examples of IDs. Finally, the study of industrial districts has been developed within a "geographical" approach by those researchers who wanted to describe the empirical structure of local agglomeration of specialised systems of production (in manufacturing and services), providing an extensive classification of the variety of existing districts, within a country or a region. Clearly when we shift from a theoretical concept to its operationalisation, either with qualitative or quantitative research, using empirical data, administrative database, and real variables, we encounter a lot of difficulties in matching the theories and the survey results. So, the analysis of spatial clustering has often resulted in an overproduction of concepts, where the individual categories sometimes overlap, and are not mutually exclusive, or where typologies are poorly described. For instance, in the Anglo-Saxon tradition, the terms ID and cluster have often been used interchangeably. And also among the more "Marshallian" researchers, like for instance Garofoli, Camagni, Sforzi, Markusen, Paniccia, Guerrieri and Pietrobelli, and Viesti, we encounter a kind of rich, and sometimes confusing, categorisation of terms like industrial districts, clusters, industrial mileux, system areas, and local production systems, many of which are extensively used as synonyms (Garofoli, 1991; Camagni, 1993; Markusen, 1996; Viesti, 2000; Guerrieri and Pietrobelli, 2001).

To return, once again, to the Marshallian terminology, in the perspective developed in this volume, we have focused our analysis around the process of territorial agglomeration of specialised industry, where a specialised production system is detachable and a district model could be identified. This means that we did not take analytically into account all possible morphological manifestations of local development. In fact, local production systems (LPSs) can be more or less developed, more or less concentrated and specialised (depending clearly on the criteria adopted to define the borders of the system and the necessary threshold of the index of specialisation), without necessarily showing the feature of being an industrial district. While each ID is indeed enclosed in the LPS category, the reverse does not hold. Whatever system of parameters one decides to employ, IDs are densely populated industry-specific organisations, composed of significant territorial concentrations of similar and complementary enterprises, specialised in one or more connected industries, which form an idiosyncratic (spatially embedded) production system. In our view this is, again, quite separate from the definition of a territorial cluster, which lacks the characteristics belonging to the "district formula": district identity, embedded institutions, and social capital, as it has been developed in the post-Marshallian tradition by the Italian economists, and among them by Becattini. While the ID concept goes back to the idea of an existing "dominant" industry (or product) specialisation in the territory, and a close "community", the concept of the TC is more elusive. It requires only the existence of one or more networks of firms composed of similar and complementary units within a coherent sectoral matrix. But again, just to clarify our view, while each ID is also comprised in the more general TC category, the reverse does not hold. Perhaps it can be argued that embryonic districts are indistinguishable from TCs, and that the upper limits beyond which we can identify not just a TC but a "mature" specific ID are still not well developed in the literature.

Districts as networks of localised capabilities and learning

Our return to this old contribution of economics was realised with the aim of re-interpreting the recent studies of the evolution of districts as a part of this ongoing interdisciplinary effort that aims to interpret the processes of local agglomeration and network formation (Belussi and Arcangeli, 1998) from a cognitive and institutional perspective. Concentration has been traditionally considered beneficial for firms that agglomerate with others in similar or related industries essentially because the creation of external static economies, as has been historically endorsed by economic geographers, and as emphasised by the "new economic geographers" like Krugman (1995) because scale economies are growing when concentration occurs. In the perspective developed in the volume, the success of the IDs is explained in relation to the (internal to the district) mechanism of creation of firm dynamic capabilities. The focus of the contributions presented is indeed on the innovative capabilities of district firms and on the processes of knowledge leakages, spillovers and socialisation activated within the various communicative channels existing in the districts. As evidenced in the cases analysed, agglomeration *per se* is not the genetic protein that infuses growth capability to the district, but the interrelated cumulative circular causation between innovation, efficiency, market expansion, and innovation. So, in order to understand the interesting features of district competitiveness, all authors dwell upon the impact of internal economies developed by the firms and their networks localised in the district area. When the innovation mechanism takes off it requires, firstly, a cognitive awareness, and secondly, a high capability of external knowledge absorption, and, finally, a high degree of experimental search and knowledge recombination. Horizontal and vertical networks provide in the IDs the necessary resources for the activation of knowledge. Within firm networks new knowledge is created, imitated and recombined. The IDs can be interpreted as a cognitive system: a social-productive system in which knowledge, social experiences, mental models and collective beliefs are accumulated in a specific space through time. They can be thought of as a "cognitive lab", where knowledge and information are elaborated in a complex way, and culture and social values are generated. The local accumulation of tacit knowledge in firms provides the necessary distinctive element fuelling ID firm competitiveness. The existence of numerous overlapping networks favours knowledge sharing and information flows. Emulative behaviours underlie the wide adoption by firms of innovative strategies. Various strategies of absorption, assimilation, sharing, transformation, creation, transfer, and diffusion of knowledge are adopted by district firms. Geographical proximity favours information and knowledge exchange among the co-localised agents. In this approach industrial districts must be interpreted as complex systems of knowledge creation, absorption, and sharing, and as nets of social interactions.

We chose to adopt those analytical categories that consider variables and events in a more complex way than is typically done by standard economic analysis. The complex nature of patterns that bind present and future allows districts (and firms) to explore a wide range of opportunities that are open to individuals, social entities, and institutions, but that are based on the existing capabilities developed in the territory through time. This bears the analytical consequence that the industrial district model, which represents a specific form of agglomeration of productive activities (both in manufacturing and in services sectors), is characterised by a multiplicity of possible evolutionary patterns of growth, innovation, and learning (Belussi and Gottardi, 2000; Belussi 2000).

Districts as machines of unorganised complexity

The industrial district model and, more generally, the wide morphology of the local systems (Becattini *et al.*, 2001), represent a new category of analysis. IDs are not simply a territorial aggregation of productive units that can be studied separately or together, as a larger entity. In the moment in which the unit of analysis moves from the *micro level* (the individual firm or the individual agent) to the *meso level* (the district, the individual community located in a specific territory), there is a discontinuity in the laws governing the district, which impedes the assimilation of the first level with the second (Biggiero, 1999). So, we can observe the "social community" and the local institutions supporting the working of the system, and the synergies emerging from the levels of productive integration, specialisation, and cooperation among the local agents and social actors as specific features that from time to time characterise each district and its pattern of development.

The origin of the variety of existing organisational forms and growth dynamics of districts lies in the complexity of the various sets of elements and the sets of the connections of these elements. A first step in this direction is in interpreting the district as a *contextualised system of local interactions and external connections* (Becattini and Rullani, 1996), at the centre of which there are regulative institutions, both public and private. Again, these systems may also be viewed as nets of trusts, organised within a specific division of labour, which gives rise to a kind of social capital embedded in the territory. Districts are composed of territorialized nets and of non local relationships (Oinas, 2002).

Within these meso-systems we find new factors of efficiency that stem from the dynamics of innovation and knowledge replication, or other factors that are linked, for instance, to the reduction of transaction costs (throughout the social technology of district institutions: trust, relational contracting, or institutional confidence). They appear here as tools of *governance*, which utilise the territory as a mean of coordination among firms and agents (workers, institutions and entrepreneurs).

In reality the district is a social construction that is able to reproduce itself and the resources that have created it. Between the culture and identity existing in each district there is a relationship of strict correspondence. In the end, the inhabitants of each district, their mental capabilities, and their relational capacities, make the difference between one place and another. So the district appears, in this sense, as a learning *system* that produces localised knowledge, related to the specific culture of the territory. The cognitive capital is accumulated in circuits of learning and experimentation, and it is influenced by the local culture.

But what occurs in a district when the process of globalisation and the use of new distance communication technologies seem to reduce, and in some cases destroy, the strategic value of the specific localisation? And what remains alive of the local division of labour when large transnational firms seem to substitute local hierarchies, or globalisation threatens the disappearance of the local technical culture, and attacks the consolidated variety of local developments?

Local nets and external linkages

Technology interacts with districts not only because they may be more or less laggard in adopting new information and communication techniques, but because, once adopted, these technologies may modify the relative value of the specific resources that fill the "competitive advantage" of the district itself. In other words, they may render less important the reduction of transaction costs associated with proximity (Gilly and Torre, 1998), they push the strategies of leader firms outwards, they render perhaps less significant the social capital that bears the local division of labour, and they may reduce the specificity of the impact of localised learning (Antonelli, 2000 and 2001).

Information and communication technologies could drastically reduce the importance of physical proximity, and deeply influence the importance of scale, thus the strategic *size* of firms and the territorial embeddedness of districts. We can thus speak about a functional (size) and a spatial dimension (among peripheral and central places) of a possible pattern of re-centralisation.

Two alternative scenarios (Laubacher *et al.* 1997) can be put forward for the long term forecast, which implies opposite models and organisational solutions.

In the first hypothesis there is a dominant role of small companies and large networks. The growth of telematic nets will occur both in districts and among large multinationals and this will bring about the total disintegration of their old Fordist structure. Networks of small scale units, coordinated by the market, will substitute hierarchies. In this scenario firms would be rather small and with a flexible organisation. Economic activity would be realised by teams of temporary coalitions of autonomous entities, operating as contractors and subcontractors, all interconnected by a large-band net. To realise a new project, for instance, a large number of small independent firms, in competition one with the other, will develop various subsystems and components. They will automatically form new electronic markets, where prices and salaries will be determined. In relation to the district, what is changing is the relationship with the territory, that becomes more crossed by virtual networks, built around a plurality of localities. Centrifugal forces could be at work. A similar perspective could also bring about a partial or complete geographical disintegration of most of the currently existing districts, considering that temporal aggregations of firms could be activated on-line and ubiquitous production networks could be formed, using favourable price differentials of factors and productive inputs (Maskell, and Malmberg, 1999). Because of the difficulty of maintaining a stable cost advantage, the long term durability of each district lies in some idiosyncratic features: the innovative strategies of local firms and their cognitive capabilities.

The second scenario assumes that, within an environment characterised by global interconnections, the increasing importance of economies of scale will favour large oligopolies, and global *conglomerates*, horizontally and vertically integrated. This dominant organisational solution will contrast the strengthening of the IDs, TCs, and LPSs model. Thus giant organisations will dominate the markets, and their employees will benefit from job security, internal training, and a sense of organisational identity.

Neither of these organisational solutions is clearly in contrast with the development of districts. As regards the first hypothesis it is to be noted that often we can observe, within districts, that the exploration of novelties goes through the autonomous formulation of new ideas by individual entrepreneurs or small firms. Small entrepreneurs are able to build networks of decentralised intelligence, capable of assuming the risk of innovation, and to dedicate specific resources (obviously *network-specific*), to organise efficient and reliable productive blocks, that allow communication and the exchange of experience and knowledge. These processes are clearly organised outside the larger power of local or external oligopolies.

However, if we admit that a process of hierarchisation of firms and networks is taking place within IDs, TCs, and LPSs, we come close to the second hypothesis. Global markets work as ideal market places for the matching of demand and supply for standardised transactions, but they are poor means of coordination of complex transactions, technologies and products. When different operators of a determined *supply chain* must be put to work together for a highly variable, uncertain, and technologically complex task, the hierarchical solution seems to be more adequate. Here relationships require an extraordinary need of stability, a high level of trust, a more direct control, and a deep understanding embedded in long-term personal contacts.

So, in both alternative scenarios we can find traces of the districtualisation of the production activity, to use the words of Professor Giacomo Becattini's contribution to this book. On one hand, through the emergence of spatially based dynamic networks of production, formed often by small firms, with a continuous creation of start-ups, and the building up of local networks of innovators. On the other hand, through a process of hierarchisation of districts, as presented in many empirical cases analysed in this volume, where global market and global hierarchies have conquered an enlarged importance, both within the canonical IDs (Paniccia, 2002) and within the high-tech local districts.

The net-evolution of districts is inserted into a post-Fordist phase of transition (Gottardi, 2000; Rullani and Romano, 1998). Clearly, networks ought not to be necessarily local networks, even if until now they have been realised above all within the limited circuits of districts. On the contrary, in the near future, more and more local networks, within the global economy, will tend to enter the *international circuits of knowledge production and division of labour*.

Within the net-evolution of districts, we must foresee a reciprocal specialisation between local and global activities. Local transactions would characterise those activities better able to utilise proximity because they are strongly interconnected. They will be sustained by local institutions (private and public), which will use their regulative power to enforce the social capital embedded in the history of the local community. Local transactions will benefit from the capacity of experimentation and learning induced by the sharing of the common experience and culture. Global transactions, indeed, would characterise those relationships that can be formalised and codified, or that have low knowledge content. However, also other activities would be organised on a larger global scale, activities that can also be complex, and non formalised or codified, but that require that actors, operating in different localities, will be able to organise themselves with the construction of a common experience. These nets will evolve during time, becoming either mercantile or hierarchical nets. However, they may evolve in other directions: that is towards the formation of *virtual nets*, where we observe the presence of stable forms of coordination and communication among distant points of the global space: among systems of power that organise the various interconnections within the different contexts. The spread of distance-relationships will create many sorts of meta districts, clusters, and local systems where many places are connected by global nets. This will also create room for the development of tacit knowledge and personal contacts. In this perspective the re-personalisation of many anonymous market relations will be experimented, but at the same time this will imply a significant shift towards the codification of much of information and knowledge that is actually embedded in a tacit form in district relationships and informal transactions.

A short overview

The book is divided into three parts. The three chapters in Part I provide a background discussion of the theory of industrial district, bringing into the more traditional Italian approach some fresh air coming from the Nordic studies based on the theory of knowledge-based organisations and industrial districts, and from the institutional perspective of the French regulation school. Part II represents the core of the book where a cognitive approach to the theory of industrial district is developed, considering the cognitive models, the learning modes, the mechanism for knowledge-building, and the processes of creation of identity and trust. Finally Part III presents some empirical evidence on the new evolutionary design of traditional and high-tech industrial districts localised in Italy and in France. The individual district case-studies are introduced by a general contribution focussed on the study and classification of local systems in Italy and on the analysis of the evolution of the district model. Part III ends with a discussion of the role of academic spin-off in connecting local knowledge in less developed areas.

Part I contains three chapters which set up the conceptual framework of an institutional neo-Marshallian theory of the industrial district economics. The nature of the ID is the theme of Chapter 1, written by Giacomo Becattini, who focuses on the districtualisation of production activities. The ID may be logically decomposed in several ways, but the first and fundamental decomposition is to be the one between the productive apparatus and the human community in which it is embedded. In fact, in the industrial district the first and the second are reciprocally interpenetrated, which means that there is, between the two, a continuous, intimate correspondence. What this particular conceptualisation suggests is the idea of a relationship between a block of economic-productive relations, spatially situated in a circumscribed territory, and another block of socio-cultural relations situated in the same territory. A relationship that is on average more intense, on the long term, than the one existing between similar blocks at higher levels (for example, at national or world level). It might also be said that the study of the district typically consists of an in-depth exploration, at local level, of the many interactive connections between the productive apparatus and the community.

The existence of a plurality of elementary processes constitutes the district and the process of districtualisation which Becattini describes with the aid of reflections made studying Prato: a) the gradual subdivision of work; b) the sedimentation of material and immaterial institutions; c) the dialectic integration between contextual knowledge and codified knowledge in the flow of productive processes, which Becattini calls the cognitive spiral of the ID; d) the formation of interface institutions between the different specialisations which play the role of versatile integrators; e) the diffusion of loyalty and senses of belonging; and f) the high level of social mobility which produces the social mobilisation of the labour market and the birth of entrepreneurs coming from the working class stratum.

In Chapter 2, Pascal Petit presents the main characteristics of the new growth regime, the post-Fordist mode of competition, in relation with the regulation school approach: the state apparatus, money as an institution, international relations, the forms of competition, and wage labour relations. Institutional changes in the 1980s and 1990s have limited the advantages of labour, enlarging the room for manoeuvre for capital, such as cuts in social budgets and support for labour market deregulation. The centrality of the new post-Fordist era is now becoming the different form of competition, with the decline of Keynesian conventions, the global opening of the markets and the extension of product differentiation, based on the opportunities provided by the so called knowledge economy. To overcome this challenge, IDs have to find the right balance between the opening to outside world and the deepening of their common co-operative advantages, based on the local social capital, and weak networks of sharing tacit knowledge. In order to meet these two challenges (internal and external), not only entrepreneurs, but also workers and the public representatives of these local communities, are crucial factors. Also, the deregulation of wage labour relations will impact on IDs, but local labour was already prepared to be very flexible. However, we can expect the growth of both menial jobs and skilled jobs for competent workers. Clearly, in the case of district relocation the former will locally disappear. In the passage from the old Keynesian policies, supporting aggregate demand, to new Schumpeterian policies, supporting the modernisation of supply by promoting process and product innovation, at a first glance IDs seem to possess the very criteria for success. This is so even if, on the one hand, the capacities of learning and development, on the demand side, should be taken into account and further developed, and, on the other hand, the improvement of logistics and services remain a crucial factor for the upgrading of the district local firms, and for allowing them to remain competitive.

In Chapter 3, Peter Maskell continues the exploration of the nature of the ID model. His paper identifies some basic explanatory problems in the literature on the ID and cluster theory. He is in favour of establishing a specific theory where learning occupies a central place. The basic requirements for such a theory of the cluster are discussed. He suggests that three separate issues need to be dealt with in order to constitute a satisfactory theory: the explanation of the existence, growth, and boundaries of the ID. District firms bear the advantages within related industries of enhancing their ability to create knowledge by variation and by the deepening of the division of labour. The interdependent development between economic activities and local institutions makes the cluster attractive to some industries and hostile to others. This activates a mechanism of specialisation. Cognitive distance tends to be small within the ID firms and this accelerates the flows of knowledge. It is proposed that the reason for the existence of the cluster can be found in the enhanced knowledge creation that takes place along two dimensions. The first is horizontal. It is by watching, discussing, and comparing dissimilar solutions, often emerging from the everyday practices, that firms along the horizontal dimension of the cluster become increasingly engaged in the process of learning, and continuous improvements, on which their survival depends. The second is the vertical dimension. The vertical dimension of the cluster consists of firms being linked through input/output-relations. Specialized suppliers and critical customers become attracted to the cluster, which, once established, makes available to them particular opportunities for growth. The vertical dimension of the cluster might, however, also be developed by the inter-firm division of labour, which tends to evolve spontaneously, when economic agents are free to pursue their own advantage, as pointed out by Adam Smith more than 250 years ago.

Part II is explicitly devoted to the study of the process of generation and acquisition of knowledge within the ID model. The contribution by Enzo Rullani in Chapter 4 stresses that the idea that development, localised and organised in territorial IDs and clusters, is not a phenomenon limited to just a few sectors and places, but one of the most widespread manifestations of economic growth. In this view, the territory is a *non-replicable* and *unique* resource. The territory can be seen as a depository of culture and habits that accumulates forms of tacit knowledge and channels of communication. Local institutions remain largely *invisible* to the outside observers and even to the local actors themselves, because they are embedded in the anthropological-social system grown up in each place. In his comparative analysis of the ID model he contrasts the ID with other types of organisations (market, hierarchy, and the virtual net) arguing that the ID advantage is based on low-cost knowledge replication in the local circuit. The advantages of local networks in exploratory learning are the following. The market is efficient in responding to a simple, well specified demand. But it is unlikely that it will venture "into the dark" to explore less probable, more difficult pathways. If there are a hundred possible alternatives, the market will tend to concentrate on those which are most obvious and attractive, neglecting the less visible. Large companies (hierarchies) are also sometimes inefficient in responding to problems of this kind. Having to choose between a hundred possible alternatives, a large company needs to invest time and money in the research process which, after a series of simulations and selections, will reduce the alternatives to just one that will be tested and put into production. The ID model, on the contrary, represents a good example of how *parallel exploratory searches* may be developed when enterprises are unable to "protect and hold" their knowledge. Within IDs, variety is generated by the exploration of many possible methods. If there are hundreds of possible alternatives, many will be explored (even though some will receive more attention than others). The research process will speed up, and, in the end, it will be clear to everyone which solutions may be best appreciated by the market. Firms will immediately bet on some "winning" solutions that subsequently will be tested by the individual consumers on the market place. This collective exploratory search implies also that the research costs will be divided up among many enterprises. As Enzo Rullani argues, IDs can be very effective in this process of exploratory learning, where the propagation mechanisms allow a rapid, and generally involuntary, dissemination of knowledge among competing enterprises. The involuntary propagation of knowledge is due to the fact that individual district enterprises are unable to hold the knowledge they possess: each producer is able, if necessary, to use, without costs and rapidly, the knowledge developed by others. These are thus significant advantages deriving from *involuntary knowledge sharing*, which is often confused with voluntary *co-operation.*

Chapter 5, by Giorgio Gottardi, starts with the observation that there has been in the last decade a technological revolution in information and communication technologies (ICT), centred on Internet applications. However there is still a great controversy about the extent to which ICT are transforming the competitiveness of individual firms, districts, and large economic regions. He argues that, on one hand, the use of ICT could undermine those economic systems that are very distant from the strategic motors where these developments are taking place, but, on the other hand, the "virtualisation" of the spatial economic relations could offer a better access to agents placed in peripheral areas. In this perspective, the assumptions of the "vanishing" of physical distance could represent a fascinating "utopia". Giorgio Gottardi analyses how IDs may manage the absorption of ICT technologies. Are IDs networks without technologies? While other researchers in the Veneto region, and particularly Micelli and De Maria, have come up with the view of historical backwardness of districts as regards the introduction of ICT technologies, Giorgio Gottardi, considering the social and economic homogeneity of IDs, stresses the thesis that they could become the privileged locus of ICT adoption.

Chapter 6, by Mauro Lombardi, has the ambitious purpose of setting the cognitive properties of the ID model. They are articulated along many features which explain the complexity of the systemic configuration and information processing of the IDs: a) the information hierarchy, which is controlled by strategic operators, where systemic information is in the hands of leader agents, and minor information is distributed more evenly; b) adaptive modularity, repeated feedback, and iterative cycles of information exchanges, which produce the emergence of technical and productive parameters allowing firm coordination; c) the passage from information to knowledge, which takes place by means of the construction of structured cognitive patterns and inference; and d) a process of reciprocal conditioning among the district agents and institutions, which influences the final configuration of the stock of knowledge of the system.

Fiorenza Belussi and Luciano Pilotti, in Chapter 7, draw their attention to the mechanism of knowledge creation within the ID model. Despite the increasing attention given to the tendency towards a continuous process of knowledge codification, they interpret the success of the Italian ID model exactly on the basis of the competitive advantage possessed by the IDs in dealing with their accumulated stock of tacit and contextual knowledge. The ID model may be interpreted as a cognitive system where a wide population of firms, entrepreneurs, and skilled workers, share competencies, expertise, social experiences, mental models, and collective beliefs. They relate the phenomenon of local agglomeration and spatial clustering to the tacit nature of knowledge (dispersed, limited, unorganised, subjective, and based on practical know-how). Another relevant point is that the authors make a distinction between three different forms of learning and knowledge creation occurring in IDs. Weak learning systems are those based mainly on the horizontal expansion of a given stock of knowledge, historically accumulated in particular localities, where tacit knowledge among agents is prevailing. Here knowledge is mainly embedded in social practices. Stable learning systems are those characterised by a significant absorptive capability from the outside circuits of knowledge matched with incremental innovations. Dynamic evolutionary systems are those IDs organised around very innovative firms. Clearly not all firms located in the area share the same innovative capability but only some strategic agents (dynamic final firms, research centres located in the district, and some specialised suppliers). Here many firms are radical innovators.

In Chapter 8, Margherita Turvani connects the analysis of the ID model to the Williamsonian approach and to the institutional view of the 'nature of the firm', proposed more than sixty years ago by Coase. Traditionally, the nature of the firm has been explained by looking at the balance between production costs and transaction costs. Does this explanation fit the 'nature' of the "industrial district"? A great number of studies at the theoretical and applied level in the economic and sociological fields of inquiry show that many different dimensions must be taken into account to explain such a specific form of organisation of firms and territories. Another relevant dimension to understand the 'nature of the district' should be included: the cognitive dimension, at the individual and at the collective level. For Margerita Turvani, its integration does not only allow a better comprehension of the working of firms but offers a key to read many distinctive features of the existence of the district, such as the externalisation of knowledge among actors belonging to a community of practice and the possibility to exchange knowledge developed in a given context. The localised, interactive, and contextual nature of knowing gives rise in IDs to a variation of knowledge that is continually renewed, but that is never as such purely self-propagated on a larger scale, since its form and contents are transformed along its growth and local diffusion pattern.

In Chapter 9, Marco Bellandi illustrates the properties of the vital industrial districts, which are characterised by processes of innovation not strictly depending on strategies of large investment in R&D. This decentralised industrial creativity is one of the building blocks of a virtuous circle of relations promoting local development. This circle has its limits and threats, against which mechanisms and agents of possible, but uncertain, adjustment are to be identified. The combination of these conditions enters into the definition of various paths of local change. He supports the view that the ID model is a fabric of multiple complementary competencies, that can be activated by different teams, and in different projects, for the development of new approaches to trade and production. From the crowd of small variations, some more important innovations can emerge periodically, especially when working experience is supported by technical education and by the access to appropriate knowledge. The most important innovations are the focal point for the aggregation of new areas of specialisation in the ID, increasing the scope of its product differentiation, the discovering of new markets, and the opening of new occasions for creative interactions. Typical examples are the development of new service activities, like customised software-house firms, which offer their services to the principal ID industry enterprises. They are able to answer the various needs of local district firms (construction of data bases, specific processing of information, creation of new pieces of software), producing at local level an interesting pattern of industry diversification. Another typical process, occurring at local level, is the development of a local specialised machinery industry, producing the new machinery for the ID specialisation sector, and continuously adapting the machinery and the mechanical instruments obtained from international suppliers of technology.

The study presented by Lucio Biggiero and Alessia Sammarra in Chapter 10 contributes to an original view of ID studies, enriching the possible theoretical tools that can be applied by researchers interested in local development issues. The key concepts proposed in this work are those of district identity, and the identification process, which both derive from social identity theory, one of the main streams in social psychology. The motivation for such an enlargement of an already rich field of research is that social psychology provides the theoretical tools to bridge between the micro (individuals), the meso (firms), and the meso-macro (district) level of analysis. District identity represents a cognitive construction of what the firms perceive as the set of central, distinctive, and enduring characteristics of the district. *Central attributes* refer to common features among district members in terms of formal and structural aspects, such as the firms' size and ownership structure. Relational features are the overlapping of social networks and communicative patterns. Informal and symbolic features are the institutional and historical background, and the social habits of the ID. *Distinctive attributes* refer to those features that emphasise similarities among district members and differences in relation to outsiders. For instance, the geographical localisation may become a salient component of district identity, making it possible to distinguish district members from organisations which are similar but that located in a different territory. Finally, *enduring attributes* refer to those features that characterise the district with some degrees of temporal stability and continuity. The emergence of a district identity presupposes a cognitive effect, the recognition that the district exists as a by-product of the actors' attempt to scan and make sense of their competitive and social environment. For the authors, two main theoretical implications are involved in the concept of district identity. The first concerns the consideration of the firms as cognitive actors; the second refers to the eventual qualification of the firms' cognitive abilities in terms of categorisation processes. A further clarification is necessary to refine the concept of district identity. District identity can be analysed at firm and collective level. At firm level, the term *perceived* district identity is used in order to underline its subjective dimension. The perceived district identity represents a sort of cognitive map that each firm in the district forms to make sense of its relevant environment. Factors facilitating the firm identification, within the ID, are the attractiveness of the district, the similarity and the interdependence among ID members. The consequences of the identification with the ID are explored in terms of trustfulness, cooperative attitudes, situated learning and attachment to the district.

In Chapter 11 we return to the solid heredity provided by the Florentine studies, which have preserved the Marshallian tradition up to the contemporary debate. More than in economics, as argued by Fabio Sforzi, the study of the ID has become a dominant issue among Italian economic geographers who, through the theoretical concepts of the ID, have redefined a large part of the scientific and methodological status of the discipline. Following Marshall's writing, the ID is defined as a local system characterized by the active co-presence of a human community and a dominant industry constituted by a set of small independent firms specialized in different phases of the same production process. An ID, therefore, enables a geographical concentration of numerous small specialized firms to organize production efficiently, in a similar manner to what happens inside a single large factory. This is made possible by the flows of external economies generated locally, among firms, and arising from the knowledge, values, typical behaviours, and institutions through which the local society acts. As stressed by the author, this is in sharp contrast to the theoretical construct of the 'growth pole' view of Perroux, where external investment through large firms becomes the motor of local development in less developed areas. Over the last twenty years, a 'new' economic geography has arisen in Italy, under the impetus of the notion of the 'ID', which opposes the *space* to the *territory,* the latter being the versatile integrator of firms, industries, and local community. The rise of the ID, as a theoretical paradigm, has compelled Italian economic geographers to reflect on the soundness of the interpretation of its origin – established in spatial economics – and to re-assess its geographical identity as the 'science of places'. This has been accompanied by the reorientation of empirical research, which has led in Italy to a new definition of a *territorial classification grid* (local systems), as an instrument of analysis which matches the theoretical construct of the ID. Sforzi concludes asserting that one may therefore say that the ID, framed by a Marshallian reading, has engendered a knowledge circuit where theory and empirical research blended and evolved together, producing that change of perspective, frequently urged, but never effectively accomplished, whereby the territory and the local system become a new unit of analysis.

The last theoretical chapter (Chapter 12), written by Maurizio Mistri and Stefano Solari, develops a sound theoretical understanding, building an institutional view of the ID model. This chapter combines some important streams of the Austrian school of thought (and particularly the Hayekian contributions) with the more conventional literature on the ID model. In economics, the best way to cope with the concepts of values, norms, culture, or any factor which structures the economic environment, is to understand them as a system of rules (institutions). Norms and institutions do not constitute an additive set since they show strong complementary effects resulting in systems of rules. A system of norms produces a structuration effect of the economic space, and emerges as interrelated regularities, in the repeated interaction of individuals. A system of norms defines the actor's available behaviours, in given situations: the domain of viable actions which are implicitly accepted in a community. A number of interrelated self-enforcing rules limit human action by inducing specific behaviour in well-defined situations. Following Aoki, the authors claim that the relationship between a system of norms, and the organizational arrangement of an economic system, can be theorized in the perspective of multiple equilibria. In their view, organizational forms match best some – formal and informal – institutions, and give birth to some self-reinforcing co-ordination schemes. In their work, the types of rules (culture, norms, reputation, loyalty, trust and confidence), the types of enforcement mechanisms, and the types of institutions they referred to, are defined and studied.

All chapters in Part III are based on the analysis of data sets or empirical cases of evolving high-tech IDs. The IDs selected offer a wide variety of the dynamic forces impacting on local development, and transforming the territorial production systems. In the case of the canonical ID, such as the textile ID of Prato, a process of complexification of the industrial structure is highlighted. Other less canonical IDs - the case of the embryonic ID specialised in furniture products of Matera–Altamura-Santeramo, the mature biomedical district of Mirandola (in the province of Modena), the packaging machinery district of Bologna - represent a new generation of Marshallian IDs, where large endogenous firms are the motor of the ID itself. In another canonical ID, the Manzano district, around Udine, the most important chair manufacturing district in Europe, we can observe an important feature: the district development is more and more interwoven with the emergence of collective actions of local actors. We have also presented two other interesting cases of IDs which prove the process of districtualisation within different local systems. The first is the car industry district of Turin, in the past exclusively dominated by a large oligopoly (the Fiat firm), and the second is a science park, a technopole in which a new swarm of start ups have created a district "atmosphere" (the Sophia-Antinopolis area). These are local systems, that, with time, have shown a mode of development that can be assimilated to the ID model.

Ivana Paniccia opens the last section, with Chapter 13. She attempts to offer, on a comparative basis, a sound measurement of the ID identification and performance for Italy. We still lack rigorous empirical research works. This chapter represents one of the first attempts to "operationalise", with empirical research, a methodology for testing ID performance, through the application of multivariate analysis. So Paniccia's work is particularly valuable. The analysis is based on a first attempt to measure the performance, as well as the socio-economic structure, of 39 Italian districts (local/subprovincial areas) in the period 1961-1996. These districts, as discussed by the author, can be defined as craft based IDs, canonical IDs, concentrated IDs, and embryonic IDs. Performance is related to the structural factors highlighted in the literature. Concentrated IDs exhibit the best performance on average, in that they score low on unemployment, and high on global productivity, and per capita income (mainly in 1991). They also achieve good export performance. As regards individual indicators, the results are again varied. Canonical IDs outperform the others with respect to the infrastructure index. Embryonic IDs are the most dynamic, and overall they exhibit a variation in export which is much higher than the national average. These results point to the fact that good performance may be achieved even if the area does not comply with all the requisites of an ideal-typical ID. The relationships between external economies and economies of agglomeration are also tentatively described. The results show how a large variety of institutional arrangements in Italian IDs is combined with a positive social or economic performance. Within the ID model, new organisational forms, such as the emergence of large firms, networks, or constellations of firms, may also generate external economies for the local environment. On the contrary, economies of agglomeration seem to have exhausted their thrust on growth in the more mature and world well-known IDs included in the sample.

Chapter 14, by Ronny Bianchi and Aldo Enrietti, is focused on an emerging ID that marks the transformation from a "hub and spoke" area, dominated by a large oligopoly (Fiat), to a more diversified district. The authors examine how changes in the production methods, and the emergence of new post-Fordist organisational forms, have affected the suppliers of Fiat and, more generally, all mechanical firms working as component suppliers and sub-contractors in the automotive district. They have studied the implications of the reorganisation of Fiat for small local firms, and especially for its sub-contractors: the ones which continued to supply Fiat and the ones which were excluded. They also offer some evidence to speculate whether firms excluded from the Fiat supply-chain have been able to survive without Fiat. Another purpose was to verify whether the region possessed an innovative/competitive capacity, independently from Fiat and, if this did exist, how it had developed, and what its decisive elements were. Furthermore, considering these factors, the authors show that, in Piedmont, in the post-war period, an industrial area had evolved into an automotive technological district.

Chapter 15, by Fiorenza Belussi, presents the results of empirical research focused on the Bologna ID, a district specialised in automatic packaging machinery. Within this technological district, a clustering of innovations has occurred; in turn, this has led to a self-reinforcing mechanism of industrial growth. The generation and diffusion of knowledge have been promoted throughout specific nets, crossing local large and small firms, suppliers, subcontractors, and local institutions. The empirical study presented aims to show the close relationship existing between the dynamics of localised technological change and the role of larger firms. The Bologna district represents an urban ID. The Bologna district is made up of a few leader firms and about 150 smaller niche producers. In addition, numerous intermediate firms that are not classifiable as packaging producers are involved in sub-contracting. Many (more or less stable) large production networks form this district. It is quite different from the typical morphological structure of the Italian industrial "districts", which are characterised by the presence of small-size firms, competing in the same global market. Here district firms work in very segmented markets (niche markets), and the typical client of the Bologna packaging machinery firm is a large multinational firm, serving mass production markets. The existing production networks are "project-based temporarily organisations", co-ordinated by final firms, on the basis of orders, which may be very complex and may correspond not just to the purchase of a single machine but to an entire packaging line (which includes the operations of preparation, making up, and packaging). While the local production chains are continually re-adjusted and changed, the district is characterised by stable, long-term, and co-operative relationships among the firms. In the district, the final firms are the principal agents that receive orders and know the market outlets. They possess not only all the technical skills of engineering and building the machinery, but also the relational skills for governing the production network (who in the district is able to perform a specific task, how, at what cost, and when). The packaging ID, in the Bologna area, is formed by about 1,000 mechanical firms, altogether employing nearly 20,000 workers.

Chapter 16, by Lucio Biggero and Alessia Samarra, is dedicated to the study of the Biomedical Valley, which is located in a small group of municipalities centred around the city of Mirandola, in the province of Modena. This ID is the most important European centre for the biomedical industry, representing an example of an innovative and high-tech Italian industrial district. The Biomedical Valley contradicts the idea that Italian IDs are only specialised in traditional sectors. It is a successful example of a recently formed high-tech and high-wages ID. In this district, medium sized companies mainly operate, some of which have been recently acquired by large European and American multinationals. The district emerged in 1963, triggered by a true Schumpeterian entrepreneur (Veronesi), who created the "first parent firm" of the district, producing the first Italian artificial kidney for dialysis. Starting from the first nucleus of founding firms, many others new firms were generated through spin-off processes. The authors analyse how the actual members of the ID perceive and evaluate the district they belong to, and verify to what extent they feel they are part of the district. In the Biomedical Valley, firms that declared a stronger identification with the district were also more likely to have a positive evaluation of the district (perceived attractiveness of the district), based on the awareness of the existence of unique tangible and intangible resources, such as specialized skills, diffused know-how, unique competencies or high status firms. The firms' evaluation of the district was also found to be positively related with the perception of a positive external image. This means that, in the Biomedical Valley, the firms that declared that they identified strongly with the district were more likely to express a positive evaluation of the district. Interestingly, the entry of multinational firms was found to be one of the facilitating factors in the process of identity formation.

The case of Sophia-Antinopolis is presented by Michel Quéré in Chapter 17. The Sophia-Antipolis park started from scratch in the 1970s, and, progressively, appeared as one of the most suited locations in Europe, where "high-tech" activities have been concentrated, mainly in the field of information and communication (ICTs), and life science technologies. The number of local firms and employment has reached respectively 1,193 and 21,535 units in the year 2000, and the number of students and researchers is around 5,000, when the initial number was zero. This ID was launched in 1972 by P. Laffite, who declared the area "the city of science, culture, and wisdom". In 1977 the project collapsed financially, and the experiment was transferred to local public authorities. In the development of the Sophia-Antipolis experiment, the author has identified the existence of a two-stage process. The first has been dominated by a very wide accumulation of external resources, be they private companies or public institutions. This first stage results in a very fragile project, because of the extremely high dependence on the external (international) environment. The internal system of Sophia-Antipolis was dominated by external relationships. During time this "satellite" ID was transformed into a proper technological ID. Consequently, it developed some localized knowledge, and the second stage was characterised by a combination of resources and capabilities accumulated locally. This transition towards a more endogenous type of economic development appeared in the 1990s, and this experiment is now facing a significant change. With the behaviour of large companies, or the recent exploding growth of local start-ups and spin-offs, the structural characteristics of Sophia-Antipolis are changing. Finally, a very interesting point lies in the observation that this transition results largely from a spontaneous order, in the sense that the ability of public policy-making to sustain this transition of the Sophia-Antipolis experiment has been extremely weak. Local policy-makers were much more concerned with the international attractiveness of the site, and with providing it with the usual physical infrastructures, than with tools and means dedicated to support the existence of a dynamic endogenous district.

In Chapter 18, Luciana Lazaretti and Dimitri Storai continue the never-ending exploration of the Prato district. However their approach is quite heterodox, post-Marshallian, and "organisational", founded on the study of organisational ecology. Having reconstructed the evolution of the industrial district through the analysis of the vital flows of the different populations of existing firms, they argue that the main feature of the evolution of the Prato district appears marked by a growing complexification characterized by an increase of the organizational variety, leading from two fundamental processes. The first is the deepening of the degree of labour division among the firms, which leads to the birth of new types of economic activities, which were previously carried on autonomously inside the different firms. The second is a process of physiological diversification that determines the appearance of firms specialized in productions not typical of the district. According to the analytical methodology used, it is possible to point out the existence of a progressive complexification in the Prato district, starting from the second half of the 1970s, in terms of the increase in the degree of organizational variety. This is represented, on the one hand, by the birth of new populations of firms, and, on the other hand, by the strong increase of the density of those already existing.

Furthermore, none of the emerging results allow the authors to foresee a reversal trend, leading towards a reduction of the variety of ID organizations (and therefore a reduction of the number in the firm population). One of the most interesting results of their research is that, in the course of the fifty years examined, none of the types of organizational populations became extinct. The district's competitive environment, therefore, turns out to be highly "self-conservational". In a historical period such as today's, characterized by strong processes of industrial restructuring, which often penalizes the small and medium size, the district formula seems to hold the roll of "socio-economic shock absorber", attenuating the process of selection of the species. This is the exact opposite of what happens in the other economic contexts.

Vito Albino and Gianni Schiuma, in Chapter 19, discuss the importance of the codification of knowledge, in one of the most dynamic districts in Italy (and in the "Mezzogiorno"), that is localised around the area of Matera-Altamura-Sateramo. The district is specialised in the production of leather sofas, and comprises about 700 firms and 16,000 employees (data refer to 2002). The most important aspect of the development of the ID, underlined by the authors, is the growth, within the district, of large firms which play a coordinating function. The ID of Matera-Altamura-Santeramo is characterised by the growth of some leader firms which, at present, occupy a leadership position within the local area and the international markets. Leader firms tend to modify the ID organisation towards a more hierarchical structure. This can be the result of two different growth processes. The first is the "internal" growth of the large firm, carried out by direct investments and/or by merging or acquisition processes. The second is related to the "external" growth of the leader firm that, focusing on its core competencies, increases the number of partnership, links, alliances, and agreements with local and external firms, owning strategic complementary competencies. In recent years a gradual restructuring process of firm networks has occurred. In particular, to improve their performance, local leader firms have restructured their inter-firm relationships. Recently, the most important firm localised in the district, Natuzzi Spa, has internalised all production phases, also acquiring many small district firms, which in the past worked for Natuzzi as subcontractors. Natuzzi has also increased its size, creating an industrial group which at the present is the world leader in leather sofa production. The implementation of many codification projects has allowed the leader firms of the ID to support the reengineering and industrialisation of the product cycle. For example, the codification of the know-how at the basis of leather cutting has allowed the development of new technology for the automatic cutting of leather. While, in the past, an expert was not able to cut leather for more than four sofas daily, today he is able to cut leather for about forty sofas. Also, the codification of the organisational knowledge has pushed the district firms towards the adoption of technologies developed outside the district area.

Chapter 20, by Roberto Grandinetti, is based on the elaboration and interpretation of many research works regarding the evolution of the chair district of Manzano, near Udine. This district comprises about 1,000 firms and 10,000 employees. Over time the chair district of Manzano has taken on all the distinguishing features of a typical Italian industrial district: a large number of production networks, an ample division of labour among small and medium-sized firms that control various stages of the value chain, a strong link between production and the local community, the creation of an ample stratum of skilled labour, a localised process of creation of technical knowledge and competence, and high export flows. Grandinetti also furnishes us with a clear anatomy of the Manzano ID, which is composed of four main types of firms: 1. small craft firms producing semi-finished goods, or carrying out specialist processing, working for local chair manufacturers on which they are almost totally dependent; 2. slightly larger firms of craftsmen, that purchase semi-finished goods, complete the work (assembly and finishing), and sell the finished product; 3. small industrial firms (between 20 and 100 employees) that are the mainstay of the district, and complete almost the entire production process alone, except for specific processing that is contracted out to craftsmen and home-workers; 4. a few quasi-verticalised medium-sized firms that, though capable of carrying out all the stages of the production cycle, often tend to decentralise some processing. One of the most important features is the great variety of entrepreneurial formulae that exist in district subcontracting. Alongside numerous firms working in just one specific stage of the production chain, there is an even greater number of firms that carry out a wide range of different activities, but they are re-combined and integrated in various ways, which are not determined only by technological requirements. Grandinetti traces the many evolutionary paths of local firms: a gradual diversification of customers and market outlets; the shift in core competences from production to planning and supply chain management, and business to business marketing; a transition from traditional methods related to process technology, based on tacit knowledge and informality, towards a model based on a higher level of codified knowledge, certified relations and processes (this in relation with the introduction of ISO9000 certification of quality); and the emergence of leader (and global suppliers), and niche firms. It is interesting to note, as widely discussed by the author, how many of the these changes have been actively promoted, supported, and sponsored by local institutions.

The twenty-first and final chapter, by Guido Capaldo, Luca Iandoli, Mario Raffa and Giuseppe Zollo, is a study of the role of academic spin-off in less developed areas, in which we typically observe weak links between the places where technical and scientific knowledge is produced and places where this knowledge may be utilised (local SMEs, and University start ups). The research of the authors aims to understand how some public actors, such as universities and research centres, can become potential sources of technological resources, through the incentive of academic-spin-offs. They elaborate a manual of good prescriptions for the management of a service centre that operates as an incubator of entrepreneurial competences, analyses barriers (cognitive, institutional and behavioural), and facilitates factors (team working, collegial support, internal and external networking, commitment to technology transfer, research orientation toward technology transfer, good experience and skills, sense of achievement). The described good practices represent the research results of a case study relative to two departments (physical science and material engineering) of the Italian universities, localised in a less developed part of Italy, where the start up ventures are studied. They conclude arguing that the best option for the university is to implement a service firm that must be partially owned by the university itself, but that must be run by an external entrepreneur (or a "surrogate entrepreneur"). Two other critical figures are indicated: the technological originator (an agent with an academic background) and the market champion (an agent involved in current business), who, of course, could be either individuals or groups.

The editors sincerely thank Ms. Alexandra Speirs for her long and invaluable assistance in the preparation of this volume.

References

Antonelli C. (2000), Collective knowledge communication and innovation: the evidence of technological districts, *Regional Studies,* vol. 34, p. 31-40.

Antonelli C. (2001), *The Microeconomics of Technological Systems*, Oxford Press, Oxford.

Becattini G. (1987), L'unità di Indagine, in Becattini G. (ed.), *Mercato e forze locali: il distretto industriale*, Il Mulino, Bologna.

Becattini G. *et al.* (2001), (eds.), *Il caleidoscopio dello sviluppo locale*, Rosenberg& Sellier, Turin.

Becattini G. and Rullani E. (1996), Local systems and global connections: the role of knowledge, in Cossentino F., Pyke F. and Sengenberger W. (eds.), *Local and Regional Response to Global Pressure: the Case of Italy and Its Industrial Districts,* Ilo, Geneva.

Belussi F. (2000), Accumulation of tacit knowledge and division of cognitive labour in the industrial district/local production system, *Working Paper On Economics and Evolution, Max Planck Institute*, #0012.

Belussi F. and Arcangeli F. (1998), "A typology of networks: flexible and evolutionary firms", *Research Policy*, no. 27, p. 415-428.

Belussi F. and Gottardi G. (2000), "Model of localised technological change", in *Evolutionary patterns of Local Industrial Systems*, Belussi F. and Gottardi G. (eds.), Ashgate, Aldershot.

Camagni R. (1993), 'Il concetto di "milieux innovateur" e la sua rilevanza per le politiche pubbliche di sviluppo regionale, mimeo.

Biggiero L. (1999) 'Markets, hierarchies, networks, districts: A cybernetic approach', *Human Systems Management*, vol. 18, p.71-86.

Garofoli G. (1991), *Modelli locali di Sviluppo*, Franco Angeli, Milan.

Gilly J. and Torre A. (1998), "Dinamica di prossimità e reti, numero monografico", *L'industria*, year XIX, no. 3.

Gottardi G. 2000 (guest ed.), "Technology Management in a Post-fordist Environment", *Special Issue of International Journal of Technology Management*, vol. 20, no. 1-2, p. 1-19.

Guerrieri P. and Pietrobelli, C (2001) "Models of Industrial Clusters' Evolution and Changes in Technological Regimes, in Guerrieri P., Iammarino S. and Pietrobelli C. (eds.) *The Global Challenge to Industrial Districts: Small and Medium-sized Enterprises in Italy and Taiwan*, Edward Elgar, Cheltenham.

Krugman P. (1995), *Development, Geography, and Economic Theory*, Mit Press, Cambridge, Mass.

Laubacher R.J., Malone T.W., MIT Scenario Work Group, (1997), "Two Scenarios for 21st Century Organisations: Shifting Network of Small Firms or All-Encompassing 'Virtual Countries'?", *Working Paper, Sloan School of Management; Massachusetts Institute of Technology*, MIT Initiative on Inventing the Organisation of the 21st Century, January 21CWP #001.

Markusen A. (1996) "Sticky Places in Slippery Space: A Typology on Industrial Districts", *Economic Geography*, vol. 72, p. 293-313.

Maskell P. and Malmberg A. (1999), 'The competitiveness of firms and regions: 'ubiquitification' and the importance of localised learning, *European Urban and Regional Studies*, vol. 6, p. 9-25.

Oinas P. (2002), Competition and collaboration in interconnected places: towards a research agenda, *Geografiska Annaler*, Special Issue "Local development: Issues of competition, collaboration and territoriality", vol. 84, no.2, p. 65-76.

Paniccia I. (2002), *Industrial Districts*, Edward Elgar, Cheltenham.

Porter M. (1998), *On competition*, Harvard Business School Press, Boston.

Rullani E. and Romano L. (1998), (eds.), *Il post-fordismo. Idee per il capitalismo prossimo venturo*, Etaslibri, Milan.

Storper M. (1997), *The Regional World*, Guilford Press, New York.

Viesti G. (2000), *Come nascono i distretti industriali*, Editori Laterza, Bari.

Part I

Towards a conceptualisation of the industrial district model

Chapter 1

From the industrial district to the districtualisation of production activity: some considerations

Giacomo Becattini

1.1 Towards a physiology of the industrialisation process

The industrial district may be logically decomposed in several ways, but the first and fundamental decomposition is to be the one between the productive apparatus and the human community in which it is, so to speak, "embedded". It is usually said, in fact, that in the industrial district the first and the second are reciprocally interpenetrated, which means that there is, between the two, a continuous, intimate correspondence as they fulfil their roles. What this particular conceptualisation suggests is the idea of a relationship between a block of economic-productive relations, spatially situated in a circumscribed territory, and another block of socio-cultural relations situated in the same territory. A relationship that is on average more intense, on the long term, than the one existing between similar blocks at higher levels (for example, at national or world level). It might also be said that the study of the district typically consists of an in-depth exploration, at local level, of the many interactive connections of that pair: productive apparatus and community.

It must be noted that each attempt to limit the analysis either to the economic-productive side or to the socio-cultural one, to work on more common disciplinary ground, however legitimate the analysis may be, makes us run the risk of taking the part for the whole; causing, so to speak, the dissolution, before our eyes, of the very object of the analysis.

Well, that type of "all-round" exploration of an industrial district, which I have carried out with reference to the Prato district after the second world war, has shed light on the existence of a plurality of elementary processes that constitute the district: a) gradual subdivision of the central productive process *(gradual, specific and self-contained division of work)*; b) sedimentation in formal and informal, material and immaterial institutions of the practices "rewarded" by the external market and emptying of the institutions corresponding to the practices "punished" by the market (*formation of local phase and service markets, consolidation of specialised communication networks, birth of co-operative networks (for buying, selling, etc.)*; c) dialectic integration between contextual knowledge and codified knowledge in the flow of productive processes typical of the district (*"cognitive spiral")*; d) formation of interface figures and institutions between the different specialisations *(versatile integrators*); e) interiorisation, by the individual agents in the district, of some consequences of the district practice *(remixing and consolidation of loyalty and senses of belonging)*; *and* f) high level of social mobility *(two-faced Janus: labour market or entrepreneurship incubator?).*

Of course I would never dream of stating that the list comprises all the elementary processes of the districtualisation phenomenon, nor even of the wider process of industrialisation, to which the first belongs; nor would I exclude the possible existence, in that list, of a few cases of double reckoning, but I believe that this decomposition of the districtual process into lots of elementary processes is, in the meantime, useful for tackling the problem from a different - and, I believe, more pregnant - standpoint than usual. Let us now examine these processes one by one, albeit briefly, also, but not only, with the aid of reflections made when studying Prato (Becattini, 1979).

1.2 The gradual subdivision of labour

The *deus agitans* of the district organisation form consists of the incorporated tendency to a growing subdivision of the "central" production process of the district (or of the "central" production processes in the case of a multi-sector district), and then of all those belonging to its filière or connected with it in various ways. It is the Smith-Marshall-Young principle of the division of labour as a wheel which, emanating from the inside of man, moves the world. With a particular qualification, however, which often passes unnoticed, and that is that, instead of referring to a general market, or referring problematically to the known levels, the macroeconomic level (the nation-state "imported" from political history) or the microeconomic level (the firm, emblematically represented by Smith's pin factory), the districtual subdivision of labour refers to a mesoeconomic unit - the industrial district - which (though often without any juridical existence) is continuously defined and redefined in its "frontier" by the comprehensive movement of socio-economic relationships. It therefore varies over time in all its dimensions, including the spatial ambit.

From the moment in which the district spreads beyond the territorial neighbourhood a new phase begins.

The continuous, growing, "local" articulation of the "central" productive process of the district and of its filière is indispensable for two sets of reasons. First of all, because it is the indispensable vehicle for the increase of productivity and innovation in the district. Secondly, because it is necessary in order to allow the interpenetration of the productive process and of daily life of the community to which it refers to be realised with an adequate "pervasiveness".

The link between the productivity of labour and its specialisation is so well established that no comment is required, but the one between innovativeness and the division of labour perhaps deserves a few lines. As the district operates on markets with a differentiated, variable demand, one of its requirements is the ability to reallocate the material and human capacity, but especially the human one, from one type of production to another.

It is thanks to this capacity that the "odd bits" of time, knowledge and working energy produced by the movement itself find, in the "operating" local production process, the "space for competition" to infiltrate. Without it, those odd bits at the disposal of the district agents would be no more than a waste of time and of productive capacity (know-how and energy), moreover, being the expression of disappointed hopes, there would be a possibility of their being coagulated in socially upsetting initiatives. Through a continuous phenomenon of internal articulation (generating phases within phases of a certain process, inducing instrumental, collateral or complementary processes, etc.), the productive process of an expanding district offers, on the other hand, the possibility of monetising the odd bits, know-how and energy made available by the "permanent socio-cultural micro-revolution" - or those that flow into the district from other areas - and transforming them into goods or services.

One of the moments of this process about which the historian can most usefully have his say is "the birth", so to speak, of the industrial district. There is an ideal point beyond which the proliferation of phase firms reaches such a level as to produce on many intermediate transformation lines - the main ones - a situation of competition which may be said to give rise to short, medium and long term local normal prices. The point, in short, where it may be said that phase markets are born. But this birth is entwined and intimately combined with another birth: the birth of values, standards and institutions congruent with the local markets. The tendencies for the formation of local phase markets, in fact, clash continuously with the tendency of the local firms to become vertically integrated and with the convenience of resorting to firms outside the district. In addition, the formation of a sense of belonging to the district finds immense obstacles in the contemporary culture of modernisation in which the district that is to be born is immersed.

To conclude, while in research it is possible to devote oneself to detecting the signs of the articulation of the typical production process in so many phase populations of production units, one must never lose sight of the fact that that articulation is not a technical phenomenon but a social one, which leads to determined modifications in every other dimension. If we really want to assign a date to the birth of Prato's industrial district, the date that springs to mind is 1958, when the verbal agreements between final producers and phase sub-contractors precipitated in a formal agreement. However, we must say that, considering the way the problem is framed here, the question of the date of birth loses a lot of its fascination: Prato, for instance, has been a kind of district for centuries.

In essence, this is what an industrial district is: a socio-economic vortex, a kind of "permanent small industrial revolution", the transforming energy of which is imprisoned - like petrol in a combustion engine - in a form which, on the one hand produces products that are sold and on the other reproduces the fundamental socio-cultural relationships of the place.

1.3 The formation of districtual institutions

The regular placing of the district product on markets outside the district and the gradual insertion of new productive agents and odd bits of time in the district process create a more and more complex socio-economic structure. For example the network of local markets which is formed in this way will be fed, on the side of suppliers, both by what we could call "autonomous" or free firms, and by firms with the "participation" (financial, due to family bonds, or other forms) of the final producer, but often encouraged to sell part of their production on the open market. In this way, in the district, as well as a scattering of small autonomous companies, ready to work for anyone, a number of more or less stable "teams" of phase producers (suppliers) are formed, each one topped by a certain final producer. In any case, the network of infra-district links of pure exchange of semi-processed materials and services is interwoven with a host of financial, family, friendships and neighbourhood ties. Well, strictly speaking, even the last of these should not be overlooked, if one wants to understand the concrete operation of this particular system over a given period of time.

The "process of converging on the district form" certainly involves long periods of time during which many of the sub-populations of phase suppliers do not present the numerosity and the other conditions necessary to guarantee conditions of competition, but if access remains free and the district's final product becomes successful on the external market, the grid of markets may tend to become closer to the canonical form of the district structure. From a certain point onward, the prices according to which certain semi-processed materials and certain productive operations are usually negotiated will form, visibly and explicitly, or in a "concealed" way, an authentic "local price system" which is, at the same time, a sub-system of the total (national and/or world) system of prices. Now, every local price system, if it belongs to a district, having to allow a "satisfactory" distribution, among the crucial sections of the productive community, of most of the net product (in the sense that, if it does not allow it, the district ceases to exist as a district), must also respond to requirements of "local equity". In this sense it may be said both that the districtual system of prices for processing and services - anchored as it is to a local socio-cultural equilibrium, univocally defined by history - is a system *sui generis,* and that it is the visible and tangible form, the prototype we might say, of each orderly market, that is each market which, in the ordinary performance of exchange, is able to allow the reproduction of its socio-cultural conditions. In its normal operating condition, the industrial district would therefore be the normal form of the capitalist market.

The growing subdivision into phases, the induction of collateral processes, the formation of open teams (but with a districtual preference) of specialised companies, the birth of a grid of internal markets and of an embryo "local price system" for numerous services and semi-processed materials, are therefore, in this interpretation of industrial phenomena, different aspects of a single process: the gradual formation, coinciding with the unfolding of a typically multi-phase productive process, by a multiplicity of autonomous agents, spatially close to one another, of an organic plurality of local institutions complementary to the central productive process, hinged - and this is the point - on a grid of local markets for semi-processed materials, machines, both new and second-hand, and operations.

But the formation of this network of internal markets, hinged, as we shall see below, on the highly singular "district market" of human resources, could not be understood if we did not consider, at the same time, the growth, around the markets, of formal institutions (e.g. handicraft associations, technical schools, technological centres, buying or selling consortiums, etc.) and informal ones (the rules of the district game) which, on the one hand, develop the institutions and the values of the place and on the other prepare the socio-cultural hinterland necessary for the grid of markets to support the population of companies in their action on the external market. It is needless to say that, in the districts, the local regulating structures develop, to a large extent, as local variations of national structures. But we must not let ourselves be deceived by this: the mere fact that the "Unione Industriale Pratese", for example, is a member of the Confindustria, is not sufficient, logically speaking, to reduce the former to a local specification of the latter. It might be logical to interpret the Confindustria as a confederation of local industrial unions. The same applies to the trade unions, the artisans' associations, etc.

1.4 The cognitive spiral

Here there are two significant theoretical operations: a) the recovery, for theorisation of the productive process, of a knowledge that was considered pacific and non-problematic: contextual knowledge, which is mostly tacit; b) the decomposition of the process of production and circulation of knowledge into clearly identified phases.

Production requires a blend of codified (or technico-scientific) knowledge and contextual (or practical) knowledge. We can imagine classifying all production types according to the codified knowledge/contextual knowledge ratio. At one end we will find an ideal production process that is entirely automated, or at least which can be reduced to an entirely specified, precise sequence of operations, and at the other a totally "formless" productive process, run by an operator, the classic *bricoleur* or do-it-yourself type, convinced that he doesn't know how to do anything well, but he "gets by" somehow, doing a bit of everything. The truth, it is needless to say, is in the middle. On the one hand, in fact, it is not true that the perfect *bricoleur* knows nothing: better than others who have been trained more than him, he knows certain recurrent operations of everyday life; on the other hand, the most automated factory in the world cannot operate unless there is someone who has accumulated experience of the typical causes of its breakdowns.

So, an area that has accumulated a certain amount of contextual knowledge (a certain know-how) based on its own, unrepeatable experiences, in the heads of its agents and in its productive and para-productive institutions - as long as it is dedicated to production types in which that contextual knowledge, suitably mixed with codified knowledge, forms the appropriate productive base - will have the possibility of ensuring a competitive advantage.

A line of historic research which could spring spontaneously from what we have described above is the research on how the know-how that characterises every currently successful district has been formed over the centuries. The simple idea would be to retrace the steps of the formation of that know-how, taking note of instances of hybridisation, diversification and dissipation. A kind of local history of productive techniques and, at the same time, of the design and forms of organisation of local production, in which particular attention is paid to the ways and concrete means adopted in each locality in order to absorb, at different times in history, the influx of codified knowledge from outside in the technological mixes adopted locally. A history that does not isolate the technical component, but ascertains the way technology has been coupled, in different periods, with the organisation of production and of companies. Fundamentally a history of the local forms of innovation, of both product and process, and of the aesthetic and functional influences on the forms of local productive tradition. A history that inextricably combines local technique and *design*, projected against the background of what is happening outside, in the wide world. A history that takes care never to be resolved either in the mechanical absorption of a general technical progress, which, like manna, falls evenly from the sky, or in a ridiculous glorification of local tradition for its own sake. Scholars must therefore watch out both for the arrival of external innovating agents and for the formation of conditions that make it possible for them to catch on.

This type of study can also stress the presence - not an obvious one in a culture obsessed with technology, such as the age we live in today - of a continuity that is deeply rooted in the comparative slowness of the "interior" evolution of local evolutionary processes which, considered in an "externally" commodity-related or technological key, may appear filled with "clamorous" instances of discontinuity. Of course, this does not exclude the possibility that psychological phenomena also have points of discontinuity. But the "loading" is slower and the "explosions" are less frequent.

The other crucial passage is provided by the capacity of social theory to decompose the process for the formation, transmission and productive use of knowledge, in such a way as to allow its empirical analysis in determined concrete contexts (e.g. territorial areas). In fact, this has already been done by Nonaka and Takeuchi, and there are already studies that apply these schemes to concrete situations. The principal defect of these studies, as far as I can see, is their isolation from the parallel, but also inter-related, process of forming values and institutions. The matter is a more general one: as always, the change of knowledge cannot be isolated from the change of all the other dimensions (e.g. ethics) of culture.

1.5 Versatile integration and productive diversification

This are the crucial points: 1) variously distributed throughout the territory, there are reserves of knowledge and business morality that can be exploited to produce certain types of goods efficiently, 2) the world market expresses ever different demands and presents unexplored potentialities. How can we mediate between the concrete lumps of know how and business morality, deposited in every single point of the territory, and the variability and unforeseeability of the world? The industrial basis, the fortune of which is based on the use of those lumps in the world market, must possess structures and men that help the suitable conversion of that potential. Hence both the structure for open teams and the generation of "versatile integrators" and, finally, the diversification of the product.

One of the characteristics that make up Italian industrial districts is the fact that they have brought economic success to populations of companies and families that canonical industrialisation had left aside. These are mostly areas accustomed to "scraping a living" (at least, that is the way things appear to "others") on the margins of the great currents of exchange, in the narrow and precarious bits of the market that have been left to them, for long periods of time, by hegemonic political power (e.g., Prato's woollen cloth with respect to its Florentine counterpart) or by triumphant mass-produced industry (e.g. Prato's carded woollen cloth with respect to worsted from England or Biella). This economic and often social and political exclusion has produced many handicaps in those areas, but, in compensation, the long permanence in difficult situations has entrenched in local manufacturers - in the "place" that then became the district - a great capacity and promptness in grasping even the slightest variations in the "humour" of consumers and in exploiting technical possibilities in an imaginative, often "improper" way that nevertheless "works". If this were not so, those communities would have disappeared from the group of candidates for district form. Well, if it is true that the present time is characterised by great turbulence and unforeseeability of demand, is this mental adaptability and trend to contaminate fields of knowledge, none of which they possess completely, *only* a handicap, as many scholars claim? Or is it not perhaps, "in this situation", *also* one of the conditions for obtaining, from a given cultural heritage, replies suited to a world that is changing so violently?

One of the characteristic features of the ideal type of industrial district is the wealth of entrepreneur figures, but also of managers and technicians at various levels, right down to the ordinary blue-collar worker, who span different specialisations, so as to mediate promptly and efficiently between the vast and variable ranges of opportunities that are continuously opened and closed, both on the production side (the "versatile integration" of district specialisations) and on the side of openings ("product diversification", aimed at exploring the most minute and unforeseeable "folds" of the human imagination). The way the district works is therefore to be seen in its routine procedure, much more than in that of a large company, offering a continuous opportunity of exercising mental versatility.

The average district operator seems to have absorbed one of the surest results of evolutionary theory: species that are too highly specialised are more exposed than others to changes in the environment. One might also say, rather imaginatively, that each category of agents in the district - final producers, phase producers, managers, intermediaries, technicians, etc. - gravitates towards the emblematic figure of the *bricoleur*. Specialists in interspecialisation, that is in all-round freedom.

A structure of the kind just described allows, when necessary, maximum productive rearrangement with minimum upsetting of the individual productive units. This is not a property of the individual firms in the district, for the use, for example, of versatile technical means (e.g. numerical control machines), but a property of the "district system", hence the designation of the industrial district as a form of *flexible integration*. *Integration* because both the firm and the team, that is the district, integrate, that is co-ordinate, the specialisations of individuals and firms; *flexible* because of the decisive role played in that integration by the network of local phase markets and of special agents (the versatile integrators), to be considered below, who each allow the rapid passage from certain mixes of suppliers and productions to others. We are still quite far away from the so-called *flexible specialisation* which is originally defined at the level of the individual company and is significantly based on technological data.

1.6 Loyalty and belonging

A task which I cannot quite explore properly, but which I "feel" is essential, is the focussing of the "senses of belonging", which are the immediate guarantees of the self-containment of the process of division of labour, on which the whole question of the industrial districts is based. If these senses of belonging are lacking, or are present in incongruous forms, although the specific economies of the district are formed, they become dispersed over a wider platform of agents and do not trigger the cumulative process that was mentioned previously. One of the best known distinctive features of the district is, in fact, the differential of trust and solidarity in business - of course not without envy, emulation and all the other human sentiments typical of the limited group - between its members and "the others". This corresponds to the above-mentioned "conditioned closing" of the district with respect to certain productive operations, to the repeatedly noted "originality" of the district's productive language and, finally, to the presence, in the district, of a widespread network of reciprocal relationships that are not formalised in precisely quantified obligations.

At a first glance, these features may seem anachronistic, almost pathetic relics of bygone periods, which connote the district's traditionalism. Many people believe that modernity is characterised by a systematic extension of trust, by a universal technical language and by the tendential resolution of all the vague reciprocal relationships in precise relationships of exchange. There is some truth in these statements, of course, but, considering the studies carried out on industrial districts, one may start by wondering whether there is any sense in thinking, for example, of the process of building up trust as a unilinear process that takes place "in a vacuum", that is excluding places and concrete human groupings. Well, what is the relationship between this process, which I imagine is tortuous and arduous, with the formation of institutions, in the broad sense, and the development of the know-how and values illustrated previously?

If in our interpretation of world change, we add to the uniforming strength of capital - which affects human experiences, breaking them up and rearranging them like pieces of Lego, and tendentially concentrating their control in only a few hands - the complex, variegated and variable dialectic of "associated human minds", as Carlo Cattaneo would have said, which I tried to "sketch", we come to confirm, at one and the same time, the continuity of the forces that change the world from the inside and the variety and unforeseeability of their developments. This is, in the end, a representation of the things that assign to a tangible and complete man, "pushed from the inside, for good or ill, towards change", the role of a motor of history.

Even the originality of the district's productive language, similar but not identical to that of other centres that produce, let us say, the "same things", must be considered in this way. It is only in the extent in which the particular features of a place's productive culture succeed in appreciating some latent need of the consumer, to which other productive cultures have less access, that they succeed - by means of their products - in reproducing themselves.

These processes of constructing local trust and reproducing the originality of the productive language are closely intertwined with the process whereby exchange relationships take over from reciprocity relationships. When certain relationships between men (e.g. working relationships) are opened to monetisation, there is a great fluidification of social movement and increased possibilities of new, unprecedented aggregations of this movement. Now, as we have already seen, for example when referring to the institutions that complement the district markets, one may think that the relationships of reciprocity, far from disappearing, are gradually redefined, assuming, if they succeed, the forms suggested and/or allowed by the dual dependence on the past of the place and on the variable conditions of the external market.

This happens all the time, of course, but what the embryo of the district theory that we have sketched can offer us in addition is the possibility of studying in an orderly fashion, in the concrete situation of a determined place, the concrete movement of those trends and of all the others mentioned previously, against the background of a local identity *considered true* - and as such a genuine historic force - by a group of men.

1.7 Two-faced Janus: labour market or entrepreneur-ship incubator?

Finally we come to the knot of all knots, the so-called "district labour market". The enigma of the district lies in the fact that the district is, at the same time, the place where "labour" is bought and sold like a commodity and the place where the minor entrepreneur class is formed (the incubator). Taking this concept to its limit, but maintaining a good degree of correspondence with known experience, we can indeed say that the same socio-cultural process produces both workers and small entrepreneurs there. And so, workers, entrepreneurs and... capital. This systematic osmosis between the two reproductive processes is what characterises the district form in comparison with canonical capitalism.

In fact, what is capital in a very general sense of the term? It is the possession of knowledge and of social relationships which have a productive importance in any given historical context. Knowing how to make, or how to sell, a certain thing and having certain social relationships (e.g. the trust of the bank manager) is all that is necessary to get a productive activity started. An environment where know-how is widespread and personal knowledge allows relationships of differentiated trust, quite apart from the money at one's disposal, is therefore an environment where "diffused capital" can be said to be produced.

This situation is quite different from the typical situation of contemporary capitalism, in which capital, passing through its monetary form, tends to build up at one end of the company while the forming, reproduction and assumption of manpower are processes completely separated from the formation of the "business capacity in command of capital", by a deep, insuperable chasm. The key to interpretation proposed here for those who base their social and economic analysis of capitalism on the dry and uncomposable contraposition of capital and blue-collar workers is therefore scandalous and unacceptable.

The social mobility of the district forms a kind of umbilical cord between local society and the formation of capital - that is between families and companies - through which pass many of the best juices, along with some of the worst of civil evolution.

A social mobility, that of the district, which, when added to its notable professional mobility, makes it a very "plastic" and flexible container of human resources which are continually qualified and requalified in response to external conditions and stimuli. In my discussion of the industrial district, these considerations always lead me to use the term "so-called" before the usual expression "labour market".

At the basis of this, as well as the "physiological conformation" of the district which we have already seen - which, by continuously breaking up the central process, continuously opens, here and there, new possibilities of access to entrepreneurial activity - I believe there is a "cultural" component, a fundamentally "optimistic" attitude, so to speak, to what is generally called the "business risk". The employee or the share-farmer, let us suppose, who, renouncing modest certainties, transforms what little he and his relatives own into capital goods (plants, machines and raw materials), which he *hopes* to use in the production of goods which he *hopes* to sell, must in fact be convinced that success is fairly probable. And yet, a "mass" conviction such as the one needed here cannot rest on subjective, accidental and fleeting opinions, but requires repeated, comparable experiences of success. So the positive attitude towards the business risk is indeed a cultural element of the district population, not in the sense of a gratuitous, widespread "optimism of will", but in that of a rational interiorisation, by the subject representing a given population, of repeated experiences of his own community or of similar communities. As a "piece of history accepted in consciousness", this conviction helps define the comparative potential of economic development (at least for the types of production suitable for the district form) of each given human grouping. Certainly, not all human groupings dominated by the trust in success will really make it, but those in which that trust is not clearly felt will certainly not make it.

This is a difficult passage of the theory of local development, because the sufficiently generalised conviction in a given community that "we can make it" does not form as long as a certain number, which cannot be specified *a priori*, but certainly not a small number of agents in the community have actually not "made it". So the weight of the past can be, and often is, decisive, both on the good side, in confirming and exasperating the record of successful districts, and on the bad side, in stressing the disadvantage of those areas that have not achieved progress. Fortunately the positive experiences are, by necessity, not only those of "face to face" experiences; more indirect and "distant" experiences can also be valid, as long as they are known and seem to be similar in the type of use and of the surrounding conditions. According to some authors, as well as local, immediate contagion, a "remote" and deferred contagion is also at work.

There is nothing better than the ambivalence of the so-called district labour market - direct bridge between subordinate and autonomous employment, seed of entrepreneurship, dynamic negation of the "dry" contraposition between "dead" capital and "living" work - for illustrating the continuous transformation of quantitative into qualitative changes which characterises the district at work.

1.8 Pulling the strings (momentarily): from the industrial district to the districtualisation process

We are now in a condition to reconsider the problem of the identification of the industrial district against this more general and intrinsically dynamic background. What does it mean when we say that a certain territorial group of firms and the respective community (or communities) forms an industrial district? It means that all those processes are jointly present, with an intensity no lower than certain levels, according to a mix which falls within a range of admissible mixes. It may also mean that, at the limit, every situation is "a bit districtual", but only some of them are districtual in such ways and extents as to be able to be defined authentic industrial districts. And also that a certain area is more or less districtual at different periods in time. And also that the mix of characters which defines the district X permeates a different territorial area as time passes. And also that the hybrid forms of transition between the very few pure forms that have been identified so far (industrial district, vertically integrated large company, polarised industrial area, etc.), are, in practice, by far the prevalent ones.

A consequence that is not rarely noted is the denunciation of the inadequacy of the concept of the life cycle of the district. In my interpretation there is no sense in speaking of a life cycle of the district, because the phenomena included there are, in reality, to be read as "transformations" of the grid of relationships between the different terms of the industrialisation process considered in its entirety over a wider area. The search for the life cycle of one of its compound forms, rather than of the cycles of typical relative positioning of the different elementary processes of industrialisation, is therefore to be rejected as oversimplified. That is, one does not look for the formation of the districts, but the laws of transformation of the overall relationships. If the life cycle of the districts is seen as an imperfect, embryo form of the second, that is all right.

1.9 An appendix: A great alliance between history and social theory?

For each historic level of knowledge, social science tells us what functions a social body must possess if it wants to survive and develop in a competitive context. History tells us: a) where and when these functions are satisfied more or less completely; b) that these functions can be combined in various ways; c) that there are phenomena that refer back badly, *prima facie*, to these functions. Social science takes these results that unblock history and, by inserting them - when it inserts them - in the system of its own categories, crystallises them in a certain way. Social science therefore keeps an eye on history; but as it always keeps a particular eye on history, which cannot go any farther than the culture of its time, it places obstacles, from then on, to the perception of what is new, in the study of both the present and the past. So the great alliance between a social science that is ready to supply more and more general and elastic schemes - though never perfectly general and elastic - and a history which systematically tries to "go beyond" what can be seen with the schemes offered by the existing social science, that is to leave room for intuition and imagination in interpretation, is the correct solution.

The study of the district, as a determined abstraction of a social compact which exists (in different examples) "in nature", allows us to identify - seeing them at work and almost "touching them with our hands" - those functions (or perhaps only some of them) and to appreciate from life their interconnections (or perhaps only some of them). In this sense the industrial district is a particularly important form for anyone studying society. But once he has done this job, in which observation of concrete districts in the field - attentive, scrupulous, repeated, carried out with all the aid offered by statistical, historiographic and investigative methods - and general reflection are alternated, the scholar can go farther, abandon the road that took him that far, and try to speak, in general, of forms of organisation of the social process as typical, self-reproducing combinations of those processes. On this new, completely theoretical plain, the industrial district is of course only one of the possible forms. But the sole fact of having "shown", if not demonstrated, that a different, unexpected, unforeseen form "is possible", is sufficient to force social science to rearrange itself. And so, nourishing history with new concepts, it forces history to modify the interpretation of industrial phenomena.

References

Becattini G. (1979), Dal "settore" industriale al "distretto industriale". Alla ricerca dell'unità di analisi dell'economia industriale, *Economia e politica industriale*, no.11, p. 1-79, reproduced as, L'unità di Indagine, in, Becattini G. (1987), (ed.), *Mercato e forze locali: il distretto industriale*, Il Mulino, Bologna.

Chapter 2

Local development in a post-Fordist growth regime[1]

Pascal Petit

2.1 Introduction

If Fordism clearly belongs to the past one still finds it difficult to characterise the New Growth Regime of the developed economies. The popularised notion of New Economy was of little help in that respect and its vagueness has been part of its success until the dramatic fall of the shares of the hi tech companies on the stock markets in 2000-2001 badly affected its credibility.

Still, major structural changes have taken place as well as large institutional changes which have all created a new context of economic development. The growth regime it led to may have specific properties, including some inner contradictions and challenges which may lead to different outcomes, but these properties have to be analysed and we cannot simply argue that we are in a transition. Growth regimes in a regulationist perspective should not be understood as stationary states with the normative long term view that neoclassical economists attach to it.

Development in capitalist economies is not self-regulated by some abstract market mechanisms, but clearly embedded in institutional contexts which are rooted in history, but are also the ever evolving results of social disputes, debates and compromises. This perspective also implies that, if growth regimes may have, in a given historical period, some common features between various trade partner countries at a similar level of technological development, national trajectories remain diverse. Growth regimes have nevertheless broad common features that help to account for some basic working of the economy as well as

[1] This contribution is a revised and extended version of a communication made at the Vicenza conference, Il Futuro dei Distretti, 4th June 1999, which largely drew on Petit (1999).

for common inner contradictions, slowly developing and fostering the social and international tensions of the period. Our purpose is certainly to give some schematic views of these new broad features at macroeconomic national levels. But growth regimes have also spatial dimensions, in the sense that the historical peculiarities they underline should also account for the basic characteristics of local development and its relations with national and international levels.

This contribution will accordingly proceed as follows. Since the characterisation of a growth regime requires an institutional approach, we shall in a first section refer to the "regulation approach" and briefly present the conceptual framework used to analyse the past Fordist growth regime.

On this basis we shall propose in Section 2.2 a characterisation of the main features of the contemporary configuration of institutions and structural changes that seem to shape the present growth regime. This macroeconomic approach of the fabric of institutions and of its influence on the working of economies will then be completed, looking in Section 2.3 at the spatial dimension of these processes. There is no deterministic relationship between local and national levels, but general trends change the perspectives at local levels in many different ways and reciprocally the national trajectories remain largely dependent on the positive network externalities and on the synergies that can stem from the various local achievements. The conclusions in Section 2.4 will try to assess some of the traits of the future of industrial districts as suggested by the previous developments.

2.2 Five structural forms to assess the institutional context

In order to assess the main features of institutional mediations in contemporary developed economies, let us first recall the theoretical apparatus which the Regulation Theory (RT) provides. RT distinguishes between five institutional packages or structural forms (also called mediations), according to the field of economic activities that they cover. One may consider them according to their anteriority in history, the kind of legitimacy they enjoy, or in terms of the kinds of decisions they involve. These five forms are the state apparatus, money as an institution, international relations, forms of competition, and wage labour relations.

The State corresponds to the set of historical institutions associated with central power in the times of absolute monarchies, such as army, police, and justice. This form obviously supersedes the others.

The second package is money. It is closely related to the previous set, but remains relatively independent as the symbol of a social link between people in a nation and between generations of citizens.

The third form, international relations, is clearly linked directly to the other two. It is the direct consequence of a nation not living in autarchy, but entertaining relationships with other nation states with similar premises. How the states coexist, and how money acts as currency, are matters of international relations.

The fourth form is more recent and dates to the time when states had to organise commerce in order to develop it. The rules of commerce, the organisation of traders, and all the institutions taking part in creating economic transactions are concerned here. We refer to this institutional package as the forms of competition.

Finally the fifth set concerns the functioning of the labour market and the organisation of work. This could be considered a subset of the forms above, dealing with labour and employment issues. However, its development during the Fordist period gave it its full autonomy.

The sequence of this presentation is telling. It hints at a historical process which gave its structure to the whole fabric of institutions that we consider. Nation states first developed state apparatuses and money, i.e. the "fonctions régaliennes" establishing modern states (Théret, 1992). They then organised their international relations on a pattern consistent with their sovereignty as nation states, giving full meaning to this notion at least in Europe[2]. Finally, during the long period of industrialisation of the 19th and 20th centuries, states were little involved, restricting their activity to setting up the regulatory framework of markets. The period has thus been characterised as a first liberal phase, market-led with little public intervention despite the hazards of the markets. (Boyer, 1988).[3]

This period ends with the economic crisis of the 1930s and the Second World War. Capitalism and the ideology of the inter-war period of free competition and non intervention were blamed for the economic disaster of those years. This led to the promotion of a modern capitalism (à la Schonfield) in the post-war period which was much more open to institutionalisation and state interventions in economic activity. This changed the structure of the institutional nexus. In the

[2] A step towards modern international relations between states which can be dated back to the 18th century; see Dehove (1997).

[3] Polanyi, in his book "The Great Transformation", rightly emphasises the discovery of the market as an organisational principle at the time.

first place the economic crisis and the war led to a revision of the boundaries between structural forms. A more clear-cut definition of the fields aimed to limit the hegemony of the dynamics of the "forms of competition". The institutional and regulatory framework of international relations and money were clearly defined from the beginning of the period with a set of regulations and international agreements, such as the international Bretton Wood settlement on the gold standard and national laws segmenting banking activities. Means for state intervention were also clearly defined. In Europe, that seems the most active field of institutionalisation in the post war period concentrated on the wage labour nexus[4]. The debates on full employment across the western world in the period also grounded the legitimacy for such activism, even if the arrangements differed among countries (Petit, 1995). These institutional adjustments show up not only in the increase of expenditure in the relevant budgets, but also in the increase in number and extent of measures organising the wage labour relation and its reproduction.

The point that needs to be stressed is that, everything else being equal, the whole post-war period of rapid economic growth took place in a specific institutional context, a configuration of structural forms, which had been radically transformed from the start, and within which the dynamics of expansion and activism (e.g. extension of coverage, new measures of protection and development) occurred in the field of the wage labour nexus. Somehow the ideal type of Fordism expressed the centrality of this last structural form, with noticeable shares of productivity gains being devoted to a more and more comprehensive reproduction of labour. By contrast, the development of this wage labour nexus underlined how it had previously been entangled with the structural form of competition (e.g. in the inter-war period and before).

Many things developed on the basis of this overall institutional context in which the wage labour nexus dynamics prevailed. Individualism developed while societies were getting richer. Inner contradictions developed, such as an on-going fully fledged internationalisation for which the old regime was ill adapted, leading to a major worldwide economic crisis in the mid 1970s. The various stages of this crisis and of the exhaustion of the old regime have been largely and differently exposed and will not be reviewed here (see for instance Mazier *et al.*, 1993). We shall jump directly to the 1990s to analyse the structure of new institutional nexus which could help to identify the emerging post-Fordist growth regime.

[4] See Delorme and Andre (1983) for an overall assessment of the evolution towards a more comprehensive implication of the state in the economy.

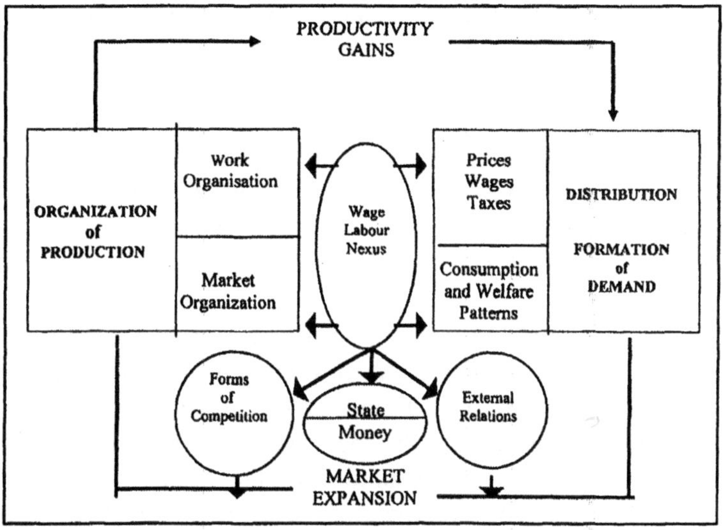

Figure 1: On the centrality of the dynamics of the wage labour nexus in the Fordist period

2.3 A new institutional configuration centred around a broad redefinition of the forms of competition

Institutional changes in the 1980s and 1990s appear as a reaction to previous evolutions, limiting the advantages of labour and enlarging the room for manoeuvre of capital. Cuts in social budgets and deregulation support the hypothesis of backlash or down-turn in the cyclical evolution of the dynamics of institutions. We argue that the new priority given to the forms of competition should be analysed in its specific aspects, and not simply considered as an approximate return to the inter-war situation as far as the dynamics of institutions is concerned.

A better understanding of the change in the forms of competition is crucial to an appraisal of what the new growth regime consists of. The assumption that one structural form plays a prevalent role and affects all other institutional changes refers mainly to the fact that the legitimacy of institutional change is borne by some principle, widely accepted at the time, which historically concerns one specific form (organise the labour market in order to socialise its risks as in the Keynesian conventions, expand and diversify market mechanisms to increase their social embeddedness, redefine the role of the nation in the international context, restore the credibility of a currency or of a group of currencies). This focus on the central legitimacy of changes in one institutional form follows simply from the observation of modern history; it should not exclude the possibility of combinations of structural forms. The hierarchy assumed between institutional changes is likely to be strongly path-dependent, e.g. largely determined by the levels attained and the trajectories of institutional transformation specifically followed by each country. This centrality finally means more a prevailing direction of changes open to on-going institutional adjustments than a definite state of reforms installed at the beginning of the period.

Still, to consider the changes in the forms of competition in the 1980s and 1990s as central to the dynamics of institutional change raises many questions regarding the actors, the purposes, the range of actions as well as the range of institutions concerned, and finally the legitimacy of these institutional changes for society as a whole. Our understanding of the forms of competition, as a structural form, ranges from regulations regarding market access and fair competition to the thick fabric of mediations which organises market transactions between economic agents, be they firms or individuals. In this last sense it concerns all economic relations between independent agents (or units), setting aside wage labour relations and the work organisation. Producers have always been interested in the evolution of the forms of competition, but markets have evolved, the norms of competition are more complex, following the differentiation of products and the opening of markets. Paradoxically the new laws of markets have increased its social embeddedness; not only the true properties of the products are accountable, but its origin and destination or long-term consequences also matter. This implies a deeper involvement of consumers in the working of markets. This process does not rely only on a new consciousness or consumerism on the part of consumers[5]; it is rooted in important structural changes. A main transformation has been the full development of wage earners' societies (Aglietta and Brender, 1984) where wage earning activities have become the overwhelming norms for a labour force which concerns men and women in much more similar ways; all of which implies a deep transformation of the domestic sphere. Just as important has been the development of fully fledged intermediations as given by all kinds of services, with a special mention for the large service networks, which are the real support of the extension and sophistication of markets we referred to. Even if these transformations are the consequences of the development of the Fordist

[5] Organised consumerism did not develop to the extent that one would have expected in the 1970s for instance.

growth regime, it has reached a stage of maturation beyond the co-evolution of mass productions and mass consumptions, which raises totally new issues and challenges.

This Second Transformation is referred to in the literature in many different ways: the clearest statements refer to new capabilities and competences in production and/or consumption: the ability to differentiate products, the development of more specified needs, elaborate consumers, and more sophisticated consumers. The terms complex economy, knowledge economy, tertiary economy all refer more or less to this greater sophistication (Foray and Lundvall, 1996). In the same vein it is often said that markets are more competitive and that non-price competition keeps growing; and the notion of reflexive economy is relevant to stress the new accountability and its social implications that are now part of the laws of markets (Lash and Urry, 1994; Callon, 1998).

One should not pretend, though, that all markets have the above characteristics and that price competition and mass production have disappeared. On the contrary, the mix of markets more or less oriented by price and non price competitive mechanisms, inducing new complex cycles and segmentations, may well be also a major characteristic of the new growth regime.

Still, the existence of highly differentiated markets, variously embedded in the social contexts of producers and consumers, is shaped by a whole fabric of mediations. These, which include all public services as well as a fair share of intermediation services (banking, communication, transport, and distribution services, all of which are under some kind of public tutelage), expand the field of institutions relevant with the change in the forms of competition under view.

The debates around the deregulation of public services, such as telecommunications or highly regulated sectors such as banking, are indicative of this broad interest as well as of the conflicts surrounding what these intermediation services should be. It is thus revealing that the "revolution" that constituted the wave of deregulation in the 1980s was headed or symbolised by two popular heads of state, Mrs Thatcher and Mr Reagan. Even more interesting is the fact that this so-called deregulation does not end up by suppressing all rules, but more in a general change of direction (re-regulation), corresponding to a shift away from regulations that segment activities towards rules that normalise and certify procedures (prudential orientation). In a way, the debate regarding the organisation of these services has just begun. Health and environmental hazards in some product markets suggest that similar sophisticated layers of institutions/mediations may be needed on a larger scale.

In all these transformations some structural changes played a major role, such as the diffusion of new information technologies[6], or the expansion of customised service activities, the accumulation of knowledge, and the internationalisation of the economies. Banking and telecommunications activities gave numerous examples of such challenges, combining the pressure of new technologies and the entry of new competitors. It is also in the field of external competitiveness that the crucial importance of non-price competition for high wage countries has been perceived most accurately. All national institutional peculiarities have been adjusted and turned into (built in) comparative advantages, once a country has been successful on world markets.

Finally it should be recalled in any assessment of institutional changes that not all markets are in the same position. Problems of information and externalities differ strongly from one to the other. Thus much more should be said about the special dynamics of financial markets and banking, where the institutional change has been significant. Specific problems of hi-tech goods and services also need to be considered, which involve cultural barriers to entry or lack of social connections (a factor of social networking which tends to gain renewed importance). This diversity and complexity is likely to fuel inner dualistic trends in our societies, as will be emphasised later.

Thus, even if we take a broad approach to the forms of competition, it remains difficult to assess the direction of this set of institutional dynamics. In the previous period, changes in the wage labour relationship could be read as a growing institutionalisation of the wage labour relation, sharing the productivity gains and organising the protection of wage earners in return for widespread diffusion of the principles of scientific work organisation. Conversely the present dynamics of the forms of competition can be seen as a trade off between a continuous extension of market mechanisms, with the new room for manoeuvre permanently given to entrepreneurs that it implies, and a constantly extending customisation of products (goods or services) and provision for consumers and citizens, with the adaptation to specific needs and the development of local synergies that it implies.

Still, this "trade off" is not as clear in the political debates and social conflicts which limit the development of a consistent institutional framework. It raises in particular two types of issues. The first is whether these developments

[6] ICTs allow sectors to change their production processes and products in ways that create pressures for deregulation, or cause shifts in the conditions of competition that lead to broad re-arrangements of activities. Most large network services underwent major transformations, as shown in intermediation activities, e.g. distribution, transport, communication and finance.

will systematically be accompanied with a weakening of the old wage labour relationship.

The second is whether such "trade off" is not biased in favour of consumers and citizens possessing a good combination of income and skill or knowledge. The old Fordist regime had learned how to cope with the hampering factor that represented income inequalities by means of fiscal redistribution and public services. The New Growth Regime may be confronted with new barriers largely raised by differences in access to information and knowledge.

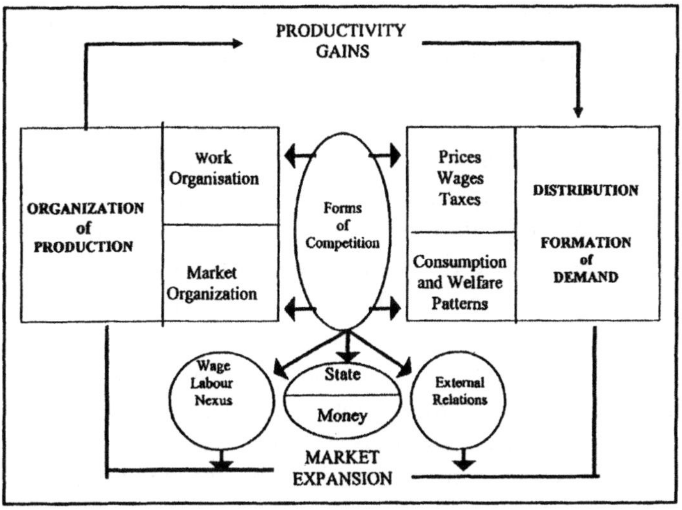

Figure 2: On the centrality of institutional changes in the forms of competition

This points at some new frontier of welfare slowly emerging from various social and political debates on access to different kinds of networks, tangible and intangible. Conversely it follows that the segmentation of demand and the lack of social cohesion may well be a major limiting factor of economic growth. A crucial issue, in that respect, depends on how the changes in the forms of competition (broadly taken) interact with the evolution of the other structural forms, and more especially with the wage labour nexus and the local conditions of organisation.

2.4 Changes in local and wage labour conditions

2.4.1 The wage labour nexus and its geography in a Fordist growth regime

The search for economies of scale, which fuelled the steady growth of productivity gains over the period led to the development of large establishments. This took place in expanding existing sites when possible, but also led to greenfield investments when constraints on space and labour force in old urban areas limited the expansion of existing sites. In all cases the driving principle of work organisation was Taylorian. It led to the specialisation of the various sites under the centralised control of the firm head office. This organisation has been the engine of growth of the Fordist model. It inspired the organisation of small and medium firms, diffusing scientific work organisation and setting the standards in terms of wages and working conditions. The trajectories of large Fordist firms thus headed the development of the wage labour nexus. At local levels this driving force combined with the development of local services and craft industries. The existence of a surplus of labour force out of agriculture or other primary industries added to the specificity of local labour markets. In this dynamic, industrial districts, as identified in Italy, appeared much as oddities, as ways to take advantage of some historical contingencies and develop local systems in some production niches. The success of these ventures was to take advantage of trust, common knowledge and shared know-how to set up rational local and co-operative systems of production. To survive, these local systems had to meet two constraints: a) to be able to upgrade the wage labour conditions in pace with the Fordist core of the economy, b) to remain competitive on product markets where price competitiveness was more and more challenged by the entries of low wage countries.

Many of these local systems did not survive and it is precisely when some local successes seem to last up into the 1980s, and rely on a strong non price competitiveness, that the notion of Industrial District caught attention; in some ways it even looked like a precursor of a post-Fordist regime, a model of flexible specialisation as Piore and Sabel (1984) rightly put it. This may oversimplify their situation. The new economic environment certainly favours some of the traits of these successful districts, but this change in the forms of competition also submits them to new challenges.

2.4.2 A new division of labour between firms

The change in economic environment, which manifested itself by a rise in information and knowledge that a firm could dispose of to set up its strategies, led firms in developed open economies to search for rents from innovation and to concentrate on their know-how. It implied externalising peripheral tasks and looking for support in their quest for innovation. Both trends correspond to a strong call for business services (of both types, highly knowledge intensive and poorly knowledge intensive), which indeed showed up in the fact that these services experienced the largest employment growth rates of all sectors in the 1980s and 1990s. This new division of labour between firms is not the only outcome of the new driving organisational principle. Firms did increase their access and monitoring of information and knowledge in engaging in all kinds of accords, agreements, joint ventures regarding both strategies to access markets and to innovate in terms of products or processes. This new context does change the spatial fabric of firms, it directly hints at some of the characteristics put forward by the Industrial Districts (IDs). Still, it also goes farther, as the search for co-operation and innovation is not limited to local partners. On the contrary, new means of communication and new sources of information and knowledge directly press towards some dismantling of local systems, opening them up to outside ventures. To overcome this challenge, IDs have to find the right balance between an opening to the outside world and a deepening of their common co-operative advantage. It requires a new consciousness of what these advantages are, and somehow the fame of the IDs, as successful local systems, did provide them with a positive representation of their joint venture, which legitimised and reinforced the social capital at the roots of their common lasting success[7].

First, the opening of IDs to the world to broaden the range of their strategies in providing information and knowledge, which in turn will comfort their rent from innovation in product markets and in production processes, has to be seen as more profitable when ventured in common.

Second, the deepening of their local co-operation has to be further developed, taking into account even more local synergies, especially in the conduct of the various learning processes they are engaged in. In order to meet these two (external and internal) challenges, the agents engaged, e.g. not only the

[7] In that respect it would be important to look at failures or short lived IDs which could not adjust either to the pressure of the Fordist regime, for those which disappeared up to the end of the 1970s, or to the pressure of the post-Fordist regime, for those which collapsed in the 1990s.

entrepreneurs, but also the workers and the public representatives of these local communities, are crucial factors.

We rapidly surveyed the challenges met by the firms themselves. The change in economic environment also transformed the strategies of workers and public representatives as we underline below.

2.4.3 On the transformation of the wage labour nexus

The search for more flexible labour markets has been an early sign of the end of a Fordist growth regime for a long time, characterised by an on-going institutionalisation of the wage labour nexus.

Still, this flexibilisation did not so much occur through a weakening of the employment protection legislation as through the combination of sectoral changes in the distribution of employment on one side and a differentiation of the labour market products on the other side. The drift towards services, where the labour conditions are quite diverse, and on average less formal, accounts largely for the sectoral reallocation of employment.

The process of jobs differentiation has some similarities with the process of product differentiation previously evoked. Firms have various labour needs according to the type of competitiveness prevailing on their product market. Workers have various competencies which are all the more praised by firms that want to have a flexible organisation of work, more relying in parts on individual characteristics and more able to change and react to changes in product markets. All these reasons explain in the first place the differentiation of labour market products. It leads also to a wider-ranging hierarchy of status. This diversity of appreciation and of situations restrains the implementation of collective agreements across the board of all workers and jobs, pushing farther the dismantlement of the wage labour nexus.

The new technologies of information and communication (ICTs) also provide new opportunities to take advantage of workers' competencies. (as opposed to skills which are more dependant on a given line of work organisation). Their propensity for networking facilitates new forms of organisation, promoting externalisation of tasks as well as the organisation of upstream (suppliers) and downstream (distributors and marketing) relations between firms. Secondly ICTs, and more precisely the diffusion of personal computers (networked on intra or inter nets), facilitate flows of codified information to be sent and received by workers, flows which are all the richer since workers have the capacity and the freedom to treat and transform this information into directly executable signals and/or into tacit knowledge that can be capitalised.

Conversely these technologies allow work to be controlled down to a very individual level. Whatever is the prevailing direction (skilled or unskilled labour with more or less room for manoeuvre regarding the treatment of codified knowledge and information), the new arrangement displays a more individualistic organisation of work.

In these organisations (competent) labour is not a substitute for fixed capital, but a complement. In other words the codified knowledge channelled by the capital equipment is complementary to the tacit knowledge that competent workers can put to use.[8]

Competencies are praised to face the shifts in production processes (externalisation of tasks, new input services, inter-firm network collaboration, delocalisation of jobs, persistence and regeneration of old Fordist type of productions). Systems of wages also tend to become more individualistic, distinguishing the interests of individual workers from more collective interests, tying bonuses and premiums to the financial results of profit centres (instead of tying wage increases to productivity goals, expressed in terms of physical indicators, as prevailed previously).

To summarise, we can say that structural changes, be they changes in the sectoral composition of employment or in technology, or in the availability of new intermediary services, helped to erode the old wage labour nexus more effectively than direct changes in the industrial relations and work organisation of each industry.

Development of the regulations and institutions of welfare systems have been influenced by similar values. Financial allocations, which leave the most room for individual decisions, are preferred to measures that distort product markets (as a minimum wage scheme could do), or that tightly monitor individual welfare choices (either by means of national systems of provision or by means of systems of allocation with strictly defined types of uses).

These transformations of the wage labour nexus have diverse effects on local labour markets.

Along with the increase in new forms of employment, with more precarious status, self employment is encouraged as well as local initiatives to develop services for specific local uses to support productive activities or help with domestic problems. Again, these developments can foster sources of competence that benefit local activities or generate menial jobs which erode the welfare of

[8] A view consistent with a feature often recalled in this book (see Belussi's contribution for instance), stressing that the spread of codified knowledge does not lead to the disappearance of tacit knowledge, but on the contrary helps to develop it.

both the users and providers of these services. More flexible production organisations, as required by the new forms of competition, are thus obtained by extensive rearrangements of inter-firm relations, as well as by more marked functional segmentation of internal labour markets. Also, the more risky environment in which firms are developing bears heavily in most-industries on wage earners, especially when firms face mergers, acquisitions, or closures.

IDs in that context are faced with changes in local labour markets, favouring the development of both menial jobs and qualified jobs for competent workers. Both trends have various effects on the dynamics of the local arrangements presiding over IDs. Much in that respect will depend on the perspectives drawn by collective actions.

2.5 Industrial districts at a cross roads

At first sight, industrial districts seem to possess the very criteria of success in a new growth regime driven by the forms of competition. These local arrangements are by essence very market oriented, highly responsive to market changes and rather product quality driven on the basis of cumulatively shared know-how. Still, the challenges that these local arrangements have to meet should not be underestimated, as we have stressed in this contribution. First, the forms of competition have changed, highly differentiated markets pressed to innovate constantly, both in product and in its provision and packaging in constantly more open competition. International access to knowledge and finance became imperative and was facilitated by the development of a whole logistic of intermediation services. This did impinge upon the local arrangements, ruined some, restructured others, and especially the hierarchy within partner firms in IDs. It forced those which survived with success to re-affirm their mutual interest in collectively meeting the challenges of the post-Fordist era. Of crucial interest in that respect have been the new forms of public interventions that developed in the new growth regime, away from the old Keynesian policies (support of rather aggregate demand) towards more Schumpeterian policies (supporting the modernisation of supply chiefly by promoting process and product innovations[9]). Though such supply side policies, even if considered in their structural dimension of innovation policies, have not sufficed so far to launch in most developed economies (not even in the US) the new growth regime on a pace of steady growth and full employment. The major deficit seems to come from a too restrictive view of the forms of competition,

[9] The Italian law of 1991 to promote innovation in SMEs is a good example of these practices.

rather exclusively supply oriented when a great deal of their dynamics should imply also restructuring and development of demand. Labour forces are also consumers and citizens in accomplished wage earner societies. Part of the differentiation of products is therefore also a broad customisation, which does not only take into account the context (and even local context), but also the capacities of learning and development on the demand side. It should thus take into account the enlarged importance of user-producer relationships - which Lundvall (1988) signalled forcefully more than a decade ago - but also the dynamics it can acquire in developing relevant learning processes. We have stressed (Petit and Soete, 1997) that such actions on the demand side were especially relevant to develop an efficient (i.e. growth enhancing) use of ICTs and of the modernised large network services they allow[10]. At local levels IDs may encounter a similar drawback. To clarify, it will also help to distinguish this issue clearly from its Fordist counterpart, stressing that the producers of some products are also its consumers. It relates more to the challenge to IDs brought by changes in local labour markets. Skilled workers have to train their competencies in keeping in touch with the external world of information and knowledge, which is more easily and efficiently done in some collective ways, beyond the limits of the firm (and its hierarchy) and within the limits of local labour markets (and the sharing of tacit knowledge they allow). Such "weak networks" (in the sense of Granovetter, 1985), with a professional blend, are important assets in the restructuring of social capital which accompany a change of growth regime. To create or maintain such an environment for skilled labour is a challenge for IDs (to which they respond differently), but it is not the only one. The change of context has also stressed the externalities, positive or negative, tied to the situation of the overall local labour market. It certainly counts for the attractiveness and quality of life of territories, but it also conditions a proper working of all these network services, which play an important role of intermediation. In other words the broad logistics of services is a crucial factor for the continuous updating of co-operation between firms now required for an ID to remain competitive. The quality of this logistic of services is closely tied with the quality of its employment. It shows how interdependent are the challenges we distinguished concerning market, production and employment strategies. It is all the more important to develop such a capacity of

[10] An example of the one sided aspect of contemporary structural policies is given by the measure of reduction of working time taken in France with a clear attention to its beneficial impact on the reorganisation of the work processes, but little or no attention is given to the use of the time freed for other activities.

adaptation that IDs may well be forced or willing to change product markets. Technologies give opportunities for such drastic shifts, some communities can seize them and their capacity to do so relies on the social capital, preserved and/or rebuilt at local level, all of which depends on the macro institutional environment provided by the new growth regime, in the making, and itself conditioned by the political capacity of countries to express a coherent project of development, as they did in other terms and contexts, in the aftermath of the Second World War.

References

Aglietta M. (1997), "Le capitalisme au tournant du siècle" postface to the new edition *Régulation et crises du capitalisme*, Odile Jacob, Paris.

Aglietta M. and Brender A. (1984), *Les métamorphoses de la société salariale.* PUF, Paris.

Boyer R. (1988), *The Search for Labour Market Flexibility*, Clarendon Press, Oxford.

Callon M. (1998), (ed.), *The Laws of Markets,* Blackwell Publishers, London.

Dehove M. (1997), "L'union européenne inaugure-t-elle un nouveau grand régime d'organisation des pouvoirs publics et de la société internationale?" in *l'Année de la régulation,* vol 1.

Delorme R. and Andre C. (1983), *L'Etat et l'économie,* Seuil, Paris.

Foray D. and Lundvall B. (1996), (eds.), *The Knowledge Based Economy*, OECD, Paris.

Granovetter M. (1985), Economic Action and Social Structure: the Problem of Embeddedness, *American Journal of Sociology*, vol. 91, p. 481-510.

Heilbroner R. and Milberg W. (1995), *The Crisis of Vision in Modern Economic Thought*, Cambridge University Press, Cambridge.

Lash S. and Urry J. (1995), *Economies of Signs and Space*, Sage Publications, London

Lundvall B.A. (1988), Innovation as an interactive process - from User-Producer Interaction to the National System of Innovation", in, Dosi *et al.* (eds.), *Technological Change and Economic Theory*, Pinter, London.

Mazier J,. Baslé M. and Vidal J-F (1993), *Quand les crises durent*, Economica, 2nd ed. Paris.

Petit P. (1995), "Technology and Employment: Key Questions in a Context of High Unemployment", STI Review, No. 15, OECD, Paris.

Petit P. (1999), "Structural Forms and Growth Regimes of the Post Fordist Era", in *Review of Social Economy,* September.

Petit P. and Soete L. (1997), "Is a biased Technological Change Fueling Dualism?", paper presented at the annual meeting of the American Economic Association, Chicago, January 1998.

Piore M.J. and Sabel C. (1984), *The Second Industrial Divide,* Basic Books, New York.

Streeck W. (1991) "On the Institutional Conditions of Diversified Quality Production", in, Matzner E. and Streeck W. (eds.), *Beyond Keynesianism, the Socio-Economics of Production and Full Employment,* Edward Elgar, Aldershot.

Théret B. (1992), Régimes économiques de l'ordre politique. Esquisse d'une théorie régulationniste des limites de l'Etat, Presses Universitaires de France, Paris.

Chapter 3

The theory of geographical agglomeration - minimum requirements and a knowledge-based suggestion

Peter Maskell

3.1 Introduction

One of the most significant consequences of the present process of globalization is the way in which it continues to turn inputs, previously crucial to the competitiveness of firms, into ubiquities. Globalization has increasingly been associated with the 'unbundling' of the previous relationship between sovereignty, territoriality and state power (Ruggie, 1993) and, as a consequence, steadily weakening nation states (Maskell, 2000), but it is arguably the many economic consequences of 'ubiquitification' that has contributed most in making globalization the favourite business buzzword at the recent turn of the century. Ubiquities are inputs equally available to all firms at more or less the same cost almost regardless of location (Weber, 1909). A large domestic market is, for instance, no longer an unquestioned advantage when global transport costs are becoming negligible; when the loyalty of customers toward national suppliers is dwindling; and when most trade barriers have eroded. Domestic suppliers of the most efficient production machinery are, similarly, no longer a solid competitive advantage, when the sales- and marketing strategies of the suppliers reach across borders, and their equipment becomes available world-wide at essentially the same cost. The omnipresence of organizational designs of proven value makes, furthermore, a long industrial track record less valuable. So when input becomes ubiquitous, all competing firms are, in a sense, placed on an equal footing. What everyone has cannot constitute a competitive advantage.[1]

[1] The role of ubiquities in changing the competitive environment is discussed in more detail in Maskell *et al.* (1998) and in Maskell and Malmberg (1999).

Firms cope with this situation in various ways. Some invest heavily in order to increase productivity, while others outsource, leaving the old industrial areas in a slowly more and more desolate and jobless state. 'Automate, emigrate or evaporate', as the saying goes. Other firms, in contrast, confront the new competitive situation by sharpening their abilities to learn and create knowledge a little faster than their competitors.

The creation of knowledge is usually seen as a process requiring dedicated investments either as pre-competitive research and education through universities etc. or at the level of the individual firm through R&D activities. At least as important is, however, the investment in incremental 'low-tech' learning and innovation (Laestadius, 1996; Maskell, 1998) that takes place when firms, also in fairly traditional industries, create strongly corroborated knowledge while handling and developing mundane day-to-day operations like resource management, logistics, production organization, personnel, marketing, sales, distribution, industrial relations, etc. (Malerba, 1992). The possessors might know little or nothing of the origin of the knowledge or how they have come to know it, but 'it's here' and 'it works' (Spender, 1996; Baumard, 1996).

However, scholars and policymakers have increasingly come to suspect that the specific spatial arrangement of economic activities might also *in itself* somehow influence the creation of knowledge and, consequentially, economic growth (OECD, 1999).

Broadly, we may recognize two major categories of agglomeration economies (Estall and Buchanan, 1961). First are those that accrue from the geographical propinquity of industries and services in general, usually referred to as 'urbanization economies' (Hoover, 1970). The second category is usually referred to as 'locational economies' and embraces those economies that arise from the geographical agglomeration of related economic activities. It is the second category of geographical agglomerations or 'clusters' that in particular have been selected in recent years by scholars from a number of different disciplines as *the* territorial configuration most likely of enhancing learning processes.[2]

Some justification for this choice has been found in empirical studies showing, for instance, how 'innovative activity, as measured by patent data, and

[2] The terms 'geographical agglomeration' or 'cluster' are used almost synonymously in the literature together with 'industrial agglomeration' or 'localization', while the term 'industrial district' initially used by Marshall (1890) for the result of locational economies is now often applied when wishing explicitly to emphasize the values and norms shared by co-localized firms (see, for instance, Brusco, 1982).

the location of high-tech industries is ... highly concentrated' (Breschi, 1995), and how the agglomeration of firms within one or a few inter-related industries in the Italian industrial districts gave rise to superior performance and some of the highest regional income levels in Europe (Bellandi 1989).[3] Today Silicon Valley and Hollywood are probably the world's best-known examples of successful non-random market-led clusters[4].

Presumably, clusters of related firms have been contributing to economic growth for quite a while but the contemporary turn towards a knowledge-based economy (Carter, 1994) in many parts of the world has certainly sharpened our interest in understanding the nature of this process.[5]

The existing literature provides two types of understanding of the phenomenon. One source of insight is to be found in ideographic, historical work on how clusters have originated and developed into fruition, occasionally accompanied by accounts of subsequent descents (Malmberg & Maskell, forthcoming). Another attempts to specify conceptually the mechanisms that provide advantages to be reaped by firms located in a cluster. The present article is concerned with the latter aspect. It suggests a way of structuring our perception regarding how the cluster might partake in knowledge creation. In dealing with this issue, it moves mainly within the world of concepts, raising a

[3] Nevertheless, it must be admitted that while the cluster discourse is characterized by an overabundance of valuable case studies the lack of solid empirical evidence across cases, sectors and countries is still profound. The present article can be seen as an attempt to provide a renewed platform for subsequent empirical investigations.

[4] In order to exclude 'random' agglomerations the number of co-localized firms must be larger than if no locational economies are present. Ellison and Glaeser (1994) notice that if firms in an industry choose locations by throwing darts on a map, only six darts must be thrown at a map of the US before it is most likely that two will hit the same state (see also Malmberg and Maskell, 1997).

[5] The swelling interest has occasioned a number of distinct schools of thought to develop including the GREMI approach (Maillat, 1991, 1998; Camagni, 1995; Ratti *et al.*, 1998), the many largely Marshallian studies of the Italian industrial districts (Beccatini, 1990; Brusco 1986, 1999; Brusco and Righi 1989; Dei Ottati, 1994a, 1994b, 1996; Bellandi, 1996; Garofoli 1992a, 1992b, 1993; Gottardi, 1996; Belussi, 1999a, 1999b), the French 'proximité' tradition (Kirat and Lung, 1999; Blanc and Sierra, 1999), an econometric type of cluster analysis (Swann *et al.*, 1998), different 'systemic' analysis (Markusen *et al.*, 1986; Saxenian, 1994; Malecki, 1991), some of which have focused explicitly on the geography of innovation (Feldman 1994; Stenberg, 1999; Breschi, 2000), as well as the cherished approach applied by Porter (1990). Until we have found the 'serene and luminous region of truth where all may meet and expatriate in common', it is impossible within the frame of a single article to take into consideration the diversity in these and other schools of thought and to pay due respect to even important distinctions and points made in this waste literature. I apologize for this when allowing myself a certain degree of simplification in the following sections.

set of questions regarding the way in which economic performance is related to space in general, and to the role of localized learning in particular.

The aim of the article is thus to investigate the nature of the cluster when knowledge creation becomes key. It does not necessarily assert that learning and innovations take place in the cluster only or deny that a good portion of all firms is happily located outside the cluster. Neither does the focus on the cluster exclude the fact that circumstances, events, and decisions in distant parts of the world heavily influence many firms today. The article merely presupposes that the cluster plays a role in knowledge creation that is by and large sufficiently important to affect what is going on in the world to warrant analysis.

The article is structured along the following lines. Section two looks briefly into previous cost based accounts of how firms might benefit when being part of a cluster. It is suggested that such approaches often fall short when addressing the more fundamental question of the cluster: the existence of many co-localized firms in related industries rather than a single, but larger entity, carrying the same tasks. It is proposed that the reason for the existence of the cluster can be found in the enhanced knowledge creation that takes place along its horizontal and vertical dimensions. In section three the learning advantages stemming from the intrinsic variation between co-localized firms with similar capabilities is discussed while section four deals with the division of labour and the interaction taking place among firms along the cluster's vertical dimension. The various factors contributing to the growth of the cluster are sketched out in section five before moving to the more detailed discussion on the boundaries of the cluster. In section six it is suggested that the boundaries can be defined by the interdependence between certain kinds of economic activities on the one hand and their appropriate institutional framework on the other. An institutional endowment favourable towards one kind of economic activity can be hostile to others. The very reasons why cognitive distance might be small within the cluster will, it is asserted, make the cognitive distance very great between clusters. When access to dissimilar bodies of knowledge is required in product innovation, too much clustering becomes perhaps a burden and further clustering ceases. The final section points to areas where future research is needed to expand and elaborate on the theory of the cluster.

3.2 The existence of the cluster

At least since Alfred Marshall's initial reflections on localized industries and the industrial district were published in his 'Principles of Economics' in 1890, scholars from a range of different fields have regularly concerned themselves with the issue.[6] The bulk of the studies in most of the next century were, however, mainly ideographic and the reasons why firms cluster were assumed or implied rather than carefully investigated and specified. It was almost as if the benefits associated with the cluster were considered self-evident enough to require little discussion (Feser, 1999). When an explanation was offered it was usually based on a model where the balance between centripetal and centrifugal forces determined the locational pattern of firms. The dispersing forces normally included the costs of congestion, or the bidding-up of prices for land and labour. The concentrating forces were, in contrast, often identified as the cost advantages in transportation or when sharing an environment made particularly agreeable by, for instance, a dedicated infrastructure, a pool of notably skilled labour, an educational system of distinctive relevance, etc.

This model largely disappeared as the swelling interest in the clusters towards the end of the twentieth century occasioned a number of novel research propositions to unfold.[7] Instead the main emphasis shifted towards explanations more or less explicitly based on transactions costs including search and information costs, bargaining and decision costs, as well as policing and enforcement costs (Babbage 1835; Dahlman, 1979).[8] As Coase pointed out:

> In order to carry out a market transaction, it is necessary to discover who it is that one wishes to deal with, to inform people that one wishes to deal and on what terms, to conduct negotiations leading up to a bargain, to draw up a contract, to undertake the inspection needed to make sure that the terms of the contract are being observed, and so on. These operations are often extremely costly, sufficiently costly at any rate to prevent many transactions that would be carried out in a world in which the pricing system worked without cost. (Coase, 1960; p. 15)

[6] Since the outstanding contributions by Marshall (1890, 1919), major works have been published by Weber (1909), Hoover (1948), Perroux (1950), Hirschman (1958), Ullman (1958), Jacobs (1961), Chinitz (1961), Greenhut (1970) and Pred (1976, 1977).

[7] Accounts of this literature can be found in Harrison (1992), Norton (1992), Baptista (1998), Storper (1995), Bianchi (1998), and Yeung (2000).

[8] Other costs of using the market include the cost of establishing the appropriate incentive arrangements (Foss, 1993). Babbage (1835) for instance observed how flour could be purchased cheaper on the market than if the government produced it themselves. Nevertheless the latter course of action was preferred rather than to carry the costs of verifying each sack of flour purchased. Information asymmetries give rise to monitoring costs that make authority more efficient than market governance.

Much in this spirit some of the recent cluster studies have emphasized how the local activity will rise and the economic growth rate increase when the co-localization of firms benefits from the information easily available on potential partners in the vicinity and, perhaps more importantly, from the ease in conducting business with such local firms. The reason for the latter is found in the behavioural constraint imposed on co-localized firms by the knowledge of the unattractive consequences of misbehaving. In a cluster it will immediately be noticed if a firm attempts to over-utilize asymmetrical information; or pass defective or substandard goods as first class; or create hold-ups in order to benefit at the expense of others in the local milieu. The information of such misbehaviour will be passed on to everyone, who in the future will tend to take their business elsewhere. Worse still, by becoming a local outcast the firm is deprived of the flow of knowledge, including its tacit parts, which can prove very difficult to substitute. Co-localized firms will therefore often benefit from the emergence of a general climate of understanding and trust where resources are committed to activities whose outcome depends upon the cooperative behaviour of others (Glaeser *et al.,* 1999). Such arrangements, it is asserted, help reduce malfeasance, induce reliable information to be volunteered, cause agreements to be honoured, place negotiators on the same wave-length and ease the sharing of tacit knowledge.

The cluster thus exists, it is often implied, because the co-location of firms cuts the expenses of identifying, accessing or exchanging products, services or, not least, knowledge between firms (Malmberg, 1996, 1997).

However, it is not always realized that such costs might be eliminated altogether by joining the different activities and placing them under one common authority or ownership. When it comes to reducing transaction costs only, the single firm is superior to all market configurations imaginable – even to the high-trust cluster. The benefits of substituting inter-firm interaction with the managerial authority of a single firm is, incidentally, one of the most significant reasons identified in the management literature for the birth and rise of the successful multinational enterprise as Teece, among others, has observed:

> Internal trading changes the incentives of the parties and enables the firm to bring managerial control devices to bear on the transaction, thereby attenuating costly haggling and disruptions and other manifestations of non-cooperative behaviour. Exchange can then proceed at lower cost and with higher returns to the participants. (Teece 1980; p. 232)

The joining together of co-localized firms in related industries under one common ownership will, in addition to possible scale economies, help both to align incentives and to diminish transaction costs. It seems to follow that no theory attempting to explain the existence of the cluster can be based on the reduction of transport, information and transactions costs only.

In order to get a grip on the problem at hand we need to start by recognizing how the continued formation and survival of the cluster attest that the total economic effect of curtailed information and transaction costs as well as of scale advantages are *inferior* to the locational economies available when *being separate firms*[9].

But what *are* then the advantages of N co-localized firms of size S undertaking related activities that are not transferable to a single firm of size S*N doing the same? This is arguably the single most important question for understanding the existence of the cluster, yet largely ignored in the conversation on the subject.

In order to structure the discussion that follows, Richardson's (1972) now classical dichotomy can be helpful when distinguishing between the horizontal dimension of the cluster, consisting of firms with similar capabilities that carry out similar activities, and the vertical dimension composed of firms with dissimilar but complementary capabilities that carry out complementary activities. Richardson explains:

> Now it is quite clear that similarity and complementarity... are quite distinct; clutch linings are complementary to clutches and to cars but, in that they are best made by firms with a capability in asbestos fabrication, they are similar to drain-pipes and heat-proof suits. Similarly, the production of porcelain insulators is complementary to that of electrical switchgear but similar to other ceramic manufacture. And while the activity of retailing toothbrushes is complementary to their manufacture, it is similar to the activity of retailing soap. (Richardson, 1972: p.889)

Complementarity signals scope for fruitful exchange while similarity in activities spells contest and market encounter. The firms in the vertical dimension of the cluster will, accordingly, often be business partners and collaborators. The horizontal dimension will, on the contrary, consist mainly of rivals and competitors. Both dimensions contain features that might contribute to

[9] The advantages of proximate specialized suppliers and customers in the cluster is in principle equally available to one big firm as to, say, twenty smaller doing similar things just as most of the advantages in relation to the skills developed in the local labour market might be just as big or small for twenty co-localized firms of a given size as for a single firm, twenty times bigger.

explaining the existence of the cluster and both will in turn be looked into below.

3.3 The horizontal dimension of the cluster

Alfred Marshall (1890) long ago hinted at an explanation for the existence of the cluster along the horizontal dimension of the cluster.[10] Marshall's reflection concerns the advantages of variation that are caused by the parallel performance of similar tasks. It is based on the conjecture that firms (i.e. owners, managers, and employees) have different perceptive powers, divergent insights and unlike attitudes. Their different valuation of the information at hand results from an idiosyncratic and at least partly tacit way in which the information is initially assembled and interpreted (Casson, 1982). Consequently, firms develop a variety of solutions as an intricate part of their daily operations when holding dissimilar beliefs about their chances of success if using one of several possible approaches to similar problems (von Hayek, 1937).

Even when trying hard it would be extremely difficult and often impossible for a single, multidivisional firm to replicate internally the process of parallel experimentation and testing of a variety of approaches that take place among a group of independent firms doing similar things in the cluster. For as Loasby points out:

> Competing visions between firms are necessary features of an evolutionary or experimental economy. But competing visions within firms, unless very carefully managed, and limited in scope, cause trouble. (Loasby 2000; p. 11)

Co-localized firms undertaking similar activities find themselves in a situation where every difference in the solutions chosen, however small, can be observed and compared. While it might be easy for firms to blame the inadequate local factor market when confronted with the superior performance of competitors located far away, it is less so when the premium producer lies down the street. The sharing of common conditions, opportunities and threats make the strength and weaknesses of each individual firm apparent to the

[10] For some reason or another Marshall's explanation never really entered the discussion of the cluster before Brian Loasby (1999, 2000) recently reintroduced it. A crude and unsubstantiated hypothesis could be that those who had been occupied with clusters had focused their attention on book IV in Marshall's 'Principles of Economics' while those who also cared to read book V did so as part of a different agenda and never felt inclined to become engaged in the cluster discourse.

management, the owners, the employees and everyone else in the cluster who cares to take an interest. Co-location, furthermore, provides firms with an arsenal of instruments to obtain and understand even the most subtle, elusive and complex information of possible relevance developed along the horizontal dimension of the cluster.

It is by watching, discussing, and comparing dissimilar solutions often emerging from the everyday practices that firms along the horizontal dimension of the cluster become increasingly engaged in the process of learning and continuous improvement, on which their survival depends. Harrison C. White saw this very clearly in his account for the essence of competition:[11]

> Markets are self-reproducing social structures among specific cliques of firms and other actors who evolve roles from observing each other's behavior. I argue that the key fact is that producers watch each within a market. Within weeks after Roger Bannister broke the four-minute mile, others were doing so because they defined realities and rewards by watching what other "producers" did, not by guessing and speculating on what the crowds wanted or the judges said. Markets are not defined by a set of buyers, as some of our habits of speech suggest, nor are the producers obsessed with speculations on an amorphous demand. I insist that what a firm does in a market is to watch the competition in terms of observables (White, 1981; p. 518)

If the firms operating along the horizontal dimension of the cluster were to be spread thinly throughout a large city among many unrelated businesses, their ability to monitor and subsequently learn from each other's mistakes and successes would be severely restricted. In the focused and transparent environment of the cluster successful experiments can be more easily distinguished from the less successful by knowledgeable local observers. Sharing a communal social culture including collective beliefs, values, conventions and language often significantly assists them in this process. Promising avenues identified by one firm become available to others. Even when carefully guarded or protected by a patent enough information often leaks out to set local competitors on the track and enable them to 'invent around' the protection (Maskell, 2001). Firms along the horizontal dimension of the cluster are constantly given the opportunity to imitate the proven or foreseeable success of others while adding some idea of their own.

[11] White's proposition can be found in several later works and his idea is at the core of Porter's (1990) concept of rivalry.

The resulting enhanced knowledge creation following from the ongoing sequence of variation, monitoring, comparison, selection and imitation of identified superior solutions is in essence why N similar firms of size S are not equal to one firm of size N * S doing the same.

The advantages suggested stem from the specific forms of knowledge creation available to the individual firm when pursuing self-defined objectives, but not to the division of a larger entity where instructions are received and actions restrained by some procedure or limitation imposed from above.

It might be worth emphasizing an essentially Darwinian feature of the process of variation; as long as the firms share a common language and certain codes that ease their interpretation of local events no trust is required as a prerequisite for learning. The sequence of variation, monitoring, comparison, selection and imitation can take place without any close contact or even an arm-length interaction between the firms. While suppliers and customers simply *need* to interact with each other in order to do business competitors do not. Most relationships in the cluster will therefore be along the vertical dimension.[12] This, however, does not imply that the firms in the horizontal dimension of the cluster never co-operate by helping each other in overcoming technical problems, by lending materials and swapping surplus capacity or by exchanging information. In fact, they may interact regularly, even intimately so, in order to forward some particular scheme (Allen, 1983). On the other hand, they might just as well hate each other intensely, never exchanging anything useful whatsoever.

The proposition put forward here simply suggests that the cluster exists because of locational economies *that are independent of the internal degree of interaction* at least in principle. The sole requirement is that *many firms undertaking similar activities* are placed in circumstances by co-locating where they can monitor each other constantly, closely, and almost without effort or costs.

Other arguments for the existence of the cluster can be found along the vertical dimension of the cluster and to these we shall turn next.

[12] This theoretical point has been supported by empirical findings (Håkanson, 1987).

3.4 The vertical dimension of the cluster

The vertical dimension of the cluster consists of firms linked through input/output-relations.[13] Specialized suppliers and critical customers become attracted to the cluster, once established, by the particular opportunities available. The vertical dimension of the cluster might, however, also be developed by task-partitioning, which tends to evolve spontaneously when economic agents are free to pursue their own advantage, as pointed out by Adam Smith more than 250 years ago:

> In a tribe of hunters and shepherds a particular person makes bows and arrows, with more readiness and dexterity than any other. He frequently exchanges them for cattle or for venison with his companions; and he finds at last that he can in this manner get more cattle and venison than if he himself went to the field to catch them. From a regard to his own interest therefore, the making of bows and arrows grows to be his chief business, and he becomes a sort of armourer (Smith, 1776; p. 119)

Some firms will thus gradually move from the horizontal to the vertical dimension of the cluster by concentrating on some particular process, where they believe they possess or might develop certain lucrative capabilities, dissimilar to others. Such distinct capabilities, once developed, will gradually be improved through a continuing process of learning-by-doing. As the cluster's vertical dimension develops and firms become more specialized they often find solutions to problems otherwise overlooked and bypassed, even when specializing in performing some particularly trivial tasks. An extended division of labour is therefore often closely associated with an acceleration of the growth of knowledge in the cluster.

The steady deepening of the division of labour is limited not only by the extent of the market,[14] but also by information asymmetries and the costs of coordination. Knowledge dispersed needs to be reassembled in order to be useful and firms need to co-operate in matching their related plans in advance since '...the one that make the heads of the pins must be certain of the cooperation of the one who makes the points if he does not want to run the risk of producing pin heads in vain' (List, 1841; p. 150).

[13] The product innovation literature has firmly established that firms learn from each other when inter-acting. See, for instance, Rosenberg (1972), Freeman (1982, 1991), Håkansson (1987), Kline and Rosenberg (1986), Hagedoorn and Schakenraad (1992), and OECD (1992).

[14] See Smith (1779), Young (1928) and Stigler (1951).

In addition, firms hold asymmetrical knowledge about products and market opportunities. These asymmetries arise as an unavoidable consequence of the way in which knowledge is produced. Inter-firm learning is therefore always subject to both thresholds, before the knowledge-bases of firms divided have grown sufficiently apart for interacting to imply learning, and ceilings, after which the cognitive distance becomes too great for firms to bridge, and learning, consequentially, will cease.

Firms in the cluster might have some advantages on both accounts compared to outsides. The spatially defined community that often emerges when related firms co-locate makes it easier for them to co-ordinate and to bridge communication gaps resulting from heterogeneous knowledge endowments (Eliasson, 1996) and understand motives and desires that in other circumstances would remain opaque. By reducing the costs of co-ordination and by overcoming problems of asymmetrical information the process of clustering tilts the balance in favour of further specialization so that a higher level of knowledge creation might be obtained. The main advantages are not the ease of intra-cluster interaction as such, as our manner of speech sometimes seems to suggest, but the deepening of the knowledge base that it enables.

The analysis so far thus suggests a reason for the existence of the cluster along the vertical dimension supplementing the one offered in the previous section on the horizontal dimension. When creating an appropriate vertical differentiation new economic activities become possible, knowledge creation is forwarded, and the resulting extension of the internal market helps make the process self-reinforcing (Young, 1928).

It follows from the concept of variation dealt with in the previous section, that if all firms in the cluster hold complementary capabilities while no two firms hold similar capabilities then all learning through variation and monitoring must necessarily cease. A continued division of labour among firms in the cluster might thus only be expedient for the overall knowledge creation up to a certain point. Beyond that, the benefits might be offset by the corresponding reduction in knowledge creation as variation is diminished and fewer possible avenues of progress are tried out in parallel. Only by a steady increase in the number of firms in the cluster would it be possible to create knowledge simultaneously by variation and by the division of labour.

3.5 The growth of the cluster

To the extent that incumbent firms in the cluster are able to reap the benefits of enhanced learning along its horizontal or vertical dimensions a non-random improvement in performance is to be expected.

In addition to the expansion of incumbents the cluster might grow by an increase in the number of firms through three different processes. First, already existing firms located elsewhere might be tempted to relocate all or a part of their activities to the cluster because of the real or imagined advantages of getting better access to the local knowledge base or to the suppliers or customers already present. As the Finnish CEO of Nokia-Mobira, J.U. Nieminen, once stoically noticed when commenting on these cluster-based advantages:[15]

> When an inventor in Silicon Valley opens his garage door to show off his latest idea, he has 50 per cent of the world market in front of him. When an inventor in Finland opens his garage door, he faces three feet of snow (van Tulder, 1988; p. 169).

Second, a dominant position will also attract entrepreneurs with ambitions to start firms in the particular industry. This is why many of the tallest wannabes within the film industry tend to end up in Hollywood and many of the world's best specialists in information and communication technology at some stage find themselves in Silicon Valley. Immigrating individuals and firms can over time have quite spectacular effects by the way they fuel the growth of the cluster.

Third and finally, new firms come into being in the cluster by spin-offs; smaller or larger groups of former employees recognize a potentially profitable business opportunity and decide to exploit it by becoming entrepreneurs themselves (Belussi, 1999b).[16]

[15] Nokia, however, stayed in Finland and has by now become the world leading developer and manufacturer of mobile or cellular phones.

[16] Dalum (1995) shows how most of the many firms currently active in the communication cluster in Northern Jutland (Denmark) can be traced back to one initial firm producing off-shore radio equipment. Similar genealogical accounts for the emergence of many Canadian clusters have been established by a variety of local organizations. The general tendency for new firms to have their main activity within similar or complementary industries to the ones already operating in the area has been shown empirically for Denmark by Maskell (1992). If all incumbents were placed in a region by industry matrix with 2.496 (12 by 208) cells, most cells would be empty but only 12 per cent of the new firms established throughout a twenty year period would be located in an empty cell.

By starting activities close to what is already going on in the cluster all new spin-offs, newcomer or local independent entrepreneur alike, can safely skip the burdensome and costly process of gathering a lot of circumstantial knowledge about the business environment otherwise crucial; when it works for the neighbour why shouldn't it also work for me? New start-ups are thus given for free, the advantages of a business environment tailored to their specific needs, even in situations when they might still be unaware of what these needs might be or how they may best be accommodated.

The availability of a suited business environment is, of course, important not only for starts-ups but also for the incumbents. By their everyday practices both new and old firms simultaneously rely on and contribute to the further development of this particular environment. It is to this issue that we shall now turn.

3.6 The boundaries of the cluster

The processes of knowledge creation along the horizontal and vertical dimensions of the cluster are rooted in the day-to-day operations of the firms but influenced by a complex set of institutions developed over time.[17] Some of these institutions are of a general nature, equally applicable and useful for promoting the economic activity in all clusters, or at least in a large number of clusters, almost regardless of the particular activities carried out by the firms located there. The emergence of general formal constraints, communal regimes of appropriation and a common climate of understanding and trust, discussed above, belongs to this category.[18]

[17] See Cannan (1912). We might follow North (1994; p. 360) in defining institutions as "humanly devised constraints that structure human interaction. They are made up of formal constraints (e.g., rules, laws, constitutions), informal constraints (e.g., norms of behaviour, conventions, self-imposed codes of conduct), and their enforcement characteristics", while explicitly adding Smith's (1997) economic (knowledge) infrastructures that result from conscious policy decisions and investment programmes and include special programs in local schools and universities, government-supported technical institutions and training centres, specialized apprenticeship programmes, etc.

[18] It might be argued that the specific way by which trust is obtained will make it differ from cluster to cluster and that very few 'general institutions' can therefore be expected to be found in practice.

Other institutions have, however, a definite scope and will differ from one cluster to the next.[19] It is reasonable to assume that the cluster's particular set of institutions has emerged as a response to the special requirements of the activities performed in the cluster.[20] There is thus a fundamental *interdependence* between the economic structure and the institutions of the cluster as they have developed over time.[21]

It has been suggested that while the cluster's particular set of activities affects what is done within and among the firms in the cluster and therefore *what is learnt*, it is the institutions in the cluster that define how things are done and consequently *how learning takes place* (Lundvall and Maskell, 2000).

Just as the set of firms undertaking similar and complementary activities differ between clusters, so do institutions. Different activities each have their own mode of learning that gives rise to different institutional outcomes. The resulting institutions in turn assist the firms of the cluster when facing the challenges and opportunities presented by changes in the outside world.

Recent research has established the close interaction between structure and institutions when investigating industrial failure in places with a particularly favourable factor endowment. Eskelinen and Kautonen (1997), for instance, demonstrate how Finland, with its bounteous supply of high-quality timber resources, high educational and training standards and a long track record of world class designers, has been losing out in wooden furniture production to countries with obviously inferior resources on some or all of these counts. The wooden furniture industry is generally characterized by

- very flexible small batch production,
- modest capital intensity,
- skilled or semi-skilled labour,
- integrated design,
- frequent contact to many different and shifting groups of customers,
- few long term contracts, and

[19] This is the line taken in much of the innovation systems literature (Lundvall 1992; Nelson, 1993).

[20] On the national level recent research has proved the existence of such a correlation between patterns of specialization in production and trade, on the one hand, and the knowledge base on the other (Archibugi and Pianta, 1992).

[21] Some argue that the differences emanate mainly from structural characteristics (Breschi and Malerba, 1997) while others look at how institutional specificities affect the location of certain industries (Guerrieri and Tylecote, 1997). See also Gertler (1995a, 1996, 1998) and Maskell and Törnqvist (1999).

- periods of high activity alternating with inactive spells of uncertain length (Maskell, 1998; Lorenzen, 1998).

However, the relevant Finnish institutions of the wood processing industry were defined, *not* by the wooden furniture industry, but by the far larger user of wood as a primary input: the manufacturing of paper and pulp. This industry, in contrast, is characterized by

- long production runs,
- extremely high capital intensity,
- few highly skilled employees
- many low or semi-skilled workers,
- no design,
- a rather stable set of very big customers,
- long term contracts dominate, and
- very structured systems of production and maintenance.

The dominance of the paper and pulp industry in Finland has led to certain highly significant idiosyncrasies in business behaviour on the timber market. As a result an institutional environment is created that is highly supportive for the paper and pulp industry, but distorting to the wooden furniture industry (Kautonen, 1996). Important institutional features hostile to the wooden furniture industry are:

- no distinction between roundwood quality classes of relevance for furniture production
- fixed standard pricing practices for roundwood
- large volume discounts on purchase of wood combined with
- long term wood contracts and upstream vertical integration
- emphasis in governmental policy and managerial ethos on technology, scale economies and process innovation rather than on market presence, design and product innovation
- output markets seen as given once and for all
- hierarchical labour relations supported by labour market agreements structured in ways penalizing small firms,

The institutions developed and refined to accommodate the needs of the firms in the paper and pulp industry created so unfavourable a business climate for the wooden furniture industry that it stagnated and declined, while the same industry blossomed in the neighbouring country of Denmark. To the naked eye Danish firms faced a considerable handicap as the insufficient local timber supply forced them to import most of the wood needed from countries such as Russia, Sweden, Poland, and – noticeably – also Finland. However, this cost disadvantage counted for very little compared to the advantages of *not* being burdened with an unfavourable institutional endowment created by a dominant industry like the Finnish paper and pulp industry (Lorenzen, 1998).

The lesson learned by the fate of the Finnish wooden furniture industry is that the more helpful an institutional endowment becomes for one type of activities the less suitable it can be for others. The significance of an appropriate *fit* between industry and institution also suggest why certain types of activity are never found in the same cluster. A cluster producing fashion wear or financial services will simultaneously develop (dissimilar) institutions that most likely will turn out to be alien to the production of ships, coal or cars.

The restrained ability to 'stretch' an institutional endowment to serve different kinds of economic activities equally well might also partake in explaining why new clusters emerge; when knowledge grows and the economic activity begins to differentiate, requirements starts to depart and new clusters are likely to be established with institutions of their own.

The boundaries of the cluster might therefore be defined by the *fit* between the economic activities carried out by the related firms of the cluster on the one hand and the particular institutional endowment developed over time to assist these activities on the other.[22] The expansion into new activities along the vertical dimension of the cluster ceases to be feasible when the fit begins to weaken.

This framework might also account for the dispersing forces at work when the additional value created from spanning across distant bodies of knowledge must justify the additional transaction costs involved. Firms heavily engaged in inter-firm innovation across usually unrelated activities and bodies of knowledge might, perhaps, be better off if not being *too* embedded in a particular cluster and, as an unavoidable consequence, be facing an even greater cognitive distance to potential partners when interaction is required.

Furthermore, the framework might provide an explanation for the demise of clusters as existing value chains at some point become fragile and new ones are being moulded. As the new vertical dimension is gradually developed the needed institutional adjustments will almost inevitably meet resistance from old incumbents struggling to survive. If some compromise is not found the resulting tension can easily lead to steady decline.

[22] If no such mechanism restricted the cluster's institutional endowment to a *certain kind of related industries only*, we would ultimately expect to end up with a single and rather large cluster containing all economic activity.

3.7 Final comments

The core of the argument presented in this article is that any economic theory of the cluster must address certain basic questions in order to be satisfactory.

First, such a theory must at the very least contain an explanation for the *existence of the cluster.* The theory must specify the process or processes that impel related firms to assemble and stay together in one place and - by doing so - make them thrive. More specifically, the theory must provide an explanation for the advantages that *many* related and co-localized firms might accrue but which are not available to a hypothetical *single* firm carrying out precisely the same activities, even if at the same location, using the same suppliers, customers and workforce.

It is suggested that the cluster exists because of the enhanced knowledge creation stemming from the variation developed along the horizontal dimension of the cluster, supported by the reduced costs of co-ordinating dispersed knowledge, of overcoming problems of asymmetrical information and aligning incentive, as well as of easing the actual transactions taking place along the vertical dimension.

Second, a theory of the cluster must include an explanation for the *growth of the cluster*. It must identify how new firms emerge and add to the strength of the cluster.

It is argued above that the cluster, once established, acts as a selection device, attracting particular kinds of economic activity comparable with the incumbents and reducing the ambiguity and costs facing local entrepreneurs when keeping close to the activities already present.

This selection device carries with it a set of constraints that might hamper future prosperity when external changes make readjustments necessary.

Third, the theory of the cluster must be able to identify *the boundaries of the cluster* by specifying why the clustering of some economic activities preclude the integration of others.

The reason put forward in this article is based on the idea of a close interdependence or *fit* between the specific economic activity of a cluster and the particular institutional endowment developed. A dawning mismatch leads to decreasing returns. Negative feedback loops start to develop.

Further work might reveal how some of the specific suggestions made in this article when attempting to flesh out a theory of the cluster are ill conceived, or that other issues than the three identified should be included. One such possible candidate will answer questions regarding the *external* fit between the characteristics of the cluster on the one hand and its broader environment on the other. The last decade's many research publications frequently perceive the cluster as the basically random outcome of present or historical processes. Relatively few have so far taken care to ponder the factors that might have made certain environments more or less suited for the emergence of the cluster.[23] The evidence available so far does not constitute the foundation for any general ex ante statements about the suitability of a given economic environment to sustain the growth of a non-random, market-led group of co-localized firms doing similar things. The amount of case studies produced during the last decade does, however, suggest that an effort to develop such statements might lead to interesting results.

There are other important aspects that require further consideration in subsequent research: The question of the *internal organization* will, for instance, be concerned with the ways that different configurations within the cluster might influence its knowledge creating abilities. The theory of the cluster might also be asked to further specify the reasons for the decline of the formerly successful cluster.

Maybe, over time, new research will also make us able to tell whether the possible mismatch between a slowly adjusting institutional endowment and the highly dynamic requirements of many contemporary industries is the primary reason why innovative firms also survive and prosper without being supported by the many proposed advantages of the cluster. The theory of the cluster will not be complete before we more fully understand the successful solitarian.

Acknowledgement

This chapter was initially published in Industrial and Corporate Change vol. 10, no. 4. 2001 under the title: 'Towards a knowledge-based theory of the geographical cluster'. The permission granted by Oxford University Press and Fondazione Associzione di Storia e Studi sull'Impresa (the copyright holders) to use the work in the current volume is gratefully acknowledged.

When writing this article I have benefited from discussions with Kirsten Foss, Brian Loasby, Michael Storper and other members of the DRUID research network. Anders Malmberg, Gabi Dei Ottati, Meric Gertler, Paivi Oinas, Edward J. Malecki, Ash Amin, three anonymous reviewers and the participants at The World Conference on Economic Geography, Singapore, December 4-7,

[23] For an exception see, for instance, Gertler (1993, 1995b).

2000 have provided valuable comments to earlier drafts. I thank The Nordic Center for Spatial Development (NORDREGIO) for financially supporting the research on which this article is based. The usual disclaimers apply.

References

Allen R.C. (1983), Collective invention, *Journal of Economic Behaviour and Organization* vol. 4, p. 1-24.

Archibugi D. and Pianta, M. (1992), *The Technological Specialization of Advanced Countries*, Kluwer, Dordrecht.

Babbage C. (1832), *On the economy of machinery and manufactures*, Charles Knight, London.

Baptista R. (1998), 'Clusters, Innovation and growth: A survey of the Literature'. In: Swann, P.G.M., Prevezer, M. and Stout, D., (eds.) *The dynamics of industrial clustering. International comparisons in computing and biotechnology*, Oxford University Press, Oxford.

Baumard P. (1996), 'Organizations in the fog: an investigation into the dynamics of knowledge'. In: Moingeon, B. and Edmondson, A., (eds.) *Organizational Learning and Competitive Advantage* Sage, London.

Becattini G. (1990), 'The Marshallian industrial districts as a socio-economic notion'. In: Pyke F., Becattini G. and Sengenberger, W. (eds.) *Industrial Districts and Inter-Firm Co-Operation in Italy*, Geneva, International Institute for Labour Studies.

Bellandi M. (1989), 'The industrial district in Marshall', in, Goodman, E. and Bamford, J., (eds.), *Small firms & industrial districts in Italy*, Routledge, London.

Bellandi M. (1996), Innovation and Change in the Marshallian Industrial District. *European Planning Studies* vol. 4, no. 3, p. 357-368.

Belussi F. (1999a), Path-Dependency vs. industrial dynamics: an analysis of two heterogeneous districts. *Human Systems Management*, vol. 18, p. 161-174.

Belussi F. (1999b), Policies for the development of knowledge-intensive local production systems, *Cambridge Journal of Economics* vol. 23, p. 729-747.

Bianchi G. (1998), Requiem for the Third Italy? Rise and fall of a too successful concept. *Entrepreneurship and Regional Development*, vol. 10, p. 93-116.

Blanc H. and Sierra C. (1999), The internationalisation of R&D by multinationals: a trade-off between external and internal proximity, *Cambridge Journal of Economics*, vol. 23, no. 2, p. 187-206.

Breschi S. (1995), *Identifying regional patterns of innovation using patent data*. Paper presented at the workshop on 'Regional Innovation Systems, Regional Networks and Regional Policy', Organized by the STEP group at Lysebu Conference Centre (Oslo), Norway, October 27-29.

Breschi S. (2000), The geography of innovation: A cross-sector analysis, *Regional Studies*, vol. 34, no. 3, p. 213-230.

Breschi S. and Malerba F. (1997), 'Sectoral innovation systems', in, Edquist C. (ed.), *Systems of innovation: Technologies, institutions and organizations*, Pinter Publishers, London.

Brusco S. (1982), The Emilian model: Productive decentralisation and social integration, *Cambridge Journal of Economics*, vol. 6, no. 2, p. 167-184.

Brusco S. (1986), 'Small firms and industrial districts: The experience of Italy', in, Keeble D. and Wever E., (eds.), *New firms and regional development in Europe*, Croom Helm, London.

Brusco S. (1990), 'The idea of the industrial district: Its genesis', in, Pyke, F., Becattini, G. and Sengenberger W. (eds.), *Industrial districts and inter-firm co-operation in Italy* International Institute for Labour Studies, Geneva.

Brusco S. (1999), 'The rules of the game in industrial districts', in, Grandori A., (ed.), *Interfirm networks. Organization and industrial Competitiveness* Routledge, London.

Brusco S. and Righi, E. (1989), Local government, industrial policy and social consensus: the case of Modena, *Economy and Society*, vol. 18, n. 4, p. 405-424.

Camagni R.P. (1995), The concept of "innovative milieu" and its relevance for public policies in European lagging regions, *Papers in Regional Science*, vol. 74, no. 4, p. 317-340.

Cannan E. (1912), *The history of local rates in England in relation to the proper distribution of the burden of taxation*, P.S. King & Son, Orchard House, London.

Carter A.P. (1994), *Measuring the Performance of a Knowledge-based Economy*, Working Paper no. 337, Brandais University, Departments of Economics, Brandais.

Casson M. (1982), *The entrepreneur. An economic theory*, Martin Robertson, Oxford.

Chinitz B. (1961), Contrasts in Agglomeration: New York and Pittsburgh. *American Economic Review* vol. LI, no. 2, p. 279-289.

Coase R.H. (1960), The Problem of Social Cost, *Journal of Law and Economics*, vol. 3, p. 1-44.

Dahlman C.J. (1979), The problem of externality, *Journal of Law and Economics*, vol. 22, no. 1, p. 141-162.

Dalum B. (1995), Local and Global Linkages, The Radiocommunications cluster in Northern Denmark, *Journal of Industry Studies* vol. 2, no.2, p. 89-109.

Dei Ottati G. (1994a), Co-operation and competition in the industrial district as an organisational model, *European Planning Studies*, vol. 2, p. 463-483.

Dei Ottati G. (1994b), 'Case study I: Prato and its evolution in a European context', in, Leonardi, R. and Nanetti, R.Y., (eds.), *Regional development in a modern European economy: The case of Tuscany*, Pinter Publishers, London.

Dei Ottati G. (1996), Trust, interlinking transactions and credit in the industrial districts. *Cambridge Journal of Economics*, vol. 18, p. 529-546.

Eliasson G. (1996), *Firm Objectives, Controls and Organization. The Use of Information and the Transfer of Knowledge within the Firm*, Kluwer Academic Publishers, Dordrecht.

Ellison G. and Glaeser E.L. (1994), *Geographical Concentration in the US Manufacturing Industries. A Dartboard Approach*, Working Paper no. 4840, National Bureau of Economic Research (NBER), Cambridge, Mass.

Eskelinen H. and Kautonen M. (1997), 'In the shadow of the dominant cluster - the case of furniture industry in Finland', in, Eskelinen H., (ed.), *Regional Specialisation and Local Environment - Learning and Competitiveness*, NordREFO Report 1979, no. 3, NordREFO, Stockholm.

Estall R.C. and Buchanan, R.O. (1961), *Industrial Activity and Economic Geography*, Hutchinson, London.

Feser E. J. (1999), 'Old and new theories of industrial clusters', in, Steiner M. (ed.), *Clusters and regional specialisation. On geography, technology and networks*, Pion, London.

Feldmann M.P. (1994), *The geography of innovation*, Kluwer, Dordrecht.

Foss N.J. (1993), More on Knight and the theory of the firm, *Managerial and Decision Economics*, vol. 14, p. 269-276.

Freeman C. (1982), *The economics of industrial innovation*, Pinter Publishers, London.

Freeman C. (1991), Networks of Innovators: a Synthesis of Research Issues, *Research Policy* vol. 20, no.5, p. 5-24.

Garofoli G. (1992a), 'Industrial districts: structure and transformation', in, Garofoli, G., (ed.), *Endogenous development and Southern Europe*, Avebury, Aldershot.

Garofoli G. (1992b), 'The Italian model of spatial development in the 1970s and 1980s, in, Benko G. and Dunford M., (eds.), *Industrial change and regional development,* Bellhaven Press, London.

Garofoli G. (1993), Economic Development, Organization of Production and Territory, *Revue d'Économie Industrielle*, vol. 64, p. 22-37.

Gertler M.S. (1993), Implementing advanced manufacturing technologies in mature industrial regions: towards a social model of technology production, *Regional Studies*, vol. 27, p. 665-680.

Gertler M.S. (1995a), 'Being there': proximity, organization, and culture in the development and adaptation of advanced manufacturing technologies, *Economic Geography* vol. 71, p. 1-26.

Gertler M.S. (1995b), Manufacturing Culture: The Spatial Construction of Capita, paper presented at the annual conference of the Institute of British Geographers, 3-6th January. (mimeo).

Gertler M.S. (1996), Worlds apart: the changing market geography of German machinery industry, *Small Business Economics,* vol. 8, p. 87-106.

Gertler M.S. (1998), 'The Invention of Regional Culture', in, Lee R. and Wills J., (eds.), *Geographies of Economies,* Hodder Headline Group, London.

Glaeser E.L., Laibson C.L., Scheinkman J.A. and Soutter C.L. (1999), *What is social capital? The determinants of trust and trustworthiness,* Working paper no. 7216, NBER, Cambridge Mass.

Gottardi G. (1996), Technology strategies, Innovation without R&D and the creation of knowledge within industrial districts, *Journal of Industry Studies* vol. 3, no. 2, p. 119-134.

Greenhut M.L. (1970), *A Theory of the Firm in Economic Space*, Appleton-Century-Crofts, New York.

Guerrieri P. and Tylecote A. (1997), 'Interindustry differences in technical change and national patterns of technological accumulation', in, Edquist, C. (ed.) *Systems of innovation: Technologies, institutions and organizations,* Pinter Publishers, London.

Hagedoorn J. and Schakenraad J. (1992), Leading companies and networks of strategic alliances in information technologies, *Research Policy*, vol. 21, p. 163-181.

Håkansson H. (1987), (ed.), *Industrial Technology Development - a Network Approach.* Croom Helm, London.

Harrison B. (1992), Industrial Districts. Old wine in New Bottles? *Regional Studies*, vol. 26, p. 469-483.

Hayek F.A. (1937), Economics and knowledge, *Economica,* vol. IV, no. 13, p. 33-54.

Hirschman A.O. (1958), *The strategy of economic development*, Yale University Press, Clinton Mass.

Hoover E.M. (1948), *The location of Economic activity*, McGraw-Hill Book Company, New York.

Hoover E.M. (1970), *An introduction to regional economics*, Alfred A. Knopf, New York.

Jacobs J. (1961), *The death and life of great American cities*, Random House, New York.

Jaffe A.B., Trajtenberg M. and Henderson R. (1993), Geographic localization of knowledge spillovers as evidence of patent citations, *Quarterly Journal of Economics* vol. 63, no. 3, p. 577-598.

Kautonen M. (1996), Emerging Innovative Networks and Milieux: the case of the Furniture Industry in the Lahti Region of Finland, *European Planning Studies*, vol. 4, no. 4, p. 439-456.

Kirat T. and Lung Y. (1999), Innovation and Proximity: Territories as Loci of Collective Learning Processes, *European Urban and Regional Studies*, vol. 6, no. 1, p. 27-38.

Kline S.J. and Rosenberg N. (1986), 'An overview of innovation', in, Landau R. and Rosenberg N., (eds.), *The positive sum game* National Academy Press, Washington D.C.

Laestadius S. (1996), *Technology level, knowledge formation, and industrial competence within paper manufacturing*, Working paper, Kungliga Tekniska Högskolan, Department of Industrial Economics and Management.

List F. (1841), *The National System of Political Economy*, Longmans, Green and Co., London.

Loasby B.J. (1999), 'Industrial districts as knowledge communities', in, Bellet, M. and L'Harmet, C., (eds.) *Industry, space and competition. The contribution of economists of the past*, Edward Elgar, Cheltenham.

Loasby B.J. (2000), *Organisations as interpretative systems*. Paper presented at the DRUID Summer Conference, Rebild, Denmark (www.business.auc.dk/DRUID)

Lorenzen M. (1998), (ed.), *Specialisation and Localised Learning. Six Studies on the European Furniture Industry*, Copenhagen Business School Press, Copenhagen.

Lundvall B.-Å. (1992), (ed.), *National systems of innovation: Towards a Theory of Innovation and Interactive Learning*, Pinter Publishers, London.

Lundvall B.-Å. and Maskell P. (2000), 'Nation states and economic development - From national systems of production to national systems of knowledge creation and learning', in, Clark, G.L., Feldmann, M.P. and Gertler, M.S., (eds.), *The Oxford Handbook of Economic Geography*, Oxford University Press, Oxford.

Maillat D. (1991), 'Local dynamism, milieu and innovative enterprises', in, Brotchie, J., Batty, M., Hall, P. and Newton, P., (eds.), *Cities of the 21st century*, Longman, London.

Maillat D. (1998), Innovative milieux and new generations of regional policies, *Entrepreneurship and Regional Development*, vol. 10, p. 1-16.

Malecki E.J. (1991), *Technology and Economic Development: The Dynamics of Local, Regional and National Change*, Longman, Harlow.

Malerba F. (1992), Learning by Firms and Incremental Technical Change, *Economic Journal* vol. 102, p. 845-860.

Malmberg A. (1996), Industrial geography. Agglomerations and local milieu, *Progress in Human Geography*, vol. 20, p. 392-403.

Malmberg A. (1997), Industrial Geography: location and learning, *Progress in Human Geography*, vol. 21, p. 573-582.

Malmberg A. and Maskell P. (1997), Towards an explanation of industry agglomeration and regional specialization, *European Planning Studies*, vol. 5, no. 1, p. 25-41.

Markusen A., Hall P. and Glasmeier A.K. (1986), *High Tech America*, Allen and Unwin, Boston.

Marshall A. (1890), *Principles of Economics* , Macmillan, London.

Marshall A. (1919), *Industry and Trade. A study of industrial technique and business organization, and of their influences on the condition of various classes and nations*, Macmillan, London.

Maskell P. (1992), *Nyetableringer i industrien og industristrukturens udvikling (New firm formation and industrial restructuring in Denmark)*, Copenhagen Business School Press, Copenhagen.

Maskell P. (1998), Successful low-tech industries in high-cost environments: The case of the Danish furniture industry, *European Urban and Regional Studies,* vol. 5, no. 2, p. 99-118.

Maskell P. (2000), 'Social capital and competitiveness', in, Baron S., Field J. and Schuller T. (eds.), *Social Capital. Critical Perspectives,* Oxford University Press, Oxford.

Maskell P. (2001), 'Knowledge creation and diffusion in geographic clusters: Regional development implications', in, Felsenstein D., McQuaid R., McCann D. and Shefer D. (eds.) *Public Investment and Regional Economic Development* Edward Elgar, London.

Maskell P. and Malmberg A. (1999), Localised learning and industrial competitiveness, *Cambridge Journal of Economics,* vol. 23, no. 2, p. 167-186.

Maskell P. and Törnqvist G. (1999), *Building a Cross-Border Learning Region. The Emergence of the Northern European Øresund Region,* Copenhagen Business School Press, Copenhagen.

Maskell P., Eskelinen H., Hannibalsson I., Malmberg A. and Vatne E. (1998), *Competitiveness, localised learning and regional development. Specialisation and prosperity in small open economies,* Routledge, London.

Nelson R.R. (1993), (ed.), *National Innovation Systems. A Comparative Analysis,* Oxford University Press, New York.

North D.C. (1994), Economic Performance through Time, *American Economic Review,* vol. 84, no. 3, p. 359-368.

Norton R.D. (1992), Agglomeration and Competitiveness: From Marshall to Chinitz, *Urban Studies,* vol. 29, no. 2, p. 155-170.

OECD (1992), *Industrial Policy in the OECD countries.* Annual Review Paris, The Organisation for Economic Co-Operation and Development.

OECD (1999), *Boosting innovation: The cluster approach,* OECD Proceedings, OECD, Paris.

Perroux F. (1950), Economic space, theory and applications, *Quarterly Journal of Economics* vol. LXIV, p. 89-104.

Porter M.E. (1990), *The Competitive Advantages of Nations,* Macmillan, London.

Pred A. (1976), The interurban transmission of growth in advanced economies: Empirical findings versus regional-planning assumptions, *Regional Studies,* vol. 10, p. 151-171.

Pred A. (1977), *City Systems in Advanced Economies. Past Growth, Present Processes and Future Development Options,* Hutchinson, London.

Ratti R., Bramanti A. and Gordon R., (1998), (eds.), *The Dynamics of Innovative Regions. The GREMI Approac,* Ashgate, Aldershot.

Richardson G.B. (1972), The organisation of industry, *Economic Journal,* vol. 82, p. 883-896.

Rosenberg N. (1972), *Technology and American Economic Growth,* Sharpe, White Plains, N.Y.

Ruggie J.G. (1993), Territoriality and beyond: Problematizing modernity in international relations, *International Organization,* vol. 47, no. 1, p. 139-175.

Saxenian A. (1994), *Regional Advantage. Culture and competition in Silicon Valley and Route 128,* Harvard University Press, Cambridge, Massachusetts.

Smith A. (1776), *An Inquiry into the Nature and Causes of the Wealth of Nations,* W. Strahan and T. Cadell, London.

Smith K. (1997), 'Economic infrastructures and innovation systems', in, Edquist, C. (ed.) *Systems of innovation: Institutions, Organisations and Dynamics,* Pinter, London.

Spender J.-C. (1996), 'Competitive advantage from tacit knowledge? Unpacking the concept and its strategic implications', in, Moingeon, B. and Edmondson A. (eds.), *Organizational Learning and Competitive Advantage,* Sage, London.

Stenberg R. (1999), Innovative linkages and proximity: Empirical results from recent surveys of small and medium sized firms in German regions, *Regional Studies* vol. 33, p. 529-540.
Stigler G.J. (1951), The division of labor is limited by the extent of the market, *Journal of Political Economy* vol. LIX , no. 3, p. 185-193.
Storper M. (1993), Regional 'Worlds' of Production. Learning and Innovation in the Technology Districts of France, Italy and the USA, *Regional Studies*, vol. 27, p. 433-455.
Storper M. (1995), The resurgence of regional economies, ten years later: the region as a nexus of untraded interdependencies, *European Urban and Regional Studies*, vol. 3, no. 2, p. 191-221.
Swann P.G.M., Prevezer M. and Stout D. (1998), *The dynamics of industrial clustering. International comparisons in computing and biotechnology*, Oxford University Press, Oxford.
Teece D.J. (1980), Economies of scope and the scope of the enterprise, *Journal of Economic Behavior and Organization*, vol. 3, no. 1, p. 223-247.
Ullman E.L. (1958), Regional development and the geography of concentration, *Papers of the Regional Science Association*, IV, p. 179-198.
van Tulder R. (1988), 'Small European countries in the international telecommunications struggle', in, Freeman, C. and Lundvall, B.-Å. (eds.), *Small Countries Facing the Technological Revolutions*, Pinter Publishers, London.
Weber A. (1909), *Über den Standort der Industrien*, J. C. B. Mohr, Tübingen.
White H.C. (1981), Where do markets come from? *American Journal of Sociology*, vol. 87, no. 3, p. 517-547.
Yeung H.W. (2000), Organising "the Firm" in Industrial Geography I: Networks, institutions and Regional Development, *Progress in Human Geography*, vol. 24, no. 2 p. 301-315.
Young A. (1928), Increasing returns and economic progress, *Economic Journal*, vol. 38, p. 527-542.

Part II

The generation and acquisition of knowledge: the cognitive approach to the industrial district model

Chapter 4

The Industrial District (ID) as a cognitive system[1]

Enzo Rullani

4.1 Introduction: facts and theories

Localised development is a fact, but is not yet a theory. The empirical evidence of the fact is not sufficient for detecting the causal chain that triggers the growth of a territorial system and determines its pace and direction.

Economic growth has a territorial texture. The idea that development is localised, and organised in territorial *clusters*, is now widely accepted. Empirical observation suggests that territorial *clustering* is not a phenomenon limited to just a few sectors and places, but one of the most widespread manifestations of economic growth (Enright, 2000).

Localised development is not restricted to some well known cases - for instance, the Italian industrial districts, the clusters of *flexible specialisation* in Spain and Germany, as discussed by Becattini (2000), Brusco (1989), and Piore and Sabel (1984). The growth of regional clusters is not restricted to mature sectors of the economy, but also occurred in high-tech sectors, for example in information technology and telecommunications like Silicon Valley and Route 128 (Zagnoli, 1991; Saxenian, 1994). Even multinational companies now carefully select the local implantation of their activities (Vaccà, 1995).

On the contrary, we have many different (and questionable) theories or ideas about how to *re-produce* or *re-direct* local development, since the casual mechanism is almost vague and seems deeply rooted in some singular features of each case history. Despite the success of territorial clustering in practice and

[1] This text summarizes some ideas presented at the Ocde Conference on "Enhancing the Competitiveness of SMEs in the Global Economy: Strategies and Policies", Bologna, 2000. The author is grateful to Sergio Arzeni for the useful questions and suggestions received.

the popularity of districts, clusters, local production systems, we are still far from achieving a satisfactory theoretical systemisation (Amin, 1993, 1988). The situation is embarrassing and, to some extent, dangerous. The danger is that it will encourage *unsuitable* or *dubious* intervention policies, which may prove to be a *waste* of public money, or cause more damage than benefits.

There are many reasons for this uncomfortable epistemic and pragmatic situation.

First of all, localised development is difficult to rationalise because it is the fruit of *on-going learning* processes which remain largely implicit and dispersed among the many elementary units of territorial clustering, without any *master mind* monitoring or directing the process as a whole (Bonomi, 1997). It is often with surprise that researchers "discover" the forms of localised development *after* they have come about.

Secondly, localised development depends on mechanisms with a high level of *complexity*[2]. Complexity is far greater than is generally taken into account by conventional economic theories. For example, it is difficult for economic theory to deal with non *linear processes*, or to take into account non-linear circuits (like feedback or increasing returns mechanisms). In global economy, some local systems may have a fast unexpected growth, while others may rapidly lose their past distinctive advantages, owing to a small change in competitive conditions. The triggering processes of this evolutionary turnaround are not well defined and their premises are under discussion (Becattini, 2000; Amin, 1998; Storper, 1995 and 1997; Rullani, 1998).

Another problem is that the natural "borders" of IDs do not correspond to administrative borders. They are often clustered around small towns, or agglomerations of towns which practically run into one another. This creates discrepancies between IDs and the local public administration whose business it is to promote local development (Sforzi, 1993). We find similar discrepancies between *national* and *local* policies. Because development has its own territorial logic, and because the competitive factors at work are related to local systems and societies, that very rarely meets the aims of national policies.

Further, the categories that economics has adopted – basically the concepts of *local clustering* and *external economies* – are not always adequate to explain the complexity of the phenomenon.

[2] Complexity is a condition produced by great variety (in space), variability (in time) and indeterminacy.

4.2 A multiplying force: the role of knowledge

Our thesis is that local development is necessarily linked to a *multiplying cognitive process*, that increases the size and quality of *intellectual and social capital* that is embedded in a particular place. Local development is due to the locally shared knowledge, which cuts the cost of each agent knowledge production and favours its application. The available shared knowledge can be obtained through co-operative attitudes among local actors, or, more often, through involuntary leakages of information and know how, that local actors are not able to hinder. What is really important, in both cases, is that knowledge can be re-used and multiplied in the local circuit.

But the district multiplier is a *necessary* condition, not a *sufficient* one. Many other premises (generally non-cognitive premises) are required for making the cognitive engine work. But nothing can replace the cognitive engine, because of the *multiplying effect due to a special property of knowledge: the possibility of costless re-production and re-creation of ideas, models, formulas, laws etc.*.

The knowledge engine produces the amplifying effect: development is not the final result of a gradual cumulative growth, but is the outcome of a non linear, discontinuous, evolutionary path.

For example, the initial know-how needed for development was rarely available locally to IDs, even if traditional connected activities were spread almost everywhere. In the great majority of the cases, the innovation that was required to break the traditional equilibrium came from outside. And, afterwards, it was amplified by circuits of learning and accumulation, which also lead to a specific local specialisation.

So, we had a combination between a *contingent* event, that could happen or not (a seed), and a *required* mechanism of amplification and diffusion (a fertile land, acting as a multiplier). IDs development can be explained by a *sufficient* chain of events, not by a *necessary and sufficient* law of causation. In other words, it is a historical path, not a casual effect.

This aspect is crucial also today, if we want to examine the possibility of replicating or extending the district model to other places.

4.3 Production of goods by means of localised resources

Territory is the place both of the *accidental context* (the singular event and the specific circumstances) and of the *necessary* cognitive engine (the multiplying effect).

There is now a tendency to interpret the territory in this second sense, as an economic resource, with its own singular "identity" and "productivity". After the 1970s, economists began to speak of (national) *capitalisms*, each different in terms of institutions, regulations, culture and productive structure. Instead of an undifferentiated space, economists began to look at the different regions and nations, each characterised by specific labour, services, visible and invisible infrastructures, and unique culture and institutions. Localised production – whether on a large scale (national forms of capitalism) or at the grass roots (local systems) – is distinguished from abstract production.

In this view, the territory is a *non-replicable* resource (the goods or services that a territory produces are limited). Besides, the territory is a *unique* resource, which stamps its specific characteristics on the resulting productive process and products, which in turn are marked by the character of the "place" where they are made. The territory can be seen as a depository of culture and habits, and accumulates forms of tacit knowledge, and channels of communication. These institutions remain largely *invisible* to the outside observers, and even to the local actors themselves, because they are embedded in the anthropological-social system grown up in each place (Brusco, 1989; Becattini, 1997).

But uniqueness is not an explication. It is, at most, a description.

The idea of clustering, which summarises the advantages of a specific spatial concentration of activities and the positive effects of external economies, tries to overcome this difficulty, because it refers to a general multiplying mechanism: the increasing returns of territorial concentration processes, through physical proximity. But it is an illusory way to escape difficulties. In an abstract world, advantages and disadvantages of concentration are symmetric forces: their convergence can supply only a new version of the equilibrium theory, not a discontinuous evolutionary path. In order to justify evolutionary surprises and multiplying effects, we need a combination of contingency and necessity, of uniqueness and reproducibility. Only cognitive processes contain both these two dimensions and can transform the one into the other, connecting their evolution along time.

The classical models of localisation (Christaller, Losch) have generated a rather geometrical structure of economic space, which does not fit at all with the increasing observed tendencies to the territorialisation of production activities. What needs to be explained is why a place has an economic, cultural and social *quality* different from that which can be observed in other places. And again, why there is not an easy way of imitating or reproducing the capabilities embedded in firms belonging to a specific ID.

To sum up, IDs and clusters are *signs* of the economic value generated by the territorial singularity of a multiplying mechanism which is fed by unique, localised, knowledge. This is what gives differential value to a specific territory, not the mere presence of some "external economies" resulting from concentration.

4.4 IDs and clusters in a networked economy

Only during the last 20-30 years, were the productive properties of local districts understood and acknowledged. The role of small firms and IDs in economic policy has changed radically compared with the long period of dominant Fordism.

Until a few years ago, we were used to seeing economies of scale and large enterprises as the precondition for acting on the global economy. Small firms were restricted to limited, often local, geographical environments.

Research into foreign direct investment, as a determining variable of the new global economy, tended to confirm this dualism: small firms were mainly exporting, while large enterprises were investing abroad. And this difference was becoming increasingly important in the process of globalisation. At the beginning, the "discovery" of IDs further emphasised this division of roles: small firms were increasingly associated with the local sphere, large companies with the global market.

However, this is not the way things really are. The error of judgement, regarding the degree of internationalisation of IDs, was based on a distorted viewpoint, which looks at *individual enterprises* rather than at the *connected system* (the complete value chain) to which they belong. Large companies contain the whole value chain within themselves. Only in a world of vertically integrated firms, their exports (or direct investment abroad) may well give an accurate idea of their degree of internationalisation. For small firms, what really matters, in terms of competition, is the degree of internationalisation of the value chain as a whole (Grandinetti and Rullani 1996).

In addition to that, today the process internationalisation is more favourable to the small firm. This is because the growth of networks, and the ICT paradigm, are lowering the entry barriers for small firms and, at the same time, are enforcing their capacity to operate at long distance. Small firms cease to belong just to a specific local setting, becoming part of a global network that connects the local sphere with the global economy. While globalisation becomes a pervasive phenomenon, small firms, through their access to global networks, enter into the whole economic circuit, assuming its standards and its rules of efficiency.

The network, which connects the local to the global, becomes the true environment in which small firms develop, grow and proliferate in each district or territory. By interacting with the networks which connect them with other local systems, or directly with the global economy, IDs do not lose their identity, but they must adapt to these new circumstances. In some cases, competing IDs can offer products or processes at lower costs; in others, they enjoy a strategic competitive advantage. With regard to the global economy, different IDs activate a *reciprocal specialisation*.

Networked economy is not ruled only by territorial heritage, but is also a normative and political construction. In a network, standards, rules, and entry conditions are shared due to a trust attitude or to some form of visible/invisible hand.

In fact, IDs are cognitive, but also regulatory machines, such as trust and confidence, are created in order to enforce the efficiency of the system. This contributes to enlarge the division of labour and, in this way, a reliable regime for *sharing* knowledge among the various actors of the system is built-in. The resulting effect is that each agent belonging to the local system earns improving returns on his investment in knowledge. Seen in this way, IDs are systems where agents take advantage of their physical proximity using their personal and direct interactions to organise an efficient division of labour in the production and use of knowledge.

4.5 Competing learning systems

IDs are therefore *learning systems* which, in the global market place, compete with *other modes of learning*. Indeed, they are not the *only way* of use knowledge to produce economic value and competitive advantages, because there are other forms of learning, whereby it is possible to differently organise the circuit of exploration, specialisation, sharing, accumulation, and division of knowledge.

For example, hierarchy competes with localised systems, through very different ways of seeding and multiplying knowledge. All the cognitive cycle of hierarchy is performed under the *power-of-command* of large multinationals, which manage their knowledge needs in the global economy without a significant recourse to personal interaction and to the advantages of physical proximity.

Another competing solution is placed at the other extreme of this ladder, relying on the "invisible hand" of the *market*, which links anonymous producers and anonymous users of knowledge and productive know-how, using contracts and prices as tools for managing the cognitive/productive circuit of exchange.

Between these "classic" rival systems (market, hierarchy, local system), we now have to consider another modality: the *virtual network*, i.e. a cognitive/productive circuit relying on virtual interactions organised throughout ICT technologies, which connects different enterprises located in various parts of the global economy.

IDs and local systems are fully part of the global economy, but they have no exclusive claim on it. Their space may be large or small, they may grow or shrink, depending on the outcome of the competitive confrontation with the other ways of organising resources, within the cognitive/productive circuit of the global economy: multinational enterprises (hierarchies), markets, and virtual networks.

Hence, IDs cannot be analysed *per se*, but their role stems from the global competitive arena, in which IDs face the alternative forms of cognitive/regulatory circuits of organisation.

Markets, hierarchies and virtual networks are partially competitive and partially complementary to IDs, because IDs can use these forms of organization to acquire knowledge, expertise and goods at low prices. For example, a decisive factor for IDs' competitiveness is the degree of *global openness and hybridisation* that each ID can achieve. The recourse to external suppliers (or markets) whenever they can offer semi-finished products, know-how or services at advantageous prices must be welcomed. And also the penetration of multinational hierarchies within IDs (or, conversely, the external investment of leading local firms moving abroad), whenever this represents the most effective way to have access to distant markets, or to acquire outside technology. The same is true for the use of virtual networks (e-commerce, metropolitan services, etc.). IDs need to utilise virtual services or distant connections, whenever this is proved advantageous.

Let us discuss in depth the four competing forms of network interaction, also as competing alternatives in the strategic function of knowledge creation, absorption and diffusion.

4.5.1 Local learning systems

Local systems make their own special contribution to the production and exploitation of economically useful knowledge, because:

1. they create an environment favourable to the experimentation of new ideas;
2. they facilitate the rapid, "sponsored" propagation of the most successful solutions, acting as "natural systems" of evolutionary learning;
3. they permit the regulated sharing of knowledge, because IDs are an environment of shared values, predictable behaviours, unspoken rules and socially controlled procedures;
4. thanks to the local regulation, knowledge can be easily shared and exchanged, fuelling a process of specialisation, which reduces the costs and risks of investment in knowledge and increases the returns on such investment. Consequently, a continuous flow of micro-investments leads very quickly to a cumulative growth of knowledge.

When considering local networks, it is important to understand how they succeed (or fail) in defending the three fronts on which they face, respectively, markets, hierarchies and virtual networks, regarded as competing alternatives in the division of cognitive labour.

4.5.2 Market learning systems

In the *market*, the production and exchange of knowledge is mediated by transactions: the buyers and sellers remain often unknown to one another, and the only bond between them is the price. In these circumstances, localisation is important only because it implies a certain amount of *costs* (transportation, communication, etc.) for moving goods, persons and information from one place to another. The market links producers and consumers localised in unspecified space: the only thing that matters is that the location must be "efficient" (i.e. not involving excessive costs).

Of course, this happens only if markets are perfect. But in practice, markets are highly imperfect, and buyers and sellers are not anonymous, but *localised.* The space that buyers and sellers share, if they come from the same district, may in fact reduce the imperfections and improve the efficiency of the transactions between them. Market and space, to some extent, co-exist. However, their co-existence is precarious and often only temporary. When the main *medium* of exchange is the market, the territorial space has no precise identity.

Why may the "use" of a specific space show a superior efficiency? This occurs when a territory is able to provide advantageous service, in terms of information, use of knowledge, reduction of uncertainty, transparency of rules and behaviours, etc. In this last case a specific economy is coupled with the corresponding territory (Maturana and Varela, 1985), and a permanent cross-fertilisation is built. However, there are indeed many cases in which the growth of market-mediated relationships gradually destroys the pre-existing territorial local culture, making the places linked by anonymous market transactions.

4.5.3 Hierarchical learning systems

In a hierarchy, the production and exchange of knowledge are organised by the "visible hand" of the proprietary power of the (often large) corporation.

For a "perfect" hierarchy, the territory simply represents a cost (measured by the distance). Typically, the hierarchy superimposes its own logic on the territory where it is settled. The territory is "utilised" if it offers some useful services, but the aim of the company is to reduce its dependence on it.

This gives rise to two possibilities:

- a type of localisation in which the large enterprise "colonises" the territory (let us think of the Fordist company town);
- a multi-localised enterprise, where the possibility of transferring activities from one territory to another is related to the costs and benefits offered by the different locations.

Usually, the hierarchical *medium* and the territorial *medium* are in competition: the former tries to mould the latter to suit its own needs, during time sometimes destroying its territory's internal logic.

4.5.4 Virtual learning systems

There is then a fourth possibility of dealing with knowledge and information exchange: the use of virtual networks. In the past, most networks were territorially based. They were localised because communicative interactions were favoured by, and sometimes almost demanded, physical proximity. Nowadays this condition is no longer strictly necessary because there are many technologies that can be developed through the access to ICT (Information and Communication Technologies). This makes possible long distant interactions, or/and real time interactions, or/and virtual interactions (Vicari, 2001).

The virtual network stimulates all sorts of interactions between a vast number of places, spread over a potentially planet-wide area. It does not suppress the specificity of the individual place, but integrates it globally within a system of competition and division of labour which is clearly based on a wider-scale.

The virtual network allows the production and exchange of knowledge between enterprises and individuals using communicative interactions, and it is more open and balanced than the hierarchical system, which is built around one dominant source of power and interest.

Communicative interactions can therefore take place effectively at different levels: *locally* (within a district or a town), and they are based prevalently on face-to-face relationships; at the *metropolitan* level (within cities/regions), and they are based on a mixture of *virtual* contacts (e-mail or tele-conferencing) and face-to-face meetings; *globally* (at national or world level), where contacts are normally virtual ones. But, in most cases, we shall see a *hybridisation* of local and virtual interaction. In relation to the issue of knowledge creation and diffusion, only the most complex and indeterminate problem-solving types of activities will privilege the local relationships (where we encounter the most direct, informal, polycentric and face-to-face exchanges).

4.6 The local knowledge factory

To identify the "engine" which generates economic value at the local level, we need to look at the way in which, in different places, knowledge is produced, exchanged and used to generate value and competitive advantage. In other words, IDs need to be regarded primarily as *cognitive systems*.

Most of the labour performed in a modern economy is *cognitive labour*, i.e. labour spent in the production, exchange and use of knowledge. Human energy effort has been in constant decline since the first Industrial Revolution. Therefore most of the working time spent in a modern economy is devoted to planning, communicating, directing machines and transformation processes, codifying and storing information, programming, inspecting and selling. Even in the so-called "manual" jobs, what really matters is not so much the (increasingly limited) muscular energy put into the process as the intellectual/mental ability to modulate the muscular energy needed for operations that have not yet been mechanised. The most time-consuming operations in these processes are those in which evaluations and decisions have to be made in complex circumstances, or when the phenomena concerned are variable and indeterminate. The kind of capital that is now vital to the generation of economic value and competitive advantage is *intellectual capital*, otherwise known as an invisible resource or *intangible asset.*

This being the case, local development must be defined in terms of the competitive advantages that arise from the production and replication of knowledge when these activities assume a *localised* form. Only by entering into the cognitive "black box", do we understand why the territory has been of increasing importance over the last thirty years, and whether it will continue to be important in the years ahead. And it is by entering into this "cognitive engine" that we can find out how and when specific forms of localised production can be reproduced, modified, and enhanced in power, and what types of *policies* need to be introduced.

Within IDs, the productive use of knowledge is based on the economies stemming from knowledge replication and from the division needed by the *innovative labour* to produce new knowledge, integrating it for economic purposes (Arora, Gambardella and Rullani, 1997). Being a resource that can be reproduced at zero cost (or almost), knowledge is not scarce, as imposed by the dominant mainstream economics for all other inputs. Its economic value may in fact be increased by its dissemination, or by extending the number of applications for which it is used.[3] However, diffusion increases its value for the society, i.e. for those who use it. This does not mean that diffusion also increases the value for the producer of the knowledge itself, or for those who have invested in it and taken the associated risk. The important element for the producer is not the social value generated by its diffusion, but the value he or she privately extracts from it.

4.7 Diffusion and retention of knowledge: a race for value

Knowledge-creating enterprises can benefit from an increased speed of knowledge diffusion, but they tend also to slow down the speed of knowledge socialisation (realised by copying, imitation, re-interpretation and development of alternative solutions).

The economic value of "proprietary knowledge" is inversely related to the speed of socialization process. Patents (for inventions), copyright (for software) and licences (for the use of solutions sold on the market) are ways of protecting the intellectual property of a person who has produced, or in some way owns, new pieces of knowledge. For our purposes, market, hierarchy, local and virtual

[3] The essential point is the gap between the cost of producing (new) knowledge and the cost incurred in reproducing the same knowledge for other, even different applications. The reproduction of knowledge occurs at very low costs if knowledge is perfectly codified. When it takes the form of tacit knowledge or if it must be contextualised before its utilisation, the costs of reproduction are higher but lower than the original production costs. According to Robbins's definition (1932), a resource is scarce if it is available in limited quantity in relation to a range of alternative uses, which compete for it on the basis of price. In this sense, knowledge is scarce in relation to the needs and means available only before it is produced. Once an effective solution to an important problem has been found, there is no longer any scarcity constraint on its possible uses. Knowledge is not consumed when it is used, because it can be reproduced at no or low cost.

networks can also be regarded as (competing) ways of extracting economic value from the process of knowledge diffusion/socialization.

The *market* is a means that favours diffusion, because it puts each producer in contact with a vast audience of potential consumers willing to pay the required price. If the market is vast and well organised, knowledge can be disseminated very rapidly, and it is possible to extract large profits. However, the market has many disadvantages:

- it is not suited for disseminating complex knowledge, which requires prior interaction and ex-post evaluation;
- it does not cover the producer against the opportunistic use of the knowledge by final users, because the intrinsic limitation of formal agreements (licences) and given the possibilities to overcome the various forms of protection (patents, copyright);
- it is not functional to have a select audience of potential users for a market that needs to be organised in advance. But a market which is pre-organised and filtered in this way is effectively a network of virtual interactions.

The *hierarchical system* makes diffusion less easy but, in return, slows down the socialisation of knowledge and makes it possible to extract value from the knowledge one possesses in a controlled way. It enables the owner to maintain secrecy or control over knowledge (even if some pieces are disseminated among selected suppliers and customers). However, the hierarchical exploitation of knowledge requires heavy investment and high risk. But, indeed, knowledge tends nowadays to lose its economic value so quickly (because technology and markets change so rapidly) that it becomes very risky and dangerous to try to "conserve" proprietary knowledge for a long time.

The hierarchical system is now losing its efficacy and it gives room to networks, which can ensure a better way of sharing the costs of knowledge creation, and that allows for a more rapid and widespread dissemination. Large manufacturers traditionally have relied on the market and the hierarchy to build their system of knowledge production. But the growing demand for variety and flexibility has required a quick adaptation of their knowledge, and the introduction of rapid learning curves. As regards this, since the 1970s, local networks have performed quite well in this respect, sometimes exhibiting even a superior efficiency. This is because they are very dynamic learning systems, able to develop small-scale innovations and to increase the search for new solutions.

4.8 The competitive advantages of local exploratory learning

Local networks have some advantages in exploration and learning when it is required to identify complex but incremental innovations. In this field, local networks perform better than markets and hierarchies in many cases that empirical studies have analysed.

The market is efficient in responding to a simple, well specified demand. But it is unlikely to venture "into the dark" to explore less probable, more difficult pathways. If there are a hundred alternatives, the market will tend to concentrate on those which are most obvious and attractive, neglecting the less visible. Large companies (hierarchies) are also more inefficient in responding to problems of this kind. Having to choose between a hundred possible alternatives, a large company needs to invest time and money in the research process which, after a series of simulations and selections, will reduce the alternatives to just one, and only this sole alternative will be tested and put into production.

The local network, on the contrary, represents a good example of how parallel efficient exploratory searches may be developed when enterprises are unable to "protect and hold " their knowledge.

In IDs variety is generated by the exploration (acting in all directions) of many possible methods, based on the contribution of different agents and on competing ideas. If there are a hundred possible alternatives, all will be explored (even though some will receive more attention than others). The research costs will be divided up among many enterprises. The research process will speed up and, at the end, it will be clear to everyone which solutions were best appreciated by the market. Firms will convert immediately to the "winning" solutions.

This is the secret of the success of many IDs in traditional sectors, high-fashion goods, special machinery, household products, furnishings and so on. A local system can be very effective in this process of exploratory learning where the propagation mechanisms allow a rapid – and generally involuntary rather than co-operative – dissemination of knowledge among competing enterprises.

The involuntary propagation of knowledge is due to the fact that enterprises are unable to hold the knowledge they possess. Each producer specialises in just one thing, with limited costs and risks, because he is able, if necessary, to use without costs and rapidly the knowledge developed by others. These are the advantages of *involuntary sharing*, which are often confused with those of *co-operation*.

4.9 Involuntary sharing or co-operation? A common misunderstanding

Involuntary sharing is the phenomenon that enables rival firms, which do not voluntary co-operate, to benefit from the collective localised process of exploration, which eventually become common knowledge.

Voluntary co-operation, on the other hand, is a phenomenon arising between suppliers and customers, in the chains of value developed within the IDs. In this case, firms deliberately organise themselves to benefit from the replication of the knowledge possessed by each specialist and by the enlarged division of labour existing in the area.

Involuntary sharing and voluntary co-operation are both important. But, to operate, they require different conditions. In the case of involuntary sharing, what matters above all is the *impossibility of closing off the cognitive circuits* at firm supply- chain level. This outcome must be regarded as socially useful, even though the individual firms would prefer to prevent any knowledge spillovers, against which - maybe without much success - they actively devote themselves.

Involuntary sharing stems from the difficulties to keep things secret. Within IDs, members of a single family may work for different firms, creating an "information bridge". In a small town, where there is a dominant industry, bars, places of recreation, schools, and generally daily life itself are means of knowledge sharing and socialisation. Everyone knows everyone else's business.

On the contrary, in local systems with a lower level of specialisation which include firms from various sectors, involuntary sharing will be much less significant, as well as within large towns, where people's personal life is quite separate from the work place.

The deliberate co-operation, which generally occurs between suppliers and customers (and exceptionally between competing firms which have a common interest), requires other conditions. What is needed are cultural circumstances and institutions which actively favour co-operation, less antagonistic industrial relations (among workers and entrepreneurs), and trustful and integrated interactions among final firms and their subcontractors.

4.10 Reusing shared knowledge

The local network is an environment that favours a high speed of knowledge and information diffusion while, at the same time, regulating the process of socialisation by setting rules for the sharing of knowledge. As we have argued above, the value of a piece of knowledge depends on the capability of its owner to profit from it more rapidly than it becomes "socialised".

In the local context, there are two important factors promoting rapid dissemination:

1. direct and informal communication based on a shared language;
2. spatial proximity, which facilitates the logistics of transferring goods, persons and information, simply and at low cost, because of the short distances involved.

Here again, it is easy to see why IDs are placed in a stronger position in comparison with other types of local network.

Communication is easy when there is a dominant industrial specialisation and a common professional language able to express all the details involved in the description of the specialised skills possessed by a large part of the local population. This common language is fed by local vocational schools, research centres, laboratories and certification institutes which set common standards. The local labour market itself is a powerful pool of communication, being characterised by a high rate of labour force turnover. This favours the formation of a common professional language and the dissemination of basic skills (blue collar workers move frequently from one firm to another, and many of them leave the established firms to become new entrepreneurs themselves).

Moreover, because the logistics of the whole supply-chain is typically concentrated just in the surrounding area, there are closed circuits supplying semi-finished items and just-in-time manufacturing processes; within them knowledge and information can simply circulate. This is more complicated for other types of local network, when supply chains are globally localised.

At the same time, the local network has a further advantage: it is able to maximise the dissemination of knowledge, while at the same time regulating its socialisation. A local system is in fact a society which exercises a *regulating function* over the behaviour of its members. There is, for example, a social punishment against opportunistic behaviours, lack of respect for contractual obligations, or lack of diligence and rigour in performing tasks (including the respect of deadlines). The existing informal *rules* make labour and business relations more reliable. They support the *sharing* of knowledge, enabling a manufacturer to keep some control over its application and use, even when it is entrusted to other parties belonging to the local circuit. These other parties will respect the rules of preventing the manufacturer from being damaged.

Local institutions can be of great help in ensuring that this kind of regulation works smoothly by consolidating the rules that govern the productive *filière*. This regards sub-contracting arrangements, minimum quality standards in products and processes, contracts guarantees, prevention of unfair competition, control over the excess mobility of firm-specific, knowledge-bearing and skill-full labour from one company to another.

In these local systems the sharing of knowledge is regulated in a more secure and targeted way, with an acceptable balance between the advantages inherent to a collective sharing of knowledge and its firm-specific and proprietary utilisation. They are able to bring down the cost of knowledge replication, speeding up the propagation of knowledge from producers to users. They greatly reduce the level of investment needed to produce novelties from the stock of existing knowledge. They increase the returns depending on specialised know-how by broadening the range of potential uses. To sum up, in a local network the efficient use of shared knowledge depends on:

- *specialisation* among the various operators (each can focus on a very circumscribed and precisely defined problem);
- *abundance of knowledge sources* deriving from the existence of a large number of potential producers and users (other specialists or systems operators);
- *regulation of the knowledge diffusion process*, with shared rules among suppliers, customers, and sometimes competitors, which achieve a reasonable compromise between competing interests;
- development of *new entrepreneurial talent*, by lowering entry barriers (for entrepreneurs, labour, services, and infrastructures).
- labour transform itself into *self-organising labour*, capable of taking an active part in building up the economic environment in which the work itself is performed (Corò, 1995; Rullani, 1995). There is no need for formal structures to direct the inputs of labour and intervene to modify programmes, as the employees are themselves capable of intelligent adaptation to circumstances. Workers switch fairly easily to the role of entrepreneur, changing their job or company without too much trouble, and add to their professional skills an autonomous capability of self governing the productive tasks, learning their jobs stage by stage, starting very often from an initial lack of academic education.

In local networks, knowledge is intrinsically a social resource. The conditions which create trust, whether they are "natural" (proximity or shared experience), institutional/cultural (social capital) or contractual (formal and informal), trigger virtuous circles of cumulative learning (Bagnasco, 1988, 1999; Trigilia, 1999; Burroni, 1999; Messina, Riccamboni and Solari 1999; Floridia, 1999).

4.11 How districts learn

Every local system is unique. The differences between one and another are enormous, because the specific differences associated with a place are augmented by those associated with the evolution of various sectors and firms that developed there.

Nevertheless, in all IDs there is a common "multiplying engine" which explains their *economic vitality* and enables many individuals, or entrepreneurial localisations, to exist as a system which owns a differential *identity*.

As we have said, this "engine" is the *cognitive system* expressed and organised by the territory or, rather, by the local economy and society. Space, economy and society are closely wedded, thanks to the many links, flows, and interdependencies created and required to organise the circuit of knowledge.

Italy is a veritable kaleidoscope of cases which, each with its special emphasis, demonstrate how initial learning processes can be amplified to the point of becoming powerful factors of territorial concentration, able to attract other similar activities (Becattini, 1997; Bonomi, 2000; Fortis, 1998).

What factors have determined this "explosion" of competitive advantages? How did a local network manage this extraordinary take off and improvement of the local living conditions?

These are cases in which some IDs starting with relative simple production processes remained quite weak concerning their capability of learning. Others show a very dynamic mechanism of knowledge creation (Belussi and Pilotti, 2000). Whatever the competitive advantages that determined their success in the past, local networks still must *learn*, change and rebuild their identity on new foundations. The competitive advantages inherited from the past are now exposed to the duress of the incoming globalisation process; this means that the productive circuits need to be renewed, and that knowledge-used must be enriched by new ICT applications. The local actors need to transform their cognitive "engine" that has served them so well up to now.

Though it might seem surprising, many IDs did not arise from pre-existing artisan activities. They have a very definite birthday and "founder"[4]. Others are

[4] For instance, the first spectacles factory, in the frame district of Belluno -Cadore, employing three people, was established in 1878 at Calalzo di Cadore, by Angelo Frescura (De Lotto, 1956). Frescura was a native of Calalzo, but left the Belluno area to work as a travelling salesman. He eventually became involved in the optics business by chance, having opened a shop selling optical equipment and fancy goods in Padua. In the past, the skills needed for making lenses had formerly been concentrated in Venice which, during the 12th-17th century, supplied the whole of Europe with glassware. However, after that, other countries (France,

based on pre-existing widespread artisan activity. Historical events seem determinant like in the case of chair district of Manzano, which developed thanks to a decree issued by the Empress Maria-Theresa of Austria in the eighteenth century. This new regulation encouraged the cutting of timber from the woods of the commune of Mariano (Gorizia), giving rise to a flourishing chair-making industry, whose products were exported to Lombardy/Venetia and the Balkans.

Then we have to note that IDs grow on the basis of two different processes. Small firms grow as a result of the *dissemination* of the knowledge produced within the district (as a whole), and often by the larger firms. From the larger firms a continuous spin-off is generated leading to the establishment of new small-scale business, either involved in subcontracting or to cover new market segments. The technology used is often fairly simple, and only a minimum of investment is required if a firm restricts itself to serving just one phase of the manufacturing process. At the same time, there is an abundant pool of dependent customers and end-users, which provide new ideas and new knowledge.

The general shortage of production capacity increases the bargaining power of even small-scale suppliers. Despite the increase of the size of the local system, IDs remain with a flexible structure, enabling them to deal with a rapidly changeable demand.

In many cases, larger firms, on the other hand, grow by applying the typical tools of rationalisation. Their production processes are based on internal scale economies, and integration, which, paradoxically sometimes recall the Fordist culture. This is of course a Fordist culture adapted to the requirements of variety and flexibility imposed by the nature of the local systems.

Typically the stock of localised knowledge is a combination of disseminated basic knowledge, coming from the ID as a whole, and firm-specific knowledge created by the larger companies, which control the more advanced technologies. The first type of knowledge can be acquired free of charge, or almost so, from the direct experience. The second is expensive to build, and requires adequate

---

England, Spain and Germany) had industrialised the process. So, the initial knowledge was brought in from outside. It was only "captured" by the local system thanks to the network of contacts created by the flow of migrant workers, who put the Cadore area in contact with the skills and products of the more highly developed European economy. The first factory specialising in the manufacture of components (supplying the whole district) dates from 1947. This provided the decisive impulse for the establishment of other small businesses, lowering the barriers to getting started in the industry, and creating a market in components and semi-finished products which could also be acquired by outsiders.

structures to cover risk and to secure intellectual property rights (designer labels, new brands, specific advertising, and direct sales structures). Often in the district pioneering enterprises contribute to the first introduction of radical technological innovations (Belussi, 1999).

With the globalisation of the 1990s this complementary relationship between small and large firms began to break down. Nowadays large companies are less in need of smaller ones for subcontracting. They can strengthen their internal organisation, or subcontract through international supply-chains. Some large companies can make the strategic decision to go for a full integration of the cycle, to have a complete control of quality and delivery deadlines. Larger companies within IDs are actually tackling two key problems: market globalisation and product dematerialisation (designer labels or *griffes*, own brands, building of sales structures or acquisition of outlets chains). Small firms, and this is an indirect effect of the very existence of the district, have became accustomed to achieve high levels of profitability with little or no direct investment in new knowledge acquisition. They also may be in serious difficulties because the growing competition from low-cost countries (Asian manufacturers and East European firms). Smaller producers of finished goods find difficulties in establishing direct and interactive relationships with distant markets. Many district firms lack organisational capabilities and they do not have the financial structures required to build it. In some IDs, the habit of the *involuntary sharing* of knowledge has undermined the incentives to co-operate in a formal way. Despite the growing district rhetoric on co-operation issues, in many cases we see a persistent individualism of the localised agents.

4.12 A glance to the future: local networks vs. virtual networks

Until recently, many local networks, taking advantage of their competitive position examined above, have dominated some segments of the international markets, competing with the other forms of organisation like "pure market" and "hierarchies".

Small firms and local networks have therefore become essential complementary partners, to a far greater degree. Markets put territories in competition with one another. Hierarchies have difficulties in "blending " their structure into the local environment.

However, in recent years virtual networks making use of opportunities for communicative interactions at long distance (ICT) have extended over many territories, with the result that network supply-chains actually cross local, metropolitan and global levels.

The decisive advantage of the virtual network is that it can dramatically cut the cost of knowledge using, producing, reassembling, and storing. This is because it enlarges the scale of replication of existing knowledge and information. Moreover, it allows a greater division of labour within a system of almost one billion potential producers, users and consumers. It also brings into play at global level national differentials (in the cost of labour, environmental protection, tax pressures, technological competence, etc.).

The virtual network relates together:

1. markets (for example, many business-to-business relationships, previously relatively unorganised, are now with the use of ICT developing into highly organised markets, organised by infomediaries which, in turn, can easily turn to non anonymous relationships);
2. hierarchies (which are trying to speed up the dissemination of knowledge by cutting off proprietary boundaries, standards, blending their organisation with closely involved suppliers and customers into a captive network). In other words, a network which retains a dominant centre but in which knowledge also flows outside the system.

Given the new opportunities offered by the global economy, there is an urgent need for local networks to enter into relationships with other territories, to achieve economies of scale (size of market) and variety (access to favourable national differentials). However, this opening up to distance relationships is perhaps more painful for local networks than for the other forms of organisation. This obligates IDs to share multiple territorial identities, and different cultures. The extension of virtual relationships can be managed only through the recourse to radical new technologies (Internet, information systems, ERP, complex logistic systems), and this contrasts with the ways in which knowledge was typically organised in the past in local networks.

In this process of transformation, a *path dependent behaviour* tends to dominate against the needs of an innovative adaptation to the economic conditions (Amin, 1978; Grabher, 1993). Local networks, left to themselves, could be paralysed by political vetoes and evolutionary inertia. To escape from *lock-in* it is necessary to have a good understanding of the possible alternatives, and follow the general interest and change the external relationships (with the market, hierarchies and virtual networks). When local networks open up to the virtual world, there is no possibility of fusion, but only of hybridisation. The recourse in favour of international outsourcing is in this context a natural passage, but this will break up some segments of the local supply-chains. The following of the logic of market advantage will tend to disrupt the pre-existing local bounds based primarily on trust. In many cases, then, the local network will be pushed towards its association with the *hierarchical* forms of organisation. It will develop large local companies that follow more directly the process of globalisation, or it will open up to external multinationals, investing into the local system to gain access to its specific resources.

The access to virtual networks will imply an effort towards a partial codification of knowledge which will be introduced within the international circuits. Only the more complex activities, related to the local mechanism of knowledge production will remain anchored to the territory. The transformation of local networks will be neither simple nor easy. It will encounter not only political resistance from interests that could be damaged, but it will also face the necessity to experiment new technological solutions and risky investment.

To keep the ship off the rocks, it may be necessary to navigate "by sight", but always steering straight for the intended landfall.

4.13 Three lines for local learning policies

Within the new forms of division of cognitive labour, the competitive repositioning of IDs depends on the evolution of local networks in their functional relation with market, hierarchy, and virtual networks. Territorial policy has to interact with the dynamic active forces underlined by this development. Consequently, any policy that is set to achieve some objectives, or to make corrections, must take into account the spontaneous forces at work and the existing evolutionary tendencies. Otherwise it is likely to be proven as ineffective. It is not a question of *laisser faire, laisser passer*, but one must avoid the error of believing that it is possible to direct the development of the territory by decree.

There are three possible kinds of territorial policy which could perform a positive role:

1. *selective policies of investment* in the territory which aim to maintain a distinctive difference and a specific competitive advantage;
2. *access* and *outsourcing policies* which provide local networks with the lower costs and greater variety available in global markets, multinational hierarchies, and virtual networks;
3. *identity policies* which give a territory a sense of cohesion, maintaining its historical continuity.

Let us take a closer look at these three possibilities.

The aim of *selective investment policies* in a local system is to augment the level of *local intelligence* and, at the same time, support the development of local networks towards highly complex activities. Only activities of this kind can compensate for the *greater cost/risk* associated with face-to-face relationships, the scale and variety of which are considerably smaller than those available in global markets, multinational hierarchies, and virtual networks. In fact, local income growth in a given territory presupposes a selection of activities, and a concentration of resources on those that are most innovative and offer the greatest added value. The only way of "keeping all going on" is to rely massively on low-paid labour or immigration. In areas of full employment (two thirds of Italy), it is not possible to have policies giving undifferentiated support to all existing activities. Policy-makers need to decide what activities are compatible with the typical higher costs of the local networks, and concentrate public support and promotion on those activities.

To facilitate a change of this kind, it is necessary, on the one hand, to encourage the delocalisation of the "less favoured" activities. On the other hand, to encourage a corresponding investment in the production of new knowledge that will enhance the quality and productivity of the remaining activities.

How can this be done? There are three broad areas for initiative. The first is done by putting more public investment into knowledge-building activities (research, education, training, entrepreneurial experimentation) and into tangible and intangible infrastructure. The second by giving incentives to individual investment in professional activities, up-grading skills, competencies and capabilities. The third is by introducing forms of *capital venturing* for small enterprises to support the implementation of good ideas.

For all those activities that are external to the *core business* of a given territory (when the comparative advantages are unfavourable), the most appropriate strategy is *outsourcing* from foreign sources (i.e. purchasing machinery, goods, services and manufacturing processes from international suppliers). By opening up to the international production, local systems can escape the risk of localism (Perulli, 1998).

The policy of *outsourcing* provides *access* to the best resources and enforces *competition*. A policy of access is, therefore, a policy of specialisation and differentiation. The aim is to set up global logistics systems, and develop standards, languages and new forms of guarantees that will enable local users to make use of knowledge, skills and resources externally produced. Hence, local networks must not only defend themselves against the global market, multinational hierarchies, and virtual networks; they must also *make use of them*.

Finally, a local system needs to maintain its *identity* over time, in spite of the changes induced by the globalisation process. Identity is promoted by local institutions and with the daily life and experience of the people who live within the system.

If a territory lacks the appropriated institutions, the task of regulating conflicts of interest and choosing among the various possibilities will be left to spontaneous development, menacing the established identities.

In Italy, IDs, for example, still lack institutions to represent their interests. There may be many administrative bodies involved in a district, but often they are in conflict with one another. In Italy, regionalism and "districtualism" have been two independent, often conflicting, phenomena, without an effective point of synthesis (Trigilia, 1989).

A policy designed to rebuild identity must be able to mend a social fabric, that is constantly torn by the selection of new local activities, where some are expelled in favour of others. In addition, there is a growing recourse to outsourcing, as the process of specialisation needs to consolidate some kind of territorial "vocation". IDs infiltrated by the global market, multinational hierarchies, and virtual networks, tend to fall apart, becoming a confused mess of conflicting interests. Recreating a territorial identity means actively contrasting the entropic nature of the spontaneous development.

Local networks can increase their capacity for competitive survival, if they are able to undertake *self-selection processes*, to develop strategies of *outsourcing* via the market or hierarchies, and to use virtual networks. They must also be able to regenerate their own *identity*, absorbing changes without excessive traumas and damage. In this respect, cultural and political influences have a very important part to play. The reasons for staying together can only be reconsidered and redefined if there is a strong social capacity for reciprocal awareness and communicative understanding.

References

Amin A. (1993), "The globalization of the economy: an erosion of regional networks?", in, Grabher G. (ed.), *The Embedded Firm. On Socioeconomics of Industrial Networks*, Routledge, London.

Amin A. (1998), "Una prospettiva neo-istituzionalistica dello sviluppo locale", *Sviluppo Locale*, vol. V, no. 8, p. 75-94.

Arora A., Gambardella A., Rullani E. (1997), "Division of labour and the locus of inventive activity", *Journal of Management and Governance*, vol. 1, no. 1, p.123-140.

Arthur B.W. (1989),"Competing technologies, increasing returns and lock-in by historical events", *Economic Journal,* vol. 99, p.116-131.

Arthur W.B. (1988), "Self-reinforcing mechanisms in economics", in, Anderson P.W., Arrow K.J. and Pines D., (eds.), *The Economy as an Evolving Complex System*, Santa Fe Institute, Studies in the Sciences of Complexity, Addison-Wesley, Redwood City, CA.

Bagnasco A. (1988), *La costruzione sociale del mercato,* Il Mulino, Bologna.

Bagnasco A. (1999), *Tracce di comunità*, Il Mulino, Bologna.

Becattini G. (2000), *Il distretto industriale. Un nuovo modo di interpretare il cambiamento economico,* Rosenberg e Sellier, Turin.

Becattini G. (1997), "Prato nel mondo che cambia (1954-1993)", in *Prato: storia di una città*, Le Monnier, Florence, volume four.

Belussi F. (1999), Path dependency vs. industrial dynamics: an analysis of two heterogeneous districts, *Human Systems Management*, vol. 18, p. 161-174.

Belussi F. and Pilotti L. (2000), Knowledge creation and collective learning in the Italian local production systems, Working Paper Dipartimento di Scienze economiche Marco Fanno.

Bonomi A. (1997), *Il capitalismo molecolare. La società al lavoro nel Nord Italia*, Einaudi, Turin.

Bonomi A. (2000), *Il distretto del piacere*, Bollati Boringhieri, Turin.

Brusco S. (1989), *Piccola impresa e distretti industriali*, Rosenberg e Sellier, Turin.

Brusco S. and Paba S. (1997), "Per una storia dei distretti industriali italiani dal secondo dopoguerra agli anni novanta", in, Barca F. (ed.), *Storia del capitalismo italiano dal dopoguerra a oggi*, Donzelli, Rome.

Burroni L. (1999), "Regolazione locale nelle regioni della Terza Italia. Una comparazione tra Veneto e Toscana", *Sviluppo Locale*, vol. VI, no. 12, p. 5-43.

Corò G. (1995), "Sviluppo locale e lavoro auto-organizzato. Il 'modello Nordest' come prospettiva strategica?", *Quaderni di analisi e di programmazione dello sviluppo regionale e locale*, Ires Toscana, May-August.

De Lotto E. (1956), *Dallo smeraldo di Nerone agli occhiali del Cadore*, Tipografia Silvio Beretta, Belluno.

Enright M.J. (2000), "The Globalization of Competition and the Localization of Competition: Policies Toward Regional Clustering", in, Hood N., Young S. (eds.), *The Globalization of Multinational Enterprise Activity and Economic Development*, Macmillan, London.

Floridia A. (1999), "Le 'risorse istituzionali' dello sviluppo locale: la Toscana e i sentieri divergenti della Terza Italia", *Sviluppo Locale*, vol. VI, no. 12, p. 79-104.

Grabher G. (ed.) (1993), *The Embedded Firm. On Socioeconomics of Industrial Networks*, Routledge, London.

Grandinetti R. (ed.) (1999), *Il seggiolaio e l'economia globale*, Cedam, Padua.

Grandinetti R. and Rullani E. (1996), *Impresa transnazionale ed economia globale*, NIS, Rome.

Maturana H.R. and Varela F.J. (1985), *The Tree of Knowledge*, New Science Library, Boston.

Messina P., Riccamboni G. and Solari S. (1999), "Sviluppo economico e regolazione politica nelle regioni di piccola e media impresa: un'analisi comparata tra Veneto ed Emilia-Romagna", *Sviluppo Locale*, vol. VI, no. 12, p. 44-78.

Perulli P. (1998), *Neoregionalismo. L'economia arcipelago,* Bollati Boringhieri, Turin

Piore M.J. and Sabel C.F. (1984), *The Second Industrial Divide,* Basic Books, New York

Robbins L. (1932), *An essay on the nature and significance of economic science*, Macmillan, London.

Rullani E. (1995), "La politica del post-fordismo: un'agenda per la sinistra", *Il Ponte* 4 and 5, p. 3-28 and 13-28.

Rullani E. (1997), "Tecnologie che generano valore: divisione del lavoro cognitivo e rivoluzione digitale", *Economia e Politica Industriale*, no. 93, p. 141-168.

Rullani E. (1998), "Riforma delle istituzioni e sviluppo locale", *Sviluppo Locale*, vol. V, no. 8, p. 5-46.

Saxenian A. (1994), *Regional Advantage: Culture and Competition in Silicon Valley and Route 128*, Harvard University Press, Cambridge, Mass.

Sforzi F. (1993), "Lo sviluppo urbano e industriale in Italia: alcune riflessioni", in, Lombardo S. and Preto G. (ed.), *Innovazione e trasformazioni della città. Teorie, metodi e programmi per il mutamento*, Franco Angeli, Milan.

Storper M. (1995), "The resurgence of regional economies ten years later: the region as a nexus of untraded interdependencies", *European Urban and Regional Studies*, vol. 3, no. 2, 191-221.

Storper M. (1997), "Le economie locali come beni relazionali", *Sviluppo locale*, no. 5, p. 5-42.

Trigilia C. (1999), "Capitale sociale e sviluppo locale", lezione inaugurale negli Incontri Pratesi sullo Sviluppo Locale, *Teorie e pratiche dello sviluppo locale*, Artimino, September.

Trigilia C. (1989), "Il paradosso della Regione: regolazione economica e rappresentanza degli interessi", Meridiana, no. 6, p. 173-189.

Vaccà S. (1995), *L'impresa transnazionale tra passato e futuro*, Franco Angeli, Milan.

Vicari S. (2001), (ed.), *Economia della virtualità*, Egea, Milan.

Zagnoli P. (1991), *I rapporti tra imprese nei settori ad alta tecnologia. Il caso della Silicon Valley*, Giappichelli, Turin.

Chapter 5

Why do Ict technologies and the Internet find it hard to spread into industrial districts and favour knowledge exchange?

Giorgio Gottardi

5.1 Introduction

It is well known that Ict and Internet technologies, namely all *www* and e-commerce applications, enormously multiply the number of relationships that any agent can establish and hold. So, the intensity of such relations tends to grow, in terms of frequency of contacts and speed of information exchange. However, in order to exchange not just data, but also information and knowledge, we need not only technological infrastructures that use the same technical standards, but also common languages and interpretative codes. This last condition constitutes a serious restraint to the globalisation process that is in principle allowed by a worldwide utilisation of the Internet. If the environment is sufficiently large, we will encounter fundamental diversities, linked to the various cultures, languages and institutions that regulate social and economic relations. Today these heterogeneities, more than the enduring lack of global standards for the transactions (e.g. contracts, reliable systems of payment, etc.) impede the international development of e-commerce. This could bring us to admit that physical proximity is now, and could remain for long, a very important factor (Storper and Harrison, 1991; Rullani, 1995). In this perspective, the assumptions about the possible vanishing of physical distance, often associated with the advantages of open networks as the Internet is, could represent a fascinating Utopia.

Then, we wonder: is it possible that the most interesting effects of the diffusion of Ict, more than a global scale, could be manifest in more limited homogeneous environments? And in this case which are the opportunities, and also the obstacles to the introduction of these technologies and to the digitalisation of relationships? This seems to be a relevant question for all localised production networks and industrial districts (IDs), because they are not just environments characterized by common institutional, cultural and interpretative codes, but also by high density of knowledge, information exchanges and economic transactions.

In order to answer this query we shall proceed in the following way. Having defined the main theoretical issues related to the analysis of how different kinds and forms of knowledge are developed, we shall deal with the problems of knowledge transfer. Then we shall discuss how this theoretical frame has a deep implication for the study of IDs . We shall develop a general descriptive model on the possible effects of digitalisation of knowledge. Finally, using empirical materials, we shall verify the real state of the art, in the attempt to offer some explanation for the gap between expectations and reality. At the end of this chapter we shall provide some observations on the theoretical problems that the digitalisation of the IDs opens, and on the policies that can tackle this issue.

5.2 Rationale

The most important problem that we want to discuss here is why Internet applications have not produced a substantial de-coupling of economic relations from their physical localisation (Foray and Mairesse, 1999; Brousseau and Rallet, 1999) in comparison with what was forecasted. Up to now, we are witnesses of a steady growth of activities related to the new economy only in areas where a strong tertiary economy and a high density of communication infrastructures already exist. There are more telephone networks (and even more Internet applications) in Manhattan island than in all sub-Saharan Africa (ITU, 1994). It is in these 'central' areas that the biggest investments are directed towards new technological infrastructures (e.g. the broad band). So, many of us now start to wonder if, tomorrow, the majority of the new economic relations promoted by Ict technologies will move outside the concentrated areas where we currently observe the main poles of development (Gomez, 1998). If these trends were confirmed we could even face a growth of existing asymmetries.

The application of the new Ict paradigm offers many possible outcomes. It could also be hypothesised that one of the more interesting effects related to the application of electronic commerce and web technologies could not be seen at all within a system that is globalised at a planetary scale, but in smaller environments. In other words, where we find the co-existence of a high density of exchanges based on innovative and dynamical transactions. These types of environments can assume different forms. One of the most frequently studied is represented by the local production networks, and, particularly, by the so-called industrial districts. With regard to that, even if we find many definitions of the notion of the ID model, many research works, particularly belonging to the Italian school, describe such environments, characterised by common cultural and interpretative codes, as a fertile area for the exchange of experiences and applicative knowledge. In these environments new knowledge is produced also in the absence of pre-determined R&D investment programmes (Gottardi,,1996); Belussi and Gottardi, 2000). IDs are typically characterised by a dynamic mix of co-operation and competition. Co-operation is not often a deliberate action among rival firms (Antonelli and Gottardi, 1991), while it can occur along the same *filière*. Competition is enforced by the presence of an internal market for subcontracting, and by a large co-localised number of competitors belonging to the same industry. It is exactly the conjunct existence of these two apparently contradictory conditions that assures long life to many IDs.

The question we are trying to answer is thus the following: could the diffusion of Ict technologies and *www* applications within these systems (which represent a quite homogeneous environment from an economic and social point of view, and are bound together by a dense 'glue' of relations and *exchanges*) be accomplished more easily and generate more clear advantages in comparison with what is occurring in relatively heterogeneous environments?

In principle these advantages could not just determine a wider openness of the local system: they could also increase and improve their internal relationships, in terms of connectivity, frequency, intensity, amplification of types of exchanges, and, as a consequence, diffusion and generation of knowledge by means of cross fertilisation.

This hypothesis implies that the application of Ict could exert a very positive impact within these local systems, with respect to more vast and heterogeneous environments. For instance, in terms of intensification of network economies, more rapid learning, and a higher rate of innovation. Looking back at the level of each local system, the consequences would consist not only of an increase in efficiency (simply measurable), but also of a more complex increase in effectiveness of the organised tasks. As a whole, these changes could be synthesised with the metaphor of a better local production organisation for a higher global competitiveness (Gottardi, 2000a; Rullani, 1995). Unfortunately, empirical evidence now provides an answer that is not at all ambiguous, but is clearly negative. What is simply verifiable is that small firms that belong to many IDs are generally late in adopting ICT technologies. And why? There are two levels of possible explanation. The first is that IDs are facing a temporary delay, which may be simply overcome under certain conditions. The second, on the contrary, suggests a 'structural' inability to adapt to the new paradigm. It is clear that the problem is not just related to the inappropriateness of technological infrastructures. The problems opened by the substitution of interpersonal relationships based on natural language, with other digitalised relationships, spread along networks, must be taken into account. In order to analyse this issue deeply we must first deal with non-technological aspects regarding the processes of knowledge generation and exchange, and above all the modalities of inter-organisational transmission.

5.3 Environments and kinds of knowledge

When we talk about the effects of Ict technologies on the processes of knowledge generation and diffusion, we do not explicitly refer to data transmission through electronic supports. Knowledge cannot be equated to data or information, and the processes at issue cannot be reduced to the problem of choosing a communication infrastructure for data or information exchange. Data may be seen as a set of discrete, objective facts about events; that may or may not be pertinent for a particular task or use; information is data whose form, content, and time are appropriate for a particular use. This can influence the judgement and behaviour of the user. Knowledge, instead, and in particular valuable economic knowledge, can be seen as a combination of information, individual perceptions, ideas, hypotheses and beliefs that govern decision and action (Demarest, 1997). Even though the amount of data and information transmitted does not have a direct relation with the amount of knowledge generated, the new technologies have a great potential in producing and managing knowledge, since they mark a significant advance in the quantity and quality of the information that can be processed. But in addition to that, when we digitalise information, we allow the separation of the information and the knowledge that it incorporates from the original source, so we render information fully storable and transferable at virtually zero cost. Two aspects assume relevance in knowledge transfer: the *type of knowledge* transferred and the applicative *context* where transfer occurs (Nonaka, 1991; 1994). Knowledge can assume, as is well known, two different forms. The basic contrast between explicit and tacit knowledge has been widely illustrated. The meaning of this distinction is that explicit knowledge represents *knowing about* (*objective* knowledge), while tacit knowledge represents *knowing how* (or *subjective* knowledge; Polanyi, 1967). This distinction emphasizes the issue of the *transferability* of knowledge and the modalities employed. Explicit knowledge comes from the rationalization of information about facts, and can be codified, for instance, in the form of formulas, designs, reports; it is therefore relatively easy to obtain, transfer, and store. On the contrary, tacit knowledge is directly connected with ideas, perceptions and experience, and therefore is quite difficult (and sometimes impossible) to codify and transfer. Knowledge of individuals and organisations can be described as a combination of those two 'ideal' forms (explicit and tacit); but the different forms require different organizational and supporting infrastructures to be managed (Millar et al., 1997; Bolisani and Scarso, 2000).

The knowledge transfer between organisations raises complex problems. Organisations may be far from one another in both spatial and cultural terms; they may work in different environments, speak different languages, have different mental models, beliefs and decisional routines. So, the exchanging of knowledge involves not only the question of explicitation/codification, but also the modalities of transmission and reception. But there is not just the problem of knowledge transferability. A correct interpretation must also investigate how it is absorbed (Davenport and Prusak, 1998; Millar et al., 1997). Therefore, the difficulties involved in the transferability depend not only on the kind of knowledge which is exchanged, but also on the 'distance' between the organisations that exchange knowledge. In the literature it has been underlined that the major difficulties are related to the cases of transferring tacit knowledge, when heterogeneous organisations are involved. This is due to the fact that tacit knowledge resides within (and may be embedded in) specific people, organizations, or social groups. To proceed to any exchange, all terms and meanings must be accurately defined or 'codified'; in addition, this encoding must be shared among the different organisations.

Our analysis also illustrates another problem: when we use an electronic network, we must not only provide an effective and efficient knowledge exchange (due to the fact that actors do not interact 'face-to-face' and they may not be co-located), but we must also substitute advantageously the pre-existing relationships. This means that we must substitute the communicative chains belonging to 'culturally homogeneous' groups of agents. In respect to this fact, when we exchange knowledge there is an antithesis between communication efficiency and the richness of the communication. While codification may enhance efficiency, by improving the velocity of transmission and increasing the number of possible contacts, it may also substantially impoverish the content of the message. This is because the process of codification–formalization inevitably compresses the richness of the content.

The growing demand of efficient exchanges and transactions among an ever larger number of different subjects makes a certain degree of de-contextualisation and codification unavoidable (naturally we cannot exclude that future developments also allow the transmission of tacit knowledge along technological networks; but this is not yet on our agenda).

The complexity of cognitive interactions (intra and inter-organisational) may be better understood if we consider that the three different sub-processes of knowledge generation, codification, and transfer in reality overlap and influence each other (Davenport and Prusak, 1998). New knowledge generation is mainly tied to the interaction and exchange among subjects with complementary competences and experience. Often the transfer generates cross-fertilisations and at the end the formation of new knowledge. Generally, valuable economic innovations can be created exactly during the diffusion process (for an accurate proof see Antonelli and Gottardi, 1991). Due to the fact that knowledge codification aims to make knowledge accessible to those who need it, it should be regarded not only as a preparatory phase to the transfer process, but, on a longer temporal scale, as a new mechanism for knowledge generation. We think that the crucial issue is not only the process of knowledge conversion, but also whether it is possible and/or opportune to codify knowledge, considering that tacit knowledge cannot be codified without losing part of its deeper meaning.

5.4 On the codification and interpretation process

Let us examine in detail the process of knowledge transfer. As we have seen above, this activity preliminarily implies a process of *codification*. Codification is not just an externalisation of tacit knowledge (that requires a specific formalisation and the use of models or analogies), but more simply a representation of knowledge through symbols or 'codes', that can also be images or terms of natural language. The process of codification has the aim of realising 'objects' or representations that translate the knowledge that must be transferred into a feasible form, suited to the channels that one intends to utilise and (possibly) useful to the scopes of the user himself; without codification the transfer is impossible.

Within certain organisations, where knowledge consists of competencies in principal difficult to transfer, codification is thus a fundamental process. The way in which codification is realised inevitably reflects the cognitive characteristics of the subject that activates this process. The description of a piece of knowledge is indeed influenced by the knowledge, values, and experiences possessed by the individual that realises this process; in particular, this regards the way in which the selection and organisation of information is organised for the realisation of knowledge representation.

Knowledge that is codified cannot reach the receiver without involving a process of *interpretation*. This process requires other conditions, and, specifically, the construction of mental models. Also when we interpret something, it is important to select the correct information and to organise it in a coherent pattern. It must be underlined here that interpretation is not a process with a deterministic end. Many interpretations are always possible. The resulting ambiguity involved is essentially related to the lack (or the difficulty found) in setting an explicit and shared code (Daft and Lengel, 1986). For this reason we often cannot avoid interpersonal communications, even in situations where advanced media are available.

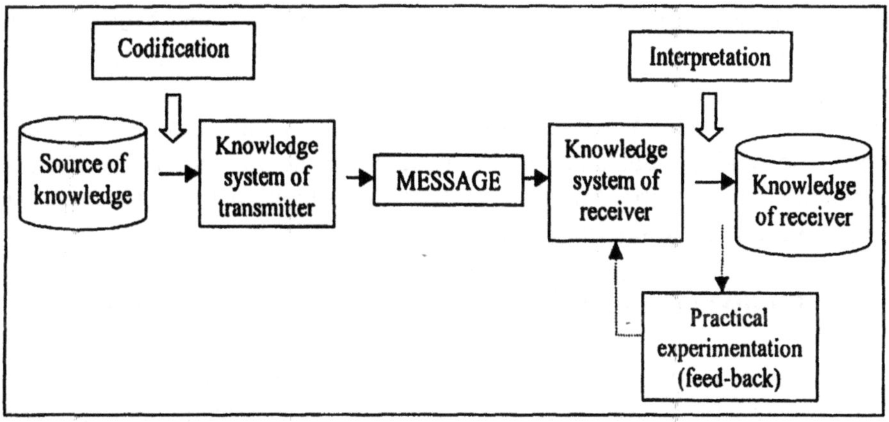

Figure 1. Modelling the transfer process
Source: Garavelli and Gorgoglione (2000).

The more complex situation is that where, in order to understand and interpret the knowledge transferred, the interconnecting actors also have to exchange their *interpretative framework*. In the case in which the processes of codification and interpretation are realised by different cognitive systems, what is involved is not only the object (knowledge) to transmit, but also the *cultural context* belonging to the receiver and to the sender of the message. In this case it is the *interpretation* which becomes part of the process that opens up major unresolved issues. When cognitive systems are different and there is a substantially divergent cultural background, it is more probable that a different interpretation and meaning is attributed to terms and symbols with which information and knowledge has been codified.

A first consequence is that the infrastructures or the digital tools able to transfer highly rich contents should be flexible and transparent with regard to languages, meanings, and interpretative context. Similar technologies exist (eg. videoconference), but they are inefficient for supporting both context communication and the transmission of formalized or codified contents. A second consequence is that, at least until now, it is justified to have doubts on the validity of the diffuse opinion that the Internet could become a universal communication tool. Maybe this could occur in the future, when a larger number of new applications (www is ten years old), and technical solutions based on a very broad band infrastructure, supporting a real multimedia communication, will have become a reality.

The experience of many years has shown that *e-commerce* systems based on formal communication and on a complete automation of transactions (as in the case of EDI) are highly efficient, but very restricted in scope. Moreover, they require a high homogeneity of organisational and strategic contexts, or at least precise agreements based on long term relationships (and this makes spot transactions impossible). An activity of knowledge codification could thus be not only too costly, but also largely inefficacious if the problem of *interpretation* is not tackled and solved. And this is a type of problem that the present process of digitalisation can deal with only in part.

5.5 Knowledge transfer into homogeneous environments

Let us start to discuss the case in which we are most interested, where knowledge transfer is organised within culturally homogeneous clusters. If we are confronted with situations where the cultural background is the same (or at least is the same with the reference to information or activity involved in the process of knowledge transfer), the interpretation of symbols and codes can be simplified. On the other hand, it is the process of codification that is now critical. The reason is the following. If the scope of codification is the fluidification of knowledge exchange among different subjects, and among them informal processes of exchanges of the *bouche à oreille* type are pre-existing (we are referring to non-codified types of communications), the process of codification *per se* constitutes a relevant cultural change. This requires modifying symbols, linguistic grammars, codes and communication models that were historically created within a cluster. Failures in transferring knowledge through the support of Ict technologies can thus stem not only from the divergence of cognitive systems, and from the possible divergence in knowledge interpretation, but they can be also created when a codification process within homogeneous cognitive systems is not completely efficacious.

Within local production systems or clusters, and within the same sector or *filière*, there are firms with similar evolutionary paths. Because they have shared the same experiences and the same cognitive systems, it is likely that they have built languages and informal processes of knowledge transfer, which are very efficient. A long time is needed to develop new technology aided modalities of knowledge transfer, which are more efficient than traditional ones. In addition, the necessary process of knowledge codification and re-codification, with more appropriated modalities for the transferring of knowledge along telematic nets could be very costly, in relative terms, when tackled by small firms with scarce resources and scarce propensity for this type of operation. At the beginning, the process of knowledge codification and re-codification could also produce less efficient results in comparison with the traditional ones; thus, we would lack a demonstrative effect that could trigger innovation.

The transfer of knowledge, in the case of Ids formed by small business, is related not to formal pieces of scientific knowledge but to a type of know-how. Therefore, the difficulties of transferring are even larger, because the process must also include a phase of *practical experimentation*, and this further delays the passage to digitalisation. This phase has the purpose, for the receiver, of acquiring contextual and applicative knowledge necessary for the concrete transferring of know-how. The experimentation phase enriches the receiver's tacit and codified knowledge, through interaction with the applicative context. Using Fig. 1, the transfer model described above must, thus, be completed with a feed-back chain. If the receiver is to valuate the efficacy of the new knowledge acquired, he would also have to adapt this new knowledge to the body of technical knowledge or experience he already possesses. Then the loop can "close" on the receiver, as shown in the figure; in all other cases the sender must be actively involved in the transfer process (Garavelli and Gorgoglione, 2000).

At present, the large part of information technology is used by firms for supporting the codification of pre-determined information. The efforts for developing new technologies, able to interpret information intrinsically characterized by ambiguity, are now growing. Hypertextual communication goes in this direction; also the new applications of *data mining* and the ontological reconstruction of contextual meaning appear useful tools. An example of this new approach is the use of broad band for transmitting moving images and for the reconstruction of virtual environments capable of reproducing contextual characteristics (prospective, weight, colour). This shows how, in these new directions, more room is given to the transmission of practical "savoir faire".

5.6 Modelling knowledge transfer into IDs

It has already been explained that the transfer of knowledge within the ID model is based on its strategic and structural process of evolution. The need of knowledge transfer is, by and large, induced by the introduction of innovations: it is the necessity to spread the use of autochthonous technological solutions, or to adapt to new, externally generated technological paradigms, that has activated and fertilised the processes of knowledge exchange. These exchanges influence the strategic choices and the sequential trial and error experiments; in the end, and in the long run, they influence the very structure of the district. This evolution goes through different learning-selection phases – as we have schematised in Fig. 2 (for a detailed description see Gottardi, 2000).

As discussed above, the transfer of any kind of knowledge requires that the actors involved be able to interpret all the implicit and explicit contents and meanings. And, because this knowledge has an economic value, we must also consider the *aim* of transfer: the exchange of valuable knowledge is useful only when it enriches and renews the knowledge base of the receiver. The consequences are, in part, twofold: on one hand, the efficiency and efficacy of the process is granted by the cultural 'proximity' and by the *congruence* of the knowledge contents which the user needs; on the other, the more extensive and articulated these ones are, the bigger is the potential value of knowledge transferred. For the firms that can have access to new knowledge, it is knowledge *variety* that may enlarge their mental models and their flexibility that they put in practice in order to face new problems.

Cultural codes and common interpretative meaning are important, as well as a high congruence with the problems of the receiver; however, knowledge transfer is more profitable when this knowledge makes a break with the old interpretative models and uses different representations of processes and facts.

The digitalisation of knowledge within IDs could thus open up significant opportunities. Benefits could be seen both within the district - in other words, within the market place - and in the value chain, and in the external relations.

Structure in *t*
⇓
Exogenous innovations
⇓
Preliminary selection
⇓
Strategy definition – learning
⇓
Localised innovations – differentiation
⇓
Co-ordination - internal selection
⇓
External market selection
⇓
Structure in *(t+1)*

Figure 2. Innovative strategies and structural change into dynamic IDs
Source: Gottardi (2000b).

5.7 Theoretical benefits and practical obstacles

Digitalisation reduces the specificity of information, increasing the number of subjects that can benefit from it, prolonging the period of utilisation (a virtual catalogue can be utilised by a vast host of operators, an on-line auction amplifies the temporal horizon of information).

Digitalisation moves the trade-off between largeness (number of subjects that use information) and deepness (richness of content) of exchanged information (see also Evans and Wurster, 1997). While today rich and specific flows can be exchanged only through interpersonal relations within limited areas, digitalisation allows the spread of complex information among an elevated number of agents.

Within a specific value chain, the digitalisation of information and knowledge could, in principle, generate the following benefits.

Both suppliers and buyers of intermediate input could be more specific and they could also make their offers/requests more complex, and/or address them to a larger number of potential clients. If the number of agents that can be brought into an electronic market place grows, it is the same market structure that changes; in particular, this increases its liquidity and competition. In other words, this enforces the efficiency of the market as a coordination mechanism of local productive relations.

The modality of value generation becomes more complex and flexible. The possibility related to the efficient use of the market allows firms to concentrate on the core business; so, firms can adopt appropriate specialisation strategies. At a meso level we see a de-construction of business and value chains of the district firms. The traditional division of labour becomes highly articulated; so, we observe a new variety of agents and strategies involved in systemic network relationships that take the place of linear-sequential transactions.

To an increased variety of business strategies corresponds an enlarged product variety. This variety derives from the process of specialisation, that creates an increased offer of modules and components.

The digitalisation of markets improves the interface between the district and the external environment. New relational potentialities are derived. The introduction of network economies is not just an internal process, pushed by the construction of *e-digital platforms*, internal market places, and *e-procurement systems*, but also an external one, which facilitates external interactions in international global trade.

As a whole, the principal expected effects are the intensification of network economies, the growth of the rate of generation and diffusion of new applicative knowledge, and an increased learning due to better interactions. On the district scale we can assume not just a simple growth of efficiency, with a modification of the firm's architecture (and in perspective with a modification of the social linkages), but also a more complex pattern of increased efficacy. The digitalisation of the traditional nets has the potentiality of opening up the system towards the external economy. Thus, district firms could participate in a major number of economic networks, benefiting from network economies, reduction of transaction costs and growth of business opportunities.

In this perspective, the district's working mechanisms could be strongly reinforced. Thus, for instance considering strong firm specialisation, the existence of a diffused small size among district firms could be appropriate. Ideally one could think of the use of second-generation flexible automation technologies, able to interact with a broad band digital network. In the cases in which this scenario could be realised, the number of *technology intensive* firms could grow. The districts could consolidate into technological poles.

Table 1. Theoretic effects of digitalisation on IDs value chain

Market effect	Value chain effect	Cognitive effect
Transaction encoding	Value chain de-construction, business de-construction	Knowledge encoding
Creation of e-*marketplaces* and vertical portals	New business and corporate strategy: (focalisation, specialisation)	New intangible assets enrich the cluster
Information diffusion rate grows	Firms differentiation	Knowledge transfer stimulates effects of hybridisation /innovation
Transaction costs are reduced	New industrial models	Cognitive heterogeneity grows
Internal market efficiency grows	New division of labour (systemic vs. linear value chain)	New specialisation and network economies take place

But we will try here to examine, first of all on a theoretical base, the obstacles that can contrast this perspective. The use of Ict technologies is advantageous only if it allows more complex information to be made known to a higher number of agents. The fundamental issue is thus whether physical proximity and interpersonal relations will continue to play a decisive role, given the fact that this has until now constituted the richness of the IDS. It has not yet been demonstrated that digitalisation could substitute the system of traditional relations advantageously (in terms of costs/benefits).

The fact that physical proximity could continue to play an important role is indeed in contrast with the diffused opinion on the de-localising power of the Internet and Ict technologies. However, up to now, these effects have been modest or non existent; more recent interpretations and forecasting on the real spatial effects of Ict technologies are in contrast with those that were initially *naif* (Brousseau and Rallet, 1999).

This problem can be reformulated in the following way: how is it possible, and when is it convenient to introduce the electronic commerce in order to support with efficiency relations and transactions within specific environments where already the existence of proximity and of a common social context render these operations very efficient? In our opinion the main obstacles derive above all from the insufficient knowledge regarding the virtual reproducibility of those mechanisms of governance and coordination which characterise the district. The feasibility of a strategy of virtualisation, through telematic nets, of social and economic interactions based on trust appears at least quite uncertain. This is because proximity substitutes sanctions on opportunism, the use of hostages, and reduces the need of formal contracts.

These difficulties can also be analysed in greater detail. Despite the apparent absence of costs in using the Internet, the diffusion of its applications through electronic nets will need big investment during a significant time span and a direct assumption of risk. Major investment are required for the development of information and knowledge codification, and its re-distribution in a digitalised form. Also the standardisation of transactions and their security, along the nets, will imply significant costs.

The so called district portals (or district market places) would have to deal with all these functions, but it is disputable whether their realisation would be simple. The implementation of *e-commerce* systems will have to face many risks, because the final outcomes are not guaranteed at all. Moreover, tacit knowledge cannot be made explicit or codified without the loss of 'valuable meanings': with the attempts of formalisation we inevitably lose something, something that might be essential. Within IDs, any exchange largely involves tacit knowledge and informal routines. Considering the actual state of the art of Ict applications, the efforts to translate district knowledge into more formal pieces of knowledge, with the scope of introducing (partially) automated communication systems, might also end up as not useful or even negative.

Contrary to what we may think, in these operations there are not many possibilities for the district small business of *free-riding* or imitation at low costs (this has often permitted a reduction of the overall investment costs). Then, as we already noted, codified communications allow a high degree of automation but, as the EDI systems have shown, such applications require that interpretative contexts, languages, and contents be largely pre-defined; this is clearly the major limit of those applications.

The inter-firm relations are based on the transfer of different forms of knowledge having very different contents. This also represents a great challenge in developing Ict technologies and *e-commerce* applications. Their potentiality strictly depends on particular communication processes; so, these applications differ from one another with regard to their technical components, and also if we consider their scope and function. A universal electronic communication tool, capable of resolving all the possible problems of communication with the same efficiency and effectiveness, is impossible or, at least, far from being available.

Within IDs, it is likely that it would be impossible to adopt standard solutions, and that one would have to implement customised projects, and search for *ad hoc* solutions. Firms belonging to IDs have developed in specific and historically determined contexts: so, even when based on the same standard components (e.g. basic software, operating systems and Internet protocols), the implementation process of each specific application will necessarily require a specific investment. Knowledge developed in IDs is a localised process.

It is also interesting to try to verify these issues at a more empirical level, in order to detect what is really occurring nowadays, and try to provide even a rough interpretation. To explore this point, we will use the main results of some empirical investigations (Chiarvesio and Micelli, 2000; Di Maria, 2000; Chiarvesio, 2001; Capitani and Di Maria, 2000). The principal reflections available underline the fact that network technologies are scarcely diffused within IDs, even considering the more dynamic districts. In the more recent period, the situation is perhaps changing in some IDs, under the combined push of the mass media, the efforts of consulting companies for Internet services, and the activity of innovation promotion organised under the competence of local collective governmental and private organisations. The largest firms (which are the local leaders of IDs) are more or less in line with the other firms existing at national level as regards the use of electronic mail, Internet services, and applications of rapid connectivity, like ISDN, while small firms come farther behind (Chiarvesio, 2001). Communication technologies are used in a 'conservative' mode: the more diffused are those which are simple to use and do not require reorganisation of the firm's work processes and relationships. The biggest absentee is electronic commerce.

Quite infrequently the languages utilised and the codes developed at local levels are standardised. Machinery, tasks, components, often have different names: they are comprehensible only to local operators but not to external ones.

The use of CAD systems for design is still limited. Quality certification is not very diffused, because what is prevailing is the self-certification based on trust and on the direct knowledge of the partner.

The low level existing in the use of electronic commerce is related to the high real effective costs, and the lack of human resources. However, this tool is perhaps not adequate, considering the characteristics of products and processes, both in relation to B2C and B2B. This point of view does not depend on a preconception, or on an obscurantist cultural position; and, indeed, many attempts have been recorded. Many firms nowadays have realised their own website, directly experimenting the fallacy of many common perceptions on the direct advantages of a better visibility and on the low managerial costs of commercial websites on the Internet. Naturally, these *e-business* systems still have many potentialities, but, on an empirical base, the solutions required are not at all trivial, neither can they be deduced from the models valid for large firms. It is obviously clear that these websites can be a basis for B2B transactions. It is not just a matter of developing new solutions, but new professional figures for the developing of *digital intermediations*, which are still lacking within IDs. The creation of electronic markets cannot be assumed as a spontaneous process. New operators are required for the digitalisation of knowledge and for the standardisation of processes and products, but they must also be able to facilitate the matching between demand and supply, and to define new rules for the *e-market places*, to guarantee its correct working, to conceive and experiment new business models.

5.8 Some concluding remarks

The hypothesis that we have tried to verify is whether the most interesting effects of the diffusion of *www* applications and e-commerce, more than within the globalised system on a planetary scale, could be manifest in more limited, but dynamic and innovative environments, where there is already a high density of transactions and exchanges. The unit of analysis chosen is the IDs, for which a significant number of researches are available, and there is a vast accumulation of knowledge about their characteristics and pattern of evolution.

IDs, many of which can be considered very successful like those in Italy, were born as 'networks without technologies' (Micelli and De Pietro, 1997). In their recent evolution they have to face a new technological variable, which could induce structures and paths with a very different outcome. Some recent analyses have dealt with this issue in a very reductive way, limiting themselves to examining only the aspect of technical feasibility, or assuming that the experiences developed in other environments (such as the construction of digital platforms, portals and electronic marketplaces used by multinational firms) could simply be transferred to other contexts.

However, the problem of digitalisation within IDs has wider dimension. The transfer of knowledge within the IDs, and the modalities with which it occurs, are the principal driver of its evolutionary processes, both in the strategic and the structural sense. Due to the impact of Ict on the processes related to information and communication, and on the creation and transfer of knowledge is formidable, it is unthinkable that the adoption of these technologies could occur softly. The analysis must necessarily overcome the technical and operational dimension, to tackle explicitly the strategic and evolutionary processes of these systems.

The outcome of the encounter of 'districtual nets' with technological nets is not to be taken for granted (Rullani, 2000). One can imagine a convergence, in which IDs gradually assimilate the organisational solutions implied by the technological nets; or alternatively, that the latter will converge towards the models of local nets adapting to the necessity of small firms. The outcomes are still not at all determined: on the contrary there is great uncertainty as to what could happen. We have chosen to analyse not only the advantages related in principle to diffusion of the Internet but also, realistically, the obstacles that retard this process.

We share a certain scepticism on the actual capabilities of IDs on the rapid adoption of the new technological paradigm. It is very difficult that the new digitalised processes of information and knowledge exchange could be activated spontaneously without the presence of promoters playing the role of *service hub*. For the same reason, the diffusion of e-commerce does not come about if it is not accompanied by the emergence of new subjects (i.e. info-mediaries and market makers), not to speak of the needed institutional adjustments. Within the IDs, subjects are needed which are able to aggregate and re-distribute knowledge. It is a process that can be long and complex. The use of Ict requires, within the IDs, the adoption of communication standards and formal procedures in order to substitute the existing routines based on the natural language, proximity, and interactions. These new standards and procedures must be used to govern complex relationships rooted in trust, reciprocity, and reputation. In our opinion, traditional communication forms cannot be simply and rapidly substituted by digitalised relationships.

IDs must thus make a cultural leap, that could be even more demanding than for large organisations, more confident in the use of technical-productive and commercial standards. Probably it is the process of knowledge codification that constitutes the more relevant cultural change, because it requires modification of the bases of communication (symbols, codes and languages), in comparison with what has been historically determined.

Nowadays all agree on the fact that the new technological paradigm will be directed above all on B2B through the electronisation of transactions. In the case of IDs, this passage will be very gradual, with a first step of using Ict technologies for the support of natural language (for example *email*) or operational routines; only after that shall we see the adoption of more symbolic representations. In our opinion, the development and use of Internet applications of e-commerce inside the IDs will depend on two conditions: 1. on one hand, the availability of a range of digital tools, applications and technologies for carrying tacit communication with a large variety of different contents (media richness); 2. on the other hand, the capability to transform tacit knowledge into gradually more formalised parts and components (e.g., following the general process of internationalisation of managerial culture).

Generally speaking, the automation of communication and transactions will not remain a realistic task, maybe with the exception of e-procurement for some categories of inputs in the MRP processes, and of totally repetitive operations.

Which could be the drivers of this change within the IDs? The first is related to the role of local leading firms, not only referring to the internal net but also in connection with outside. These firms will stimulate the adoption of telematic nets, as in the case of the diffusion of EDI: this technology has been promoted by final leaders for pushing their subcontractors to adopt the relative standard. In comparison, the Internet offers the advantages of a more technological flexibility deriving from the use of a non-proprietary standard.

The second possibility is that small innovative newcomers will start-up autonomously, bringing new business ideas based on digitalisation; however, we are of the opinion that this path is less likely to occur. A third opportunity could be created by new co-operative projects with different local agents, both public and private (including Universities and Research Centres). These projects could be directed towards the design and installation of vertical portals for the b2b or other net services.

References

Antonelli C. and Gottardi G. (1991), The Interaction between the Generations and the Diffusion of New Technologies, *Economics of Innovation and New Technologies*, vol.1, no. 4, p. 309-325.

Bolisani E. and Scarso E. (2000), "Electronic communication and knowledge transfer", *International Journal of Technology Management*, Special Issue on Technology Management in a Post-Fordist Environment, vol. 20. no.1-2, p. 116-33.

Brousseau E. and Rallet, A. (1999), (eds.), Technologies de l'Information, Organisation et Performances Economiques, *Commissariat Général du Plan*, Rapport du group de travail, Paris.

Capitani G. and Di Maria E. (2000), "Le nuove tecnologie dell'informazione e della comunicazione come fattore strategico di sviluppo locale", in, Micelli S., Di Maria E, *Distretti industriali e tecnologie di rete, progettare la convergenza*, Franco Angeli, Milan.

Chiarvesio M. (2001), "Un portale per il distretto industriale della sedia", in De Toni A, Grandinetti R. (eds.), *Conoscenze, relazioni e tecnologie di rete nelle filiere distrettuali*, Franco Angeli, Milan.
Chiarvesio M. and Micelli S. (2000), "Reti e nuove tecnologie nei distretti industriali del Nord-Est. Risultati dell'osservatorio Tedis", in, Micelli S. and Di Maria E, (eds.), *Distretti industriali e tecnologie di rete, progettare la convergenza*, Franco Angeli, Milan.
Evans P. and Wurster T. (1997), "Strategy and the new economy information", *Harvard Business Review*, no. 5, p. 70-82.
Foray D. and Mairesse J. (1999), (eds.), *Innovations et performances des firmes*, Editions de l'École des Hautes Etudes en Sciences Sociales, Paris.
Daft R.L. and Lengel R.H. (1986), "Organisational Information Requirements, Media Richness and Structural Design", *Management Science*, vol. 32, no. 5, p. 554-71.
Davenport T.H. and Prusak L. (1998), *Working Knowledge: How Organisations Manage What They Know*, Harvard Business School Press, Boston.
Demarest M. (1997), "Understanding knowledge management", *Long Range Planning*, vol. 30, no.3, p. 374-84.
Di Maria E. (2000), "Vertical Communities e distretti virtuali", in Micelli S. and Di Maria E. (eds.), *Distretti industriali e tecnologie di rete, progettare la convergenza*, Franco Angeli, Milan.
Garavelli A.C. and Gorgoglione M. (2000), *Attori processi e tecnologie del trasferimento di sapere e saper fare nelle imprese: un modello di analisi*, Tecnologia, organizzazione e fattore umano, Associazione Italiana di Ingegneria Gestionale, Turin, 10 November.
Gottardi G. (1996), Technology strategies innovation without R&D, and the creation of knowledge within industrial districts, *Journal of Industry Studies*, vol. 3, no. 2, p. 119-134.
Gottardi G. (2000), "Innovation and the creation of knowledge in Italian industrial districts. A system model", in Belussi F., Gottardi G. (eds.), 2000, *Evolutionary Patterns of Local Industrial Systems. Toward a Cognitive Approach to the Industrial Districts*, Ashgate, Aldershot.
Gomez M. (1998), *Non redistributing prices and exclusion in the evolution of Internet*, Uranga University of the Basque Country, Bilbao, Departamento de Economia Aplicada I, mimeo.
von Hayek F.A. (1948), *Individualism and Economic Order*, University of Chicago Press, Chicago.
ITU (1994), Report on the State of World Communications", *International Telecommunication Union*, Newsletter, no. 1, p. 9-12.
Millar J,. Demaid A. and Quintas P. (1997), "Trans-organisational innovation: a framework for research", *Technology Analysis and Strategic Management*, vol. 9, no. 4, p. 399-418.
Micelli S. and De Pietro L. (1997), "Reti senza tecnologie: la diffusione delle tecnologie nelle imprese del Nordest", *Economia e società regionale*, no. *2*, p. 5-27.
Nonaka I. (1991), "The knowledge creating company", *Harvard Business Review*, Nov.-Dec, p. 96-104.
Nonaka I. (1994), "A dynamic theory of organizational knowledge creation", *Organization Science*, vol. 5, no.1, p. 14-37.
Polanyi M. (1967), *The Tacit Dimension*, Doubleday Anchor, Garden City, NY.
Storper M. and Harrison B. (1991), "Flexibility, Hierarchy and Regional Development: The Changing Structure of Industrial Production Systems and their Forms of Governance in the 1990s'", *Research Policy*, vol. 20, p. 407-421.

Rullani E. (1995), "Distretti industriali e economia globale", *Oltre il Ponte*, no. 50, p. 5-61.

Rullani E. (2000), Tecnologie della conoscenza e distretti industriali: due linee di evoluzione, in, Micelli S. and Di Maria E. (eds.), *Distretti industriali e tecnologie di rete, progettare la convergenza*, Franco Angeli, Milan.

Chapter 6

Cognitive models, efficiency, and discontinuities in the evolution of Industrial Districts and Local Production Systems

Mauro Lombardi

6.1 Introduction

In this study, a research hypothesis on Local Production Systems (LPS)[1] is developed on the basis of theoretical statements introduced in Lombardi (2000a). Starting from the results of empirical inquiries carried out recently by many scholars, we continue the objective of delineating a long-term trajectory of their development (Sections 6.1-6.4) on the one hand, and on the other, investigate potential future trajectories by analysing emerging properties during the ongoing transition process (Section 6.5).

Notwithstanding several important specific aspects, long-term dynamics of LPS in Italy showed common general features. Since the 1980s, similar trends have appeared, even though in different gradations, as a reaction to discontinuities related to fundamental – i.e. technological, organisational, institutional – aspects that are inherent in changes of leadership (Pilotti, 1999, 2000).

In order to understand the genetic processes of changes and emergent phenomena, a precise theoretical perspective is adopted here that is based on the

[1] In this paper, we shall use the concept of local production systems, which is a more general term than industrial districts basing ourselves on the properties present in various typologies of local productive micro-universes. Our choice is based on the concept of Local Systems (LS), composed of sub-systems (LPS, institutional infrastructures, cultural and political shared values, etc.), and on different mixes of links between sub-systems and differences in the composition of each sub-system foster varied forms of LS.

analysis of three components: 1) systemic configuration, that is the working model of a set of entities which interact in a way that is not simple (Simon, 1962); 2) mechanisms of information processing and of the creation and diffusion of knowledge; 3) relationships between environmental changes, information flows, and organisational models.

6.2 Evolutionary discontinuities

A fundamental turning point, that was fully realised starting in the 1980s, was certainly the one caused by the changes in the competitive environment. New factors (the quality of products, strategic abilities to project into the markets) became relevant in a world marked by growing complexity and incessant turbulence. The growing complexity identifies a situation within which there emerges a technical-productive dynamics towards products based on technical and productive competencies, which are different from those of the past, and sometimes stem from a multiplicity of sources that are external to the local context. In other words, local "micro-universes" have to absorb or use knowledge that is no longer accumulated only inside them. The incessant turbulence synthesises the innovative dynamics of international demand, which requires commodities that are characterised by new technological mixes, while competitors, able to offer products with radically different components (technologies and properties), appear on the international markets. There have been conspicuous clusters of process and product innovations, which require a higher technological and scientific level of both input and output. Global competitors (firms and countries) have become prominent, some of which are able to compete in matters of price and quality, while others adopt "aggressive" strategies in order to win over and keep markets.

These phenomena, which are related to the so-called "globalisation of the markets", are tightly entwined with new types of consumer behaviour: consumers have become more careful in evaluating the price/quality ratio and in searching for first-hand and specific solutions in order to satisfy their needs. During recent decades, LPS have shown a considerable ability to obtain "access to final markets in which the capacity of timely adaptation to change in the quantity and the quality of demand is highly requested" (Bellandi, 2000).

The changes in the competitive environment have triggered processes of transformation in the basic components of traditional LPS.

A first significant point is the following: skills and technical-productive competences, accumulated within traditional production cycles, have had to cope with the need to control global information flows relative to science, technology, economics, and finance. In fact, in order to perform effectively in markets characterised by innovative dynamics, strategies are necessary that are not so much adaptive or merely responsive as they are able to forestall and foster trends. This is impossible without adequate mechanisms of control of new information flows.

A second important point, and one that relates to the preceding point, is that in many LPS, a process of erosion of the traditional "core competences" developed during the last decades has begun.

The factors pointed out – above all – the need for higher technological and scientific levels of input, have led to a substantial change in the economic space for products based on practical and hands-on knowledge, and incremental innovations (Corò and Rullani, 1998).

In fact, we must be aware that the traditional technical-organisational configuration of LPS is characterised by the existence of a "multiplicity (...) of different nuclei of specialised know-how and approaches to production and innovation" (Bellandi, 1996; p. 358), while a main system can be identified within this multiplicity of nuclei of competences (Bellandi, 1994), that are professional and entrepreneurial, around which the process of division of labour among units with different degrees of specialisation is developed (Bellandi, 2000).

The changes in the competitive environment and the needs related to global information flows have altered the division of labour and the pre-existent specialisation[2], by triggering a transition process during which LPS have undertaken strategies of response and structural adjustment as regards exogenous shocks: diversification and a rise in the quality of the products; restructuring of relations between firms; progressive internationalisation of the production cycle; changes in the division of labour, and distribution of the competences during the various production phases. It must be emphasised that the following phenomena have occurred, in different degrees, in LPS:

1. a reshaping of the local production potentials, with a reduction in the number of firms and employees;
2. an increase in the imports of components and half-finished products from other (domestic and international) LPS;
3. changes in the mixes of technologies used, raw materials worked, and goods produced.

[2] "Thus, the shrinkage of the Prato textile system has been accompanied by a complex redefinition of the division of labour among local firms" (Dei Ottati, 1996; p. 44).

Economic literature on the evolution of LPS, as analysed in Lombardi (2000b), has made it possible to point out a morphological homogeneity between the evolutionary profiles of different Italian LPS. It is characterised by three general trends which are the outcome of the discontinuities that occurred during the past decade: A) firms and LPS interact with markets differently from the past, on the basis of non-traditional mechanisms of acquisition and processing of information; B) there are a new cognitive division of labour and a multiplicity of technical-organisational models; C) networks of firms organised in hierarchical forms have appeared and have become stronger.

Since points A and B are analysed in sections 6.2, 6.3 and 6.4, let us direct our attention to C, which emerges in different forms within the Italian Regions. In the Marshallian industrial districts (MID), that are more greatly present in Tuscany than elsewhere, leader firms (above all woollen mills) appear which tend to become consolidated in groups of businesses through participation in, e.g., woollen mills in joint-stock companies (phase firms) (Ciappei and Mazzetti, 1996). In Emilia-Romagna, there are networks of firms with "local roots, but which are open and integrated to some extent with external entities, both Italian and foreign" (Messina, Riccamboni, and Solari, 1999). Lastly, the Veneto region is characterised by network-firms that are inserted within a fabric that is less open to external economic influences.

The previously-emphasised morphological homogeneity should not lead us to disregard the profound differences between "two models of local development" (Burroni, 1999a). In fact, a comparative analysis of LPS in Tuscany and in the Veneto shows that systems of small firms prevail in the former, while systems of small and medium-sized enterprises prevail in the latter. The same inquiry has pointed out a growing variation in economic structures: 1) in the Veneto, there is a more structured model that is characterised by the presence of leader firms; 2) in Tuscany, a configuration similar to the MID ideal type remains.

During the past two decades, the Region with MID has exhibited weaker dynamics: Tuscany has lower values in its economic performance indexes (employment and valued added by the industry and tertiary sector) (Burroni, 1999b). It must be emphasised that Veneto, on the one hand, and Tuscany and Emilia-Romagna, on the other, are characterised by two different models of regulation or local governance. In the first, a "model with low involvement" has prevailed (Burroni, 1999b), within which institutions indirectly invest, by giving economic and financial support to projects by firms or associations of firms. In the other two Regions, local institutions have centred their actions on the involvement of economic agents (first of all, associations of firms) and, at the same time, have implemented strategies aimed at directly providing public services (infrastructures, professional training, trading services). In the first-named Region, the distinction between the political and economic spheres has become greater; in the other two, integration between the two spheres has continued. Two models – an "aggregative" one and an "integrative" one – have thus been embedded within the institutional configurations described (Messina, Riccamboni, and Solari, 1999).

The preceding reflections have highlighted structural differences and common features, viewed as trend factors which foster specific interactions within local contexts. The exogenous discontinuities and the changes in the basic components of internal structures described inspire us to proceed to an analysis of the evolution in the properties of LPS, in an attempt at defining possible trajectories for the near future.

6.3 System properties: self-containment, information hierarchy, distributed and parallel information processing

In this section, we shall deal with some general properties of systems of small firms, in relation to exogenous and endogenous evolutionary discontinuities.

In general, a system can be described on the basis of precise variables, by means of which it is possible to characterise a set of components and the relationships between them. Complex systems, which are "made up of a large number of parts that interact in a non-simple way" (Simon, 1962; p. 468), are of particular interest as far as the purposes of this study are concerned.

Three types of explanations of complex systems have been proposed (Allen and Lesser, 1991). The first one consists of models based on equilibrium, which is arrived at by means of processes that are not subject to investigation. The prototypical example is a system described with simultaneous equations, that represent optimising types of behaviour of agents (individuals and firms) belonging to the same system. The second one, which is expressed in dynamic systems, is developed through models based on equilibrium, that are viewed as the result of (internal) variations caused by exclusively exogenous changes. In this case, the dynamics and the behaviour of the system are explained by assuming the homogeneity of the components: all entities are ranked together with the "average type", which is a logical, mathematical or statistical construction that works as a "representative agent" (Allen and Lesser, 1991; p.

163). These authors emphasised that this type of explanation model is not able to go beyond an analysis of linear mechanisms, while "learning, adaptations and changing responses are treated as exogenous factors". At the same time, this theoretical approach cannot "capture" non-linear phenomena, chaotic dynamics, or structural and organisational changes within the systems. The third type of explanation, which we call evolutionary, treats these aspects without assuming equilibrium and by using very different concepts[3].

For the purposes of this study, we wish to stress that an analysis of the behaviour and the dynamics of complex systems cannot disregard two essential aspects: 1) the relationships between the internal milieu and the external environment; 2) how the impulses are transmitted within the inner circuits and spread outside of them.

On the basis of the preceding reflections, our analysis of LPS is developed here by means of a conceptual framework that is centred on three factors: A) a distinction between "inside" and "outside" (that is, a study of the system's boundary); B) the typology of the interactions between components; C) mechanisms of information processing and of the creation and diffusion of knowledge.

As for A, we have to argue a basic property of LPS: self-containment, based on the distinction between the internal milieu and the external environment. This property is important, above all if the boundary is subject to more or less frequent alterations. The question becomes more complex in the case of LPS. This is because individuals and firms are, themselves, sub-systems that belong to local systems within which other entities – such as institutions, associations, banks, and agencies which provide firms with services – act . On the other hand, the single firm can be viewed in different ways. Eliasson (1989; p. 27) considers the production process as a "knowledge machine which uses information" and the firm as an "experimental machine" (Ibid.; p. 282). Cyert (1988; p. 87-88) analyses it as an "adaptive mechanism", and Loasby (1976) sees it as a "learning system", while Chandler (1992) emphasises the concept of "hierarchies of organisational routines" as basic components of firms. Lastly, Cyert and Kumar (1996) stress the role of the firm (business) as an organisation searching for information, with two joint aspects: an efficient search for flows of information, and adaptation to them.

As far as the local systems of small firms are concerned, a large number of economic, sociological and political-science studies systematically describe the

[3] Interesting analyses of evolutionary theory are to be found in Nelson (1995), Dosi and Nelson (1994); Max-Planck-Institut (1999).

important role of the local institutions that perform functions which are decisive for the systems of enterprises: the providing of services for the productive apparatus; the facilitating of credit access, especially on the part of small firms; the development of infrastructural offerings. As for the point discussed here (self-containment), another aspect is also relevant, i.e. the function of the institutions in producing and consolidating the "sense of belonging" of the agents operating in a given territory. The creation of a sort of "collective identity" is fundamental, because it enables the development and reinforcement of interactions between the subjects on various levels. In this way, endogenous dynamics are developed, which lead to peculiar forms of exercising the basic functions for LPS and to the emergence of socially-shared values and socially-accepted rules of behaviour. Particular and incessant information exchanges are, thus, the basis of the whole process.

We have thus defined the self-containment property of LPS, but a complete explanation of it requires a discussion of the B and C factors as well.

Socially-shared values and socially-accepted rules of behaviour are among the fundamental conditions which define the boundary of a LPS, but two other factors contribute: 1) the type of relationship between the internal milieu and the external environment; 2) a model of information distribution (how it is processed).

As regards 1, the territorial agglomeration of a myriad of productive units of commodities and services, over a more or less extensive period, is necessarily linked to how impulses from the market arrive, are filtered, and then spread within a given area. If we consider the market as the external environment, the transformation of a limited "micro-universe" of small firms into a real system depends, precisely, on how the signals extracted from the market are translated into specifications for the different phases of the production cycle. In other words, the process ranges from the demand parameters to the factorisation of a product into components[4], so that the production sequence is consistent with the starting parameters. Depending on who extracts the signals and how the factorisation process is realised, different principles and organisational structures emerge within local systems. We should add that the technical-organisational configuration of a local context is linked both to the type of information and to the way in which it is spread. We now arrive at factor 2. One of the most basic elements in the existence of a system is the difference between system information, which is inherent in its basic structure (Langlois, 1983), and

[4] Here, we use the idea of "factorisation of problems into semi-independent subparts" in a more extended manner (March and Simon, 1993; p. 214).

parametric (operational) information, which is related to the performing of functions and activities within specific contexts. Given the nature of information, the typology of the system is linked to the asymmetry of the distribution of the information, in the sense that two aspects become relevant: on the one hand, the aspect of how system information is controlled, and on the other, the forms and the limits within which operational information is created and spread. With reference to LPS, we can analyse the factors previously described (the relationship between the internal milieu and the external environment; the distribution of information) in general terms.

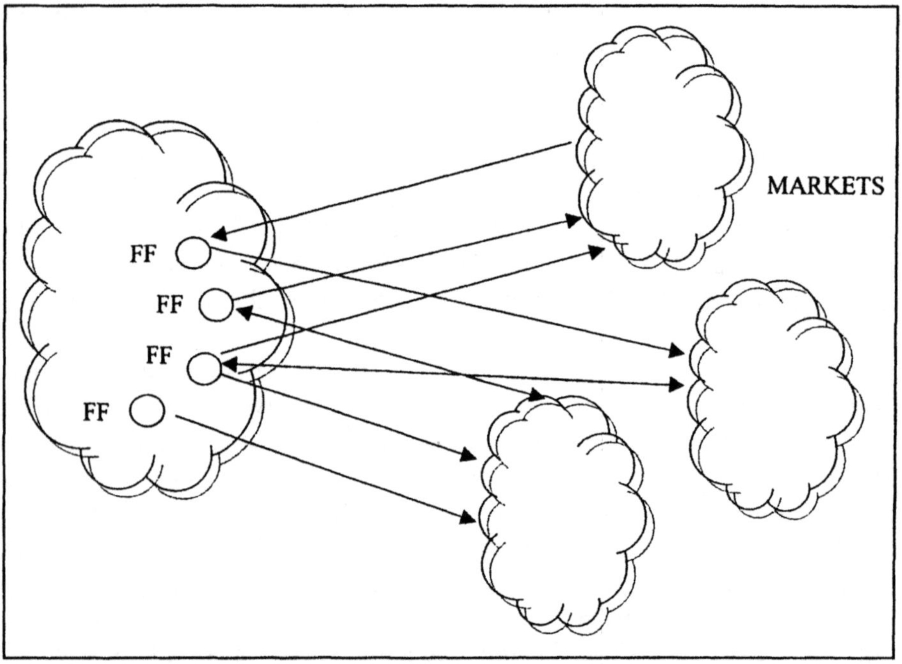

Figure 1: Markets: different sources of demand parameters (system information).
FF = final firms

Over several decades, in MID in Tuscany, strategic operators (called final firms) have appeared which are able to acquire and control market information: opportunities, predictable trends, demand features (fig. 1). They transform these into job orders and delivery notes for one or more production phases of a cycle that is fragmented and dispersed within a given territorial area (fig. 2,3). While operational functions are being performed and technical-productive problems are being solved, there are no limits to spreading information in view of the frequent and intense relationships among small firms, that are basically without restraints as far as belonging to groups or privileged relations with particular purchasers are concerned. Technical information flows can develop in any direction, and are associated with a knowledge of work costs and charges within a stratified local system.

The "finishing line" of the analysis that we have made is the following: systems information is firmly controlled by strategic operators, and parametric information is widely distributed between a myriad of units.

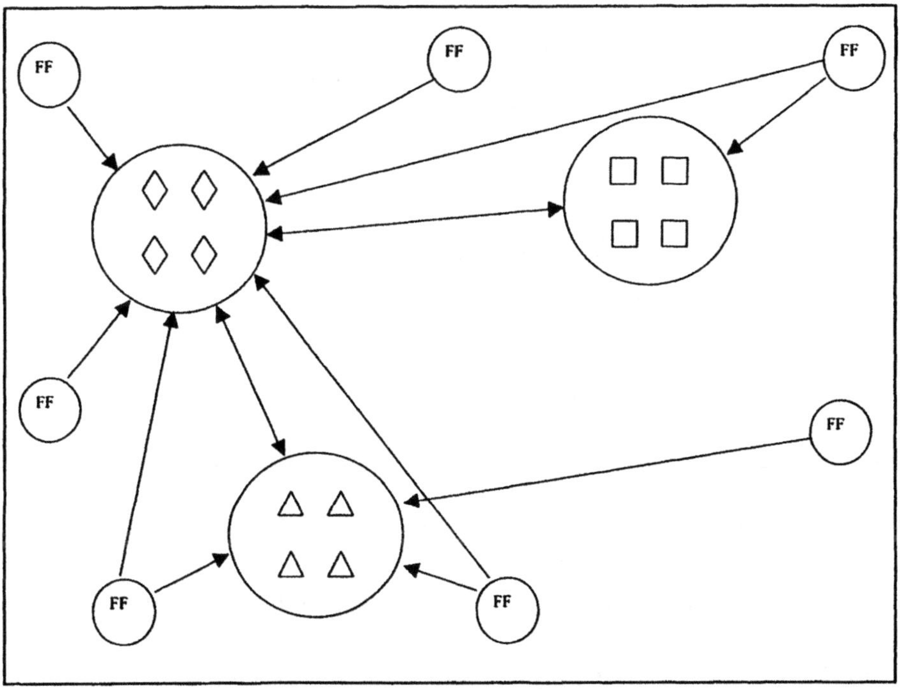

Figure 2: From FF to first layer groups of phase firms (groups of dispersed production phases). Flows of operational information.

On the basis of these conclusions, we can emphasise the existence of an internal milieu that is based on high-frequency interactions between local units. The result of these interactions is the factorisation of the local production potential, in order to match the demand parameters. The competitive success of LPS has thus been based for a long time on the ability of the internal milieu to produce quick and effective reactions to the impulses deriving from the external environment and described by means of information parameters in the possession of final firms.

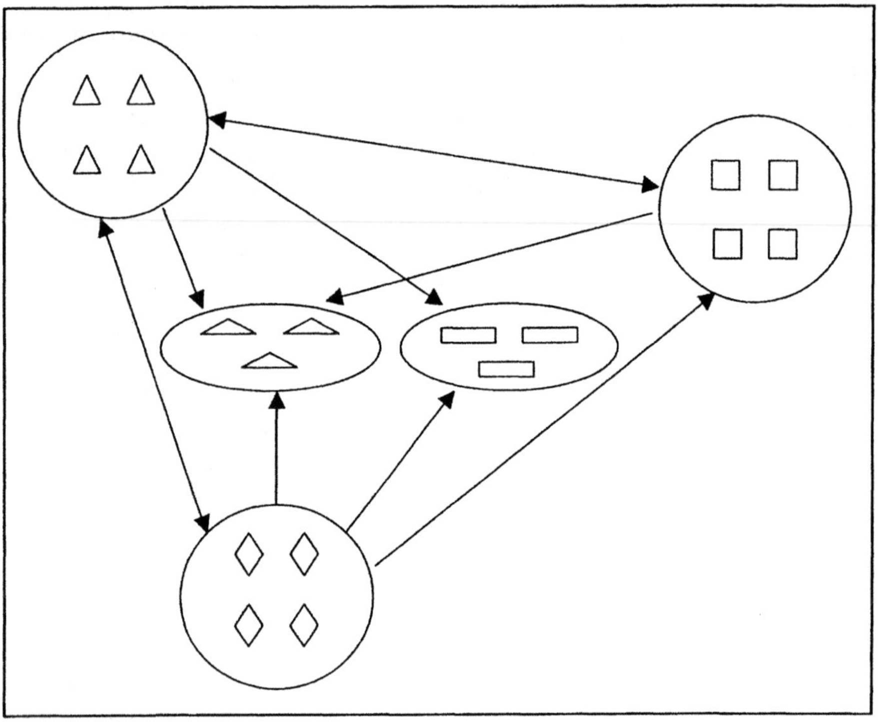

Figure 3: From the first layer to the second layer of groups (phase firms). Flows of operational information.

At this point, we can synthesise the traditional configuration of LPS with:

Proposition 1:

LPS are characterised by an information hierarchy: system information, which is controlled by strategic operators, and parametric information which is widely distributed within the local context.

Therefore, the double environment (internal and external) corresponds to a double level of information flows. However, it is also necessary to identify another aspect: in addition to the basis information asymmetry, there are the so-called "minor" information asymmetries – in the sense that, within the local context, distribution of technical-productive information is carried out on the basis of relationships of competition and cooperation. Thanks to these, widespread knowledge is acquired during performance of the work cycles. Since the strategic workers enlarge the aggregate of firms to which the job orders are assigned, above all during the expansion phases, the development of technical-productive information takes place, in fact, during distributed and parallel flows, with a level of parallelism that is directly connected with the business cycle. This also involves another characteristic of LPS, i.e. redundancy, which can be understood in two ways: redundancy of production sources, in that the local production potential (physical and human capital) is only partially employed (Bellandi, 1994; Bellandi, 2000); and information redundancy, in the sense that parallel cycles of exchanges between units lead to the production of converging information and knowledge (production-phase and product parameters).

6.4 Evolutionary features: bounded endogenous variability; adaptive modularity

Information hierarchy, distributed and parallel elaboration of information, and redundancy are the basic features of LPS, but it is necessary to stress an important characteristic of the tradition local information flows. These are essentially limited, since the "yeast" within the local context is directed at solving technical-productive problems in response to exogenous parameters that have been obtained from the external environment. The nature of the economic-productive process is fundamentally adaptive, in view of the peculiar configuration of the relations between internal milieu and external environment. In this type of scenario, the possibility that technical-productive research would undertake innovative trajectories is rather remote. It is therefore not surprising that the traditional LPS model (MID, systems of small firms in general) has been characterised by essentially incremental innovations. In fact, traditional LPS react to exogenous disturbances or impulses, while the endogenous capacity for change is substantially limited by the particular operative horizon within which the myriad of units operates. The distributed and parallel elaboration of information and redundancy can be represented as a model of an organised system in flexible modules that are greatly interactive with each other and are capable of identifying and selecting appropriate forms of behaviour for the final demand parameters. In fact, small firms located in a given territory constitute groups of stratified phases within the overall production sequence. They increase, accumulate and improve their own competences in the activity of solving problems, that is, in pursuing the objective of tuning the productive capacity with the demand parameters. The production units can therefore be considered as adaptive agents with goal-oriented behaviour. In this way, the entire local system swarms with goal-oriented sub-systems, which develop complementary competences that are homogeneous and yet alternative, through continuous interactions.

This widespread adaptive modularity is at the basis of a particular specification process for the parameters inherent in the production cycle. They derive from repeated feed-back and iterative cycles of information exchanges which are realised in the performing of changeable production routines. The parameters are not determined from above, but emerge from below, thanks to a real and proper evolutionary dynamic that we can define as "the evolutionary emergence of the parameters".

We can synthesise the conclusion to the analysis with:

Proposition 2

Adaptive modularity, repeated feedback and iterative cycles of information exchanges produce an evolutionary emergence of technical-productive parameters[5] *within LPS.*

At this point in our study, two themes have to be considered: 1) what are the limits of this type of dynamics?, and 2) how does the coordination come about between n-modules of competences, operating in a distributed and parallel manner, until they generate significant levels of redundancy?

As regards point 1), the analysis carried out has the logical implication that this is dynamics with precise characteristics: it has endogenous sources, but at the same time it is bounded, since it must respond to exogenous parameters and constraints that are synthesised in what we have defined as "system information". Nevertheless, the bounded endogenous variability is realised by means of intense communications flows.

As regards point 2), it must be observed that, in the absence of any sort of constraint, an aggregate of interacting n-entities has a very high probability of producing chaotic and unpredictable types of behaviour. In LPS, this is excluded from the very beginning, because they originate with general restraints on the dense communications patterns. In this way, a powerful inner drive acts within the interactions, both between groups of firms and also within a single group, in the direction of the emergence of socially-accepted models and rules of behaviour. The powerful drive is provided by the necessity for the local system that abounds in competence modules, to satisfy the exogenous demand parameters. In other words, determined patterns of behaviour must emerge, so that the economic-productive potential can acquire a satisfactory degree of

[5] There is a clear convergence between the view presented here and the theories of collective and localised learning (Maskell and Malmber, 1999).

fitness with respect to a particular competitive environment. The "finish line" of the analysis is, therefore, that an apparatus such as that of LPS shows a spontaneous coordination, which derives from the unusual configuration of both endogenous and exogenous information flows.

The conclusion to this section is thus:

Proposition 3

The bounded endogenous variability and the iterative cycles of information exchanges produce a powerful inner drive towards a spontaneous coordination of the information flows between units belonging to a local context[6].

In the light of the reflections made, it may be useful to develop the analysis, in order to study more profoundly the mechanisms that lead from information processing and the creation of knowledge to the prevalence of patterns of behaviour.

6.5 Information, knowledge, and P-cognition

We assume the concept of complexity as proposed by Hodgson (1997; p. 669). It "refers to the density of structural linkages and interactions between parts of an interdependent system, and is not necessarily or primarily a problem of extensiveness or scale."

This definition, which is coherent with the one taken from Simon, makes it possible to put the problem of information correctly in perspective. If the myriad of local units interacted in a completely free manner, without restrictions, it would be very difficult for the economic-productive potential to avoid a combinatorial explosion, that is, an enormous amount of alternatives to weigh up and select in the search for a solution to a technical-productive problem: the number of combinations and information exchanges could be so high that a high cost is required in terms of the resources employed (e.g. times for decision-making processes and for their being put into effect). Within this perspective, the role of information and of its transmission mechanisms is an essential one.

Hence, the importance of characterising in a precise manner the concept of information, which "is not an entity but is rather a relationship between entities (or states)" (Clark, 1991; p. 99). It is not even a set of entities or of definite objects, but rather a more or less complex association between sets of entities. It

[6] Our analysis reinforces the thesis that "firms and regions…are ensembles of competences that emerge from social interaction" (Lawson, 1999; p. 158).

is "inherently relational, a mapping between two or more sets of events" (Buckley, 1983).

If information is relational, the information processes require organisation because, in "an unstructured system, the probability of message transmission is small and as a result the system displays no coherence. Conversely, the greater the degree of organization, the more information the system possesses" (Clark, 1991; p. 99).

Having concluded that information is relational and connected both to the organisational structure of a system, and also to the degree of coherence characteristic of the latter, it is necessary to make a close examination of these two aspects in relation to traditional LPS. Within the latter, the danger of a combinatorial explosion is avoided by the fact that, in the particular configuration of the relationship between internal milieu and competitive environment, the information flows (described in section 2) develop, in fact, as if they had "focusing devices" inside them which are implicit in the bounded endogenous variety. Above and beyond this aspect, the local system must also resolve two problems of a general nature: that of overall coordination within the local micro-universe and that of system governance. Indeed, an LPS is made up of many levels or sub-sets: types of individual and social behaviour, the production sphere, the political and ideological dimension, institutional space.

Witt (2000) has analysed the same themes from the point of view of the firm, but their general valence is clear, since they arise when information and knowledge are dispersed among many agents (in our case, individuals and firms).

To deal with the two problems, we must take into consideration the passage from information to knowledge.

As far as individuals are concerned, we must first distinguish between sense data and information (Hodgson, 1997): the former are the (visual, acoustical, and somatic in general) signals of a multiple nature, which arrive in the brain from the outside world. For the mass of sense data not to become a chaotic cloud, said data must be organised on the basis of associations (structured or not) between sets of data or impulses. We thus have information. The same thing must happen for the systems made up of many interdependent parts (such as LPS): these are constituted on the basis of the fact that the myriad of impulses are – so to speak – "channelled" into specific information regularities, synthesised for example in the congruence of parameters that represent production phases or routines. Only in this way can computation problems which cannot be solved by entities (individuals or firms), that are provided with bounded rationality, become "solvable": i.e. they can be solved in reasonable times with reasonable costs. Thanks to a distributed and parallel information-processing procedure, the noise and confusion (an enormous field of signals and impulses) can be progressively reduced, leading to the definition of given characteristics and properties of the goods produced. Within this perspective, adaptive modularity is one of the essential ingredients in the search for solutions to specific technical-productive problems. The reflections made can help in understanding in what way LPS can be represented as cognitive laboratories (Becattini and Rullani, 1993; Belussi and Gottardi, 2000), by pointing out the cognitive machine incorporated inside it. Development of the analysis on this level can derive benefit from the use of concepts inferred from the evolutionary theory of human cognitive processes (Comsides and Tooby, 1992), in particular the basic idea that individual organisms (and their minds) are an aggregate of information-processing mechanisms, selected by natural selection by means of adaptation processes, research activity, and the discovery of "statistical regularities" in an environment which would otherwise be a source of unsolvable computational problems.

The same thing occurs for the n-entities that act in an LPS, the operativity of which depends on the capacity to match the exogenous impulses filtered by the strategic operators. From this derives the centrality of a particular type of cognitive process within LPS, the basis of which is fundamentally an associative dynamics between units with a double characteristic: 1) it is spontaneous and endogenous; and 2) it is subject to constraints (see section 3).

The production routines and phases interact within the framework of a process of reciprocal adaptation and tuning between the flows of information generated on the inside, so as to intercept the exogenous impulses. For the process to be efficacious, it is necessary that the crucial passage from information to knowledge take place: in the production cycles scattered throughout the territory, it is important that the associative dynamics produces cognitive patterns. So, what is knowledge, and how is it produced in LPS?

Recently, several authors (Ancori, Bureth and Cohendet, 2000; p. 266) have defined knowledge as "a structure comprised of four different [and interwoven] layers". Previously, Nonaka and Takeuchi (1995) indicated components that differentiate knowledge from information: beliefs, commitment, action, and meaning.

In this study, we have adopted a point of view that is compatible with those indicated, assuming that the passage from information to knowledge depends on the attitude in constructing representations by means of selective actions performed by a decision-making unit (entity, organism) on a set of potential states and lines of orientation. On the basis of a precise definition of the evolutionary theory of cognitive processes (Margolis, 1987; Cosmides and Tooby, 1992; Clark, 1991), we consider that this hypothesis is well-founded: by considering that the human brain "is [an] adjuster, not [a] calculator" (Margolis, 1987), an interacting entity within a complex system (like LPS) is "pattern seeking", analogously with what happens in human beings, whose evolutionary bases lead them to "seeing something, rather than nothing" (Margolis, 1987; p. 39). Therefore, the P-cognition theory formulated by Margolis, according to which knowledge is developed on the basis of a cognitive cycle in a spiral form (cues trigger the identification of a pattern, which in turn interacts with the context, selecting in this way the alternatives available for other impulses, which give a new boost to an endless process), may be usefully applied to LPS. The cognitive process, seen in terms of the "tuning of patterns and cues" (Margolis, 1987; p. 71)[7] seems particularly useful for understanding the cognitive dynamics in local systems, that are characterised by the properties analysed in sections 2 and 3 (distributed and parallel information, processing, adaptive modularity, associative dynamics in tuning modules of competences and exogenous parameters). The passage from information to knowledge resides in the construction and use of patterns. Increasingly, scholars have emphasised the importance of patterns in cognitive processes and in decision-making processes in general (from Hayek, 1952, to Loasby, 2000). Paraphrasing what Loasby (2000) has affirmed for organisations, we can affirm that in LPS, interacting n-entities continually try to tune decision-making patterns with patterns of exogenous impulses. Essentially speaking, therefore, the modules of adaptive competences, operating in local contexts, develop associated cognitive repertories through sequences of patterns and cue impulses. Two relevant points must be emphasised. In reality, the ability to extrapolate patterns is a decisive component in all cognitive processes: from directly somatic and sense processes (Cosmides and Tooby, 1992) to theoretical and scientific ones (Nightingale, 2000). The fundamental nucleus of this ability lies, in fact, in its aptitude for identifying relations of similarity between sets of elements of different natures (Hofstadter, 1995). It must be pointed out, however, that there is a fundamental difference between extrapolative activities which systematically attempt to find unpredictable final results and extrapolative activities which lead to adaptive changes. In LPS, we are above all in the presence of the second type, given the bounded endogenous variability. In other words, the ability to recognise a more limited number of relations or contexts of similarity derives from the particular

[7] "Cues" prompt a pattern which, by interacting with the rest of the context, provides an amended set of cues, which prompts the next pattern" (Margolis, 1987; p. 2)

configuration of the information flows and from the relations with the outside environment. One of the final results is precisely the fact – consolidated in the empirical literature – that traditional LPS have long been able, above all, to create and absorb incremental innovations.

We can synthesise the conclusions of the analysis made with:

Proposition 4

In LPS, the passage from information to knowledge takes place by means of the construction and use of structured cognitive patterns.

6.6 Cognitive models and patterns of behaviour

At this point, it is necessary to examine the nature and characteristics of cognitive patterns in much greater detail. On this subject, the literature abounds with concepts (frames, schemata, attunements, scripts, templates). In economics, Nooteboom (1999) has proposed a multi-level theory of learning that connects individuals, organisations, and industries. His vision of the learning cycle is centred on the concept of "scripts", derived from studies of artificial intelligence and consisting of a "sequence of nodes [....] correspond to events and actions" (Nooteboom, 1999; p. 134). In Artificial Intelligence scripts are conceived as "models of mental constructs". A converging analysis has been developed by Lombardi, by applying the concept of frames to the study of innovative processes (Lombardi, 1990) and of "Chandlerian" models of a firm (Lombardi, 1991). Frames, like networks of nodes and relations between them, are data-structures which represent prototypical situations that are related to different reference contexts (Minsky, 1974). They are multi-layered structures, and each layer can have a different power in representing real contexts: higher levels have more general and stable prototypical elements; lower levels have assigned more and more feeble values until terminal "slots", which must have been filled with specific items or data.

Whichever concept is accepted, few doubts exist that agents have the ability to represent the conditions of the world and to act on the basis of representation (Searle, 1998). Johnson-Laird (1983) distinguished three types of representations in mental life: mental models, propositional representations, and images. We shall dwell on the first ones, which human beings construct in order to search efficiently within a world that otherwise abounds in possibilities to the point of appearing confused and a source of combinatorial explosion. Since they cannot elaborate exhaustive mental models, agents think up operative models, namely structures of relationships that are similar to the real reference processes.

The formal and informal inferences of agents depend on mental models, which play a fundamental part in the representation of the world (objects, states, sequences of events, acts).

At the basis of the formation of these models are several characteristics of mental processes: A) recursiveness, i.e. the verifying of iterative cycles of sequences of elements or states; B) "parallel swarm scanning", i.e. explorative activity in many directions and at many levels (Hofstadter, 1995). In other words, we have a multi-dimensional space made up of specialised micro-explorations (information flows at various levels) and of properties emerging at a macroscopic level (concepts, visions, patterns)[8]; C) the pervasiveness of entwinings involving "top-down" processes (guided by theory) and "bottom-up" processes (guided by data) (Hofstadter, 1995). The application of these concepts to information processing and to the creation of knowledge in LPS appears to be rather interesting. We can easily find the properties in their configuration, that was previously represented in terms of adaptive modularity of competences, associative dynamics, bounded endogenous variability, evolutionary emergence of parameters, and extrapolation of patterns.

The analysis of cognitive processes supplies useful elements for dealing – in relation to LPS – with the problems of coordination and governance, as posed by Witt in reference to organisations. Mental models carry out an essential function, because individuals construct them in order "to make sense out of the world around them" (Denzau and North, 1994; p. 4), especially in environments characterised by uncertainty. Furthermore, ideologies "evolve from such constructions (…) and the external institutions are the external (to the mind) mechanism [that] individuals create to structure and order the environment" (Denzau and North, 1994; p. 4). The production and diffusion of knowledge therefore follows paths and uses instruments that are different from those of information. It is also necessary to observe that mental models are often formulated and expressed in explicit terms, but - so to speak - they work tacitly: it is a structured knowledge which is communicated by means of processes of cognitive convergence within the collectivity of agents (individuals and firms) on the basis of iterative cycles of exchanges of input or of spiral cycles of cues and patterns.

[8] In this regard, it is useful to recall the metaphor proposed by Hofstadter on the columns of ants guided by a swarm of explorers (Hofstadter, 1995; p. 248).

This view of individual and social cognitive processes has interesting implications for the debate over the codification of knowledge and, therefore, the distinction between tacit and codified knowledge [9].

For the purposes of the present study, it is interesting to emphasise, above all, four points relative to tacit knowledge: 1) it is necessarily based on mental (structured) models, abounding in dense entwinings between explicit and implicit rules for the production of information and knowledge; 2) it circulates through the spiral process of cues and patterns (Margolis, 1987); 3) it is not an exclusive characteristic of LPS, but rather an essential property of all cognitive processes within both hierarchical organisations (Senge, 1990; Nonaka and Takeuchi, 1995) and self-organised systems such as MID (Becattini and Rullani, 1993); 4) it is also fundamental in the dynamics of very formalised knowledge, as Nightingale (2000) has demonstrated by means of his analysis of scientific research in the field of chemistry.

In the theoretical perspective used here, cognitive patterns assume the function of "attractors in a very high dimensional phase space" (Witt, 1991; p. 90), and cognitive frames can be "socially-shared interpretation patterns" (Witt, 2000; p. 744). Since these socially-shared creations may belong to a set of units which operate within an organisation, an investigation into the connections between frames and cognitive leadership becomes significant (Witt, 1998). Shared interpretative patterns are also a necessary characteristic, above all of systems composed of interacting n-entities for which the problem of coordination is crucial. Thanks to their incessant dynamics, mental models are an important instrument for realising cognitive coherence, without which there would be no self-organised internal milieu on the basis of dynamic interactions with parameters extracted from the external environment (proposition 3).

The existence and communication of mental models help in understanding how modules of competences can be developed endogenously within the framework of a local context and, at the same time, how they remain anchored to particular points of reference (bounded variability) (proposition 2). Within this perspective, mental models perform the function of shifting cognitive equilibria, which are realised by means of processes coordinating economic processes and the knowledge that is dispersed through "research programmes which provide a framework to guide agents in formulating their expectations and plans" (Loasby, 1991; p. 46). A dynamic cognitive equilibrium therefore constitutes the outcome of several of the n-dimensions and components that feed the evolution of LPS.

[9] On this topic, see the special issue of the journal *Industrial Corporate and Change*, 2/2000.

Thus, we can formulate the following:

Proposition 5

Mental models are shifting cognitive equilibria that derive from the processes coordinating economic activities and knowledge dispersed within LPS.

Other elements that can be synthesised in the problem of governance are also important to the functioning of LPS: which factors induce individuals and firms to establish relations of competition, cooperation, and reciprocal trust even in situations of potential conflict?

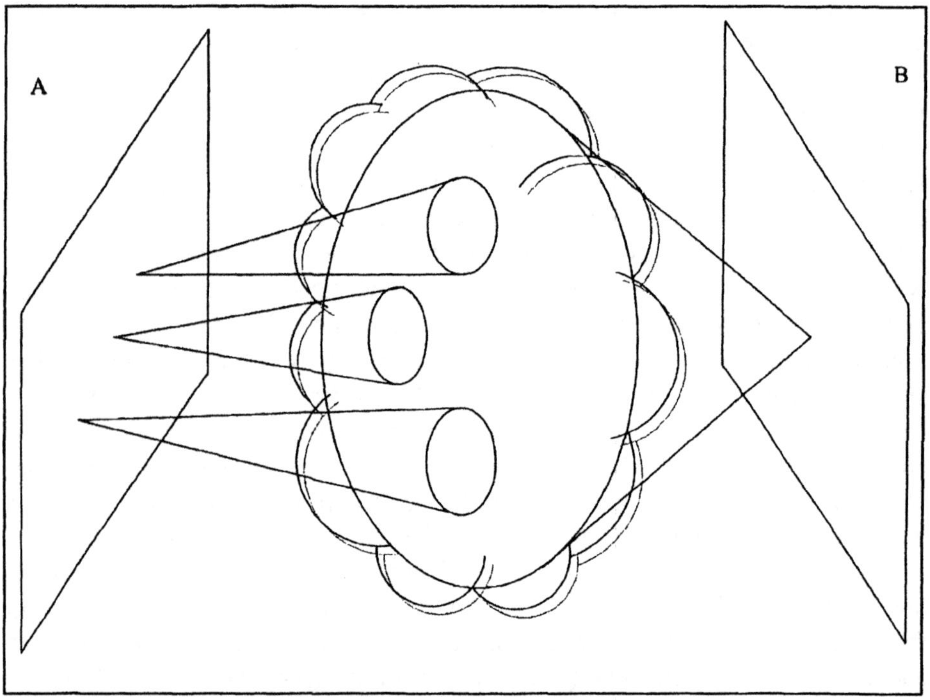

Figure 4: A: Internal cognitive frames, Mental models, Cognitive equilibria
B: Institutions as external scaffolding and system functions

The realisation of a shifting cognitive equilibrium constitutes a peculiar solution to the coordination problem; but the dynamics of LPS depends also on the complementary solution to the governance problem which, in extending Loasby's view, we can delineate in the following terms: "In each system, equilibrium may be defined as the persistence of established theories and policies" (Loasby, 1991; p. 52).

The theoretical approach adopted makes it possible to see cognitive patterns as fundamental components of the ideologies and institutions that prevail in a given area, even if it is important to point out one essential element: it is the peculiar cognitive dynamics that causes the evolution of a local system not to be reproduced in an identical manner within two contexts that are similar from a structural point of view. While performing analogous functions from the point of view of the system, the configuration and the equilibrium attained may be locally different. Our initial observations on the differentiated role of institutions in the regions of Tuscany and Veneto constitute a confirmation of the hypothesis put forth and, at the same time, show that – from many points of view – local institutions perform a crucial function.

Above all, they are a fundamental part of the "environmental scaffolding" that "helps to reduce the scope of individual choice" (Clark, 1997; p. 282). In contributing to avoiding the risk of combinatorial explosion, "such scaffolding plays a dual role: it both restricts and expands our intellectual horizon" (Clark, 1997; p. 282). By this route, individuals and firms, both characterised by bounded rationality, receive indispensable support when they tackle situations that are marked by uncertainty[10].

[10] Situations characterised by cognitive incompleteness can be described as follows. Agents have a complete list of the conditions of the world relevant to the problem of decision-making, but they do not know which is the true one, and are capable only of assigning distributions of probability to the events (Arrow, 1974; p. 33). Langlois (1984) defines this type of context as "perfect structural knowledge", because the subjects have a "certain knowledge of the structure of the problem", combined with an imperfect parametric knowledge, in that they lack information relative to operative parameters useful for selecting choices. In these conditions, the traditional model of rational decision, based on the maximisation of the expected utility, is fully applicable. Situations characterised by the impossibility of defining a complete set of events to which to assign probability distributions are defined by Langlois "imperfect structural knowledge". This first type of context corresponds to "weak substantive uncertainty" (Dosi and Egidi, 1991) and it is always accompanied by a gap between the cognitive and computational capacities of the agents, on the one hand, and the complexity of the problem relative to the choice to be made. "Strong substantive uncertainty" (analogous with imperfect structural knowledge) characterises in particular manner non stationary environments, marked by the producing of endogenous and exogenous shocks (technological changes, new products, changes in taste and demand, changes in markets, etc.). In general, however, economic dynamics can be described on the basis of the uncertainty that is generated in them: 1) *parametric*, when new factors emerge that do not bring up the basic decision-making picture for discussion; 2) *strategic*, which corresponds to contexts in which it is necessary to examine consolidated operational models, as a result of more or less great discontinuities induced by types of behaviour of other agents or by significant variations in the performance of economic activities; 3) *structural*, which occurs in the case of fundamental and irreversible changes in the

In the second place, we must keep in mind that institutions, habits and rules are closely and mutually entwined. Habits and rules are essentials in the evolutionary process of institutions, since "rules are conditional or unconditional patterns of thought or behaviour which can be adopted either consciously or unconsciously by agents" (Hodgson, 1998; p. 185). The milieu within an LPS is, therefore, an aggregate of beliefs, commitments, trust and mutually-entwined mental models. The particular profile or institutional model that emerges is nothing more than the representative vector of all these components (fig. 4).

In fact, there exists a third and relevant function of institutions, in that these are a fundamental factor in the development of the joint action which, combined with local external economies, is at the basis of "collective efficiency" or the competitive advantage of an LPS (Schmitz, 1999). From the point of view adopted in this article, the marriage between joint action and local external economies signifies that institutions perform important system functions. In section 1, we have seen how the institutional model has penetrated into (and is, therefore, closely connected with) the model of local development, by following evolutionary trajectories that may also be path-dependent.

We can synthesise our conclusions with

Proposition 6

The properties of the local milieu and the flows of interactions, which develop contemporaneously at many levels determine a process of evolutionary reciprocal conditioning between units belonging to the system, institutions, and the global configuration of the system.

The dynamics of a local "micro-world" may be profitably interpreted with the metaphor of the "stigmergic algorithm": it is one in which actions are strongly determined by external structures that are themselves the operands or objects of the actions" (Clark, 1997; p. 279).

The "finish line" of the reflections developed in this section is that, in LPS, peculiar cognitive equilibria are realised and, therefore, also different solutions to the problems of governance and co-ordination, on the basis of the evolution mutually entwined with cognitive equilibria and institution models.

The moment has come to reflect on the effects brought about in this mix by the evolutionary discontinuities described in the first section.

---

traditional operative contexts (radical technical-productive innovations) (Langlois and Robertson, 1995).

6.7 Conclusions

We shall take up again several of the properties of LPS, above all adaptive modularity and bounded endogenous variability in the process of tuning of the modules of competence, in order to tally with the changing demand parameters. The latter is a typical situation of "parametric change, that is, change of certain known variables within a known framework" (Langlois and Robertson, 1995; p. 555)[11] What happens, however, in the face of systematic or radical innovations?[12] How do we evaluate the evolutionary changes indicated in section 1?

First of all, it appears certain that the relationship with the external environment has profoundly changed: strategies that anticipate demand and active projection on the markets are becoming ever more necessary, while high standards of quality must be pursued. In these conditions, a rationality that is purely adaptive with respect to information parameters derived from outside becomes ever less efficacious, while at least three types of effects on the traditional configuration of LPS are produced:

1. Information flows change substantially.
2. The reciprocal adaptation between distributed modules of competences is less suitable for satisfying requirements of greater quality control and for the creation of technological and financial resources of a higher qualitative level, in order to cope with new competitive factors.
3. The degree of self-containment of LPS is reduced by the new strategies of the final firms, which resort to competences outside the local contexts in a search for lower costs and for particular competences.

Point 3 shows driving forces which alter the traditional evolutionary profile by acting on the basic properties of LPS, starting from the change in the boundaries of the system, that is, of the separation between high-frequency

[11] "For example, it may be highly uncertain which grade of cloth or which style of tile will be demanded this season, but it is well known to all what it means to produce a grade of cloth or a style of tile." (Langlois and Robertson, 1995; p. 555)

[12] The expression "systemic innovations" indicates that "the change in one part of the system (one stage of production, for example) necessitates corresponding variations in other parts" (Langlois and Robertson, 1995; p. 554). Henderson, Clark (1990) distinguish between four types of innovations: *incremental*, that generate variants of an existing product, by utilising its potentials wholly or in part; *radical*, based on different scientific or engineering principles, and capable of opening up new markets for potential applications; *architectural*, that change the connection modalities between components of a product, leaving unchanged a fundamental nucleus of knowledge and of project-related elements (*core design concepts*); *modular*, that transform knowledge relative to the fundamental nucleus of knowledge of a design, without altering the architecture of the product.

interactions (within the framework of the internal milieu) and the low-frequency ones (with the external environment). At the same time, the needs for greater control of the qualitative and technological parameters specific to the various phases lead to establishing more stable relationships of collaboration between units[13], thus producing mechanisms of selection and then of reduction in redundancy and of parallelism in the internal milieu. The limitation of the processes of reciprocal adaptation and the creation of more closely coordinated sequences of productive phases, belonging both to the inside and to the outside of the local contexts, has other profound implications. The new patterns of interaction influence the processes of creation and diffusion of mental models. The further consequence is that the rules of behaviour and traditional institutional models cannot remain unchanged: in fact, it is probable that on the one hand the socially-shared values and rules will be weakened and that, on the other, the mix of competition and collaboration will change. Competition tends to be extended in an environment that is outside the local one; cooperation generates incentives towards the creation of real and proper sub-systems of integrated firms.

The points described delineate a horizon characterised by a mix of phenomena that are typical both of the situations defined by Langlois and Robertson (1995; p. 555)[14] as characterised by "strategic uncertainty", and also the conditions of structural change (radical or systems innovations).

Impulses tend to be diffused in a selective manner. If, about two decades ago, the activity configuration of n-small firms could be represented as a propagation network with many "ups and down", in today's scenario the local production potential tends to assume a composition "with watertight compartments"(fig. 5), with the transmission of impulses among the less formal components which are more bound to contingencies that are programmed to some extent. A new division of work thus tends to be affirmed, with a potential multiplicity of regroupings of the production phases.

[13] An example of this process can be identified in what happened in Prato, starting at the end of the 1980s: a progressive acceleration and a consolidation of the phenomena of the formation of formal (vertical) groups of firms occurred, with the reorganisation on more stable bases of inter-firm relationships that previously had been very much "looser" and less selective (Dei Ottati, 1996). During the same period, a great reduction in the number of operative firms occurred, together with a very limited development in innovative services for the industry (Pratofutura, 1996).

[14] "This would typically involve rearranging capabilities in fairly drastic ways, but within known boundaries."

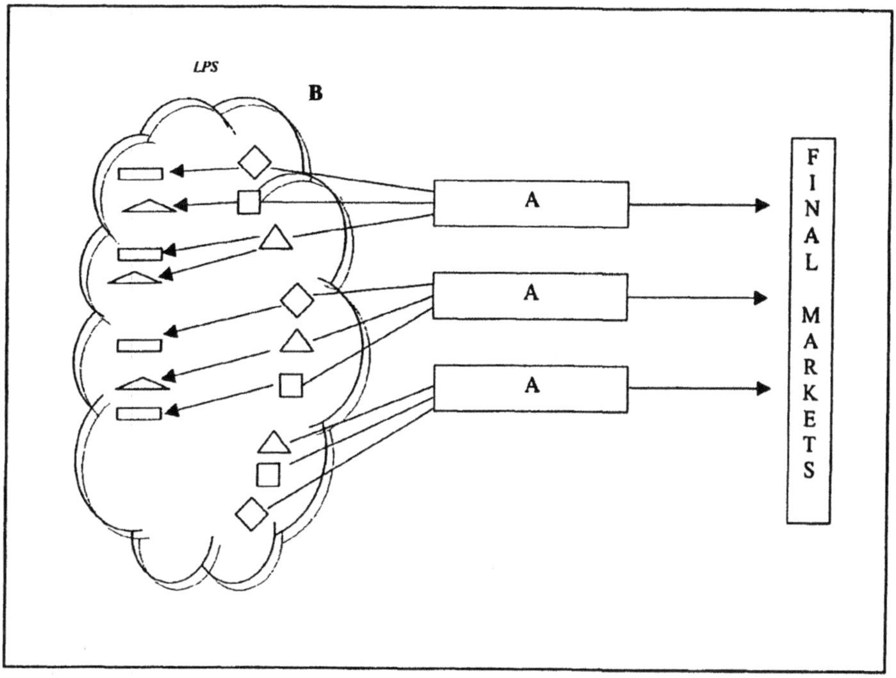

Figure 5: A: More integrated groups. Multilayered and verticalised LPS.
B: Segmented techno-productive configuration

The possibility that groups may be formed of firms linked in a stable manner or of companies with a strong tendency to vertical integration of the production cycle has relevant effects from the point of view of problems of coordination and governance. As far as the former is concerned, the presence of several groups or vertically-organised companies implies that a scenario marked by the coexistence of a multiplicity of mental models responding to different types of business logic can be opened up. In this case, there will no longer be socially-shared values, but rather an aggregate of group-shared values.

The dynamics of information processing and of the creation of knowledge may be able to demonstrate a sort of "chemical composition" that is profoundly different, the outcome of which will be cognitive equilibria that are drastically changed with respect to those of the past. The change may be described as a passing from a sense of belonging to a general community to a sense of differentiated belonging, depending on the sub-system in which it is inserted. We must keep in mind, in fact, that "the organisation of the Prato industrial district involved a shift towards a more conscious and organised form of coordination, but this continued to occur among mutually agreeing and largely autonomous partners" (Dei Ottati, 1996; p. 48). At the same time, we must not overlook the fact that, on the one hand, sub-systems have shrunk their own borders within the local context and, on the other, have extended them by including outside units.

A fundamental point on which to focus our attention is the following: the evolutionary trajectories that are realised in the local systems depend also on the particular cognitive equilibria and institutional models that have prevailed in the past. This helps us to understand the reasons for the diverging paths registered in the local systems in Tuscany and in Veneto (see section 1). On the other hand, it is not arbitrary to hypothesise that real evolutionary lock-in phenomena occur when local systems assume configurations (a mix of cognitive equilibria and institutional models) that are unfit for coping with the effects of evolutionary discontinuities. Furthermore, it is likely that the development trajectories followed in the past can, in some cases, produce path-dependence, in the sense that it becomes difficult to adopt radically different paths of development on the basis of the particular configuration of the local economic-productive potential.

The analysis developed leads us to believe that the bases for attaining collective efficiency by means of external economies and joint action and, therefore, through the exercising of systemic functions on the part of institutions, have progressively eroded (see section 1). In the new scenario that is outlined, the latter must redefine their own role within the processes of governance, because the alterations in the traditional mixes of cognitive equilibria and institutional models bring out another relevant problem. A sort of dilemma rears up in the light of the institutions, which can choose among at least three action alternatives:

1. to identify and carry out new systemic functions in contexts that are characterised by strong distinctions between sub-systems;
2. to become real and proper "strategic agencies" (Bellandi, 2000), capable of measuring themselves with a plurality of strategies that at times are cooperative, at others are competitive, and at still others are conflictual. It is evident that relevant problems arise which are inherent in decisions under conditions of radical indecision;
3. to reduce the institutional space to a minimum, by exercising very general functions and, thus, leave the field free for carrying out new driving forces.

The future development of LPS is open, therefore, to many solutions, starting from the radical changes in the forms of collective learning that have been realised in past decades.

However, the possibility that the changes in the competitive scenario favour the occurrence of something similar to the cycle theory of integration and disintegration, formulated by Nooteboom (1999) in regard to firms, cannot be excluded.

References

Allen P.M. and Lesser M. (1991), Evolution Human Systems: Learning, Ignorance and Subjectivity, in Metcalfe J.S. and Saviotti P. (eds.), *Evolutionary Theories of Economic and Technological Change*, Harwood Academic Press.

Ancori B., Bureth A. and Cohendet P (2000), The Economics of Knowledge: The Debate about Codification and Tacit Knowledge, *Industrial and Corporate Change*, vol. 2, p. 255-287.

Arrow K. J. (1974), *The Limits of Organization*, Norton & Company, New York.

Becattini G. and Rullani E. (1993), Sistema locale e mercato globale, *Economia e Politica Industriale*, vol. 80, in G. Becattini, *Il Distretto Industriale*, Rosenberg & Sellier.

Bellandi M. (1994), Le logiche del cambiamento economico locale, in Bellandi M. and Russo M. (eds.), *Distretti Industriali e cambiamento economico locale*, Rosenberg & Sellier, Turin.

Bellandi M. (1996), Innovation and Change in the Marshallian Industrial District, *European Planning Studies*, vol. 4, no. 3, p. 357-368.

Bellandi M. (2000), The Experience of Italian Industrial Districts: An Industrial Economics Reading, *Economic Analysis*, forthcoming.

Belussi F. and Gottardi G. (2000), (eds.), *Evolutionary Patterns of Local Industrial Systems*, Aldershot, Ashgate.

Burroni L. (1999a), Mutamenti nell'organizzazione produttiva della Terza Italia. Una comparazione tra Veneto e Toscana, *Sviluppo Locale*, vol. 11, p. 33-67.

Burroni L. (1999b), Regolazione locale nelle regioni della Terza Italia. Una comparazione tra Veneto e Toscana, *Sviluppo Locale*, vol. 12, p. 5-43.

Chandler A. D. (1992), Organisational Capabilities and the Economic History of the Industrial Enterprise, *Journal of Economic Perspectives*, vol. 3, p. 79-100.

Ciappei C. and Mazzetti G. (1996), *Lo sviluppo a rete delle imprese tessili del distretto pratese*, Regione Toscana, Iris.

Clark N. (1991), Organisation and Information in the Evolution of Economic Systems, in Metcalfe J.S. and Saviotti P. (eds.), *Evolutionary Theories of Economic and Technological Change*, Harwood Academic Press.

Clark A. (1997), Economic Reason: The Interplay of Individual Learning and External Structure, in Drobak J. N. and Nye J. V. C. (eds.), *The Frontiers of the New Institutional Economics*, Academic Press.

Corò G. and Rullani E. (1998), (eds.), *Percorsi di internazionalizzazione. Competenze e auto-organizzazione nei distretti industriali del Nord-Est*, Franco Angeli, Milan.

Cosmides L. and Tooby J. (1992), The Psychological Foundations of Culture, in, Barkow J.H., Cosmides L. and Tooby J. (eds.), *The Adapted Mind*, Oxford University Press, New York.

Cyert R. M. (1988), *The Economic Theory of Organisation and the Firm*, Harvester-Wheatsheaf, London.

Cyert R. M. and Kumar P. (1996), Economising by firms through learning and adaptation, *Journal of Economic Behaviour and Organisation,* vol. 29, p. 211-231.

Dei Ottati G. (1996), Economic Changes in the District of Prato in the 1980s: Towards a more Conscious and Organised Industrial District, *European Planning Studies*, vol. 4, no. 1, p. 35-52.

Denzau A. T. and North D.C. (1994), Shared Mental Models: Ideologies and Institutions, *Kyklos*, vol. 1, p. 3-31.

Dosi G. and Egidi M. (1991), Substantive and procedural uncertainty, *Journal of Evolutionary Economics*, vol. 1, no.1, p. 145-168.

Dosi G. and Nelson R.R. (1994), An Introduction to evolutionary theories in economics, *Journal of Evolutionary Economics*, no. 4, p. 153-172.

Eliasson G. (1989), The Dynamics of Supply and Economic Growth, in, Carlsson B. (ed.), *Industrial Dynamics*, Kluwer Academic Publisher, London.

Eliasson G. (1990), The firm as a competent team, *Journal of Economic Behaviour and Organisation*, vol. 3, p. 275-298.

Hayek F. A. (1952), *The sensory order*, Routledge, London.

Henderson R. M. and Clark K. B. (1990), Architectural Innovation: The Reconfiguration of Existing Product Technologies and the Failure of Established Firms, *Administrative Science Quarterly*, vol. 35, p. 9-30.

Hodgson G. M. (1997), The ubiquity of habits and rules, *Cambridge Journal of Economics*, vol. 21, p. 663-684.

Hodgson G. M. (1998), The Approach of Institutional Economics, *Journal of Economic Literature*, March, p. 166-192

Hofstadter D. (1995), *Fluid Concepts and Creative Analogies*, HarperCollins Publisher.

Johnson-Laird P.N. (1983), *Mental Models. Towards a Cognitive Science of Language, Inference and Consciousness*, Cambridge University Press, Cambridge.

Langlois R. N. (1983), Systems Theory, Knowledge and the Social Sciences, in Matchlup F. and Mansfield U. (eds.), *The Study of Information*, New York, John Wiley & Sons.

Langlois R.N. (1984), Internal Organisation in a Dynamic Context: Some Theoretical Considerations, in Jussawalla M. and Ebenfield H. (eds.), *Communication and Information Economics*, Amsterdam, North Holland.

Langlois R. N. and Robertson P.L. (1995), Innovation, Networks, and Vertical Integration, *Research Policy,* vol. 24, p. 543-562.

Lawson C. (1999), Towards a competence Theory of the region, *Cambridge Journal of Economics*, vol. 23, no. 2, March, p. 151-166.

Loasby B.J. (1976), *Choice, Complexity and Ignorance*, Cambridge University Press, Cambridge.

Loasby B. J. (1991), *Equilibrium and Evolution*, Manchester University Press, Manchester.

Lombardi M. (1990), Problems in representing knowledge within the theory of Innovations, in, Egidi M., Lombardi M. and Tamborini R. (eds.), *Knowledge, Uncertainty, and Economic Decisions*, Franco Angeli, Milan.

Lombardi M. (1991), Technological Evolution and Economic Processes: the Firm as a visible *Mind,* in *Annali di studi e storia dell'impresa*, Il Mulino, Bologna.

Lombardi M. (2000a), The cognitive approach to the study of local production systems, in Belussi F. and Gottardi G., (eds.), cit.

Lombardi M. (2000b), The Evolution of Local Production Systems: the emergence of the "invisible mind" and the evolutionary pressures towards more visible "minds", Paper presented at the REN Conference (Regional Enterprise Networks), Prague, 22-23 October 2000.

March J.G. and Simon H. A. (1993), *Organisations*, Blackwell Publisher, Cambridge Mass., [First Edition 1958].

Margolis H. (1987), *Patterns, Thinking, and Cognition*, The Chicago University Press, Chicago.

Maskell P. and Malmberg A. (1999), Localised Learning and Industrial competitiveness, *Cambridge Journal of Economics*, vol. 23, no. 2, March, p. 167-185.

Max-Planck-Institut, (1999), *Proceedings of the Symposium Assessing the Potential of the Evolutionary Approach to Economics"*, Jena.

Messina P., Riccamboni G. and Solari S. (1999), Sviluppo economico e regolazione politica nelle regioni di piccola e media impresa: un'analisi comparata tra Veneto e Emilia Romagna, *Sviluppo Locale*, vol. 12, p. 44-78

Minsky M. (1974), A Framework for Representing Knowledge, in Brachman R.J. and Levesques H.J., *Reading in Knowledge Representation*, Kaufman, New York.

Nelson R. (1995), Recent Evolutionary Theorising About Economic Change, *Journal of Economic Literature*, vol. 1, p. 48-90.

Nightingale P. (2000), Economies of Scale in Experimentation: Knowledge and Technology in Pharmaceutical R&D, *Industrial and Corporate Change*, vol. 9, no.2, p. 315-355.

Nonaka I. and Takeuchi H. (1995), *The Knowledge Creating Company*, Oxford University Press, Oxford.

Nooteboom B. (1999), Innovation, learning and industrial organisation, *Cambridge Journal of Economics*, vol. 23, 2, p. 127-150.

Pilotti L. (1999), *Leadership* allargate, *meta-organizer* e apprendimento localizzato nei distretti industriali italiani, *Piccola Impresa Small Business*, vol. 2, p. 3-32.

Pilotti L. (2000), Evolutionary and adaptive local systems in Northeast Italy: strategies of learning, leadership, and co-operation, in Belussi F., Gottardi G. (eds.), cit.

Pratofutura (1996), *I servizi per le imprese nel distretto tessile pratese*. Unione Industriali Pratese.

Schmitz H. (1999), Collective efficiency and increasing returns, *Cambridge Journal of Economics*, vol. 23, p. 465-483.

Searle J.R. (1995), *The Construction of Reality*, Allen Lane, London.

Simon H.A. (1962), The Architecture of Complexity, *Proceedings of the American Philosophical Society*, vol. 196, no. 6, p. 467-482.

Witt U. (1991), Reflections on the Present State of Evolutionary Economic Theory, in Hodgson G. M. and Screpanti E. (eds.), *Rethinking Economics*, Edward Elgar, Cheltenham.

Witt U. (1998), Imagination and Leadership -The Neglected Dimension of an Evolutionary Theory of the Firm, *Journal of Economic Behaviour and Organisation*, vol. 35, p. 161-177.

Witt U. (2000), Changing Cognitive Frames – Changing Organisational Forms: An Entrepreneurial Theory of Organisational Development, *Industrial and Corporate Change*, vol. 4, p. 733-755.

Chapter 7

Knowledge creation and codification in Italian Industrial Districts

Fiorenza Belussi and Luciano Pilotti

By means of the old, we come to know the new (Confucius)

7.1 Introduction

The aim of this paper is to present an overview of the various mechanisms of knowledge creation, diffusion, and assimilation which are visible within the Italian local production systems. The term "local production system" adopted in this paper concerns that class of phenomena related to the spatial concentration of specialised industry, often defined in the economic and business literature as: industrial districts, multi-district areas, system-areas, local systems governed by leading networks firms, "clusters", "milieux innovateurs", and so on. The paper discusses the evolutionary pattern of tacit and codified knowledge within the institutional complexity of the Italian local systems, where not only firms are active agents in knowledge producing and distributing, but also local institutions contribute to the process of socialisation of information and knowledge.

Our analysis is set in an ongoing theoretical dispute related to the understanding of the "meaning" of knowledge and the "nature" of scientific or technical knowledge. A contrast can be stated between the concept of *a priori* knowledge (propositional knowledge), and an idealisation of knowledge in a cognitive perspective as a "*situated*", and subject-related, fact (narrative knowledge). In the first case we use the term knowledge as a *de facto* synonym of information. This is what we find in orthodox textbooks: an "algorithmic" model of knowledge production and utilisation, where knowledge concerns exclusively the generation of articulable pieces of information, thus with explicit (not tacit) knowledge. On the other hand, using an other epistemological approach, we can focus our attention on the state of the agent's entire cognitive context, and his or her limited capabilities. Here knowledge refers to an interpreted set of data, information, propositions (whose meaning has a value only for a given subject). In this second frame, the introduction of the concept of tacit knowledge is crucial in order to allow the social mechanism of use, generation, assimilation, and transmission of knowledge (either scientific, or economically significant). The idea of tacit knowledge refers to the fact that "we know more than we can tell", as Polanyi (1967) stated, in his famous book, and that not all knowledge can be simply articulated and codified. Here agents have limited cognitive capabilities, bounded learning abilities, path-dependent mental models, and they must use their interpretative frames to evaluate market and technological opportunities. Agents are not endowed with all the knowledge they need, nor can they find it as a blueprint on the shelves, ready-made on demand. In order to use knowledge productively they must develop filtering mechanisms, selection and elaboration procedures that capture and utilise the relevant useful knowledge dispersed in the economic environment. As firstly acknowledged by Hayek, the type of knowledge that economic agents utilised is not just codified (scientific) knowledge but "unorganised knowledge". The activity of knowledge exploration, recognition (the knowledge of knowledge), interpretation, ordering, validation, and connecting perceptual elements with objective facts is central to the working of the economic system. If we embrace this perspective, the mental processes of individuals become relevant, such as their cognitive frames, their modalities of learning, their ability to conceptualised complex phenomena, and their capacity of storing knowledge (knowledge accumulation). More precisely, the whole set of knowledge that individuals use is not just the type of codified scientific knowledge described by the simple assumptions that we find in the neoclassical doctrine. The art of knowing[1] for practical purposes (setting up a production process, experimenting a novel process, inventing a new product, finding a new client, entering a new market, etc.), is related to the knowledge one needs to do something, and, by contrast, is characterised by skills, competencies, and the use of others' knowledge, and therefore it is made up of unavoidable elements of tacit knowledge. This vision follows the Hayek central problem for economic

[1] As Polanyi (1958) said in the Preface of his book "I regard knowing as an active comprehension of the things known, an action that requires skill. Skilful knowing and doing is so performed by subordinating a set of particulars, as clues or tools, to the shaping of a skilful achievement, whether practical or theoretical" (p. vii).

analysis: "how a solution is produced by the interaction of people, each of whom possess only partial knowledge" (Hayek, 1945; p. 530).

Using the seminal contribution of Nonaka, the idea developed within the paper is that Italian local production systems may be categorised on the basis of the prevalence of a certain type of knowledge. So, we will distinguish the Italian local production systems in three main categories:

a) local production systems where tacit knowledge prevails, and learning takes place mainly through socialisation;
b) local production systems based on a balance between tacit knowledge and codified knowledge where learning appears to be a "pure" interactive process among localised agents: here we find absorption of external knowledge (internalisation) and recombination;
c) local production systems where learning is based on more formal innovative activities (R&D type) and the process of externalisation takes place because firms localised in the area share the feature of being Schumpeterian innovators, thus they feed the global circuits of knowledge with novelties and new original pieces of codified knowledge.

Within this analytical framework - which is based on the Italian tradition of research on industrial districts [2], we focus the attention on the increasing variety and diversification of evolutionary patterns of local production systems[3].

The first section of the paper intends to focus on the theoretical framework discussing the formation of an endogenous growth mechanism. Section 7.2 discusses the various mechanisms of knowledge creation. Section 7.3, using the conceptual framework discussed, which connects learning activities and the use of tacit knowledge, presents the three main categories of local production

[2] See, for example, Becattini (1979;1987;1989), Gandolfi (1988), Brusco (1989), Gobbo (1989), Pyke, Becattini and Sengenberger (1990), Sforzi (1990; 1995), Pyke and Segenberger (1992), Moussanet and Paolozzi (1992), Falzoni Onida, and Viesti (1992), Amin and Robins (1990), Amin (1991; 1993), Schmitz (1992), Capecchi (1990), and Hirst and Zeitlin (1992), Nuti (1992), Harrison (1990), Lombardi (1994), Bellandi and Russo (1994), Gottardi (1996), Cossentino, Pyke and Sengenberger (1996), Garofoli (1983; 1992; 1995), Anastasia, Corò, and Crestanello (1993), Dei Ottati (1996); Bellandi (1993); Mistri (1993); Bramanti and Senn (1994), and Carminucci and Casucci (1997). For the international context, see Sabel and Zeitlin (1982), Sabel (1989), and Benko and Lipietz (1992).

[3] Analyses of local production systems have shown that there is great diversity among these systems in terms of morphology (with greater or lesser concentration of industrial structures), levels of competitiveness, innovative capabilities, and above all evolutionary tendencies. Moreover institutional, cultural and social factors have been highlighted as contributing to the generation and consolidation of these structures. For an up-to-date bibliography see Belussi and Gottardi (2000).

systems typical of Northeast Italy. The aim of the third section is to propose a conceptual framework for the formulation of public policies that, stressing the importance of collective learning, identifies the room for active policy. Section 7.4 deals with some provisional conclusions.

7.2 The formation of an endogenous growth mechanism

Some characteristics of the diffusion of technical change and know-how within the Italian local production systems can clarify the process that has been frequently described as the endogenous mechanism of growth of local production systems[4]. Empirical studies indicate many factors that have contributed to the consolidation of local production systems.

The "laws" of motion of local production systems have been discussed by Marshallian theorists only in terms of external economies, "industrial atmosphere", and reduction of transaction costs. In the same way, this tends to differentiate the "Northeast" approach (Rullani, Di Bernardo, Pilotti, Gottardi, Biggero, Corò, and Anastasia) from the "pure" Marshallian School of Florence (Becattini, Dei Ottati, and Bellandi). In another places (Belussi, 1996; 1999c), we have emphasised the dynamical aspects related to the various models of learning existing within the local production systems, focusing on three main factors which influence the local development patterns.

A) Firstly, the processes of inter firm "cognitive" division of labour, which is related to the intra-district division of innovative labour (where only few agents, and sometimes large firms belonging to the local context, are truly Schumpeterian). On the contrary, when Bellandi (1989) speaks of the diffused innovative capacity of industrial districts, he has in mind only the important, but not strategic aspects of learning by doing and by using.

B) Secondly, the related specialisation of knowledge, which guides the mode of resource co-ordination of the local systems, whose knowledge connections are very often external to the local system. So, in many cases, local production systems are not self-sufficient "knowledge communities" (on this issue see also Loasby (1998) who manifests a more traditionally Marshallian perspective).

C) Thirdly, we must focus our attention on the process of accumulation among local agents of embedded "contextual knowledge". In our interpretation the process of accumulation of contextual knowledge refers to technical capabilities and know-how that stem from both tacit (difficult to transfer)

[4] On the theorising of the fundamental dynamic structure of local clusters see Brenner (2001).

knowledge and the absorption of available codified knowledge[5]. This gives rise to the growth, during time, of "knowledgeable" agents[6] (individuals, firms and institutions) and, relating to this, to a stock of "contextual knowledge". It is important to stress that, contrary to what Marshall thought, "contextual knowledge" is not "in the air", (external observers crossing the systems do not have access to it) but is ingrained in knowledgeable agents. So, we must assume also that it is difficult to transfer, or to imitate. Within the Italian local production systems it has been spread out mainly in two ways: a) thanks to a diffused social system of births of new firms, founded by the employees of the most technologically dynamic firms, and b) within the matrix of the inter-firm relations of subcontracting. The existence of this form of social capital, which we have called "contextual knowledge", bears important consequences for the economic analysis. In turn, it is precisely the existence of this distinctive character of embedded capital that brings about the competitiveness of the major number of the Italian local production systems and that contributes to explain the phenomenon of the territorial agglomeration of specialised firms in Italy. Contextual, or "situated" knowledge, is not simply to be copied from outside. Spatial differences are not washed away by the formation of global markets. Agents, therefore, possess specific long-lasting competencies that influence their level of competitiveness (Maskell, 1999; Antonelli, 2000).

A recursive sequence of a cumulative growth-inducing mechanism can be described, and the various stages of growth of a typical local production system can be modelled.

1. Typically each system starts with a small group of firms endowed with some artisan skills. There appears, at the beginning, the distinctive competence that can be mobilised by local productive forces. Initially, firms are characterised by being "phase-enterprises". The governance of the local production cycle is highly decentralised among small entrepreneurs. Cost competitiveness (and self-exploitation by the self-employed small entrepreneurs) is typically the principal attribute of the start-up. Once the local system is able to capture a specific segment of demand, the growth-mechanism starts.
2. The first earnings are invested in the modernisation of production processes. This tends to maintain at the minimum level all the production costs (the predominant small size allows for a significant reduction in internal

[5] One of the authors has discussed this issue at length in a previous paper (see Belussi, 2000a).

[6] On this topic see also the contribute of Vicari (1998) which mainly focuses on the them of firms' creativeness.

organisational costs, as discussed by You, 1995). So, the shares of the national or international market initially acquired tend to expand. Demand growth increases the returns from the further division of labour among firms that has now been made possible. Specialisation increases economies of scale and may induce the generation of new knowledge (with the introduction of incremental innovations). In turn, this renders the local production system more competitive. The proximity of agents forms an integrated system where interactions are fluid. Over time, many channels (both informal and institutionalised) are created through which information and knowledge circulate quickly within the productive matrix of subcontracting and specialised firms.

3. The capability of combining "dispersed pieces of knowledge", within these channels, is intensified when the proximity of agents allows for repeated interactions. Here starts the evolutionary pattern of diversification among the various local production systems, depending on the type of learning they are able to develop. In one case, it may be a simple type of learning by doing or using, that mainly influences costs. In another, a more complex form of learning might be envisaged, that has an intrinsic technological content, via the interactions built with suppliers, providers of machinery, technical consultants, demanding clients, and suppliers of intermediate components, that may radically change the product or its performance. If the cumulated effect of changes is radical, the competitiveness of the local systems may be affected and its (national or international) market shares will grow again.
4. A higher volume of production allows a greater division of labour among enterprises[7], and the sequence recursively starts over.

7.3 The various mechanisms of knowledge creation, diffusion and assimilation

Let us turn to the various mechanisms of knowledge creation, diffusion, and assimilation, which can be found within production systems that we called Italian "local production systems". On the basis of empirical research conducted by the Italian local production systems, they may be classed into different

[7] See the arguments put forward by Rullani (1993) "If the small specialised enterprise belongs to a quite large system, the relevant scope economies ... should be measured according to the size of the value-chain of the whole territorial system. In other words, what is important for the generation of value and competitive advantages, is not the size of the enterprise, but ... the level of the efficiency of the local system." (p. 35).

categories[8] in relation to the prevalence of the pool of knowledge they have access to (tacit or codified[9]) and considering the various processes of learning.

The important notion of tacit knowledge

It can be stated that knowledge is a very complex subject to analyse. Technological knowledge involves various degrees of complexity, specificity, openness, cumulativeness, opportunity, transferability, appropriability, and tacitness. Moreover, the creation of new pieces of "useful" knowledge in firms, that support the innovation process, requires a great deal of both endogenous and exogenous sources of knowledge and expertise.

Knowledge itself (Arrow, 1994) is an input into the production of other goods, but it is also an output. It is a function of resources devoted to knowledge as well as a function of the existing knowledge. In addition, it is not just the average level of knowledge that is relevant, but also its distribution (diffusion) among agents. In some circumstances knowledge is of such a form that it can be easily transferred. Personal knowledge or tacit knowledge related to abilities, routines, know-how, or specific practical skills, deriving from experience cannot be simply codified, and, therefore, transferred from one place to another or from one agent to another agent. Knowledge in "practice" tends to be highly tacit in nature, while codified knowledge, abstract knowledge[10], or "scientific knowledge", related to the theoretical understanding, and to scientific principles, has the characteristic of being fully codified[11].

[8] We can assume that all knowledge developed and transferred among local agents bears the characteristics of being "contextual knowledge": a collective good whose generation and expansion is the result of a process that combines pieces of information and knowledge that are owned by a variety of parties and that cannot be traded as such (Antonelli, 2000).

[9] We address the issue of the importance of tacit knowledge not in its pivotal role of (new) knowledge generation in high-tech sectors, as in Senker (1995), or MacKenzie and Spinardi (1995). The main purpose of the paper is to reassess the importance of tacit knowledge in modern economic systems, and within the Italian local production systems.

[10] Please note that the concept of abstract knowledge refers to the generality of knowledge more than to its codifiability. In other words, some pieces of codified knowledge are not abstract (and general). The routines for flying in a specific aeroplane might be codified, but they cannot be applied in a vast range of different models. So, the concept of general knowledge is more restricted than that of codified knowledge, being a sub-set of the category of codified knowledge. (I owe the clarification of these aspects to a personal communication with Alfonso Gambardella).

[11] As Hayek (1945) rightly observed, scientific knowledge "is not the sum of all knowledge". A little reflection will show that there is also a type of "unorganised knowledge" dispersed among many actors: the knowledge of particular circumstances of time and place.

Using the seminal contribution of Polanyi and Nonaka, we can trace a clear-cut distinction between the two forms of knowledge mentioned above. Codification refers to a form of "objectivated" knowledge (a set of justified true beliefs[12]), thus an explicit form of knowledge that is related to the scientific results of basic research and innovative activity (a body of facts, information, principles and practical understanding of science)[13]. Intuitively, this is opposed to tacitness[14], a subjective (both individual and collective shared) property of knowledge, linked to the abilities that individuals possess on the basis of pieces of knowledge developed through practical experience (unarticulated mental models, intuitions, skills).

This distinction may be even clearer if we recall the Lundvall and Johnson (1994) metaphor. The latter type of knowledge has to do with "knowing how[15]", and includes some forms of active participation (knowing in action) and relational capability, thus, "know-who" of the knowledgeable agent. The former is related to a passive and merely conceptual understanding of "knowing what", and "knowing-why" (speculative awareness on the state of the world).

Furthermore, it is also true that to use codified knowledge we need a certain amount of tacit knowledge and ability. In the history of technology we may observe a path of evolution of many specific pieces of knowledge that have become over the time fully codified (this includes many craft products that were

[12] Using a type of connectionist approach we will clearly distinguish the concept of knowledge from that of information. While the latter is related to human reasoning, perception, and interpretation, the former is indeed compressable in symbols and codifiable types of messages. This distinction seems very important nowadays, also within the context of cognitive economics and Artificial Intelligence studies. For a review that relates Hayek's work to connectionism see Smith (1997) and Parsons (1997).

[13] In turn, explicit knowledge may be classified in two ways. As *disembodied*, if it refers to the progress of science and technology (laws, formulas, meaningful sets of information articulated in clear language, including numbers or diagrams, or scientific discoveries related to the state of organic or inorganic substance: new compounds, new materials, etc.). Or as *embodied*, if it lies within technological "tools" such as scientific instrumentation, new machinery, or new information and communication technology with an enlarged computational capability, etc.

[14] It is important to point out that tacit knowledge is not closely related to craftsmanship. Polanyi, the author who developed the concept of tacit knowledge, based his theoretical framework on the analysis of the activities performed by a group of scientists (Polanyi, 1958). See also Ryle (1949).

[15] The concept of know-how is close to the notion of tacit knowledge. Routinised activities and skills may be learned and reproduced without someone being able to go back to their scientific rationales. In regard to this, a few examples were reported by Nelson and Winter (1982). In short: the physiology of muscular movement may be unknown to an athlete without affecting his or her performance.

industrialised under the constraint of technical progress[16]). However, also the amount of tacit knowledge we commonly use has grown up during time.

In any case, the role of tacit knowledge in firms is central to the absorption (Cohen and Levinthal, 1989 and 1990) and the practical utilisation of external knowledge. And, because the amount of external knowledge that firms require to operate is undoubtedly greater than the knowledge that they can reasonably afford to produce in-house, tacit knowledge becomes a strategic element in firm organisation. More specifically, in the interpretation proposed here, which is based also on Belussi (2000a), economic agents within LPS, in order to have access to the existing stock of knowledge, and to put "codified knowledge in action", need a great deal of cognitive capabilities and informal skills.

New ideas, and new technical solutions, are continuously generated in the world, but the process of knowledge transfer is much more complex than the one portrayed by standard theory (Metcalfe and De Liso, 1995). In order to have access to novelties, firms must be aware of them, and then they must be able to assimilate and absorb the new knowledge generated outside. The application of new knowledge requires acceptance, adoption and inter-firm diffusion. All these sub-processes are related to informal learning capability (Gilbert and Cordey-Hayes, 1990). Thus, the successful management of knowledge transfer needs, firstly, a re-contextualization of external knowledge (for "reconverting" and decoding the innovation), and, secondly, an operational encoding of this new knowledge into the internal firms' capabilities and organisational routines. During these different steps tacit knowledge is retained, accumulated, and spontaneously created within organisations (Pavitt, 1984).

Within firms, different degrees of tacitness are embodied in knowledge utilised, in relation with the various activities. Let us take the two extreme cases. Research and development departments deal typically with codified pieces of knowledge that are articulated, and controlled by scientific instrumentation, formalised theories, formulae, and verifiable procedures. The more we move

[16] History is full of disquieting anecdotes. Expert systems have proved to be a complete disaster in running financial activities in stock markets, and during the black October of 1987 they nearly provoked a catastrophe, but automatic pilots are commonly used as a support and extension of human senses. In the ceramic industry, in Sassuolo, the knowledge of skilled pluggers is still far superior to any computer-controlled program. At the end of a highly automated production cycle, human abilities govern the "firing" times and the weight of specific "ingredients" (personal communication related to an interview with the owner of the largest firm of the Sassuolo district, Marazzi spa). The use of computer programs for translations has produced very poor results, while the codification of technical designs has gone further, etc.

away, the degrees of codification decline, and tacit knowledge grows in intensity. Production activities normally[17] contain high levels of tacit knowledge.

To conclude, technical knowledge is partially tacit and ingrained in the skills of individuals. If we assume the importance of tacit knowledge as a crucial aspect of the activity of invention and innovation, we can also theorise that all activities related to the introduction of technical change are in some ways "context-dependent", or localised, as discussed many years ago by Stiglitz (1987), and more recently by Antonelli[18].

Is the "codification-trend" a plausible hypothesis?

The discussion about codification versus tacitness has brought about some important contrasts. David and Foray (1995) put forward the thesis that the codification of knowledge is central to the modern process of dissemination, transfer and retention of knowledge. A similar point is also outlined in Arora and Gambardella (1994). This is because the scope of what can be codified seems to be continually expanding, thanks to the new advances produced by scientific and technological progress. Cowan and Foray (1997), linking codification with the dynamics of firm information structures, reinforce this thesis, up to the point of

[17] Given the obvious difficulty in measuring how much knowledge exists, either codified or tacit, these reflections maintain a certain degree of abstraction and inaccuracy. (One deals with affirmations that have no possible practical demonstration but that have strong theoretical and practical implications in terms of technology policy).

[18] It is useful at this point to introduce the concept of localised technical change (Antonelli, 1999). Antonelli's perspective sets the theoretical basis for understanding the role of tacit elements in the contextualization of knowledge. "Technological change is inherently localised in that it consists of changes in the technical capability of structures that are limited to a well defined set of characteristics...Localised technological change builds upon structured information ..that, as a public good, is available to everybody with low, though not negligible costs, of imitation and acquisition. Tacit knowledge instead is the result of lengthy learning processes; it is idiosyncratic and specific to the organisation and business environment of the firm". .. Technological change is more or less localised according to the mix of internal and external, codified and tacit knowledge on which it depends, but neither form may be dispensed with... It consists of specific pieces of technological know-how obtained by means of learning by using and doing. It incorporates the experience and skills of labour as well as the opportunities of improving products and production processes generated by highly circumstantial factors and events... Since localised knowledge is mainly tacit, because it is implicit and embedded in the memory of organisations and in the economic, regional and industrial environment of each firm, it is difficult to learn, imitate, transfer, adopt and use. It is more proprietary and its use is more excludable than ...in the Arrowian tradition." (p.5-6).

proposing a topography for knowledge activities in which the unclear concept of "codebox" is presented (Cowan, David, and Foray, 2000).

On the contrary, more cautious remarks can be found in the contributions of Lundvall (1996a and 1996b), Senker (1995), Dosi *et al.* (1988), Breschi and Malerba (1997), and Johnson and Lundvall, (2001). Many arguments oppose the codification trend. The introduction of the "information technology paradigm" has not just increased the stock knowledge we can use; it has (above all) increased, to the nth degree, the availability of data; in turn, this has also dramatically increased the circulation of unnecessary information. The overload of information will oblige people to make ever greater use of their tacit knowledge, to select the relevant information they can utilise. There is a spiral of conversion of tacit into codified knowledge, but more tacit knowledge is needed to handle this new codified knowledge. With the contemporary trend towards a post-Fordist society, based on skilled work, or "knowledge workers", it is the entire labour market that in fact is moving towards activities, jobs, and tasks, where learning attributes and tacit knowledge are becoming more important. Lundvall (1996b) has recently pointed to the dawn of the Learning Economy. Today we find ourselves in an economy in which the competitiveness of individuals, firms, and entire systems of innovation reflects their ability to learn. The learning economy places emphasis on interactions (user-producer relationships), and on tacit knowledge sharing and networking. The social dimension of learning is of paramount importance. The knowledge of "human knowledge" (Tamborini, 1997) finds active elaboration of mental models[19], feedback mechanisms of knowledge verification and testing (based on the individual or on the collective experience), where, to allow the passing of the message, common languages, meanings, metaphors, heuristics, visions[20], beliefs and conventions must be shared among actors. In practice, as argued by

[19] As argued by Denzau and North (1994), people act in part upon the basis of myths, dogmas, ideologies, and half-backed theories. In conditions of uncertainty, individuals' interpretation of their environment will reflect their knowledge. Individuals with common cultural backgrounds and experience will share their knowledge. Shared mental models guide choices and the evolution of political-economy. Mental models, institutions and ideologies all contribute to the process by which human beings interpret and order their environment.

[20] In his compelling article on the existing different theories of the firm, Fransman (1994), whose aim is to describe the influence exerted by these visions on the firm internal process of knowledge and information, presents what he calls the Ibm paradox. It is a clear case of a firm clinging to the mistaken belief in the ability of the mainframe computer to sustain profitability at least until 1991, despite the information which it possessed (and processed), contradicting this belief.

Lundvall, a symbiotic relation is established between the two form of knowledge. Codified knowledge may be utilised only through recourse to tacit knowledge. Individuals learn only when tacit knowledge is embodied "in actions"[21]. Tacit knowledge is used to frame the perception of their reality (Schön, 1979), to structure their behaviour, to select and encode the relevant information they need, to filter and re-assemble knowledge (abstract and tacit), to receive signals and elaborate their content, etc. Within the economic organisations, therefore, individuals experience a continuous elaboration and exchange of codified and tacit knowledge. Finally, we can observe that economic agents do not just act, they *inter*act, and establish new organisational routines (Lane, Malerba, Maxfield, and Orsenigo, 1996). Industrial networks, and the emerging knowledge-based organisations, must be regarded as the classical locus where these interactions generate a new stream of "localised" technical change (Antonelli, 1999).

The connection between learning and tacit knowledge

The growth and diffusion of knowledge, both tacit and codified, take places in firms and organisations through learning activities. Economic literature has tended to separate the growth in the stock of knowledge from its diffusion[22] (Davies, 1979; Mansfield, 1961; Mahajan and Wind, 1986; Rogers, 1962). This distinction, goes back to Schumpeter, who has marked the impact of "heroic" innovators. However, most advances in knowledge are not achieved at once. They come about by the slow adaptation of potential users (Stoneman, 1983). If we share the vision that technical change is the outcome of many building blocks, embodying many sources of new ideas, and involving many agents, tacit and codified knowledge become necessary to disentangle the main research objectives, and to implement the discovery process. If this is so, the growth of knowledge may be portrayed essentially as an interactive process between the use of tacit knowledge and learning.

As a consequence: firstly, tacit knowledge is crucial for the innovation process; secondly, learning activities in firms require tacit knowledge both to absorb the existing external knowledge (from the historically created pool of knowledge) and to create new knowledge.

Arrow's (1962) formulation of learning contains implicitly the acknowledgement of the role of tacit knowledge. Where does the better

[21] The classical reference here is in Argyris and Schön (1974). For an excellent survey see Tsoukas (1996).

[22] For a recent survey see Lissoni and Metcalfe (1994).

performance of firms come from? In Arrow's view, learning (improvements in firm knowledge) is related to the accumulated experience of firms, and experience, as we know, is based on tacit elements, and often on the results of unplanned search procedures. However the use of the term learning puts us in an analytical dilemma. Are increasing returns from learning just portrayed as Q/t or are qualitative changes involved in the production function of the firm? A way to solve this problem is offered by the work of Nonaka (1993) and Nonaka and Takeuchi (1995), who have modelled in the following ways the processes of learning among individuals within organisations:

a) socialisation, where there is a passage among individuals of tacit knowledge to tacit knowledge, through a process of collective sharing of knowledge
b) externalisation, in which tacit knowledge is transformed into codified knowledge, a necessary passage in order to diffuse knowledge in a global circuit outside the group that possesses them;
c) recombination, that implies the reuse of various types and sources of tacit knowledge and codified knowledge for the creation of new knowledge, through the use of the firm networks and of other inter-linkages;
d) internalisation, that describes the process through which firms absorb external knowledge transforming it into pieces of tacit knowledge.

Socialisation (learning as knowledge transfer from one agent to another)

Socialisation views a process of inter-active learning mainly related to the transmission of simple instructions from skilled workers to apprenticeship through (intra-firm socialisation) transfer of mainly tacit knowledge, or it refers to the process of labour mobility within a given district. Moreover, a privileged channel of knowledge transfer within the Italian local production systems is attached to the technical specification provided to subcontractors by final firms (district socialisation). Knowledge is transferred among production networks, where firms co-operate in relationships more or less at arm's length. In this modality learning refers mainly to the transfer of tacit knowledge. Socialisation of knowledge may be a planned process, or not a deliberate activity, which occurs as a spillover of a specific subcontracting relationship. It requires differences among agents (one must know and the other not). It is also portrayed as a decentralised process of diffusion of knowledge. Its activation avoids the degradation of the existing stock of knowledge but it does not increase in the firm - or within a certain local production system - the level of the existing stock of knowledge. The stock of knowledge, and learning opportunities are bounded[23] by the absence of relevant innovations (once the ability to perform a certain task is settled, it is transferred to those who do not know it).

Craft skills and simple routines form the typical main component of this process. Tacit knowledge of this type is embedded in particular Italian local systems where craft skills of labour force still seem to be crucial. One thinks for instance of some sectors such as clothing, footwear, furniture, etc. Here firms base a relevant part of their competitiveness, especially in the most qualified segment of demand, on traditional craft labour force skills. Given the fact that, in the nineteenth century, many skills have been lost because of technological progress due to the introduction of mass production, large-scale industrialisation, and the standardisation of labour tasks, the formation of the Italian local production systems has allowed the maintenance of these tacit skills. So, while world-wide loss of skills and unlearning processes (a forgetting of knowledge) have occurred, in the cases reviewed in section 7.3, abilities, skills, and tacit competencies have been kept alive thanks to a mechanism of knowledge sharing. This happened - and is still happening - through tasks and skills imitation, in various ways such as: a) on the shop-floor with side by side imitation, b) with specific job training, c) through the exchange of experience along the subcontracting activity or, again, d) after the working hours, within the process of "communitarian socialisation" where working experiences, opinions, and visions on the world are exchanged through informal communications, within associations, political meetings, friendship contacts, family life, and even "strategic osterie" (wine-bars).

Recombination (learning as knowledge improvement)

We are now focusing our attention on the search procedures conducted by individuals and firms, and on the results of the activity of exploiting and exploring new productive routines (Nooteboom, 2000). When the combination of different ideas produces new knowledge, we increase the scaling up of the existing stock of knowledge. So, this type of learning implies marginal modification of product and machinery, new methods for doing the same thing with more efficacy, incremental innovation, re-design, up to radical modifications (and Schumpeterian innovations)[24]. Within the economic literature, the word learning has been used at least in three different analytical

[23] On this issue see also Young (1993).

[24] The taxonomy of innovation we are using implicitly is quite consolidated in the innovation literature. See for example Freeman (1982).

contexts, where both tacit and explicit knowledge is involved. A. In Arrow sense, learning activities are supposed to be the involuntary result of experience: one looks more to the economic impact than to the process of new knowledge creation. For Arrow (1962), learning is something that is internal to the firms, where improvements come only when the same individual (or plant) is exposed to the repetitiveness of tasks. B. For Rosenberg (1976, 1982), it is the utilisation of new machinery that drives to learning activities in firms. Learning is related to a flow of marginal or significant improvements that go often unnoticed by economists, that sometimes are also related to the introduction of a new division of labour among firms[25]. In the Arrow-Rosenberg approach the recombination of knowledge that occurs during learning can in any case be labelled as incremental innovation. C. The exchange of information and knowledge described in the von Hippel-Lundavall approach represent, again, another point of view, or interpretation of learning activities. The "user-supplier paradigm" proposed by von Hippel some years ago (1988) and by Lundvall (1988), is presented to the audience of innovation economists as a fundamental tool for the introduction of technical change in specific sectors (medical instrumentation, food processing and dairy products, and so on). Here, learning is approaching the Schumpeterian spirit of being a "gale of creative destruction". The recent article of Lane, Malerba, Maxfiel, and Orsenigo (1996) takes a similar stance: all learning activities, depending on exchange of knowledge due to social interaction, because they increase the total amount of knowledge possessed by agents are considered a type of "generative learning".

But, clearly, not all learning is definable as "generative": in some cases it can produce original new knowledge and new incremental or radical innovations, in other cases, learning activities are just spreading in the environment the existing knowledge, among more numerous agents, through voluntary spillovers and involuntary leakage. In other words, learning depends on the existing "knowledge gaps" among the agents involved within the activated network. If some agents have "accumulated" during time a relevant stock of knowledge they may play the role of "activators" of knowledge/competencies within a specific system of relationships. But we have already discussed this case in the previous section. Returning to our main central question, which types of learning characterise the model of Italian districts/local systems? Are they "specialised" in learning by doing/using activities as assumed by Bellandi (1989; 1992), or do they show also the presence of more radical types of learning, which implies the

[25] As also discussed by Chandler (1990), the process of inter-firm division of labour is an element that should be seen as a dynamic process in a historical perspective.

introduction of significant/radical innovations as discussed for instance in the evolutionary models of LPS based in the Northeast regions described by Belussi and Arcangeli (1998) and Belussi and Gottardi (2000)?

Internalisation (learning as absorption capability)

This type of learning is clearly very important and mirrors a process of knowledge diffusion among, firms, territories, and countries. As emphasised in the cognitive model of Nonaka and Takeuchi (1995), here a process of re-contextualization of external (typically codified) knowledge takes place[26]. Here external can refer to an individual firm or, as in Becattini and Rullani (1996), to an entire territory. This is what occurred in the Italian local production systems, especially during the 1960s and 1970s. The stock of knowledge grew incrementally, with the ongoing industrial activities. Firms learned to react and adjust to market signals, to co-operate in dense but territorially dispersed networks and to modify products and processes slightly. When new pieces of knowledge were "captured" by the local "first movers" in new knowledge absorption, very soon new knowledge spread among all agents localised in the specific LPS. Social and territorial proximity favoured imitative procedures. The absorption of external knowledge was also performed by local institutions such as training centres, entrepreneurial associations, laboratories for the testing of products and materials. Very often firms in industrial districts/local production systems performed "vicarious" (low-cost) forms of learning. Even if generally disregarded by the literature on economic innovation, these low-cost sources for innovation (exhibitions, technical meetings, and so on) turned out (Istat, 1995) to be very important not just for "districtualised" firms but for all Italian firms[27]. As discussed by Becattini and Rullani (1996), the capacity of having access to a circuit of codified global knowledge through communicative codes, represents a strategic resource, that allows local-system firms to face the international competition of low labour cost countries (p. 171).

[26] However, even if it has not been considered by Nonaka and Takeuchi, we must also include the possibility of absorbing tacit knowledge (and know-how) from outside. Let us take the case of firms employing outside people with specific skills, or locating productive units in specific places in order to have access to a certain type of knowledge (Science Parks, areas specialised in specific sectors, and so on).

[27] For a deeper discussion see Belussi (1999c).

Externalisation (learning as the capability to produce new relevant pieces of knowledge)

The externalisation process focuses primarily on the creation of new knowledge that is subsequently (under different property-right regimes) released to outside agents (either internally to the district, or, more generally, into the codified "global circuit" of knowledge). It also describes the behaviour of some local production systems during the 1990s, reflecting their "maturity" as organisational form. Only the last evolutionary stage of LPS growth sees them as territorial structures populated with innovative firms. In other words, they were able to increase greatly the stock of existing external knowledge. The level of innovative activity is for the Italian LPS an indirect proxy of their evolutionary capability, as analysed in Belussi and Gottardi (2000). As we will discuss in the next sections, only few LPS show, on closer scrutiny, a strong evolutionary pattern of change. The technological dynamism of local systems is certainly related to their competitive success on the markets, as strong exporters. In the most technologically dynamic local systems, high levels of knowledge creation occur in parallel with an institutional complexity of the industrial structure (emerging leading firms, hierarchisation of the local industrial structure, and presence of large dominant networks). A cognitive division of labour tends to emerge more clearly among firms. Some firms become the privileged *locus* of innovation activities (final assemblers-final firms and/or specialised suppliers). So, the division of labour that crosses the entire local industrial structure is not understandable in terms of "task decentralisation", but in term of specialisation of "innovative labour". Final firms (or hub firms) concentrate their activity on the strategic core of new product design and engineering (together with marketing and post-sale services). Manufacturing activities (routinised activity) are assigned to local small subcontractors (and nowadays they are often delegated to external international supply chains, delocalised in low labour cost countries). The cognitive division of labour mirrors a process of "system evolution", where the benefits of a Smithian extended division of labour interact with a Darwinian process of differentiation, which generates firm variety. As noted by Loasby (1996), the growth of knowledge within the organisation of industry points out a cumulative interaction between differentiation and integration: this is an important element that Marshall added to Smith's work.

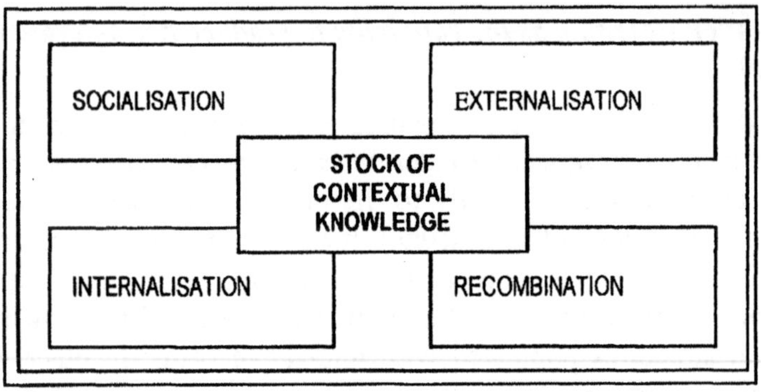

Figure 1: A model of learning and accumulation of "contextual knowledge" based on Nonaka

The formation of the stock of "contextual knowledge"

The dynamics of these four forms of learning provide the basis of the formation of the stock of knowledge embedded in Italian local systems that we can defined as "contextual knowledge". If the Nonaka processes of learning can be conceptualised as flows, contextual knowledge embedded in a territory has the nature of a stock.

As a variable, it is time-dependent, and it is correlated to the presence of technologically dynamic firms. It is formed by elements of codified knowledge (absorbed also from the outside), and tacit knowledge (developed slowly within the production process in practical experience, and internally to the networks of relationships built up by firms).

Contextual knowledge may be described as the social output of a historical process of accumulation of technological capabilities and skills. This occurs only if in a specific territory the mechanism of mobilisation of knowledge is activated (for this reason, the presence of contextual knowledge is path-dependent; it is linked with a territory, and with the "productive" knowledge existing among its firms, in a given specific industrial specialisation).

However, in order to grow, local industrial systems required favourable economic conditions, as for instance in Italy in the post-war period, when the common market was created. They also required some perceptive entrepreneurs, capable of capturing specific segments of the local (and external) demand. Then the model of a local production system could take-off. Knowledge creation and propagation occurred as a consequence of the development of "productive and contextual knowledge" within local firms. But the creation of contextual knowledge is at the same time a cause and effect of growth. In other words, a circular loop between growth and knowledge was at that time at work. Local production systems were accelerators of new technologies in the presence of production network externalities[28] (Belussi, 1996).

The evolutionary path of growth that originates within the local production systems started with the growth of a restricted number of firms: the founders of the local production systems. In these firms knowledge and technical skills become consolidated, and contextual knowledge was created. Knowledge propagation was achieved via the entrepreneurialisation of technical and professional people within the founder firms. Their level of professionalism allowed them to leave their employer and become small independent entrepreneurs. The industrial structure of LPS expanded through a process of firm scission.

Subsequently, new waves of spin-offs occurred, populating the district with small innovative producers: the frequency of contacts, and the number of exploratory searching led in the end, due to the law of large numbers, to improved products or processes: thus, often, original innovations were created[29].

The fact that the development of these idiosyncratic local production systems is territorial-specific bears an important consequence. While all resources are considered by standard economics as "ubiquitous" and footloose, one of the few remaining genuinely localised factor, in the increasingly slippery globalisation of the economic space, is precisely the 'stickiness" of some forms of knowledge and learning processes (Malmberg and Maskell, 1999). The historical persistency of a localised "stock of knowledge", shared by local agents (individuals, firms, and institutions) is the characteristic that reproduces during

[28] Within empirical literature on LPS we can find an ample range of gains associated to "externalities": a) externalities related to the rapid adoption of innovation (the "bandwagon effect"), as discussed by Gilly and Grossetti (1993), Gordon (1992), Longhi and Quéré (1993); b) externalities deriving from the co-ordination of investments among the firms which participate in the same production network (Di Bernardo, 1991; Harrison, 1991, 1994); c) externalities linked to the reduction of transaction costs deriving by a better co-operative relationship (see the work on trust), as argued by Dei Ottati (1994); d) externalities stemming from the sharing of information and knowledge, as argued by Antonelli (1994 and 1996), Audretsch and Feldman (1994); e) externalities related to the positive impact of research centres, training institutions and tailored specialised services, as discussed by Bartolozzi (1993), Bartolozzi and Garibaldo, (1995), Cooke (1994); and f) externalities deriving from the presence of specialised suppliers (this is a crucial point for the competitiveness of final firms).

[29] R&D-dependent" radical innovations " are not typically produced within the LPS, where firms are often small sized. Radical innovation would require the specialisation of dedicated resources to invention and innovation. But a great deal of innovative activity is generated through learning by interaction (Belussi, 1988).

time systematic forms of competitive advantage. This strategic (but immaterial) resource is essentially territorial-specific[30]. Contextual knowledge can thus be defined also as a cognitive form of social capital, historically accumulated in the local production systems model, and interpreted as a sunk investment. Only agents operating in the local production system have access to it, and they may further enlarge and exploit its profitability through strategies of entrepreneurial growth. This process, thus, is highly path-dependent, and built up upon a nucleus of "original" pre-existing local skills and competencies.

"Contextual knowledge" is clearly local, and it is bound to the spatial boundaries of each specific production system. Contextual knowledge therefore cannot ever be completely "globalised". Spatial proximity and social mechanisms of sharing knowledge facilitate its local "transmission". Over long-distance (which is relational as well spatial), frictions dominate[31]. It is obviously true that some innovations introduced by the firms of LPS can be clearly codified, and imitated elsewhere (through the process of externalisation that we have described above). So, some codified elements of contextual knowledge are more at "risk". But, differently to what Nonaka claimed, the conversion of tacit knowledge into "external" knowledge is far more difficult. It follows that, on the whole, the contextual knowledge of the various local production systems can never go over the local production system walls, and thus, if some flows may reach the global circuit or direct competitors, external knowledge, defined as a stock variable, will never be completely externalised.

The verification of this thesis can clearly only be indirect. Indeed, the areas of specialisation on which local production systems have competed have not changed very much over the last two decades (Quadrio Curzio and Fortis, 2000). During this time, local production systems have proved to be quite stable structures, and not footloose organisations. They have deepened their roots in their territory, which is also a community of people, sharing local traditions, habits, language and entrepreneurial visions. The process of globalisation has passed over them.

Yet during the 1980s Italian local production systems were already global in their market outlets. At the end of the 1990s, recent empirical research shows that the process of economic globalisation has not undermined at all, as a whole,

[30] A similar concept is expressed by Antonelli (1996), when he defines the structure of percolation processes as "localised knowledge"; also Maskel and Malmberg, (1999) have emphasised the increased importance of "embedded tacit knowledge" for the industrial competitiveness of the local production systems based on small enterprises.

[31] A similar perspective is also developed in Breschi and Malerba (1997).

their competitiveness. Many of them were still quite high exporter systems. The export flows calculated by a recent empirical survey show that Italian LPS were typically exporting something like 40-50% of the total output produced in their area[32], and in some cases these percentages were set at remarkable peaks.[33]

7.4 An empirical taxonomy of local production systems based on the mechanism of knowledge development

Looking back at the extraordinary commercial success of the Italian model, and its local production systems, it should be said that, in the past, too much attention (and speculation) has been concentrated on terms like flexibility, small firms, and etc. They were the magic words capable alone of highlighting the - perhaps still not completely explained - competitive power of these types of spatial economic systems (sometimes even now considered as "backward" by orthodox economists). Our attempt has been, on the contrary, to capture the existence of the "district effect" phenomenon[34]. Our tentative explanation is based on the evolutionary mechanism of "knowledge" development. Behind the extraordinary competitive success of the Italian LPS is at work an endogenous mechanism of knowledge creation which, in different ways, has implies that collective learning processes could take place (Lombardi, 2000a). So, we adopt a vision in which the Italian LPS are interpreted as "growth machines": a dense network of localised inter-related activities, that bears some synergetic properties (Lombardi, 2000b; Biggero, 1998 and 1999). Here, the formation and accumulation of "contextual knowledge" has sustained the development of local firms. This, in turn, through a process of firm scission has activated the growth of the industrial structure, given rise to the interplay between the technological

[32] Considering the 50 product groups "made in Italy" localised in LPS (Montedison, 1998), we observe that these groups are responsible for a huge positive balance of trade (in 1995: 148.015 billions of lire, which surpassed the total net balance of 67.550 billion of lire; 1996: about 125.000 and a net balance of 39.000). Export flows were 154.294 billion lire in 1995, and 104.318 billion in the first nine months of 1996. For 21 products typically manufactured in LPS, the Italian firms are first placed and Italy has the best international trade balance, and for other 8, Italy ranks as the second or third exporting country.

[33] See, for instance, the packaging machinery district in Bologna (Belussi, 1999a), where about 95 % of the total output is exported, or the Montebelluna district specialised in ski boots (Pilotti, 1998), which supplies 75% of the international markets.

[34] Using quantitative analyses, both the work of Signorini (1994), Bagella and Becchetti, 2000), and that of Paniccia (2001; see Chapter 13) have tried to measured the existence of a "district effect".

dynamism and the local economic development, characterised by entrepreneurial vitality. An important role, in our explanation, is played by the local accumulation of tacit knowledge. The presence of tacit knowledge is generally ignored in a large part of studies regarding the analysis of LPS. We believe that a proper understanding of the tacit elements of knowledge is decisive in the understanding the issue of the dynamics of local districts.

Having said that, we must now clarify the evolutionary trajectory of the LPS, that envisages a process of interplay between the creation and diffusion of knowledge, thus the learning mechanism at work, and the development of the local production structure. Using an evolutionary approach (Witt, 1992, 1993; Nelson, 1995), we can build at least a general framework, able to take into account the various mechanisms of learning already discussed. Utilising the wide range of empirical research produced in these years, and quoted in the next pages, on the basis of a qualitative judgement and abstraction, we have "taxonomised" the evolutionary pattern of the Italian LPS in the following way.

In Fig. 2, we have sketched a map of the typological characteristics of the local production systems, based on the elements of connection between learning and tacit knowledge that refer to the Nonaka model[35].

1. The first cluster represents systems where tacit knowledge among agents predominates. Tacit knowledge is mainly embodied in the skills of the labour force and in craft production. The evolution of the systems can be described mainly as a horizontal expansion of a given stock of knowledge, accumulated in particular localities by "historical accidents". Although this was the situation of many local production systems at the stage of their formation, this ideal type of district is still a living model. Knowledge development is organised essentially around the sharing of tacit knowledge among agents.

Even when some forms of learning by doing and learning by using can be registered, learning is bound by the absence of the creation of new knowledge. A type of instructive learning prevails; and learning activities have occurred (and are occurring) through a process of socialisation of skills. No relevant innovations can be detached if we exclude, naturally, the acquisition of new machinery from external suppliers. The number of the local production systems sharing these features is very low nowadays. Notably, we find here some production systems specialised in traditional skill-intensive industry sectors. It is the case of some clothing and knitwear districts (for instance the area of Carpi), or of some traditional footwear districts, like the urban area of Naples, or the

[35] See also the new contributes that Nonaka has presented on the issue of tacit knowledge (Nonaka, Umemoto, and Senoo, 1996; and Nonaka and Konno, 1998).

concentration of small firms of Casarano (in the province of Bari). Of course, in sectors specialised in artistic productions, this model is still viable, like the famous glass making district of Murano, which is localised in an island near Venice[36].

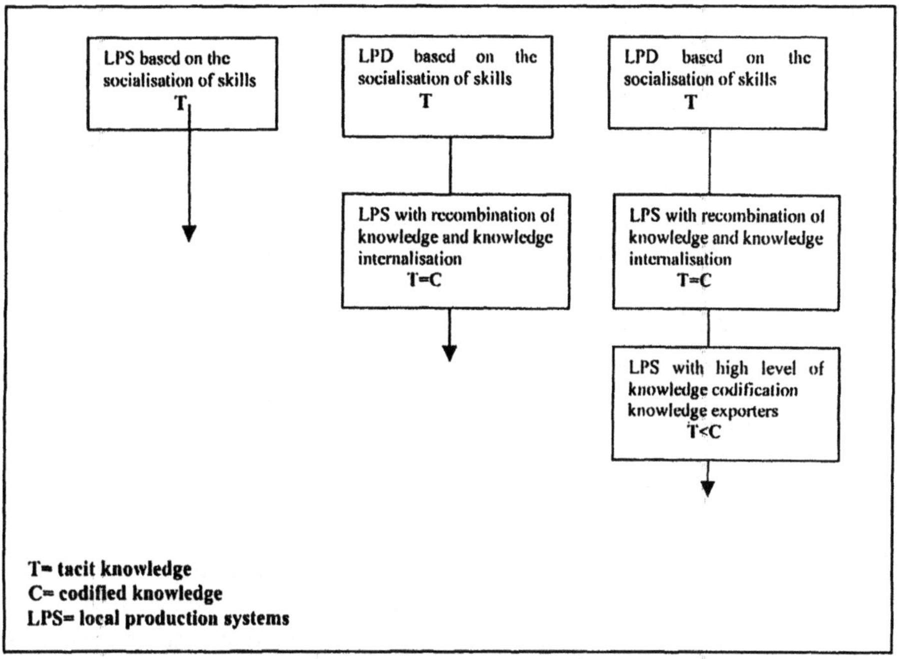

Figure 2: An evolutionary model of local systems based on the accumulation of tacit and codified knowledge

2. The second category describes those systems where the specialisation of the local manufacturing structure has gone farther, and it has activated a process of absorption of external knowledge, and a recombination of a different sort of knowledge coming from internal or external source (through networks and agents searching procedures). As empirical research has proved, in these systems tacit knowledge is still relevant but also codification knowledge is well developed. The stock of knowledge of these systems sees a balance between tacit and codified knowledge. In this cluster, the sources of innovation are more formalised: innovation activity mainly occurs in engineering departments and through local interactions with suppliers, sophisticated clients, local experts,

[36] For the analysis of the Carpi district see Bigarelli (2000); for Casarano see D'Ercole (1992); for the footwear district of Naples see Amin (1988).

consultants, suppliers of machinery, and local institutions. Product innovation is frequent.

Here many actors contribute to the absorption and socialisation of knowledge, such as training schools, universities or special services (supplied by the local authorities) for small firms (real services to firms).

This cluster is represented by a vast array of niche specialisation:

a) many mechanical engineering sectors (water tap fittings in Varese[37]; packaging machinery in Bologna[38]; the "stainless steel" valley of Conegliano, near Treviso, specialised in small electric household appliances[39]; the heating-thermal-mechanic sector of heating appliances near Verona[40]);
b) sectors related to the second transformation of plastic and other metals[41] (plastic frames for eyeglasses in Cadore, near Belluno; jewellery in Vicenza, Arezzo and Valenza Po, near Alessandria; tiles in Sassuolo; marble in Valpolicella, near Verona),
c) furniture sectors (leather upholstery producers[42] in the district of Matera - Altamura-Santeramo, in the South of Italy; the chair sector of Manzano[43], near Udine; the furniture sector of Livenza[44], near Venice),
d) textile, leather and other fashion-related sectors (Prato[45] and Biella[46] for wool, Como for silk[47]; Castel Goffredo for women's stockings; Arzignano near Vicenza[48] and Santa Croce near Florence for leather; Treviso for knitwear[49]; sophisticated women's footwear on the Riviera del Brenta, near Venice[50]).

3. In the third cluster the dominant feature is the prevalence of codified knowledge (however, as in all local systems, tacit knowledge is still important). From an evolutionary point of view this cluster represents the last stage of growth. Firms localised here are technologically dynamic, in many firms the

[37] Cf. Bozzi and Bramanti (1994), Bramanti and Senn (1994).
[38] Cf. AA.VV (1980); Capecchi (1990), Belussi (1999; see Chapter 15).
[39] Soli (1998).
[40] Belussi (1979).
[41] Cf. Moussanet and Paolozzi (1992). For Sassuolo see Marchi (1999).
[42] Cf. Belussi (1999b; see Chapter 19).
[43] Cf. Grandinetti (1999; see Chapter 20), Tamisari (2000).
[44] Cf. Guerra (1998)
[45] Cf. Dei Ottati (1995 and 1996) and Chapter 18.
[46] Cf. Lucchini and Martini (1992).
[47] Cf. Papadia (1992).
[48] Cf. Dutch and Pezzi (2000).
[49] Cf. Belussi (1992), Belussi and Festa (1990).
[50] Cf. Belussi (2000b).

research activity is formalised in the R&D department. Firms frequently patent, and original innovations have sustained the competitiveness of leading firms. This portrait does not suit all firms that populate the local system, but is related only to a small group of innovating firms. However, they feed the local system absorbing knowledge from outside, combining new pieces of knowledge, and transferring it to the net of their subcontracting and specialised suppliers. During time the knowledge creation process has enlarged the local stock of knowledge. Using the Nonaka model, in this cluster the socialisation process of tacit knowledge is related to high-tech competence and skills. The system produces a stream of innovations and original knowledge, which also feed the global circuits. Here we assume the presence of systematic radical innovations. The trade balance of knowledge sees the LPS as exporters of knowledge incorporated in goods or disembodied. Anyway, flows of acquisition and flows in which knowledge is exported are common. The clustering of innovations and the total amount of knowledge embedded in local firms allow us to define these local systems as technological districts. In Italy we do not find many high-tech systems, such as in the USA and the UK (Silicon Valley, Route 128, Cambridge area, etc). Some recognised examples are the Montebelluna area[51], near Treviso, specialised in ski-boots production, and the biomedical system of Mirandola[52], near Modena.

7.5 Conclusions

In this paper we have tried to show how local production systems are diversified. Despite the increasing attention given to the role played by knowledge in the economy, the current debate among economists and technologists demonstrates a quite contradictory nature. On one hand, as discussed by Cowan and Foray (1997), Cowan, David and Foray (2000), or Arora and Gambardella (1994), among others, the process by which knowledge and information evolve in the economy is described as a tendency towards a continuous process of knowledge codification. This is seen as the natural trajectory of scientific progress, and as a direct consequence of the development of the ICT technological paradigm , and the notion of tacit knowledge is simply rejected as "a vague bit of fashionable economic jargon" (Cowan, David and Foray, 2000; p. 212).

[51] Cf. Albertini and Pilotti (1996), and Corò, Gurisatti and Rossi (1998).
[52] Cf. Biggiero (2000; see Chapter 16).

Quite in contrast, as discussed also by Johnson and Lundvall (2001) we have tried to underline, in the cases under consideration, the exceptional importance of tacit knowledge, which is related to the characteristics of learning, the notion of localised technical change, and the evolution of competencies and routines in firms. In this work we have discussed the first elements of a line of inquiry for the verification of the "real" level of innovativeness existing within the Italian LPS systems, that "taxonomise" three possible modalities and levels of learning.

This clearly appears to be related to the internal, or endogenous, technological dynamism of LPS. However, again, this is an insufficient element to predict their competitiveness in the long-term, and how the different LPS can, or may, shift from one stage of learning to a different one. This still seems to us a much more complex issue, related to the dynamics of local entrepreneuriship. But, clearly, the ways in which local entrepreneurs interpret the range of market opportunities has more to do with the development of specific entrepreneurial capabilities (Teece and Pisano, 1994; Carlsson and Eliasson, 1994), imagination and leadership (Witt, 1996, 2000), alertness (Kirzner, 1979), and their ability to explore demand needs through marketing and new models of business (Nooteboom, 1999). Nor must we undermine the role of local and external institutions in shaping and enhancing contextual capabilities (Belussi, 1999c). This is clearly the subject of another (not yet written) paper.

References

AA. VV. (1980), *Macchine, scuola, industria*, Il Mulino, Bologna.

Albertini S. and Pilotti L. (1996), (eds.), *Reti di Reti*, Cedam, Padua.

Amin A. (1988), Specialisation without growth: small firms in an inner-city area of Naples, mimeo, University of Newcastle upon Tyne.

Amin A. (1991), These are not Marshallian times, in Camagni (eds.), *Innovation networks: spatial perspectives*, Belhaven Press, London.

Amin A.(1993), The difficult transition from informal to Marshallian district, mimeo, University of Newcastle upon Tyne.

Amin A. and Robins K. (1990), Industrial districts and regional development: limits and possibilities, in Pyke F, Becattini G. and Sengenberger W. (eds.), *Industrial Districts and Inter-firm Co-operation in Italy*, Genevre, Ilo.

Anastasia B., Corò G. and Crestanello P. (1993), Problemi di individuazione dei distretti industriali, *Oltre il ponte*, no. 42, p. 24-61.

Antonelli C. (1994), Technological districts, localised spillovers, and productivity growth. The Italian evidence on technological externalities in the core regions, *International Review of Applied Economics*, vol. 8, no. 1, p. 31-45.

Antonelli C. (1986), Technological districts and regional innovation capacity, *Revue d'Economie Régionale et Urbaine*, no.5, p. 695-705.

Antonelli C. (1999), *The Microdynamics of Technological Change*, Routledge, London.

Antonelli C. (2000), Collective knowledge communication and innovation: the evidence of technological districts, *Regional Studies*, vol. 34, no. 6, p. 535-547.

Arora A. and Gambardella A. (1994), The changing technology of technological change: general and abstract knowledge and the division of innovative labour, *Research Policy*, vol. 23, p. 523-532.

Argyris C. and Schön D. (1974), *Theory in practice*, Jossey-Bass, San Francisco.

Arrow K. (1962), The economic implication of learning by doing, *Review of Economic Studies*, vol. 29, p. 155-73.

Arrow K. (1994), The production and distribution of knowledge, in Silverberg G. and Soete L. (eds.), *The Economics of Growth and Technical Change*, Edward Elgar, Cheltenham.

Audretsch D. and Feldman M. (1994), R&D spillover and the geography of innovation production, Discussion Paper, FS IV 94-2, Berlin, Wissenschaftszentrum.

Bagella M. and Becchetti L. (2000), (eds.), The Competitive Advantage of Industrial Districts, Theoretical and Empirical Analysis, Physica, Heidelberg.

Bartolozzi P. (1993), Le politiche industriali e di sostegno all'impresa PM-Artigianato delle Autorità regionali in Emilia Romagna, Marche, Toscana, Veneto, Friuli, mimeo, Ires Bologna, Emilia Romagna.

Bartolozzi P. and Garibaldo F. (1995), (eds.) *Lavoro creativo e impresa efficiente. Ricerca sulle piccole e medie imprese*, Esi, Rome.

Becattini G. (1979), Dal settore industriale al distretto industriale. Alcune considerazioni sull'unità di indagine in economia industriale, *Economia e Politica Industriale*, no. 1, p. 7-21.

Becattini G. (1987), *Mercato e forze locali: il distretto industriale*, Il Mulino, Bologna.

Becattini G. (1989), *Modelli locali di sviluppo*, Il Mulino, Bologna.

Becattini G. and Rullani E. (1996), Local systems and global connections: the role of knowledge, in, Cossentino F., Pyke F. and Sengenberger W. (eds.), *Local regional response to global pressure: the case of Italy*, Ilo,Geneva.

Bellandi M. (1989), Capacità innovative diffuse e sistemi locali di imprese, in, Becattini, (ed.), cit.

Bellandi M. (1992), The incentives to decentralised industrial creativity in local systems of small firms, *Revue d'economie industrielle*, no. 59, p. 99-110.

Bellandi M. (1993), Structure and change in the industrial district, *Studi e discussioni*, no. 85, Department of Economic Science, University of Florence.

Bellandi M. and Russo M. (1994), (eds.), *Distretti industriali e cambiamento economico locale*, Rosenberg & Sellier, Turin.

Belussi F. (1979), *Il settore termomeccanico a Verona*, Grafiche Cierre, Verona.

Belussi F. (1988), (ed.) *Innovazione tecnologica ed economie locali*, Franco Angeli, Milan.

Belussi F. (1992), (ed.), *Nuovi modelli di impresa, gerarchie organizzative ed imprese rete*, Franco Angeli, Milan.

Belussi F. (1996), Local systems, industrial districts and institutional networks: towards a new evolutionary paradigm of industrial economics?, *European Planning Studies*, vol. 4, no. 3, p. 5-26.

Belussi F. (1999a), The generation of localised technological change through communication processes. The case of the packaging machinery industry in the Bologna district, Inloco, mimeo.

Belussi F. (1999b), Path dependency vs. industrial dynamics: an analysis of two heterogeneous districts, *Human Systems Management*, vol. 18, no. 2, p. 161-174.

Belussi F. (1999c), Policies for the development of knowledge-intensive local production systems, *Cambridge Journal of Economics*, vol. 23, no. 6, p. 729-747.

Belussi F. (2000a), Accumulation of tacit knowledge and division of cognitive labour in the industrial district/local production system, *Working Paper On Economics and Evolution, Max Planck Institute*, #0012

Belussi F. (2000b), (ed.), *Tacchi a spillo. Il distretto calzaturiero della Riviera del Brenta come forma organizzata di capitale sociale*, Cleup, Padua.

Belussi F. and Festa M. (1990), L'impresa rete del modello veneto: dal post-fordismo al toyotismo? Alcune note illustrative sulle strutture organizzative dell'indotto Benetton, *Oltre il Ponte*, vol. 31, p. 3-81.

Belussi F. and Arcangeli F. (1998), A typology of networks: flexible and evolutionary firms, *Research Policy*, vol. 27, p. 415-428.

Belussi F. and Gottardi G. (2000), Model of localised technological change, in, Belussi F. and Gottardi G. (eds.), *Evolutionary Patterns of Local Industrial Systems*, Ashgate, Aldershot.

Benko G. and Lipietz A. (1992), (eds.) *Les régions qui gagnent*, Presses Universitaires de France, Paris.

Bigarelli D. (2000), Il distretto di Carpi negli anni '90, *Small Business*, Special Issue "Il futuro dei distretti", Belussi F., Gottardi G. and Rullani E. (eds.), vol. 2, p. 23-39.

Biggiero L. (1998), Italian industrial districts: the triple helix of problem solving, *Industry and Higher Education*, vol. 4, p. 167-184.

Biggiero L. (1999), Markets, hierarchies, networks, and districts: a cybernetic approach, *Human System Management*, vol. 18, no. 2, p. 71-86.

Biggiero L. (2000), The location of multinationals in industrial districts: knowledge transfer in biomedical", *Journal of Technology Transfer*, vol. 27, p. 111-122.

Bozzi C. and Bramanti A. (1994), Struttura ed evoluzione di un settore dinamico: il caso degli antifurti nella provincia di Varese, *Quaderno no. 2, Collana Osservatorio*, Bocconi University, Milan.

Bramanti A. and Senn L. (1994), Cambiamenti strutturali, connessioni locali e "strutture di governo" in tre sistemi locali della Lombardia del nord-ovest, Quaderno Opes no. 5, Bocconi University, Milan.

Brenner T. (2001), Self-organisation, local Symbiosis of firms and the life cycle of localised industrial clusters, paper presented at the seminar "Localised Industrial Clusters", 31st January, Max Planck Institute for Research into Economic Systems, Jena.

Breschi S. and Malerba F. (1997), Sectoral innovation systems: technological regimes, Schumpeterian dynamics and spatial boundaries, in Edquist C. (ed.), *Systems of Innovation*, Cassel, London.

Brusco S. (1989), *Piccola impresa e distretti industriali*, Rosenberg & Sellier, Turin.

Capecchi V. (1990), A history of flexible specialisation of industrial districts in Emilia Romagna, in Pyke F, Becattini G. and Sengenberger W. (eds.), *Industrial Distinct and Inter-firm Co-operation in Italy*, Ilo, Geneva.

Carlsson B. and Eliasson G. (1994), The nature and importance of economic competence, *Industry and Corporate Change,* vol. 3, no. 3, p. 687-711.

Carminucci C. and Casucci S. (1997), Il ciclo di vita dei distretti industriali. Ipotesi teoriche ed evidenze empiriche, *L'Industria*, no. 18, p. 283-315.

Chandler A. D. (1990), *Scale and Scope. The Dynamics of Capitalism*, The Bellknap Press, Cambridge Mass.

Cohen W. and Levinthal D. (1989), Innovation and learning: the two faces of R&D, *Economic Journal*, vol. 99, p. 569-96.

Cohen W. and Levinthal D. (1990), Absorptive capacity: a new perspective on learning and innovation, *Administrative Science Quarterly*, vol. 35, p. 128-52.

Cooke P. (1994), Building a 21st century regional economy in Emilia Romagna, paper presented at the conference "Industrial districts and local development in Italy: challenges and policy perspectives, Bologna 2nd-3rd, May.

Corò G., Gurisatti P. and Rossi A. (1998), Il distretto sport system di Montebelluna, in Corò G. and Rullani E. (eds.), *Percorsi locali di internazionalizzazione. Competenze e auto-organizzazione nei distretti industriali del Nord-est*, Franco Angeli, Milan.

Cossentino F., Pyke F. and Sengenberger W. (eds.), (1996), *Local and regional response to global pressure: the case of Italy and its industrial districts*, Ilo, Geneva.

Cowan R. and Foray D. (1997), The economics of codification and the diffusion of knowledge. *Industrial and Corporate Change*, vol. 6, no. 3, p. 595-622.

Cowan R., David P. and Foray D. (2000), The explicit economics of knowledge codification and tacitness, *Industrial and Corporate Change*, vol. 9, no. 2, p. 211-253.

David P. and Foray D (1995), Accessing and expanding the science and technology knowledge-based economy, *STI Review*, no. 16, p. 69-92.

Davies S. (1979), *The Diffusion of Process Innovation*, Cambridge University Press, Cambridge.

Dei Ottati G. (1994), Trust, interlinking transactions, and credit in the industrial districts, *Cambridge Journal of Economics*, vol. 18, no. 6, p. 529-546.

Dei Ottati G. (1995), *Tra mercato e comunità: aspetti concettuali e ricerche empiriche sul distretto industriale*, Franco Angeli, Milan.

Dei Ottati G. (1996), Economic changes in the district of Prato in the 1980s: towards a more conscious and organised industrial district, *European Planning Studies*, 4, p. 35-52.

Denzau A. and North D. (1994), Shared mental models: ideologies and institutions, *Kyklos*, vol. 47, no. 1, p. 3-31.

D'Ercole M. (1992), Il distretto di Casarano, in Onida F. Viesti G. and Falzoni M. (eds.), *I distretti industriali: crisi o evoluzione?*, Egea, Milan.

Di Bernardo B. (1991), *La dimensione di impresa: scala, scopo, varietà*, Franco Angeli, Milan.

Dosi G. *et al.* (1988), (eds.), *Technical Change and Economic Theory*, Pinter, London.

Dutch E. and Pezzi A. (2000), I distretti industriali di fronte alla sfida della globalizzazione: sono in grado di apprendere da altre realtà?, *Small Business*, Special Issue, Il futuro dei distretti, Belussi F., Gottardi G. and Rullani E. (eds.), vol. 2, p. 141-150.

Falzoni A., Onida F., and Viesti G. (1992), (eds.), *I distretti industriali: crisi o evoluzione?*, Milan, Egea.

Foray D. and Freeman C. (1993), (eds.), *Technology and the Wealth of Nations. The Dynamics of Constructed Advantages*, Pinter, London.

Fransman M. (1994), Information, knowledge, vision and theories of the firm, *Industrial and Corporate Change*, vol. 3, no. 3, p. 713-757.

Freeman C. (1982), *The Economics of Industrial Innovation*, Pinter, London.

Gandolfi V. (1988), *Aree sistema: internazionalizzazione e reti telematiche*, Franco Angeli, Milan.

Garofoli G. (1983), Le aree-sistema in Italia, *Politica ed economia*, vol. 11, p. 57-70.

Garofoli G. (1992), Les systèmes de petites entreprises: un cas paradigmatique de développement endogène, in Benko G. and Lipietz A. (eds.), cit.

Gilbert M. and Cordey-Hayes M. (1990), Understanding the process of knowledge transfer to achieve successful technological innovation, *Technovation*, vol.16, no. 6, p. 301-312.

Gilly J.P. and Grossetti M. (1993), Organisations, individus et territoires, le cas des systèmes locaux d'innovation, *Revue d'Economie Régionale et Urbaine*, no. 3, p. 449-468.

Gobbo F. (1989), (ed.), *Distretti e sistemi produttivi alla soglia degli anni '90,* Franco Angeli, Milan.

Gordon R. (1992), State, milieu, network: systems of innovation in Silicon Valley, paper presented at the workshop "Systems of Innovations", October 5-6, Bologna.

Gottardi G. (1996), Technology strategies and innovation without R&D, *Journal of Industry studies*, vol. 2, p. 119-134.

Grandinetti R. (1999), *Il seggiolaio e l'economa globale, La transizione evolutiva del distretto friulano della sedia attraverso i risultati di un'indagine di campo*, Cedam, Padua.

Guerra P. (1998), I sistemi produttivi mobilieri del Livenza e del Quartier del Piave, in Corò G. and Rullani E. (ed.), *Percorsi locali di internazionalizzazione. Competenze e auto-organizzazione nei distretti industriali del Nord-est*, Franco Angeli, Milan.

Harrison B. (1990), Industrial districts: old wine in new bottles?, Working paper no. 90, Carnegie Mellon University, Pittsburg Pa.

Harrison B. (1994), *The lean and the mean*, Basic Books, New York.

Hayek F. (1945), The use of knowledge in society, *American Economic Review*, vol. 35, p. 519-30.

von Hippel E. (1988), *The Sources of Innovation,* Oxford University Press, Oxford.

Hirst P. and Zeitlin J. (1992), Specializzazione flessibile e post-fordismo: realtà e implicazioni politiche, in Belussi (ed.), cit.

Istat (1995), Indagine sull'innovazione tecnologica, Anni 1990-92, Rome, *Notiziario*, series 4, Sheet 41, June.

Johnson B. and Lundvall B. (2001), Why all this fuss about codified and tacit knowledge, paper presented at the DRUD Conference, January 18-20, Copenhagen.

Kirzner I. (1979), *Perception, Opportunity and Profit. Studies in the Theory of Entrepreneurship*, University of Chicago Press, Chicago.

Lane D. *et al.* (1996), Choice and action, *Journal of Evolutionary Economics*, vol. 6, p. 43-76.

Lissoni F. and Metcalfe J.S. (1994), Diffusion of innovation ancient and modern. A review of the main themes, in Dodgson M. and Rothwell R. (eds.), *The Handbook of Industrial Innovation*, Edward Elgar, Gower House.

Loasby B. (1996), The organisation of industry and the growth of knowledge, *Lectiones Jenenses*, Heft 7, Max Planck Institut zur Erforschung von Wirtschaftssystemen, Jena.

Loasby B. (1998), Industrial districts as knowledge communities, in Bellet M. and L'Harmet C. (eds.), *Industry, Space and Competition*, Edward Elgar, Cheltenham.

Lombardi M. (1994), Meccanismi evolutivi nella dinamica dei sistemi di imprese, Ricerca di Base, *Quaderni dell'Università Bocconi,* no. 3.

Lombardi M. (2000a), The cognitive approach to the study of local production systems, in, Belussi F. and Gottardi G. (2000), (eds.), *Evolutionary Patterns of Local Industrial Systems*, Ashgate, Aldershot.

Lombardi M. (2000b), Learning and organisational structures, paper presented at the Padua Seminar, "Evolution, complexity, and institutions", 7th December.

Longhi C. and Quéré M. (1993), Systèmes de production et d'innovation, et dynamique des territories, *Revue économique*, vol. 44, no. 4, p. 713-725.

Lucchini A and Martini A (1992), in Onida F., Viesti G., and Falzoni M. (eds.), *I distretti industriali: crisi o evoluzione?*, Egea, Milan.

Lundvall B. (1988), The user-supplier relationship, Aalborg Working paper, no. 5.

Lundvall B. (1996a), The social dimension of the learning economy, *Druid working papers*, 96-1, p. 22.

Lundvall B. (1996b), Information technology in the learning economy- challenges for development strategies, memorandum for United Nations Commission for Science and Technology, mimeo.

Lundvall B. and Johnson B. (1994), The Learning Economy, *Industry Studies*, no. 2. p. 23-42.

Lundvall B. and Borrás S. (1997), *The Globalising Learning Economy Implication For Innovation Policies*, DGXII, Brussels.

MacKenzie D. and Spinardi G. (1995), Tacit knowledge, weapon design and the unconvention of nuclear weapons, *American Journal of Sociology*, vol. 101, no.1, p. 44-99.

Mahajan V. and Wind Y. (1986), *Innovation Diffusion of New Product Acceptance*, Ballinger Pub., Cambridge.

Malmberg A. and Maskell P. (1999), Guest editorial: localised learning and regional economic development, *European Urban and Regional Studies*, vol. 6, no. 3, p. 5-8.

Mansfield E. (1961), Technical change and the rate of imitation, *Econometrica*, vol. 29, p. 741-766.

Marchi G. (1999), *Reti e sistemi di piccole imprese. I produttori di macchine e impianti nel distretto ceramico di Sassulo*, Franco Angeli, Milan.

Maskell P. (1999), Knowledge creation and diffusion in geographic clusters, Druid, Copenhagen, mimeo.

Maskell P. and Malmberg A. (1999), Localised learning and industrial competitiveness, *Cambridge Journal of Economics*, vol. 23, p.167-185.

Metcalfe J.S. and De Liso N. (1995), Innovation, capabilities, and knowledge: the epistemic connection, mimeo, University of Manchester.

Mistri M. (1993), *Distretti industriali e mercato unico europeo*, Franco Angeli, Milan.

Montedison Ufficio studi-Cranec Università Cattolica di Milano (1998), Il ruolo dei distretti industriali nel "made in Italy", paper presented at the conference "Distretti industriali: la via italiana al lavoro e allo sviluppo", Milan, 17th February.

Moussanet M. and Paolazzi L. (1992), (eds.), *Gioielli, bambole, coltelli. Viaggio nei distretti produttivi italiani*, Il Sole 24- Ore, Milan.

Nelson R. and Winter S. (1982), *An Evolutionary Theory of Economic Change*, Harvard University Press, Cambridge, Mass.

Nelson R. (1995), Recent evolutionary theorising about economic change, *Journal Economic Literature*, vol. 33, March, p. 48-90.

Nomisma (1991), *Strategie e valutazione nella politica industriale*, Franco Angeli, Milan.

Nonaka I. (1993), On a knowledge creating organisation, mimeo, paper presented at the Associazione nazionale Formatori, Parma, October 23rd-30th.

Nonaka I. and Takeuchi H. (1995), *The Knowledge Creating Company*, Oxford Univ. Press, Oxford.

Nonaka I. and Konno N. (1998), The concept of "Ba": building a foundation of knowledge creation, *California Management Review*, vol. 40, no. 30, p. 40-54.

Nonaka I., Umemoto K., and Senoo D. (1996), From information processing to knowledge creation: a paradigm shift in business management, *Technology in Society,* vol. 18, no.2, p.203-218.

Nooteboom B. (1999), Innovation, learning, and industrial organisation, *Cambridge Journal of Economics*, vol. 23, p.127-150.

Nooteboom B. (2000), Managing exploitation and exploration, paper presented at *Workshop on Cognitive Economics*, Alessandria, 15-18th November.

Nuti F. (1992) (ed.), *I distretti dell'industria manifatturiera in Italia,* vol. 1 and 2, Franco Angeli, Milan.

Paniccia I. (2001), Italian industrial districts: evolution and performance, paper presented at the Padua Seminar, "Evolution, complexity, and institutions", 8th February.

Papadia M. (1992), Il distretto serico di Como, in Onida F., Viesti G., and Falzoni M. (eds.), *I distretti industriali: crisi o evoluzione?,* Egea, Milan.

Parson S. (1997), Hayek and the limitation of knowledge: philosophical aspects, in Frowen S. (ed.), *Hayek: Economist and Social Philosopher,* Macmillan Press, London.

Pavitt K. (1984), Sectoral patterns of technical change: towards a taxonomy and a theory, *Research Policy*, vol.13, p. 343-373.

Pilotti L. (1998), I distretti innovativi del nord, *Sviluppo e Organizzazione*, no. 187, p. 15-32.

Polanyi M. (1958), *Personal knowledge*, Routledge, London.

Polanyi M. (1967), *The Tacit Dimension,* Anchor Books, New York.

Pyke F, Becattini G., and Sengenberger W., (1990), (eds.), *Industrial Districts and Inter-firm Co-operation in Italy*, Geneva, Ilo.

Pyke F. and Sengeberger W. (1992), (ed.), *Industrial Districts and Local Economic Regeneration,* Geneva, Ilo.

Revue d'Economie Régionale et Urbaine, (1993), numéro spécial, "Economie de proximités", no. 3.

Quadrio Curzio A. and Fortis M. (2000), (eds.), *Il made in Italy oltre il 2000*, Il Mulino, Bologna.

Rogers E. (1962), *Diffusion of Innovations,* The Free Press, New York.

Rosenberg N. (1976), *Perspectives on Technology*, Cambridge University Press, Cambridge.

Rosenberg N. (1982), *Exploring the Black Box: Technology, Economics, and History,* Cambridge University Press, Cambridge.

Rothwell R. (1994), Industrial innovation: success, strategy, trends, in Dodgson M. and Rothwell R. (eds.), *Handbook of Industrial Innovation,* Elgar, Gower.

Rullani E. (1993), L'impresa minore nella teoria economica e nelle osservazioni empiriche, *Rivista italiana di economia, demografia e statistica,* 47, no. 3-4, p. 29-63.

Ryle G. (1949), Knowing how and knowing that, in *The Concept of Mind*, Hutchinson University Library, London.

Sabel C. (1989), Flexible specialization and the re-emergence of regional economics, in Zeitlin J. (ed.), Local Industrial Strategies, *Economy and Society*, vol. 18, no. 4.

Sabel C. and Zeitlin J. (1982), Alternative storiche alla produzione di massa, *Stato e Mercato,* no. 5, p. 213-58.

Schmitz H. (1992), Industrial districts: models and reality in Baden-Württemberg Germany, in Pyke F. and Sengenberger W. (eds.), *Industrial Districts and Local Economic Regeneration,* Geneva, Ilo.

Schön D. (1979), Generative Metaphor: a perspective on problem setting, in Ortony A (ed.), *Metaphor and Thought*, Cup, New York.

Senker J. (1995), Tacit knowledge and models of innovation, *Industrial and Corporate Change*, vol. 4, no. 2, p. 425-447.

Sforzi F. (1990), The Italian districts in the Italian economy, in Pyke F., Becattini G. and Sengenberger W. (eds.), *Industrial Districts and Inter-firm Cooperation in Italy*, Geneva, Ilo.

Sforzi F. (1995), Elenco dei sistemi locali di piccola e media impresa 1991, paper presented at Artimino 11th September, "Lo sviluppo locale", organised by Iris.

Signorini L. (1994), The Price of Prato, or measuring the ID effect, *Papers in Regional Science*, vol. 73, p. 369-92.

Silberstone A. (1989), (ed) *Technology and Economic Progress,* Macmillan, London.

Smith B. (1997), The connectionist mind: a study of Hayekian psychology, in Frowen F. (ed.), *Hayek: Economist and Social Philosopher. A Critical Retrospect*, Macmillan, London.

Soli M. (1998), Il sistema produttivo inox di Conegliano, in Corò G. and Rullani E. (eds.), *Percorsi locali di internazionalizzazione. Competenze e auto-organizzazione nei distretti industriali del Nord-est*, Franco Angeli, Milan.

Stiglitz J.E. (1987), "Learning to learn, localized learning and technological progress", in: Dasgupta P. and Stoneman P. (eds.), *Economic Policy and Technological Performance*, Cambridge University Press, Cambridge.

Stoneman P. (1983), *The Economic Analysis of Technological Change*, Oxford University Press, Oxford.

Tamborini R. (1997), Knowledge and economic behaviour. A constructivist approach, *Evolutionary Economics*, vol. 7, p. 49-72.

Tamisari M. (2000), The evolution of the chair manufacturing industry in Friuli-Venezia-Giuilia, in, Belussi F. and Gottardi G. (eds.), *Evolutionary Patterns of Local Industrial Systems*, Ashgate, Aldershot.

Teece D. and Pisano G. (1994), The dynamic capabilities of firms: an introduction, *Industrial and Corporate Change,* vol. 3, no. 3, p. 537-56.

Tsoukas H. (1996), Forms of knowledge and forms of life in organised contexts, *Warwick Business School* Working Papers, no. 171, p. 1-44.

Vicari S. (1998), *L'impresa creativa*, Etas, Milan.

Witt U. (1992), (ed.), *Explaining Process and Change*, The University of Michigan Press, Ann Arbor.

Witt U. (1993), (ed.), *Evolutionary Economics*, Edward Elgar, Cheltenham.

Witt U. (1996), Imagination and leadership – The neglected dimension of the (evolutionary) theory of the firm, *Max Planck Institute Working Papers*, no. 5.

Witt U. (2000), Changing cognitive frames - Changing organisational forms. An Entrepreneurial theory of organizational development, *Max Planck Institute Working Papers*, no. 7.

You J. (1995), Small firms in economic theory, *Cambridge Journal of Economics*, vol. 19, p. 441-462.

Young A. (1993), Invention and bounded learning by doing, in Grossman G. (ed.), *Economic growth: Theory and Evidence,* Edward Elgar, Cheltenham.

Chapter 8

Cognitive economies and the "nature of the district"

Margherita Turvani

8.1 Introduction

In "The nature of the firm" Coase quotes D.H. Robertson who notes that it is possible to find "islands of conscious power in this ocean of unconscious co-operation like lumps of butter coagulating in a pail of buttermilk" (Coase, 1937; p. 386). Coase asks why is it so? As we know, he contends that the balance between the costs of producing in house and the costs of using the market explains the existence of firms "coagulating transactions" within the market.

In this paper I want to reframe the same argument at a different scale, and ask why it is possible to see firms lumping together in a specific "context" that is named a district? Theoretical and applied research on the industrial district shows that this "context" is thick; it is made up of economic transactions which evolve together with the social and territorial relations (Becattini, 2001). These features of the district attract the interest of many scholars and policy makers, trying to find out the conditions for existence and for replication of such experiences. In most cases their analysis is guided by two different frameworks: one to look at the working of the firm, and another one to look at the network of relations among firms and, more generally, among different agents and institutions operating at the economic and social level, within the specific context of the industrial district. I contend that the same reasons that explain the "nature of the firm" and its existence should be transferred to understand the "nature of the district" and the emergence of industrial clusters. A cognitive perspective for looking inside the "black box" of the firm is an appropriate tool to deal with "the nature of the district", describing it in terms of the knowledge creation process that takes place within and among firms, relying on the development of "cognitive economies" for individuals and for firms. The paper is organised as follows. Section 8.2 offers a brief survey of the theories of the firm with respect to the role of information processing and knowledge creating processes within organizations. Section 8.3 describes how individual cognition develops at the individual level and in a specific context to highlight the social dimension of the knowledge creation process and the role of organizations in creating the social community which sustain the production of knowledge. Section 8.4 discusses the most common understandings of the "nature of the district" showing that they are characterized by the isolation of an entity, the firm, within a "context" which defines its outside or environment. Section 8.5 applies a cognitive perspective to the functioning of firms showing how it improves our comprehension of the dynamic boundaries between the firm and the its outside "context". Section 8.6 shows how these dynamics within the context of a process of a collective generation of knowledge give rise to the emergence of distinctive "lumps" of knowledge, which by means of entrepreneurial action gets organized in firms (Section 8.7). Conclusions follow.

8.2 Information, knowledge and the nature of the firm

To clarify how a distinction between information and knowledge is useful to discuss the nature of the firm I refer to the definition - utilized by Nonaka and Takeuchi (1995) - of information as a good capable of producing knowledge[1].

In economics the two terms are frequently confused. Knowledge is simply made up of information and the latter is described as a set of data about the states of the world. This set is closed-ended and can potentially be acquired. This overtly contrasts with the open-ended nature of knowledge, the processes of discovery and creation of new forms of knowing (Loasby, 1991).

If information is capable of giving rise to knowledge, this cannot or does not happen in a way that can be determined *a priori*. In the case of knowledge being effectively built, its relation with a specific closed-ended finite set of data may not necessarily be reconstructed. Interpretative ambiguity and uncertainty become features of the process which gives rise to knowledge. Fransman (1994; p. 716) remarks, "knowledge (belief) will be used to interpret incomplete information," and the results of this process will always be subject to verification, revision, and comparison. It is precisely for this reason that knowledge is an unfolding open-ended, incomplete and local process (Hayek, 1952; 1945).

[1] This definition has been developed by the philosopher Dretske (1981).

This approach calls into question the interactive dimension in individual and social cognition. Information and knowledge are not freely combinable and cumulable and the media through which knowledge flows become relevant.

Information and knowledge play different roles in the theories of the firm. Some theories focus on the information costs encountered in completing transactions, both internal and with the market; other theories see the firm as a collection of forms of knowledge. In the first case, individual agents manage a variety of information problems, all implying costly solutions: organizations are seen as information processing devices and the firm is depicted as a nexus of contracts.[2]

In the second case, rationally bounded agents are unable to use optimising calculation and they develop local scale problem-solving techniques. Organizations are thus described as a set of local forms of know how, fostered by the recurrent activity of problem solving. The representation of these forms of knowledge and the way in which they are generated, selected, and modified becomes the field of research. Following Marshall's contribution, knowledge becomes a strategic factor giving way to the development of a different perspective to understand the nature of the firm. The resource-based view of the firm (Penrose, 1959) and the studies on bounded rationality and organizational learning (Cyert and March, 1963) offered the tools to study why firms differ in knowledge management and development, and why these differences persist through time, building up a firm's specific strength or weakness.[3] Differences are lasting and self-renewing because knowledge and skills are accumulated through learning, and learning "happens" in a specific time and space (Nonaka and Takeuchi, 1995). Organizational learning (March and Simon, 1958; Senge, 1990; Argyris and Schön, 1996) explains why knowledge and skills are not easy to imitate, transfer and trade: learning is not simply a process of internalisation of information, but it is an interactive process implying individual participation in communities of practice (Lave and Wenger, 1991).

[2] In particular both Alchian and Demsetz's work of team production (1972), the principal-agent theory by Jensen and Meckling (1976) and the pioneering work of Coase (1937) on which Williamson (1985) develops the economics of transaction costs. In these views hierarchy solves information problems relying on more or less complex contractual arrangements.

[3] There is a vast literature on the subject. For a survey, see Foss (1993) and Montgomery (1995). For an examination of the points of contact between the strategic approach and the approach based on resources, see Mahoney and Pandian (1992) and Prahalad and Hamel (1990).

8.3 Cognition in context

Cognitive processes reside in the mind of the individuals, but the interactive process between individual and the community to which he belongs moulds his cognitive development and the outcome of cognition.[4]

Economics is open to a new frontier of research when Simon (1986) elaborates his work by building on a critique of the paradigm of perfect rationality. His model of procedural rationality takes into account the cognitive limits of individuals and Simon describes how the alternatives are formulated, compared, and assessed while learning plays a key role in improving the cognitive performance of individuals. Cognitive activities rely on the use of symbols, and symbolic structures are images of reality; they are more or less accurate transcriptions of the outside world. These complex structures are decomposable, and the mental activity of problem solving is made up of a set of elementary information-handling processes. Simon's view develops out of cognitivism (Piaget, 1969), in which perception works to connect the individual to his environment. The environment is the "outside" of the individual and is the source of new perceptive stimuli; individual cognition, by means of a process of assimilation, accommodation and feedback, works for the combination of the old and the new mental structures.

The study of human cognition offers social sciences hints for moving attention from individual behaviour to the analysis of relations connecting individuals. Collective mental representations and inter subjective interpretations ask for an explanation: connectionism, studying the mind of individuals using the connections between neurons as a unit of analysis, and attributing the emergence of cognition to the plasticity of the system of connections, offers a useful theoretical support. The study of the cognitive structures (frames, mental models) that work as perceptive filters and link the individual to his environment becomes the clue because they limit and select the mental activities that the individual performs. Cognitive structures evolve because, when they are in action they can no longer be the subject of cognitive activity and knowing consists of creating new, more complex structures; the system of relations among individuals shows similar dynamics (Hutchins, 1995).

Social interaction is a clue to understanding the development of interpretative patterns and common frames used by individuals and it gives rise to shared forms of behaviour in a social environment and in problem-solving activities

[4] This theme was developed by Rizzello and Turvani (2000) in a paper discussing the relevance of the cognitive sciences for institutional economics.

(Denzau and North, 1994). Despite the highly idiosyncratic subjective nature of cognitive processes, the development of intense communication, favoured by proximity (not necessarily only spatial but also cultural) leads to commonality in developing cognition, which takes on the character of dependency on the context in "local and idiosyncratic" forms. As Witt (1998) observes with reference to the studies carried out by Bandura (1986), interestingly, these mechanisms produce shared behavioural models.

> "The significance of such models lies in their vicarious nature, i.e. the fact that their rewarding or non rewarding consequences can be grasped by observation rather than own experimentation. Within intensely communicating groups, learning by observation then means that the group members tend to focus on much the same model. Consequently, their individual learning processes produce correlated results which, via the prevailing social models, can intuitively be grasped by newcomers" (Witt, 1998; p. 5).

These mechanisms are important because they offer a vicarious method (i.e., using other person's experience) to appraise success and failures of deviant behaviour, i.e., behaviour diverging from that commonly established in a specific context. In such a way, individuals may receive information to evaluate their own behaviour in different possible alternative situations, previously not under consideration. A re-framing of choices may take place allowing a smooth transition from an older social model to more innovative ones (Witt, 1998; p. 6).

Complex institutional and organizational systems (the outcome of more or less intentional efforts) help individuals and groups in their cognitive activity, supplying aids for decision-making and learning. In this vein, North (1990) connects the organization of the economic process with the individual cognitive activity; the action of competing agents is the driving force in economics, but action depends on the agents' perception of the possible outcomes. Yet, perception critically depends on the available information, and the way such information is processed. Given the cognitive abilities of the agents, information is often incomplete, and is elaborated on the basis of past experience, creating the condition for a "path dependent" evolution (Arthur, 1994). Institutional context and inherited rules, therefore, affect individual cognition, because individuals perceive the environment according to the mental structures they use to interpret the world. These structures are rooted in the past and in culture and, more importantly, they are developed in specific contexts, contributing to limit and to model the possibilities of perception and choices for agents (North, 1990). In this way, differentiation and heterogeneity are continuously recreated, preserving the use of different mental structures and models leading to the elaboration of different interpretations out of a similar set of information. Subjective interpretation plays a major role in the agents' ability to frame a problem and to attribute meaning and therefore it affects decision making. Just as institutions and organizations limit possible alternatives, so, too, mental structures limit perception and the set of decisions at the individual and at the social level (North, 1994; p. 25).

8.4 The firm, its context, and the district

8.4.1 The context is made up of external economies

Theories of the firm offer different perspectives to look "inside" the black box; the "inside" is seen as a nexus of contracts when organizations serve for agent's information problems while the "inside" is a collection of capabilities when organizations serve for agent's cognitive activities. In this case, cognition takes place within the boundaries of a firm, i.e., "in context" and firms "are social communities" specialised in speed and efficiency in the creation and transfer of knowledge" (Kogut and Zander, 1996; p.503).

Outside the boundaries of the firm, of course, there is the world of which the firm is a constituent part, made up of markets, firms, nature, society, etc. Each firm relates with its "outside" in many different ways and it is embedded in a thick web made of material and intangible flows.

Most typically, economists describe what may accrue from the relation existing between what is inside the firm and what is outside, in terms of external economies[5]. Marshallian industrial district comes out of this relation between the firm, as a unit, and what is outside it.

Bellandi (1989) discusses in detail the notion of the industrial district as may be understood reading Marshall. "Marshall does not strictly define the industrial district in relation to small or medium-sized firms... However it is these firms, which most interest him when discussing industrial district. He points to "those very important external economies which can often be secured by the

[5] Let's recall the definition of internal and external economies given by Marshall: "economies dependent on the resources of the industrial houses of business engaged in it, on their organization and the efficiency of the management" are called internal economies; while "these dependent on the general development of the industry" are called external economies. (Bellandi, 1989; p.138).

concentration of many small business of a similar character in particular localities: or, as is commonly said, by the location of industry." They are an agglomeration of local, external economies, that is the economies of production and transaction costs which a firm enjoys and which derive from the size of the system to which it belongs. More precisely, in Marshall's thought, the economies come from the working of the firm within the thick local texture of interdependence which binds together the small firms and local population" (Bellandi, 1989; p.138).

As noted by Bellandi, external economies, i.e., economies dependent on the general development of the industry, become even more relevant when activated in, and associated with, co-localization of firms, giving rise to economies of agglomeration which develop in transaction dealing, skill formation, and innovation processes. Bellandi goes on discussing Marshall and concludes

> "Agglomeration effect results from the fact that standardizations are particular to local producers... and depend on the possibility of easy adjustment or setting up of networks of transactions among firms" (Bellandi, 1989; p.142).

Proximity reinforces agglomeration by the accumulation of skills because, according to Marshall, if many men are employed in the same locality and in the same tasks, they will educate each other. They develop a sense of responsibility and carefulness too and "local industry" becomes a "property of all".

Furthermore, agglomeration produces advantages in terms of innovation capacity of firms not only because good ideas may easily circulate but because there is a special positive attitude toward improvements and innovations: "Industry's secrets are ceasing to be secrets: they are, as it were in the air and children are unwittingly learning many of them. Work well done is rightly appreciated, inventions and improvements in machinery, in process and the general organization of the business have their merits promptly discussed: if one man starts a new idea, it is taken by others and combined with suggestions of their own; and thus it becomes the source of further ideas" (Marshall, *Principles*; p. 225, cit. in Bellandi, 1989; p.145).

All these factors refer to the existence of external economies, which are set in motion by cooperation and competition, and which sustain co-operation and competition within the industrial district (You, 1994). The co-existence of co-operation and competition among firms, localized in proximity, favours individual success, and the origin of the district may be found more in close proximity of firms. It provides "opportunities for entrepreneurs to specialize and for the district, as a whole, to secure economies of scale denied to isolated individual firms because of internal restriction on growth. They can therefore afford to stay small and concentrate their initiative and inventiveness on what they do best" (You, 1994; p. 264).

The existence and the nature of the district rest on external economies which fill up the outside of the firm; the nature of the firm does not matter in explaining why firms lump together because the firm is more a black box than a living organization.

8.4.2 The context is "the institutional setting"

New institutional economics develops a theoretical framework to explain the existence of various discrete organizational forms (Williamson, 1996). It focuses on transaction costs and its determinants at the micro and macro level, and it contends that rules, laws, culture, of beliefs and values, all affect the cost of transacting and shape the possible governance structures. This perspective may be used to develop an explanation of the existence of the industrial district, one of the many forms of organization for economic transactions. The existence of positive transaction costs threaten not only the economic but also the social and political system, since when there are transaction costs, individuals can exploit them to the detriment of others. Of course, the state plays a leading role in determining transaction costs: supplying a more or less stable system of property rights and a more or less adequate system to make them effective through laws and law enforcement, the state offers the rules of the game for economy and for society (North, 1994)[6]. Yet, the rule to which they conform are much more articulated: conventions, cultures, beliefs and all sorts of informal rules prevail in determining everyday social and economic life. Within these systems of rules transactions are organized; markets, hierarchies, and hybrid forms are developed. Having identified how transactions differ, and the features distinguishing alternative modes of governing transactions, transactions take place within the most appropriate governance structure, in order to minimize transaction costs.

From the methodological point of view, institutional economics develops its analysis at two different levels, that of institutional environment and that of

[6] This definition has been given by Nobel Prize Douglas North: " Institutions are the humanly devised constraints that structure human interaction. They are made up of formal constraints (rules, laws, constitutions), informal constraints (norms of behaviour, conventions, and self imposed codes of conduct) and their enforcement characteristics. Together they define the incentive structure of societies and specifically economies" (North, 1994; p. 4).

organization, or institutional arrangements, to use Williamson's words (Williamson, 1996). Within this framework, the "industrial district" as much as networks of firms are put into the catalogue of hybrid forms of organization which develop, thanks to low level of transaction cost, within a favourable institutional environment. Much effort must be poured into the analysis of the "environment", made up of both formal and informal institutions which carry the burden of minimizing the transaction costs within the complexity of transactions taking place in the industrial district. A favourable "institutional environment" is required, and should be promoted to recreate the frame supporting the replication of successful "industrial district" experiences. Given the proper institutional environment, the industrial district will emerge as the most suited organizational form for the completion of transactions. The approach is static in its essence, and the feedback from the economic transactions to the social and institutional relations is not taken into account (Turvani, 1998). Again the virtuous cycle connecting the firm with its outside and vice-versa, for example the flow of Marshallian externalities, is overlooked.

8.4.3 The context is "the district"

The specific feature of the most acknowledged theory of the nature of the district is firm embeddedness i.e., the firm and its outside are subsumed in the "district" and melt into an organic whole. The geographical aspects of agglomeration lose relevance with respect to the territorial dimensions of development (Gambarotto, 2002). Since the end of the 1970s and the early 1980s, the credibility of this approach has grown, building on the work on "the Third Italy", and it has been developed along different streams of research on the role of small and medium sized enterprises in the economic development process (Bagnasco, 1977; Becattini,1979; Brusco, 1982; Fuà and Zacchia, 1983; etc.). [7]

This vein of studies assumes the Marshallian framework describing how the progressive division of labour is accompanied by complementary forms of integration, giving rise to the peculiar features of dense externalities within the "district". Moreover, in the same Marshallian vein, the socio-economic relations that characterized the territory and the district are extremely relevant (Becattini, 2001). These specific conditions shape transactions, lower transaction costs, and

[7] To offer a complete picture of the literature on industrial districts and on the various forms of localized productive organizations and system is outside the scope of this work. For an appraisal of such literature, see for example Maillat (1998), and Gambarotto (2002).

keep the average dimension of firms small. A great deal of attention is, therefore, devoted to the comprehension of all those mechanisms that sustain exchange, such as community life, trust and general "atmosphere" (Dei Ottati, 1995).

A distinctive feature of this research approach on "district" is the combination of the "economic" elements, as developed by Marshall, with a strong twist in favour of sociological analysis. It results in a very high level of embeddedness of firms within their environment, which becomes the centre of the analysis. Descriptive studies offer insights for policy recommendations, even though each reality "uniqueness" and its organic, historical process hamper the possibility of replication.

In the Marshallian vein, the organic evolution of the district rests on the processes which mobilize and create specific resources at the local level, while the "post-Fordist" vein emphasizes the role of technological progress which creates the conditions for the permanence of district-like forms of organization, which appears as a prototype of a flexible production system.[8]

The technological and the socio-economic features are integrated in the study of the milieu innovateur (Camagni, 1991) and the innovation and learning processes that take place within a territory are under scrutiny. Networks and interrelations between different entities (individual, firms, institutions) constitute the substance of the milieu, made up of different and overlapping "spaces" for communication. Relational aspects are core features of the geographical approach (Storper, 1997), which weld the socio-economic and the technological dimensions.

> "The guiding metaphor is the economy of relations, the economic process as conversation and coordination, the subject of the process not as factors but as reflexive human actors, both individual and collective, and the nature of economic accumulation as not only material assets, but as relational assets. Regional economics in particular, and integrated territorial economies in general, will be redefined here as a stocks of relational assets." (Storper, 1997; p.28. cit. in Gambarotto, 2002).

Emphasis on the relational aspects in all these approaches leads to a common trait which unifies each of them, in various forms and degrees: embeddedness of

[8] This line of research was first introduced in Piore and Sabel's analysis (1984) and was also adopted by the "French regulationist" school, and then widely applied and developed for the Italian case by Rullani (1998).

firms within their environment. A major consequence is that each firm loses its character of uniqueness, being a repository and creator of unique knowledge.

Departing from the archetypal Marshallian district based on the "faceless" concept of external economy, the necessity to give the territory a "face" results in the idea of "thickness" of social and economic relations. A colourful description of what is going on in a district is obtained but indeed "thickness and density" come to substitute "agglomeration" (Malmberg and Maskell, 2001). A second major consequence of blurring the boundaries between the firm and its environment is the definition of a (seemingly) powerful concept: social capital. This notion is able to condense a variety of explanations, enabling the researcher to mentally "locate" those factors that "make the difference" to explain development in specific location. It is indeed a too parsimonious explanation, used to fill up the gap between explanation offered by theory (especially economic theory) and the real world.[9]

8.4.4 Knowing in firms

Processes of learning and cognition are core activities in firms and organizations. Within each firm, individual cognitive processes are developed by interacting with other minds, and the features of such collective production of knowledge have been previously described. The way in which individuals learn and know in specific contexts by interacting, for example working in a social community like the firm, has implications that go beyond the boundaries of each firm. Understanding how these processes take place within each firm may be used as a guide in explaining why firms learn and develop knowledge by lumping together. A better understanding of what is, perhaps very incompletely, understood among economists about individual mind functioning, in the context of the firm, suggests framing the understanding of "the nature of the district" by utilizing what we have learnt about the "nature" of the firm.

We know that firms are structures, which govern the production of knowledge and its utilization in novel combinations in production processes; we know that knowledge generation is an unfolding open-ended process going on by means of subsequent structuring; we know that the outcome of such process

[9] Social capital, made up of fiduciary relations and duties that foster each other, is a novel "invisible hand", that substitutes the market in explaining how cooperation and competition trigger development. The working of civil society, after all, becomes the engine of growth: "to focus on social capital means, then, to recognize that the propensity and the capabilities to cooperate that members of a society express, may deeply affect the character of economic and political development that the community may achieve" (Mutti, 1998; p. 12).

assumes specific and local features being developed in a "context". This knowledge is usefully transferable and it can be applied to the analysis of the functioning of the district.

The interactive contextual nature of knowing gives rise to knowledge that is renewed but is never purely self-propagating on a larger scale. The growth of knowledge, like the growth of an organism, does not follow a logic of "more of the same", since its form and contents are transformed (Winter, 1982).

Knowledge proceeds along growth patterns with the generations of new forms and structures by which individuals are able to select, interpret and organize the full range of stimuli coming from the environment; only within this framework is information meaningful, i.e., it is information and not simply raw data. Sociality and the existence of a community of practice are necessary for cognition, as recent findings in cognitive studies have shown, and there is a process of selective perception and retention progressively building up mental structures. These are used for knowing, and they evolve by filtering and selecting signals, following the path drawn by past experiences (Hayek, 1952; Lave and Wenger, 1991; Rizzello and Turvani, 2000).

These micro processes happen in "firms" as much as in "districts", because they take place in the individual minds and shape organizational behaviour in each and any "context". Theories of the firm, such as the evolutionary theory, the resources based theory, and the knowledge-based theory, describe the working of firms in terms of individual cognitive activities, in terms of routines, or of core capabilities. These structures sustain persistent patterns of behaviour, and evolve through time according to different and past experiences of firms. Firms then remember and know by doing, and their knowledge preserves a character of tacitness for the people holding it. (Penrose, 1959; Nelson and Winter, 1982; Nonaka and Takeuchi, 1995; Teece *et al.*, 2000).

The knowledge that organizations handle and continuously create resides in the complex of routines, and in the pool of capabilities making up the firm. It is through this lens that organizations approach messages coming from within the organization itself and also from outside it. Routines serve as such repertoire of capabilities i.e., they describe what the firm knows and how it knows it; but they are also the frame by which the outside is understood. They can be interpreted as the firm's cognitive apparatus and, at any time, they are the specific organization cognitive model, frame, or map (Langlois, 1983, 1995). Organizational routines do not only filter what may be "seen" in environments, but shape the firm's absorptive capacity (Cohen and Levinthal, 1993) which is the ability to understand and take advantage of ideas (or knowledge) developed elsewhere. Perceptual abilities of firms play a major role not only in guiding firm's strategy, but in determining firm's success in the initiatives undertaken. Firms are endowed with a system of routines that shape their perceptual abilities in two broad areas: "(1) the ability to see ways of gaining and improving competencies and (2) the ability to recognize opportunities for applying those competencies in a way that generates value. At the risk of oversimplifying a bit, let's call those area operational perception and market perception" (Langlois, 1995; p. 8).

Perceiving is not a trivial activity, given ambiguity and uncertainty and many forms of misperception may occur. It may result both from thinking that an opportunity exists when, by ex post consideration, it does not; but it may result in lacking the perception of an opportunity, which, ex post, will turn out to be profitable.[10] Langlois (1995) applies this line of reasoning to the effects of misperceiving opportunity within firms operating in the development of the computer industry, and concludes: "What determines the ability of an organization to perceive opportunities? It is my contention that perceptual ability is a matter of the "fit" between the environment and the organization as a cognitive apparatus.... If an organization is to be able to classify information in the right (the most useful) categories, it must posses a structure of categories that is somehow "like" those of the outside world, or at least that small piece of the outside world relevant to it" (Langlois, 1995; p. 10).

This conclusion may be at once trivial and profoundly complex; yet it fits well with our description of how organizations develop knowledge, relying on the mechanisms of individual cognition; it also fits well with evidence about the development of knowledge along scientific paradigms, and furthermore with the evidence and researches on localized technological change (Antonelli, 1999).

8.5 Collective knowing

Firms do not know in a vacuum; they know in "context" and this is recognized in the economic literature on innovation; technological knowledge cumulates systemically, it is created using pre-existing knowledge, and it is highly indivisible. Producing technological knowledge is a collective process in which each agent has unique yet complementary information (Antonelli, 1999; Richardson, 1972). This perspective describes knowledge as a collective good, given its features of radical indivisibility and inputs complementarity in its production.

[10] Langlois treats these possible misperceptions as errors of type one, and errors of type two, according to statistical testing of hypotheses.

> "When these conditions hold, the productivity of a given technological knowledge will be much higher if everybody would be willing to put in a common pool all its knowledge that is complementary to many other. ...We may say that such generation of knowledge is an archetype of a network externality effect, where resources productivity is much higher as many more agents are participating in the net" (Antonelli, 2001; p.229).

The possibility for each agent to take advantages in producing such knowledge as a collective good, is strongly related to the existence of a reliable net of communication. Information flows and the features of the network need to be analysed in this perspective, referring to communication theory, while individual positional advantages, with respect to the flows of information, influence his ability and performances (Antonelli, 2001).

This framework offers a deeper understanding of what constitutes the "atmosphere" of the district; many networks for communication are overlapping in a specific territorial environment, and a picture of what is a "thick" set of transaction is neatly emerging, both at the economic and the social level. Networks enabling communication include local labour markets, user-producer relations, technological outsourcing, subcontracting and relations between small and leading firms, cooperative projects between firms, relations with research institutions of various kinds, financial markets, and so on. The overall image is that of an innovative milieu but firms' knowledge is the engine of growth; if overlapping networks of this kind are existent, positive feedbacks will come into play; development at local level will appear as a self reinforcing virtuous circle. The "district" is the outcome of the previously described processes of collective production of knowledge.

Each firm enjoys a wide range of externalities in the form of economies in technological knowledge production: the overall costs that each firm needs to bear to innovate, both in terms of production and transaction cost, are lower when efficient communication networks exist between the firm and its environment (mainly made up of other firms) or, to put it in another way, with any given level of resources, such economies allow each firm to reach a higher level of innovative performances. These conditions are very likely to happen in "proximity" and because of "proximity" (Nelson, 1987; Antonelli, 2001).

Firms know in "context" and, as we saw, the collective dimension of knowing is a source of economies for firms. Furthermore, firms know and remember by doing (Nelson and Winter, 1982), and this happens in the "context" provided by the specific set of overlapping networks which feed firms with information. Each of them elaborates on similar sets of information and, according to its specific capabilities or cognitive maps, is able to produce new forms of knowledge. Communication among firms is crucial, and the variety of networks overlapping within the district contributes to "fine tuning"; yet, communication should not be seen as a way to re-allocate information and distribute signals. Communication involves deeper processes, such as recognizing differences in cognitive structures, transfer of knowledge from one context to the other, comparison and convergence of cognitive models, and so on. Social learning will occur (Bandura, 1986) welding together the interdependence between subjective perceptions and experiences at the individual level with the process of learning and knowledge growth at the macro level. Knowledge is dispersed, local and specific, and competition spreads it and helps the formation of "opinions" in markets (Hayek, 1946). Social learning then develops on individual cognition and its subjective and idiosyncratic feature: this specificity triggers novelty. Experience and insight preserve uniqueness, and each learns by experiencing and getting insight.

These features of the knowledge creation process were analysed a long time ago by Adam Smith, as Loasby noted (1999). Adam Smith observed that any process of division of labour is accompanied by a process of division of knowledge:

> "Not only does the division of labour encourage invention and other improvement; different kinds of division encourage different kinds of invention and improvements. Smith (1976b; p. 20-1) identifies three categories. The first is produced by workmen, as their "attention comes naturally to be directed towards some one very simple object"; the second by those who have developed the making of machinery into a "peculiar trade"; and the third by those "who are called philosophers or men of speculation, whose trade it is, not to do anything, but to observe everything; and who upon that account are often capable of combining together the powers of the most distant and dissimilar objects" (Loasby, 1999; p. 71).

Smith is describing a process in which the division of knowledge and the division of labour spur each other and new branches of knowing and doing develop out of this process; while developing our knowledge through experience, by specializing in peculiar trade and focusing our attention, we also open our mind to insight, putting together most distant and dissimilar objects, giving rise to novelty. Selective perception, working both as a filter and as a focusing device, shapes the outcome of mental processes. Accordingly, knowledge cannot be directly deduced from a closed set of information, but it comes out of cognitive activity pursued by individuals who elaborate information in an increasingly idiosyncratic and local way.

The very same processes take place in firms, and knowledge supporting decision-making in the firm is characterized by the strategic assessment of the available information, by interpreting it in relation to the set of relations that the firm is able to "see" or "perceive" between its own behaviour, the behaviour of other firms, and of all the other agents involved. The knowledge the firm requires is, thus, the outcome of a process of elaborating elementary information, organized around an interpretive model, connecting the firm's objectives and their context.

The calculation of the outcome of the decision-making process, relying on a complete set of information, is a fiction and as Knight (1921) pointed out, decisions in a firm are based on judgment, and this judgment is a constituent element in knowledge-production processes. A vast scope for judgment and entrepreneurial activity, using Knight's words, may be envisioned, even in everyday activities of the firm, to size opportunities in the market and in the operational field.

8.6 Entrepreneurial activity, the firm, and the nature of the district

Entrepreneurial activity, the ability to seize opportunity putting together "the most distant and dissimilar object", to use Smith's words, is the ability to derive profit out of novelty, i.e., innovating in product, processes and organization. Relying on our description of individual cognition and on Adam Smith's hints, it is possible to assume Langlois' conclusion with confidence: "when operational and/or market environment is highly focused, those organizations will do best whose cognitive apparatus allows them to perceive the finer details and to solve the smaller but important puzzles" of everyday activities of problem solving (Langlois, 1995; p. 11).

There are good reasons for firms to lump together creating districts, and we have described some of them; but there must be a good reason for each specific firm to join the district. This reason has to be found in the entrepreneurial judgment of a good opportunity for profit.

This trivial observation takes us back to the problem we have started with: why do we see "island of conscious power"? Richardson (1972) explained that transaction cost alone cannot explain the emergence of firms in a "world of transactions", but that production costs and transaction cost comparison need to be taken into account. In his view, firms exists always in a dense network of cooperation and affiliation by which firms are interrelated: cooperation is just one way in which activities can be coordinated, the other being direction and market transactions (Richardson, 1972, 1998). He develops the analysis of capabilities, i.e. the appropriate knowledge, experience and skill that sustain efficacious performance of firms; each firm will tend to specialize in a set of activities requiring similar capabilities, because the unified control of very dissimilar activities may hinder productivity, unless firms grow to reach the appropriate level of economies of scale.

Economic literature on districts has shown that these conditions do not apply when firms are in close proximity, and each new firm is able to take advantage of the tight productive connections in the local market. In this case, as Loasby observes, vertical integration is not necessary to align complementary activities and "by avoiding it firms at each stage of production process are better able to preserve the distinctive outlook of their separate trades, which, as Adam Smith noted, enables them to perceive opportunities that would be overlooked within a single business" (Loasby, 1999; p. 74). Each kind of knowledge, then, may be deepened, economies may be obtained in innovative activities, even if there is no formalized research going on within the firm, and rents may be appropriated more easily.

It is important to note that the possibility of relying on a very high and progressive division of labour offers some other cognitive economies: new knowledge in any of the sub fields and applications has a positive consequence for the firm producing it, but it also has a value for many other firms, operating in similar activities, which may rely on vicarious learning (Bandura, 1986), avoiding expensive blind alleys and concentrating on profitable opportunities. Absorptive capacity is also enhanced when entrepreneurial activity is facilitated by the high division of labour/knowledge, both because the entrepreneur's "attention comes naturally to be directed towards some one very simple object", and because entrepreneurs operate in a "peculiar trade". But entrepreneurial activity may imply radical novelty "combining together the powers of the most distant and dissimilar objects", opening new lines of business in the district, progressively diversifying its original vocation: in this case the operating of forms of vicarious learning may spread and foster diversification (Witt, 1998).

All these kinds of activities implying entrepreneurial judgment may be found in any firm as much as in any district. Judgment and uncertainty absorption performed by the entrepreneur allow the formation of new firms, because by centralizing the assumption of responsibility for uncertain outcomes it is possible for firms to function as an administrative and planning unit (Penrose,

1959; Richardson, 1998; Turvani, 1995). Each firm judges opportunities to make profit, and the perception of what is a good opportunity is, of course, developed in "context". Each firm seeks sources of economies in sizing technological and market opportunities, and these are exploited by entrepreneurs whose cognitive models and judgment capabilities are shaped and focused according to the context in which they are developed (Foss and Lorenzen, 2002). Entrepreneurial action is the lifeblood of the industrial district experience, and much of the entrepreneurial action is put forward by individual entrepreneurs: if misperceptions of opportunities in technology and market fields depend on the fit between the organization and its environment, then in the "district" context, entrepreneurial action implies a lower level of uncertainty absorption or, to put it in another way, entrepreneurial initiatives are favoured even if, on average, a lower level of profit may accrue to them. Furthermore, the prevalence of small size enterprises, which is correlated with the high level of firms' birth rate (entrepreneurial initiatives), shapes the quality of human resources. Not only is labour mobility very high, but the knowledge that human resources hold is both highly tacit, idiosyncratic, and complex, given the low level of division of labour within each firm.

The district becomes the place in which the subjectivity of individual cognition is welded with the construction of shared mental models: the way of knowing within the firm feeds in and is nurtured by knowing taking place among firms.

8.7 Conclusion

The attempt to describe some features of the microfoundations of knowledge dynamics within the firm suggested some implications for understanding the nature of specific forms of organization of firms and territories. Accepting the view of Nonaka *et al.* (2000; p. 8) that "to understand the true nature of the knowledge creating firm, viewing a firm as black box, a set of transactions or a collection of resources is not enough", the necessity to understand the features of the creation of knowledge, and the forms in which it takes place within the firm is not questionable. The creation and utilization of knowledge is a core source of a firm's capability to search and exploit profit opportunities and indeed, at any given time, the knowledge that firms hold provides the future opportunities to innovate.

Yet firms differ in knowledge management and development, and these differences persist through time, and build up a firm's specific capabilities. The understanding of the mechanisms of cognition is useful to describe how new opportunities for action emerge with the generation of new knowledge and how persistence and change co-exist in the knowledge-creation process in the individual mind and in the organization. The reason for this must be sought in the social dimension of individuals' cognitive activity, which simultaneously open up opportunities for change and innovation but also preserves a degree of inertia. The mechanisms underlying the knowledge-creation process work in interactive contexts, suggesting a picture of the individual as "a dynamic being and the firm as a dynamic entity that actively interacts with others and the environment" (Nonaka *et al.*, 2000; p. 2). In this picture a social dimension of knowledge-creation appears; it grows on communication, imitation, and comparison and it encourages both the development of interpretative patterns and common frames and the emergence of novelty, fostered by the day-by-day knowledge-creating process at the individual and at the organizational level.

Adam Smith's insights help to reframe this picture on the district scale. The progressive division of labour in firms and among firms goes hand in hand with the redefinition and the renewal of the division of knowledge: novelty is the outcome, and it opens up new opportunities for the use of entrepreneurial judgment, together with the permanence of shared forms of behaviour in the social environment and in problem-solving activities.

References

Alchian A. and Demsetz H. (1972), "Production, Information Costs and Economic Organization", *American Economic Review*, vol. 62, p. 777-795.

Antonelli C. (1995), *The Economics of Localized Technological Change and Industrial Dynamics*, Kluwer Academic Publisher, Dordrecht.

Antonelli C. (1999), *The Microdynamics of technological change*, Routledge, London.

Antonelli C. (2001), *The Microeconomics of technological systems*, Oxford University Press, Oxford.

Argyris C. and Schön D. (1996), *Organizational Learning II. Theory, Method, Practice*, Addison-Wesley Publishing Co., New York.

Arrow K. J. (1974), *The Limits of Organizations*, Norton, New York.

Arthur B. (1994), *Increasing Returns and Path Dependence in the Economy*, Michigan University Press, Ann Arbor.

Bagnasco A. (1977), *Le tre Italie*, Il Mulino, Bologna.

Bandura A. (1986), *Social Foundation of Thought and Action. A Social Cognitive Theory*, Englewood Cliff, Prentice Hall.

Becattini G. (1979), Dal settore industriale al distretto industriale, *Rivista di Economia e Politica Industriale*, vol. 7, no.1, p. 7-21.

Becattini G. (2001), *From the industrial district to the districtualization of production activity: some considerations,* mimeo, Florence.

Bellandi M. (1989), The industrial district in Marshall, in E. Goodman and J. Bamford (eds.), *Small firms and industrial districts in Italy*, London.

Belussi F. and Pilotti L. (2000), *Learning and Innovation by networking within the Italian Industrial Districts,* International Seminar on Technological development in Industrial Networks, mimeo, Urbino.

Brusco S. (1982), The Emilian model: productive decentralization and social integration, *Cambridge Journal of Economics,* vol. 6, no. 2, p. 167-184.

Camagni R. (1991), (eds.) *Innovation networks: Spatial Perspectives*, Belhaven, London.

Coase R. (1937), The Nature of The Firm, *Economica*, vol. 4, p. 386-405.

Coase R. (1991), "*The Institutional Structure of Production*", Alfred Nobel Memorial Prize Lecture, Stockholm.

Cohen M. and Levinthal D. (1993), Absorptive capacity: A New Perspective on Learning and Innovation, *Administrative Science Quarterly*, vol. 35, p. 128-152.

Cyert R. and March J. (1963), *A Behavioral Theory of the Firm*, Englewood: Prentice Hall.

D'Andrade R. (1995), *The Development of Cognitive Anthropology*, Cambridge University Press, Cambridge.

Dei Ottati G. (1995), *Tra mercato e comunità: aspetti concettuali e ricerche empiriche sul distretto industriale,* Franco Angeli, Milan.

Denzau A. and North D. (1994), Shared mental Models: Ideologies and Institutions, *Kyklos*, Vol. 47, no. 1, p. 3-31.

Dretske F. (1981), *Knowledge and the Flow of Information*, The MIT Press, Cambridge MA.

Foss N. J. (1993), "Theories of the Firm: Contractual and Competence Perspectives", *Journal of Evolutionary Economics*, vol. 3, p. 127-144.

Fransman M. (1994), "Information, Knowledge, Vision and Theories of the Firm", *Industrial and Corporate Change,* vol. 3, p. 713-757.

Fuà G., Zacchia C. (1983), *Industrializzazione senza fratture*, Il Mulino, Bologna.

Gambarotto F. (2002), Le relazioni economiche istituzionali nello spazio delle economie locali:una ipotesi teorica, in Camagni, R., Fiorentini, R. and Mistri, M. (eds.) *Auto-organizzazione ed apprendimento strategico*, Cedam, Padua.

Hayek F. (1937), Economics and Knowledge, *Economica*, vol. IV, no. 13, p. 96 – 105.

Hayek F. (1945), The Use of Knowledge in Society, *American Economic Review*, vol. 35, p. 519-530.

Hayek F. (1946), The meaning of competition, *in Individualism and Economic Order*, Routledge, London.

Hayek F. 1952, *The Sensory Order. An Inquiry into the Foundations of Theoretical Psychology*, Routledge & Kegan Paul, London.

Hutchins E. (1995), *Cognition in the Wild*, The MIT Press, Cambridge MA.

Jensen M. and Meckling J. (1976), "Theory of the Firm: managerial behaviour, agency cost, and capital structure", *Journal of Financial Economics*, vol. 3, p. 305-360.

Kirzner I. (1979), *Perception, Opportunity, Profit*, Chicago University Press, Chicago.

Knight F. H. (1921), *Risk, Uncertainty and Profit*, Chicago University Press, Chicago.

Kogut B. and Zander U. (1996), "What Firms Do? Coordination, Identity, and Learning", *Organization Science*, vol. 7, p. 502-518.

Lamberg A. and Maskell P. (2001), *The Elusive concept of Localization Economies*, AAA annual Conference, New York, February.
Langlois R. (1983), System Theory, Knowledge and the Social Science, in Machlup F. and Mansfield U. (eds.), *The study of information*, Wiley, New York.
Langlois R. (1995), Opportunities Seized and Missed in the History of the Computer Industry, in Raghu G., Praveen N. and Shapira Z. (eds.), *Technological Entrepreneurship: Oversights and foresights*, Cambridge University Press, Cambridge Mass.
Langlois R. and Foss J. (1999), Capabilities and Governance: the Rebirth of Production in the Theory of Economic Organization, *Kyklos*, vol. 52, no. 2, p. 201-18.
Lave J. and Wenger, E. (1991), *Situated Learning: Legitimate Peripheral Participation*, Cambridge University Press, Cambridge.
Loasby B. J. (1999), Industrial District as Knowledge Communities, Bellet M. and L'Harmet, C. (eds.) *Industry, Space and Competition*, Edward Elgar, Cheltenham.
Loasby B.J. (1991), *Equilibrium and Evolution*, Manchester University Press, Manchester.
Lorenzen M. and Foss N. (2002), Cognitive, coordination, Institutions, and Clusters: an exploratory Discussion, in, Brenner T. and Fornahl D. (eds.), *Cooperation, Networks and Institutions in Regional Innovation Systems*, Edward Elgar, Cheltenham.
Mahoney J. and Pandian R. (1992), "The Resource-based View within the Conversation of Strategic Management", *Strategic Management Journal*, vol. 13, p. 111-126.
Maillat D. (1998), From the industrial district to the innovative milieu: Contribution to an analysis of territorialized productive organisations, *Recherches Economiques de Louvain*, vol. 64, no. 1, p. 111-129.
March J.G. and Simon H. (1958), *Organizations*, Wiley, New York.
Marshall A. (1890), *Principles of Economics*, Macmillan, London.
Montgomery C. (1995), (ed.), *Resource-based and Evolutionary Theories of the Firm: Towards a Synthesis*, Kluwer Academic Publishers, London.
Mutti A. (1998), *Capitale sociale*, Il Mulino, Bologna.
Nelson R. and Winter S. (1982), *An Evolutionary Theory of Economic Change*, The Bellknap Press of Harvard University Press, Cambridge (Mass).
Nelson R. (1980), Production sets, Technological Knowledge and R&D, *American Economic Review*, vol. 70, no. 2, p. 62-67.
Nonaka I. and Takeuchi H. (1995), *The Knowledge Creating Company*, Oxford University Press, Oxford.
Nonaka I., Toyama R. and Nagata A. (2000), "A Firm as a Knowledge-creating entity: A New Perspective on the Theory of the Firm", *Industrial and Corporate Change*, vol. 9, no. 1, p. 1-21.
North D. (1990), *Institutions, Institutional Change and Economic Performance,* Cambridge University Press, Cambridge.
North D. (1994), Economic Performance Through Time. *American Economic Review,* vol. 84, no. 3, p. 359-68.
Penrose E. (1959), *The Theory of the growth of the Firm*, second ed. 1995, Basil Blackwell, Oxford.
Piaget J. (1969), *The Mechanism of Perception*, Routledge & Kegan Paul, London.
Prahalad C. and Hamel, G. (1990), "The Core Competence of the Corporation", *Harvard Business Review*, vol. 68, no. 3, p. 79-91.
Polanyi M. (1962), *Personal Knowledge*, Harper & Row, New York.

Piore M. and Sabel C. (1983), *The Second Industrial Divide*, Basic Books, New York.

Richardson G. (1972), The organization of industry, *Economic Journal*, vol. 82, p. 883-896

Richardson G. (1998), *Production, Planning and Prices*, mimeo, Oxford.

Rizzello S. and Turvani M. (2000), *Institutions Meet Mind: the Way Out of an Impasse*, Constitutional Political Economy, vol. 11, no. 2, p. 165-180.

Rullani E. (1998), Networks and Internationalization: Managing Complexity through Knowledge, in, Zan L., Zambon S. and Pettigrew A. (eds.), *Perspectives on strategic change*, Kluwer Academic Publisher, Dordrecht, p. 107-42.

Senge P. (1990), *The Fifth Discipline: the Age and Practice of the Learning Organization*, Century Business, London.

Simon H. A. (1986) "Theories of bounded rationality", in, McGuire C. and Radner R. (eds.) *Decision and organization: a volume in honor of Jacob Marschak*, University of Minnesota Press, Minneapolis.

Smith A. (1795), *Essays in Philosophical Subjects*, reprinted 1980, W. Wightman (ed.) Oxford University Press, Oxford.

Storper M. (1997), *The Regional World*, Guilford, New York.

Teece, D., Pierce L. and Boerner C. (2000*), Richard Nelson's Dynamic Capabilities, paper presented to the Conference for Honoring Nelson,* New York.

Turvani M. (1995), The core of the firm: the issue of the employer-employee relationship, in J. Groenewegen J. (ed.) *Transaction Cost and Beyond*, Kluwer Academic Publisher, Boston, p.189-208.

Turvani M. (1996), Il mestiere delle istituzioni, in, Williamson O.E., *I meccanismi del governo*, Franco Angeli, Milan.

Turvani M. (1998), "Black boxes, grey boxes: the scope of contracts in the theory of the firm", in, Nielsen K. and Johnson B. (eds.), *Evolution of Institutions, Organization and Technology*, Edward Elgar, Cheltenham.

Williamson O.E. (1985), *The Economic Institutions of Capitalism,* Free Press, New York.

Williamson O.E. (1996), *The Mechanisms of Governance,* Oxford University Press, Oxford

Williamson O.E. (1999), *Human Actors and Economic Organization*, mimeo, Haas Business School, Berkeley.

Winter S. (1982), "An Essay on the Theory of Production", in, Hymans S. (ed.), *Economics and the World Around It*, Michigan University Press, Ann Arbor.

Witt U. (1998), *Cognition, Entrepreneurial Conceptions and the Nature of the Firm Reconsidered,* mimeo, Max Planck Institute, Jena.

You J. (1994), Competition and co-operation: toward understanding industrial districts, *Review of Political Economy,* vol. 6, no. 3, p.259-278.

Chapter 9

Paths of local learning and change in vital industrial districts

Marco Bellandi[1]

9.1 Introduction

Vital industrial districts are characterised by, among other complementary conditions, processes of innovation not strictly dependent on strategies of large investment in R&D (Section 9.2). This decentralised industrial creativity is one of the building blocks of a virtuous circle of relations promoting local development. This circle has its limits and threats (Section 9.3), against which mechanisms and agents of possible but uncertain adjustment may be identified (Sections 9.3-9.4). The combination of these conditions enters into the definition of various types of path of local change (Section 9.5). The last section suggests that the circle defines vital industrial districts as a model of learning region. The model may also be applied to the interpretation of some of the learning processes that act in combination within dynamic local economies more or less dissimilar from industrial districts.

[1] This paper supplies a synthetic illustration of views included in Bellandi (1992), (1996), and (2000). The author wishes to thank for their comments Antonio Calafati and the audience at the *Max Plank Institute for research into economic systems International workshop on "The evolution of industrial districts" (Jena, October 8-9, 1999)* where this retrospect was presented.

9.2 Decentralised industrial creativity and external economies of scope

Spaces of competitive advantage for many industrial districts have been defined with respect to specific but not unusual conditions in contemporary markets and technologies. Final markets should be characterised by a highly differentiated and variable demand. Productive processes should combine an important role of know-how embedded in different specialised activities, with the possibility of defining the output of each activity as an economic good for specialised firms. High performance within such spaces manifests itself as the realisation of economies of scope, which are partly external to single firms, but internal to teams of actors working within the industrial district.

The basic element for such economies is that the cost and time of resetting specialised capacities for a differently specified run of production are easily smaller within small firms than within operative units of large firms. The external economies issue from appropriate combinations of such basic versatility. Different production projects can be realised with the help of different and variable teams of small firms specialised in different complementary activities.

The point is that *this same fabric of multiple complementary competencies is the support for the development of different approaches to trade and production problems within the industry*. And the exchange among specialists for trade and production necessities generates continuous opportunities of interaction between those same approaches.

Creative solutions easily spring up from the bottom, especially when they can be fruitfully applied for enlarging the capacity of adaptation and customisation of teams of companies. They are, for example, incremental innovations springing from a relative wealth of original business and technical approaches, interacting and embedded into the life (educational, working, etc.) experiences of large groups of producers and traders: this is what I call decentralised industrial creativity.

The main geographical, cultural, organisational features of the district make easier, in general terms, the formation of large pools of specialised competencies, their interaction, and the introduction of new solutions. Many aspects of this supporting relationship have already been discussed in the literature. Just to recall a couple of basic aspects, local proximity makes face-to-face contacts easier, and they are a suitable means for communicating the tacit aspect of learning embedded in production and trade experiences. Secondly, the stable relations between a population of small specialised firms and a local community of people support the early learning of the "mysteries" of the industry - in Marshall's words. This goes together with the constitution of shared views on the accumulation of human capital and conventions of labour, business, and politics.

The appropriate working of decentralised industrial creativity may be prevented by various specific situations, such as those recalled in the next section focussing on learning and the accumulation of localised productive knowledge in a long run perspective.

9.3 Reproduction of localised knowledge

The long run viability of an industrial district requires, in any given period, that its social and business relations support the interaction of a differentiated pool of know-how embedded into the life experiences of a large group of local producers. But it also requires that both the pool and the supporting relations be reproduced, and possibly regenerated with variations, in the face of new challenges from the external and the internal contexts.

A general framework which assists investigation of this long run issue is suggested by Becattini and Rullani (1996) when they borrow from I. Nonaka the idea of a circle of complementary relationships between knowledge embedded in specific contexts and knowledge easily transferable among contexts. According to this idea, new productive knowledge is developed within specific contexts of action, by combining new generic opportunities from transferable knowledge and the contextual operative know-how. More value from new productive knowledge can be obtained by the replication of its use beyond the specific native context. This demands a codifying process which translates parts of contextual knowledge into a transferable form, e.g. information, codes, products, relatively easy to understand and use. Finally, the variety of the pool of transferable knowledge increases, and these permit new opportunities of creative combination with different pools of contextual knowledge - modified themselves as a result of the previous step of interaction.

Considering the circle, Becattini and Rullani (1996) sustain that:

a) even in the face of the strong contemporary tendencies toward globalisation, embedded aspects of productive knowledge keep a necessary function;
b) teams of small companies, growing around the accumulation of embedded aspects of productive knowledge in industrial districts, are not necessarily

weaker in terms of effective creativity than large organisations, even if the latter are stronger in managing the codified aspects of productive knowledge.

A fundamental problem in this model concerns the decreasing returns of a path of local development based on the deepening of productive knowledge within a given set of products. We can refer to the Charles Babbage variation on the third principle of Adam Smith about the division of labour. Practice brings perfection, in the sense that even more detailed vertical specialisation is developed, specific routines are established, machines and automatic devices are introduced: finally, transferable knowledge is substituted for contextual know-how. That is good for cost efficiency, and the viability of an industrial district also needs a certain amount of it.

If the development of new productive knowledge follows only this path, then in the long run highly skilled workers are substituted for others with a lower level of human capital, and labour for machines. The weight of factors not specific to the district - and easily transferable from/to the outside - increases. Unit labour cost falls in correspondence with an enlarged productive capacity, and in this sense there are increasing returns with respect to the scale of the output. However the capacity of a given production process to generate income for local factors decreases. In this last fundamental sense, there are decreasing returns for the local system.

This local endogenous contradiction has been considered in various other ways by many authors: see Asheim (1996) for a discussion. Other difficulties result as by-product of a prolonged period of growth. Local wealth and greater openness to external mobility and cultural exchange bring about a decreasing propensity towards human capital investment in manufacturing and craft-based competencies. The grasp of the ethic of work and independent activity may decrease. Heavy flows of immigration may steeply increase the local supply of cheap but not skilled labour.

Paths of local decreasing returns characterise specifically mass production and standardised demand on stable markets. But the problem can manifest itself in various forms also in the case of markets with variable and differentiated demand: for example, as the hollowing out of the manufacturing capacity of the district, or as the overall down-sizing of the district with the specialisation of its industry to the service of ever smaller niches of high quality products (Sabel and Zeitlin, 1985).

Finally, the long run viability of industrial districts needs the capacity to generate new sets of specialised activities. They are possibly complementary to the old ones, but they should require new or adapted processes of learning from locally widespread production and trade experiences. New opportunities for the constitution of economies of scope and scale are their result. This capacity may be supported by some, not purely random, local processes.

An organic process is based on the same source of decentralised creativity, that is *complementary multiplicity* of activities embedded in the district. From the crowd of small variations, some more important innovations can emerge periodically, as a result of decentralised industrial creativity, especially when the working experiences are supported by technical education and by the access to appropriate codified knowledge. The most important innovations are the focal point for the aggregation of new nuclei of specialisation within the district, which increase the scope of its product/market combinations, enlarge the multiplicity, open new occasions for creative interaction. A typical example is the development, from activities of service and adaptation of industrial machines and machines used in the principal industry, of a localised industry for the production of specialised mechanical instruments. More recent examples concern the development of localised industries for the production of customised software from the activities of software maintenance and adaptation for the principal industry.

9.4 Mechanisms and agents of reaction to radical challenge

A second traditional, but less organic mechanism of long run adaptation is based on the function of multiplicity for generating substitution opportunities. In the presence of the crisis of important parts of the principal industry, a set of other activities, already nested within the district, can grow supported by a not too difficult adaptation or re-training of local factors made redundant by the crisis within the principal industry. The techno-organisational configuration of the principal industry is changed as a consequence, but the economic and social viability of the district is supported.

Jane Jacobs (1970, p. 88) offers a historical example of this type of change referring to Birmingham history: "In the seventeenth and eighteenth centuries, the city had enjoyed a large trade in shoe buckles, but the shoelace put an end to that. A rising button industry had more than compensated for the loss. Some of the button makers used glass decoratively and this had afforded opportunity to makers of bits and pieces of coloured glass who, working from this foothold, had managed to build up a considerable local glass industry."

Mechanisms of adaptation and change need implementation by agents of change. It is easy to think that a decisive function should be played by the action of R&D centres (collective or private, local or multinational) embedded in the district. Though useful, this action has important difficulties. Too much reliance of local producers on practical knowledge and its creative contents can hamper the efforts of contact coming from the R&D centres. Or, instead, technocratic myths can spread in the district and generic codified knowledge is abruptly imposed on production processes, dissipating the local fabric of practical knowledge. Furthermore, embedding can make too easy the local imitation of results of heavy investments in private R&D, reducing the incentives for those investments.

A more traditional but consistent agency in trespassing the borders of a given vertical division of labour is represented by entrepreneurs who have a large capital of relations, internal and external to the district. They should also have the control of an adequate base of financial and organisational capital, for supporting new local productive experiences (Becattini and Rullani, 1996).

The relational capital helps the exchange of knowledge among independent producers. The access to a pool of organisational and financial capital responds to the necessities of the implementation of important local innovations. It is often difficult to obtain the productive and commercial contributions of other agents, when these contributions request, from them, heavy investments whose returns are conditioned by the uncertain success of the entrepreneurial project. The entrepreneur is ready to bet on the quality and success of his/her idea. However he/she is not able, without an expensive effort, to persuade other producers with different experiences. It follows that independent producers will not be inclined to make the necessary investments. Centralisation helps in these regards (Silver, 1984).

This internal solution also brings about a set of possible problems. First of all, an industrial district, at a certain moment, may lack leading entrepreneurs. Secondly, if centralisation is realised by means of an extensive integration of activities (under one or a few large companies, possibly external giants with their pools of managerial skills), a large part of the previous experience of widespread entrepreneurship and market relations can be wiped out from the district. This situation would tend, even after the success of the innovative idea, to hold up the entry of independent specialised suppliers, the entry of newcomers competing with the incumbent firm of the systemic entrepreneur, and an appropriate re-constitution of market rules.

However, other forms of organisational centralisation exist, that allow the coexistence of an appropriate degree of systemic orientation of investments and the survival, within the endowment of factors of the production system, of a rich enough fabric of experience in market relations (see for example Dei Ottati, 1996). One such form is represented by *business leaders*, at the helm of a few more structured firms working within the district, who are able to support partially systemic projects, but who also break the ice for a larger population of competitive and innovative followers. Another form is *community leadership*, constituted by a set of inter-linked Unions, Business Associations, academic and political leaders, aiming at the constitution of collective real services and training centres. This form is productive if the risk of transforming the industrial district into a local system dependent on public flows of finance can be avoided.

9.5 Paths of local change

The industrial and social configuration of a district tends to change as a consequence of the introduction of new sets of specialised activities in which local productive knowledge can fruitfully accumulate. This happens even when the old sets are maintaining a strong central position within the district; but it is especially the case after periods characterised by substitutive multiplicity.

For example, the joint effect on the sector profile of the district can be illustrated by the following three paths of change.

- A first possible path is characterised by the preservation of a strong local set of complementary production and trade activities, possibly with an increasing importance of non manufacturing productive competencies (Sforzi, 1997), and with this of an adapted or even entirely new sector identity. That is the district form and its internal logic are preserved, as in the story recalled in the previous section.
- A second possible path is characterised by a growing local complexity which, going beyond certain thresholds, brings about the separation of the fabric of social and business relationships into a diversified set of sector fields, connected perhaps by the reference to a common metropolitan basis of urban and service activities. The apparent loss of a strong local sector identity, and the transition towards an urban metropolitan form, is the result. This second path can be easily exemplified with reference to old European industrial districts, like Lyon in France, Sheffield and Birmingham in the U.K., etc.
- The third path is more hypothetical: it is the so-called post-Fordist transition of industrial districts. The sector identity, that is a strong fabric of complementary production activities, is lost, but a strong local capacity of creative interaction is preserved thanks to a shared set of pro-active values, life experiences, languages (Rullani, 1995).

The second and the third type show possible paths of exit from the district form. Other aspects of paths of exit can be suggested by some reflections in Section 9.4 above, particularly those related to the possible persistence of centralisation after a period of deep adjustment.

9.6 Some conclusions: Industrial districts and learning regions

The local logic of industrial districts, during periods in which they are able to preserve or possibly re-define a clear and viable sector identity, suggests a model of dynamic district, featured by its strong core of complementary production activities, within the principal industry and in related activities. This core systematically supports the interaction of local know-how, codified productive knowledge, and resources of creativity - say cognitive interaction - both inside the core itself and more generally throughout the local system (see figure 1).

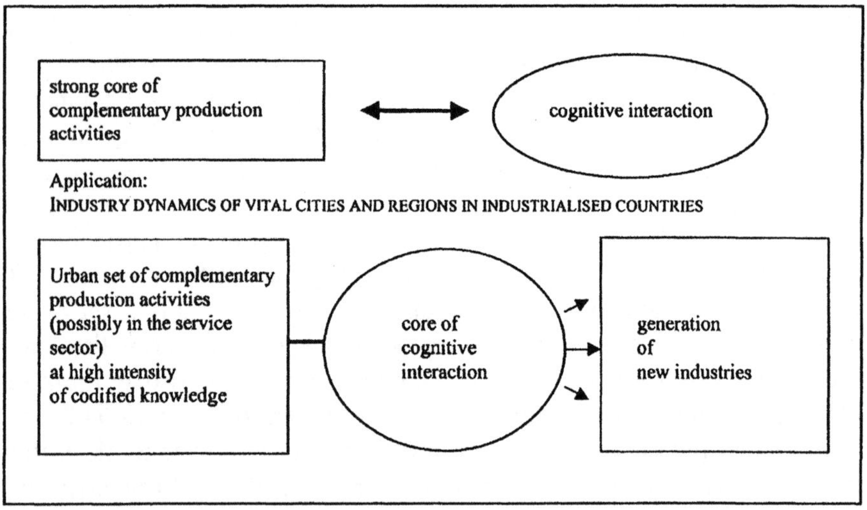

Figure 1: The model of the dynamic industrial district as a theory of learning region

According to this view, the model of the dynamic district is a theory of a "creative milieu" (Becattini, 1991), or of a "learning region", that is of a territory in which learning processes are embedded into the network of relations which identifies the system itself (Asheim, 1996).

There are of course other theories of learning region: for example, the technological district, the regional innovation system, and the region moved by an innovative milieu (Antonelli 1995; Cooke 1996; Maillat, 1996). In these other theories the capacity to foster innovation, especially systemic innovation, seems to be related to a large range of different sector experiences which are put together.

For example, considering the theory of the innovative milieu, the core of the local change process is a set of innovators. Their different, possibly global, business and technical experiences combine with shared social experiences in a common region. Such similar and common experiences support cognitive interaction and entrepreneurial action in the same region. When, as happens in large dynamic cities or industrial regions, a set of different industries is available within the region itself, then the formation of innovative *milieux* can be easier.

It is worth considering the application of the district model of learning region to the interpretation of a rich dynamic city and its metropolitan industrial area, if that city is defined as a place where a great local stock of embedded activities grant the local agents great opportunities of adding new complementary activities and creating new division of labour (Jacobs, 1970; p. 58). Is it possible to identify there a strong core of complementary activities systematically supporting internal cognitive interaction, and orientating similar relations out of the core, but inside the territorial relations of the city? Quite simply, the core can be identified with (some parts of) the dominant urban and services industries, when they are characterised by an intensive use of complementary codified knowledge (Pozzana, 1994; p. 86).

Accordingly, the manufacturing sector's diversification of the city would look more like a joint result of cognitive interactions supported by the knowledge intensive production core, than a decisive factor of the city's dynamics. A different position is presented by Glaeser *et al.* (1992).

The application of the district model to cases of big cities, or by extension to big industrial regions (in industrialised countries) implies a strong level of abstraction, and the theories mentioned above certainly play a primary role. And even some aspects of the dynamics of industrial districts may be fruitfully considered from "non district" points of view.

References

Antonelli C. (1995), *The Economics of Localised Technological Change and Industrial Dynamics*, Kluwer, Dordrecht.

Asheim B. (1996), "Industrial districts as 'learning regions'. A condition for prosperity", *European Planning Studies,* no. 4, p. 379-400.

Becattini G. (1991), "The Industrial District as a Creative Milieu", in, Benko, G., Dunford, Mick, eds. *Industrial change and regional development: The transformation of new industrial spaces,* Pinter, Belhaven Press, London and New York.

Becattini G. and Rullani E. (1996), "Local Systems and Global Connections: The Role of Knowledge", in F. Cossentino, F. Pyke, W. Sengenberger (eds.), *Local and Regional Response to Global Pressure: The Case of Italy and its Industrial Districts*, International Institute for Labour Studies, Geneva.

Bellandi M. (1992), "The Incentives to Decentralized Industrial Creativity in Local Systems of Small Firms", *Revue d'Economie Industrielle,* no.59, 1er trimestre, p. 99-110.

Bellandi M. (1996), "Innovation and Change in the Marshallian Industrial District", *European Economic Planning,* no. 3, p. 357-368.

Bellandi M. (2000), "Paths of an Evolving City: Alternatives and Relevance in Local Policy", *European Planning Studies*, no. 6, vol. 8, p. 709-808.

Cooke P. (1996), "Regional Innovations Systems: An Evolutionary Approach", in Braczyk H., Cooke P. and Heidenreich R. (eds.), *Regional Innovation Systems*, Josey-Bass, Chicago.

Dei Ottati G. (1996), "The remarkable resilience of industrial districts of Tuscany", in Cossentino F. and Sengenberger W. (eds.), *Local and regional response to global pressure: The case of Italy and its industrial districts,* Research Series, International Institute for Labour Studies Geneva.

Glaeser E.L., Kallal H.D., Scheinkman J.A. and Shleifer A. (1992), "Growth in Cities", *Journal of Political Economy,* no. 6, p. 1126-1152.

Jacobs J. (1970), *The Economy of Cities*, Vintage Books, New York.

Maillat D. (1998), "From the Industrial District to the Innovative Milieu: Contribution to an Analysis of Territorialised Productive Organizations", *Recherches économiques de Louvain,* no. 1, p. 111-129.

Pozzana G. (1994), "Distretti industriali, città, sistemi metropolitani", in Bellandi M. and Russo M. (eds.), *Distretti industriali e cambiamento economico locale*, Rosenberg & Sellier, Turin.

Rullani E. (1995), "Distretti industriali ed economia globale", *Oltre il ponte*, no. 50, p. 5-61.

Sabel C.F. and Zeitlin J. (1985), "Historical Alternatives to Mass Production: Politics, Markets and Technology in Nineteenth-Century Industrialization", *Past and Present*, no. 108, p. 133-76.

Silver M. (1984), *Enterprise and the Scope of the Firm. The Role of Vertical Integration,* Robertson, Oxford.

Sforzi F. (1997), "The Tuscan Model: An Interpretation in Light of Recent Trends", Leonardi R. and Nanetti R. (eds.), *Regional Development in a Modern European Economy. The Case of Tuscany*, Pinter, London.

Storper M. and Walker R. (1989), *The Capitalist Imperative*, Blackwell, New York.

Chapter 10

Social identity and identification processes: enriching the theoretical tools to study Industrial Districts

Lucio Biggiero and Alessia Sammarra

10.1 Introduction: the rationale for dealing with identity and identification

Becattini's (1979, 1989, 1990) "simple" redefinition of the research object from "industry" to "district", that is from a product/market-based definition to a territory-based one, allowed for an enlargement of the scope of economic analysis which has shown complex interactions between social and economic aspects of regional studies, and specifically of industrial districts, in the realm of industry studies. In order to explain such a complexity and its evolutionary character, many scientific disciplines entered the field, with their theoretical tools and empirical methodologies. In a sort of "snowball effect", sociologists, management scholars, geographers and historians developed researches or suggested ideas and hypotheses concerning IDs, besides heterodox industrial economists. Consequently, specialists in the field had to progressively take into account the new perspectives and studies which were emerging in an unending body of literature.

Therefore it is with caution and understanding possible negative reactions of some readers, that we propose other theoretical tools. They belong to a rather different discipline from the ones "traditionally" employed in these studies, especially when carried on by industrial economists, heterodox or orthodox as they may be. The discipline is social psychology (Jones, 1998), and the theoretical tools are derived from Social Identity Theory (SIT), and already partially developed in the field of management studies under the heading of organisational identity and identification processes.

The rationale for such a further enrichment of an already rich field of research is to explain thoroughly those behavioural factors of competitiveness that scholars traditionally recognise as distinctive features of IDs such as trust, co-operative attitude, and citizenship behaviours. Of course they are not the only sources of district competitiveness. Low wages and low prices are often mentioned as sources of competitiveness of Italian districts (Bagella and Becchetti, 2000), although it seems that these factors were mainly relevant in the early stages of districts' development ('70 and '80), and they were limited to low-tech districts (Brusco, 1986; Brusco and Paba, 1997). Other important sources of district competitiveness are also scale and scope economies (Belussi, 1992), agglomeration economies (Bagella and Becchetti, 2000), technological and organisational innovations (Albertini and Pilotti, 1996; Belussi and Gottardi, 2000; Brioschi and Malerba, 1997), and institutional settings (Amin and Thrift, 1994; Albertini and Pilotti, 1996; Belussi and Gottardi, 2000; Breschi and Malerba, 1997). However trust, co-operative attitude, and citizenship behaviours are stimulating a growing interest in recent studies. In particular, trust is viewed as the determinant of low transaction costs within the district (Biggiero, 1999, 2001a; Dei Ottati, 1994). Growth of social capital, of territory value, and processes of collective learning (Keeble and Wilkinson, 1999) are indicated as the engines of district competitiveness (Brusco and Paba, 1997). Solidarity mechanisms (Brusco and Paba, 1997; Brusco and Solinas, 1997), identification and attachment to territory (Bagnasco, 1977; Camagni, 1991; Trigilia, 1989), and some kinds of formal (institutional) and informal (tacit) norms of behaviours (Amin and Thrift, 1994; Goodman and Bamford, 1989; Grabher, 1993; Pyke, Becattini and Sengenberger, 1990; Pyke and Sengenberger, 1992) are also mentioned as distinctive features of IDs.

Although trust is investigated also in industrial organisation approaches (Williamson, 1985, 1993), in sociology (Gambetta, 1988) and in management studies (Lane and Bachmann, 1997; Kramer and Tyler, 1996), and co-operative attitude in game theory (Axelrod, 1984), we believe that a full and thorough understanding of these behavioural factors of competitiveness can be really obtained only by putting social psychology into play. Indeed, the emergence of trust, identity and social norms is a central issue in social psychology. The study of IDs can substantially benefit from taking that perspective into account since there is complementarity rather than juxtaposition with the management and economic approaches to trust (Lane and Bachmann, 1997; Kramer and Tyler, 1996), with neoinstitutionalism (DiMaggio and Powell, 1983; Zucker, 1986) and regional systems of innovation (Braczyk *et al.,* 1997; Cooke *et al.,* 1997;

environment, but are largely the outcome of each firm's strategic choices, dynamism, and entrepreneurship.

However, as the classic perspective came to simplistic and often stereotypical interpretations by neglecting the role of individual firms in favour of considering the district as a macro-entity, a theoretical explanation exclusively centred on the role played by individual firms, mostly presented as leading actors or strategic centres, risks falling in the opposite reductionism, which fails to recognise the existence of the district as something more than just a sum of individual firms or, at most, one of the production networks led by few focal firms. When we choose the district as the research object, then that one is the whole, that is the macro-system unity under investigation (Biggiero, 1999). Inside districts can be identified elementary (micro-units of analysis, which are the individuals (workers, entrepreneurs, etc.), and also a meso level, which, in this case, is represented by single organisations (firms, associations, consortia, etc.).

A deeper understanding can only be achieved through a bifocal analysis capable of linking the role of the local context with the behaviours of the single firms which contribute, with their actions, to shape the district environment and to generate the cultural, social and knowledge resource-base of the territory. In this vein, simplistic interpretations risk appearing as the classic perspective, exclusively centred on the analysis of the district as a whole. They forget the individual firms that actually form and foster the district. The interpretation centred on the examination of focal firms and their ability to manage exchange relationships neglects the role of the district as repository of the intangible resources, generated from the idiosyncratic interaction between economic actors and their local environment.

In order to reconcile the duality of these interpretations, we need to explore further the behavioural and cognitive dimensions of agglomeration processes that lead to both the formation and maintenance of IDs and to other forms of local production networks. In this perspective, some authors claim that the existence of localised production networks lies in the potentiality to reduce inter-firm co-ordination costs, and to enhance co-operation, knowledge transfer, and innovation based on identity creation and participation at the local level (Storper 1993; Rullani, 1997). Sammarra and Biggiero (2001a) developed this suggestion, introducing the concept of *district identity as the missing link between the classic perspective, based on the macro-level analysis focused on the district as a whole, and a meso-founded approach which takes the single firms as units of observation*. We argue that district effectiveness in enhancing co-operation, co-ordination, knowledge transfer, and innovation depends mostly on how firms perceive and evaluate their space and how they evaluate their belonging to the local territory. Furthermore, we claim that identification with the district involves qualitative changes in the way firms interact that contribute to enhance the reservoir of social knowledge available within the network.

10.3 The concept of district identity

SIT (Tajfel, 1978, 1981; Tajfel and Turner, 1979; Turner, 1982, 1985; Turner *et al.*, 1987; Abrams and Hogg, 1988) and its recent applications in the field of management and organisations (Albert and Whetten, 1985; Ashforth and Mael, 1989, 1996; Dutton *et al.*, 1994; Peteraf and Shanley, 1997) provide a fundamental contribution which allows the concept of identity and identification to be extended to the study of IDs (Sammarra and Biggiero, 2001a).

Paralleling Albert and Whetten's definition (1985), *district identity* represents a cognitive construction of what the firms perceive *as the set of central, distinctive, and enduring characteristics of the district* (Sammarra and Biggiero, 2001a). *Central attributes* refer to common features among district members in terms of formal and structural aspects, such as the firms' size and ownership structure; in terms of relational features, such as overlapping social networks and communication patterns; and in terms of informal and symbolic features, such as institutional and historical background and social habits. *Distinctive attributes* refer to those features that emphasise similarities among district members and differences in relation to outsiders. For instance, geographical localisation may become a salient component of district identity, making it possible to distinguish district members from similar organisations, such as other small and medium firms specialised in the same economic sector, but located in a different territory. Finally, *enduring attributes* refer to those features that characterise the district with some degree of temporal stability and continuity. The emergence of a district identity presupposes a cognitive effect, the recognition that the district exists as a by-product of the actors' attempt to scan, and make sense of, their competitive and social environment. It also provides an evaluative effect based on the appreciation of the district's attributes, upon which a collective identity may be built.

As follows from the above definition, two main theoretical implications are involved in the concept of district identity. The first concerns the consideration of the firms as cognitive actors; the second refers to the eventual qualification of the firms' cognitive abilities in terms of categorisation processes (Sammarra, 2000).

With respect to the first implication, two approaches can be taken, both based on the idea that organisations are cognitive systems, and that they can be considered as collective minds expressing intentions and looking for identities. The two approaches differ essentially – but not only - on the methodological side, because of the difficulty to carry on empirical researches which treat and investigate organisations as collective minds. The first one holds that it is possible, although difficult, to understand and describe organisational identity by searching among formal and informal communication flows, artefacts and symbols. Ashforth and Mael (1996) argue the existence of organisational identity as the "character" or the stylised narratives about the "soul" or essence of the organisation, advances by stakeholders as claims about meaning (p. 21). The notion of "character" is necessarily inclusive and fuzzy, encompassing aspects of mission, values, ideology and beliefs, norms, competencies, and customary ways of doing things" (p. 23). "Thus, in a real sense, organisational identities exist because they *must*" (p. 21). Authors looking at organisations as cognitive systems and sense makers share the same perspective (Gioia and Sims, 1986; Weick and Bougon, 1986; Weick and Roberts, 1993; Weick, 1995; Biggiero, 2000). There are many ways to carry out such investigations, they are very time consuming, and usually involve ethnographic methodologies. In any case, these kind of researches are still at an early stage in the field of organisation studies.

The second approach tries to be more efficient by simplifying the investigation. This means that, under some assumptions and circumstances based on the low size and low complexity of the organisation, its mind can be equalised to that of its representative. According to this simplified approach, whenever we refer to organisations as cognitive systems, we actually consider cognition by a firm to mean cognition by entrepreneurs, individual top managers, or by top management teams. Indeed, in the latter two cases, the issue of how individual identities "merge" in a single one is posed again, and leads to the same problem and assumption of a collective mind and identity. This is, for instance, the line followed by Peteraf and Shanley (1997) in applying SIT to strategic groups.

Both approaches are juxtaposed to methodological individualism and its application in sociology and economics, even though the former is more radical and explicit in sustaining the existence of collective minds and cognitive systems other than individuals. Since we also adopted the simplified approach in our empirical research (Sammarra and Biggiero 2001b; Biggiero and Sammarra, 2002), in the rest of this essay we stick to it, while, in the next section, we discuss some implications of the former (more radical) one for current economics and management theories.

As concerns the hypothesis of firms' cognitive abilities in terms of categorisation processes, several studies in the strategic field show that the managers' mental models of their relevant competitive environment are built through processes of cognitive categorisation (Rosch, 1978; Porac *et al.*, 1989; Porac and Thomas, 1990, 1994; Porac *et al.*, 1995; Borroi *et al.*, 1998). Cognitive categorisation means that, to make sense of their external environment, managers code the similarities and differences between organisations in a summary fashion, by clustering attributes that co-occur across an organisational field into general categories. Firms differ in how well they represent a category ideal, leading to a core-and-periphery structure. Core firms are those which are closer to the category ideal, and they are used as referents in similarity judgements to clarify ambiguity and define category boundaries (Porac *et al.*, 1995; Reger and Huff, 1993). The empirical studies on the Scottish knitwear industry (Porac *et al.*, 1995), then replicated in the Carpi textile-clothing ID (Borroi *et al.*, 1998), have confirmed that managers and entrepreneurs form their own vision of their relevant competitive environment in terms of cognitive categories in order to simplify it.

In this vein, we define the *perceived* district identity as a cognitive category that managers and entrepreneurs in the district form through categorisation processes aimed at defining the district's boundaries and features on the basis of a set of central, distinctive, and enduring attributes which co-occur among the firms in the district. District identity is in this sense a prototype, that is a subjective representation of the defining attributes of a social category - the district in our case - that capture the context-dependent features of group membership, often in the form of representation of exemplary members (Hogg *et al.*, 1995). The cognitive construction of the perceived district identity, in the managers' and entrepreneurs' minds, has important consequences for the firms' behaviour and strategic choices. First, the district's boundaries are defined so that it is possible to classify other organisations as in-group and out-group members, that is to distinguish firms that belong to the district (the in-group) from the external firms (the out-group). Second, as a consequence of the categorisation process, the behaviour of organisations perceived as group-members becomes more predictable. However, the focus on a selected set of central and distinctive attributes suggests that there will neither be unanimity among observers regarding the group's membership, nor unanimity of expectations about the members' behaviours.

A further clarification is necessary to refine the concept of district identity. District identity can be analysed at firm and collective levels. At firm level, we use the term of *perceived* district identity to underline its subjective dimension. The perceived district identity represents a sort of cognitive map that each firm in the district forms to make sense of its relevant environment. The set of attributes chosen to define the district identity is not objectively given: each firm selects its own set, according to the subjective categorisation processes followed by the entrepreneur or management team. This means that the selected set of attributes can differ among firms and, consequently, that different firms can hold different perceived district identities, that is different views about what is central, distinctive, and enduring about the district they belong to. Building from this definition of perceived district identity, we posit that, at collective level, it is possible to have multiple district identities when different conceptualisation exists regarding what is central, distinctive, and enduring about the district. As Albert and Whetten (1985) pointed out even within a single organisation, it is possible that multiple identities coexist. If all the members and units in the organisation share a common view about what is central, distinctive, and enduring about the organisation, then, at collective level, the organisation has a mono identity. If the members conceptualise differently the organisation they belong to, or, stated in other words, if they have different perceived organisational identities, then, at collective level, the organisation has a multiple identity. By the same token, we state that the *collective* identity of the district derives from, and is composed of, the perceived district identities held by the single firms, and that, at collective level, there can be a mono identity or multiple identities of the district. The first case occurs when the perceived district identities held by the single firms are consistent and coalesce, at collective level, into a unique and shared district identity. The second case happens when the beliefs embedded in the perceived district identities are many and different, and form a multiple and differentiated district identity at the collective level. At first sight, this seems the case of IDs, where trade associations, trade unions, public institutions, small firms, and large companies are likely to attain and perceive (partially) different identities. In the long run, a reciprocal feed-back mechanism is likely to emerge between district identity at firm and collective level, according to a circular process of influence: the collective identity of the district is forged by the process of convergence or divergence among the perceived district identities held by the single firms and, in turn, the perceived district identities held by the single firms evolve and change under the influence of the collective district identity.

In a dynamic process, considering the peculiar structural features of IDs, it might reasonably be argued that the perceived district identities held by the single firms are likely to converge into a widespread and uniform collective identity. The reason is that physical proximity and a high degree of network density, due to the division of labour among district firms and also to the existence of overlapping social ties outside of the firms, facilitate the enactment of social influence based on formal and informal communication within the district. Moreover, other mechanisms of social contagion are related to work flows, which are usually contained within the districts' borders: both entrepreneurs and workers normally live in the same community where they work; frequently also where they have grown up and it is likely that they have been trained in the same schools and that they have gone through common previous work experiences in the same firms. Even the specialisation in the same economic sector can contribute, in addition to the other factors so far discussed, to promote the convergence of the perceived district identities held by the single firms towards the formation at collective level of a shared and consistent district identity.

However, it is also possible that the perceived district identities held by the single firms evolve over time according to a differentiating process under the pressure of divergent forces, such as the entry of new actors within the district coming from different sectors or regions, which may hold very different cognitive maps concerning how to compete in the district. Also the diversification into new economic specialisation or the evolution towards *pluricultural districts*, that are districts specialised in more than one economic sector, can prevent the process of convergence towards a unique and shared district identity at the collective level, because the presence of different entrepreneurial and cognitive maps reduces the potentiality to construct cognitive categories based on a similar conceptualisation of the district.

10.4 Can districts have identities?

The concept of district identity introduced in the previous section implies the following hypotheses:

1. the district and its members hold their own specific identities;
2. the attributes of such identities are comparable;
3. the subjective perceptions of single district members do really matter.

The first hypothesis, which is also the most complex, reminds to an old dispute within social sciences. It could be synthesised in this way: are cognitive traits like intentions, decisions, preferences, and thus identities, attributes limited to individuals, or can they be extended to groups of individuals, like organisations and, consequently, to groups of organisations like networks and districts? Following methodological individualism (Donzelli, 1986; Hayek, 1976; Rothbard, 1979), neo-classical economic theory (Schmalensee and Willig, 1989), and partly also sociology (Coleman, 1990), answer negatively. Therefore they deny that identity can be assigned to an organisation and, thus, to an inter-organisational entity such as a district. According to this perspective, firms have only one "intention": profit maximisation. In the more extensive (because comprehending also non profit organisation) and recent version of that paradigm – the agency cost theory - the intention is to maximise expected subjective utility of the principal and the agents (Jensen and Meckling, 1976; Fama, 1980). Identity and other attributes are just factors which contribute to the maximisation of the utility function (Akerlof and Kranton, 2000). Identity-related behaviours are motivated by the payoffs people can derive from their own actions and from others' actions. In a more or less explicit way, identity and identification processes are variables in the utility or production function, reflecting the well known – and also extensively criticised (Simon, 1983; Biggiero and Laise, 1998) - framework of neo-classical economics.

Management, organisation science, and social psychology hold the opposite viewpoint: beyond the rejection of profit (or utility) maximisation as the sole goal, organisations are seen as cognitive systems (Tenbrunsel *et al.,* 1996), which have an identity (Albert and Whetten, 1985; Ashforth and Mael, 1989, 1996; Kogut and Zander, 1996; Whetten and Godfrey, 1998). In this view, Kogut and Zander state that "what makes a firm's boundaries distinctive is that the rules of co-ordination and the process of learning are situated not only physically in locality, but also mentally in an identity" (1996; p. 515).

Cybernetics (Biggiero, 1999, 2001b; Foerster, 1982; Maturana and Varela, 1980, 1987), theories of complexity (Biggiero, 2000, 2001a; Casti, 1994) and the developments of sociology (Eve *et al.,* 1997; Luhmann, 1984, 1990) and of management and organisation sciences, which explicitly apply complexity theories to human organisations (Biggiero, 2000; *Organization*, 1998; *Organization Science*, 1998), share this second perspective. They redefine the concept of emergent properties of complex systems, which is inherently incompatible with methodological individualism. Indeed, if there were no emergent properties, that is if a system (an organisation) was simply a sum of single parts (individuals and things), there would not be any reason to consider it as a unity, as a whole system (Flood and Carson, 1988; Klir, 1991). Besides complexity deriving from computational problems, its behaviour would be predictable and understandable by knowing that of its parts. There would not be any need for the holistic (systemic) view, and, thus, any stance for collective mind or system's cognition. Even if we cannot deeply address this issue here, we want to mention briefly three new arguments in favour of emergent properties: 1) largely unpredictable effects of interactions between the parts (individuals, groups and things) of a system (Biggiero 1997); 2) impossibility to perfectly (or sufficiently) know the initial state and the dynamics of a system and of its parts (Biggiero 2001b; Peitgen *et al.,* 1992); 3) feedback effects between the system and its parts (Flood and Carson, 1988; Klir, 1991; Lange, 1965).

With our explicit reference to organisational identity and to social psychology, we argue that it is meaningful to assign identity and intentions to an organisation or a district, hence considering them as cognitive systems (Gioia and Sims, 1986; Luhmann, 1984, 1990; Teubrunsel *et al.,* 1996) and sense makers (Weick, 1995). The view of organisations as neural networks allows merging of the social networks approach and the constructivist approach: organisations are at the same time computing and cognitive systems (Biggiero, 2000). In fact, organisations can be seen as cognitive maps, that is as cognitive networks. Further and consistently, according to a social constructivist perspective, 'if we treat interrelations as a variable and interrelating as a process, this suggests a way to conceptualise collective mind. Dispositions toward heed are expressed in actions that construct interrelating. Contributing, representing, and subordinating, actions that form a distinct pattern external to any given individual, become the medium through which collective mind is manifest' (Weick and Roberts, 1993; p. 364). 'The collective mind is "located" in the process of interrelating..., and is manifest when individuals construct mutually shared fields' (Weick and Roberts, 1993; p. 365).

In this perspective, the most interesting issue becomes a purely empirical one: how can we define and measure inter-organisational (or district) identity and identification? In the case of a self-employer the solution is obvious. In the case of a small artisan organisation, one can reasonably argue that - given the dominant role of the entrepreneur as decision maker and, thus, sense maker - the identity and the perceptions of the entrepreneur well represent those of his/her organisation. However, in the case of a larger organisation, more complex analytical and methodological tools should be built. Especially for large organisations we suggest two other solutions, supposedly more effective, but also more complex and time consuming, and thus, more problematic to use for extensive empirical research. They consist of deriving organisational identity from the analysis of external and internal communication, which take place through written documents and public sentences, and through artefacts and symbols (Gagliardi, 1990; Monge and Contractor, 2000). Of course, such an analysis should take into account formal and informal (emergent) communication, as well as intentional and unintentional meanings and messages. Studies in the field of organisational ethnography, narratives, symbolic interactionism have developed many methodologies in this direction. The challenge is to apply them to the study of identity and identification processes between (groups of) organisations, although it is clear that this kind of approach would require longitudinal studies and is more difficult to adapt to extensive researches.

However, since this field is at an early stage of development, and our case study is fortunately concerned almost exclusively with small firms, we have neglected these methodological problems, leaving them for a future agenda.

The third hypothesis argues that what matters is the subjective perception that each organisation has regarding its own and the district's identity, and its subjective comparison in terms of similarity, attractiveness, etc., regardless of the realism of such perceptions to the eyes of external observers (e.g.: researchers). In this regard Ashforth and Mael (1996) state that 'Identity is socially constructed – and as an intersubjective phenomenon, the perceived organisational identity may bear little relation to the "objective" or impartially measured organisational identity' (p. 23). For instance, a district member can erroneously perceive that, in comparison with other communities or/and districts, one of the salient attributes of the district identity is a high level of social solidarity. Let suppose that this perception of a higher level of social solidarity is not supported by a careful analysis carried on by an external observer on the basis of some indicators (parameters) of the level of social solidarity within that district. This result would not change the fact that the district member will continue to avoid opportunistic behaviour towards other members, as long as his/her positive perception of the district identity is maintained. Furthermore, it could be argued that, if this perception of the district persists over time in its members' minds, the effect of self-fulfilling prophecies could generate the "real" phenomenon that is it, stimulating the growth of social solidarity within the district.

10.5 Identification with the district: concept and antecedents

At the individual level, social identification is the perception of oneness with or belonging to some human aggregate (Tajfel, 1982). Identification is an extreme form of relational modelling that allows the individual to define his/her identity in relation to the characteristics of perceived social groups. This means that the identification with a social group, such as a political party, a religious group, or the organisation one works for, is essentially a cognitive process, which leads the individual to define and describe himself/herself on the basis of the defining attributes of that social category. The concept of social identification has recently been applied to work organisations (Ashforth and Mael, 1989, 1996) and to inter-organisational groups such as strategic groups (Peteraf and Shanley, 1997).

In their study on organisational identification, Dutton *et al.* (1994) already pointed out that the identification is not a binary variable and that what is important is the degree of identification. The authors define the degree of identification with the organisation in relation to the degree of similarity or overlap between the perceived organisational identity and individual identity. Drawing upon these recent studies, we define social identification with the district as a cognitive process (figure 1). Namely we posit that an organisation (firm or institution) identifies with the district when it defines its own organisational identity, selecting the same attributes that it views as central, distinctive, and enduring about the district. The higher the overlap and correspondence between the perceived district identity, and the organisational identity, the higher the degree of identification with the district (Sammarra, 2000).

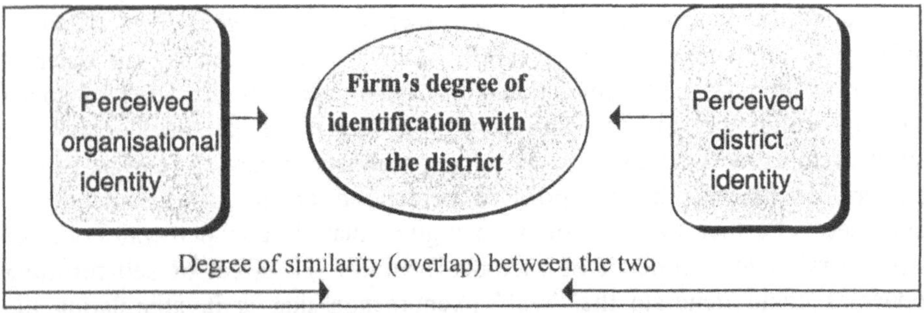

Figure 1: Identification with the district

The possibility to measure the degree of identification is a genuine empirical issue, which has been faced by different authors at various levels of analysis. Therefore, it assumes specific characters when referred to the identification between an individual and a group, between an individual and an organisation (Ashforth and Mael, 1989, 1996; Bergami and Bagozzi, 2000), between an individual and a social set (Abrahams and Hogg, 1988), between an organisation and a strategic group (Peteraf and Shanley, 1997), and between an organisation and its inter-organisational network or district (Biggiero and Sammarra, 2001b)[1]. In the last two cases, the challenge is to apply and extend to the field of management and organisations empirical research methodologies largely developed by social psychology, and to enrich and integrate them with those tested by organisation scientists.

To explain the process of social identification with the district, we identify three antecedents: 1) the perceived attractiveness of the district; 2) the degree of perceived similarity among district members; 3) the degree of perceived complementarity/interdependence among district members' (Sammarra, 2000; Sammarra and Biggiero, 2001a).

The first antecedent operates because, according to SIT, social identification is driven by the psychological need of *self-enhancement*. At individual level, this means that people are more likely to identify with prestigious social groups (Ellemers, 1993) and organisations (Mael and Ashforth, 1992) in order to satisfy the basic need to see themselves in a positive light (Tajfel, 1978). In their study on strategic groups' identity, Peteraf and Shanley (1997) argue that this same mechanism applies also to the inter-organisational level. Specifically, the two authors suggest that the presence of high-status firms within the strategic group affects the perceived prestige of the group, and thus reinforces the adoption of mimetic activity, which facilitates the process of identification with the group based on self-enhancement needs. Applying the same explanation to the case of IDs, it follows that the process of identification with the district is mediated by the perceived attractiveness of the district. Moreover, it might be reasonably expected that *the higher is the perceived attractiveness of the district, the higher is the firm's identification with the district*.

The concept of perceived attractiveness is composed of two related conceptual dimensions: the attractiveness of the perceived identity and the attractiveness of the perceived external image (Ashforth and Mael, 1989; Dutton *et al.*, 1994). The first refers to the evaluative judgement that any organisation makes of the district (*how the firm perceives the district)*. The perceived attractiveness of the district is higher the more the firm is aware of the existence of unique tangible and intangible resources within the district, such specialised skills, diffused know-how, unique competencies, high status firms. The perceived external image of the district refers to what the firm believes is the evaluative judgement that others make of the district *(how the firm believes that the district is perceived by others)*. The perceived external image of the district

[1] The operationalisation of the level of identification with the district is more extensively discussed in Sammarra and Biggiero (2001b).

is higher, the more the firm believes that the district is positively evaluated by others, that is when the firm believes that the district has a good reputation in the national and international markets, or when it feels that the local population and institutions recognise that the district plays an important role for the local development and well-being.

The second antecedent is based on the fact that, according to the SIT, social identification occurs through self-categorisation. This means that, in order to identify with a social group, it is first necessary to recognise the existence of the group as a social category, which is to put together into the same category elements that have the same characteristics/attributes selected to define the category. This does not imply that all the members of the category must be perceived to be identical, because categorisation and depersonalisation occur only with attributes selected to define the social category (Ashforth and Mael, 1996; Brewer, 1993). However, it might be argued that the higher the similarity among members, the higher the possibility to categorise them, including oneself, into the same social group. On the basis of this logical consideration it is reasonable to expect that the more a firm views itself to be similar to other firms, the easier it is to classify them, including oneself, in the same social category. Consequently, it is likely to expect that *the perceived similarity with other firms of the district positively influences the firm's identification with the district.*

Moving beyond the SIT model, we argue that the self-categorisation and self-enhancement mechanisms are not the only drivers for social identification, because personal interaction and inter-personal influence are also important factors (Biggiero and Sammarra, 2001). Social interaction affects the process of social identification in two different ways. First, as already acknowledged within the SIT framework, constant social interaction facilitates mutual observation and comparison, and consequently it facilitates the perception of existing similarities among group members. Second, social interaction may extend the basis for self-categorisation, highlighting social attributes that are not perceivable in the absence of direct and constant interaction. This second effect shifts the cognitive basis of categorisation from attributes that are easy to observe, such as race in the case of individuals, and size in the case of firms, to attributes deriving from social interdependency and similarity, such as shared goals and mutual needs. Furthermore, we argue that people are motivated to form social groups, and eventually to identify with them, not only when they perceive they are similar to others, but also when they feel they have common objectives and goals with others, and when they need to rely on them in order to accomplish salient objectives or activities. This third drive for social identification occurs not only within social organisations, but also within IDs. One of the structural features of

IDs is the high level of vertical specialisation and inter-firm division of labour, which implies the existence of strong vertical interdependencies among the district members. Moreover, direct social interaction plays a primary role among district members and cannot be considered a secondary or marginal factor when explaining the emergence of group phenomena. According to these considerations, it is reasonable to expect that *the higher the perception of existing complementarities and interdependencies with other firms in the district, the higher the firm's identification with the district.*

Table 1: Factors facilitating firms' identification with the district

FACTORS ENHANCING FIRMS' IDENTIFICATION WITH THE DISTRICT	
Attractiveness of the district	Attractiveness of the perceived district identity ▪ Presence of high status firms ▪ Presence of unique tangible and intangible resources (infrastructures, natural resources, specialised skills, competencies, know-how)
	Attractiveness of the district image ▪ Market reputation ▪ External legitimisation (administrative and governmental policies)
Similarity among district members	▪ Firms' structural characteristics ▪ Adopted technologies ▪ Entrepreneurial background
Interdependence among district members	▪ Division of labour and specialisation ▪ Shared goals and needs

10.6 Identification with the district: consequences for the firms' behaviour and the district's competitiveness

According to the SIT, social identification leads the individual to adopt pro-social attitudes towards other group members as an affirmation of the common social identity. In the SIT framework many group phenomena such as internal cohesion, co-operation, solidarity, and positive evaluation are explained as a consequences of identification with a social group. Therefore, the link between identification and organisational behaviour is not new in organisation theory. March and Simon already acknowledged that when the individual identifies with the organisation "he acquires an attachment or loyalty to the organisation that automatically - i.e. without the necessity for external stimuli - guarantees that his decisions will be consistent with the organisation objectives" (1958: 278). The SIT explains this behavioural effect as a consequence of the self-categorisation and stereotyping mechanism implied in social identification, which enhances the perceived similarities with other group members and the individual's attitude to conform to the norms and values of the group. This means that, as a consequence of the psychological need for consistency, organisational identification motivates individuals to align and conform their own behaviours and beliefs to the organisation's norms.

Kogut and Zander (1996) remark that the development of a shared identity among organisational members lowers the costs of communication and learning, because it establishes explicit and tacit rules of co-ordination, and influences the direction of search and learning. According to the authors, an important finding in experimental psychology is that learning through identifying is more powerful than attempts to "teach" individuals via incentives and propaganda' (1996: 510). Shared identities facilitate what Lave and Wenger (1991) call "situated learning", because identity implies an adherence to a symbolic coding of values and rules, and the formation of convergent expectations. As Kogut and Zander notice, this effect of identity on learning has important implications for understanding the transfer of knowledge within firms' boundaries, but also for explaining 'why geographic proximity, such as Silicon Valley, appears to be associated with rich contextual environments for the spawning of new innovations' (1996; p. 508).

Ashforth and Mael (1996) move further the analysis of the consequences of social identification. The two authors maintain that organisational identification has a positive influence on commitment, because the greater the members' identification, the tighter the perceived association between personal success and organisational success. Furthermore, identification facilitates the adoption of pro-social behaviours towards the other organisational members, enhancing internal cohesion, co-operative attitude, solidarity (Ashforth and Mael, 1989; Dutton *et al.*, 1994), and positive evaluation (Turner, 1982; 1984). Among all the forms of pro-social behaviours, Ashforth and Mael (1996) emphasise the concept of *depersonalised trust*. According to the two authors, when the individual identifies with the organisation he is likely to engage in trusting behaviours towards other organisational members as a result of their common social identity, independently of prior interaction with each other or previous direct knowledge: the process of categorisation in in-group and out-group members alone provides valid and reliable information about other members' behaviour.

The notion of depersonalised trust is very interesting when it is applied to the analysis of inter-organisational relationships, because it explains the emergence of trust even before and/or in the absence of prior direct interactions as a consequence of the process of identification. This means that the emergence of trust is based on belonging to the same group: the common social membership replaces the need of direct inter-personal relationships and/or reputation to reduce uncertainty. In this vein, we can say that the process of de-personalisation implied by social identification turns trust from a source of personal capital into a source of social capital. Personal capital means that trust is based on personal contacts and that inter-organisational relationships are founded on direct knowledge and personal trust among managers and/or entrepreneurs. Through the enactment of identification processes trust is based on the perception of belonging to the same social group, transforming the very concept of trust into social capital of the group. Applying Ashforth and Mael's (1996) concept of depersonalised trust, we argue that the identification with the district allows the firm to perceive the other members of the district as reliable on the basis of group membership, even in the absence of previous direct relationships and/or reputation. According to these considerations, it is reasonable to expect that *the identification with the district will have a positive effect on the firm's attitude to trust other district members.*

According to the SIT, social identification leads to the adoption of co-operative behaviours towards other group members. In their study on strategic group, Peteraf and Shanley (1997) applies this same explanation at the inter-organisational level, maintaining that the identification with the group facilitates the adoption of co-operative strategies with other organisations of the group. According to this argument, we posit that *the identification with the district enhances the firm's co-operative attitudes towards the other district members.* It is important to notice that this hypothesis does not assume that co-operation is context-dependent, and that all the firms of the district automatically adopt a co-operative attitude as a consequence of a shared and homogeneous culture. Different firms can have higher or lower co-operative attitudes according to their own processes of identification, that is, according to their own way of perceiving and evaluating their belonging to the district in terms of self-categorisation processes. This is a micro-founded explanation of the emergence of trust and co-operation within IDs, which is consistent with the idea that district firms are not necessarily homogeneous and that they can adopt different behavioural patterns according to each firm's way of defining and locating itself with respect to the external environment.

To the extent that the firms identify with the district, they become more attached to it, because they feel they have more at stake in the district, developing a stronger interest in continuing their membership. This means that, as a consequence of the identification with the district, firms tend to attribute more importance to the district for their own success and are more inclined to undertake collective actions and promote projects aimed at the growth and strengthening of the district as a whole. According to these considerations we expect that *the higher the firm's identification with the district, the stronger the firm's attachment to the district.*

Table 2: Behavioural effects of firms' identification with the district

CONSEQUENCES OF IDENTIFICATION WITH THE DISTRICT	
Trustfulness	▪ Depersonalised trust ▪ Convergent expectations
Co-operative attitudes	▪ Pro-social attitudes towards the in-group ▪ Adherence to social norms and values
Situated learning	▪ Adherence to symbolic coding ▪ Shared systems for meaning and knowing
Attachment to the district	▪ Collective actions ▪ Preserving membership

According to the SIT, social identification not only modifies the individual's behaviour towards the in-groups, enhancing internal cohesion, co-operation, and collective actions, but also the attitude towards the out-groups, increasing the perceived competition with other groups and favouring a negative and biased evaluation of external members. These negative consequences of social identification on individuals' behaviours and attitudes have been observed in many experimental settings in which it has been proved that the perception of existing social categories and social identification are sufficient motivations to adopt discriminatory behaviours towards the out-group. It is reasonable to expect that the firm's identification with the district is also associated with the emergence of discriminatory behaviours towards the out-group, because firms outside the district are not likely to appear as reliable and skilled as the other firms of the district. These considerations suggest the following proposition:

identification with the district will increase the firm's attitude to adopt discriminatory behaviours towards external members.

10.7 Conclusions

This study presents the challenge to introduce a new theoretical perspective to study IDs: social psychology. The motivation for such an enlargement of an already rich field of research is that social psychology provides the theoretical tools to bridge between the micro (individuals), the meso (firms) and the macro (district) level of analysis. The key concepts proposed in this work are those of district identity and identification processes, which both derive from social identity theory, one of the main streams in social psychology.

We claim that an ID, as each sufficiently stable social group, can have its own identity, that is, the set of shared perceptions about the central, distinctive, and enduring characteristics of the district. Indeed, the existence of a district identity was already more or less explicitly assumed in most of the traditional approaches to IDs, although under theoretical frameworks and languages very far from those of social psychology. However, traditional approaches lack a systematic and clear definition of district identity and of its interrelations with other key aspects, like district strategies and structures, while social psychology provides the theoretical tools to develop these concepts more systematically.

An analogous lack concerns the analysis of the process of social identification, which occurs among members of an inter-organisational group. Building upon theoretical and empirical works on social and organisational identification, this paper explores whereby firms within an ID come to view themselves as members of a particular social entity, and adopt pro-social behaviours towards co-members as a consequence of their identification with the group. The study presents two sets of propositions concerning both antecedents and consequences of social identification at the inter-organisational level. The perceived attractiveness of the district, the degree of perceived similarity among district members, and the degree of perceived complementarity/interdependence among district members' are the antecedents, while firms' attitude to trust and co-operate with other district firms, their attitude to adopt discriminatory behaviours towards external members, and their attachment to the district are the consequences of identification processes.

Although in an embryo stage, we believe that this theoretical approach can significantly improve current understanding of IDs, although substantial work must be devoted, in the future, to develop the empirical foundation of the identity approach to IDs.

References

Abrams D. and Hogg M.A. (1988), Comments on the Motivational Status of Self-esteem in Social Identity and Intergroup Discrimination, *European Journal of Social Psychology*, vol. 18, p. 317-334.

Akerlof G.A. and Kranton R.E. (2000), Economics and Identity, *Quarterly Journal of Economics*, vol. 3, p. 715-753.

Albert S. and Whetten D.A. (1985), Organizational Identity, in L.L. Cummings and Staw B.M. (eds.), *Research in Organizational Behavior*, vol. 7, p. 263-295.

Albertini S. and Pilotti L. (1996), *Reti di reti*, Cedam, Padua.

Amin A. and Thrift N. (1994), (eds.), *Globalization, Institutions and Regional Development in Europe*, Oxford University Press, Oxford.

Ashforth B.E. and Mael F.A. (1989), Social Identity Theory and the Organization, *Academy of Management Review*, vol. 14, p. 20-39.

Ashforth B.E. and Mael F.A (1996), Organizational Identity and Strategy As a Context for the Individual. In L.L. Cummings and Staw B.M. (eds.), *Research in Organizational Behavior,* vol. 13, p. 17-62.

Axelrod R.M. (1984), *The Evolution of Cooperation*, Basic Books, New York.

Bagella M. and Becchetti L. (2000), (eds.), *The competitive advantage of industrial districts. Theoretical and empirical analysis*, Springer, Heidelberg.

Bagnasco A. (1977), *Tre Italie. La problematica territoriale dello sviluppo italiano*, Il Mulino, Bologna.

Becattini G. (1962), *Il concetto d'industria e la teoria del valore*, Boringhieri, Turin.

Becattini G. (1979), Dal "settore" industriale al "distretto" industriale. Alcune considerazioni sull'unità di indagine dell'economia industriale, *Economia e Politica Industriale*, vol. 1, p. 1-79.

Becattini G. (1987), (ed.), *Mercato e Forze Locali: Il Distretto Industriale*, Il Mulino, Bologna.

Becattini G. (1989), Sectors and/or Districts: Some Remarks on the Conceptual Foundation of Industrial Economics, in, E. Goodman and J. Bamford (eds.), *Small Firms and Industrial Districts in Italy*, Routledge, London.

Becattini G. (1990), The Marshallian Industrial District as a Socio-economic Notion, in, F. Pyke, G. Becattini and W. Sengenberger (eds.), *Industrial District and Inter-firm Co-operation in Italy*, ILO, Geneva.

Belussi F. (1992), (ed.), *Nuovi modelli di impresa, gerarchie organizzative e imprese rete,* Franco Angeli, Milan.

Belussi F. and Gottardi G. (2000), (eds.), *Evolutionary Patterns of Local industrial Systems*, Ashgate, Aldershot.

Bergami M. and Bagozzi R.P. (2000), Self-categorization, affective commitment and group self-esteem as distinct aspects of social identity in the organisation, *British Journal of Social Psychology,* vol. 39, p. 555-577.

Biggiero L. (1997), Managerial Action and Observation: A View of Relational Complexity, *Systemica*, vol. 12, p. 23-37.

Biggiero L. (1999), Markets, Hierarchies, Networks, Districts: A Cybernetic Approach, *Human Systems Management*, vol. 18, p. 71-86.

Biggiero L. (2000), Complexity and Organization, in, M. Zeleny (ed.) *The IEBM Handbook of Information Technology in Business,* Thomson, London.

Biggiero L. (2001a), Self-organizing processes in building entrepreneurial networks: A theoretical and empirical investigation, *Human Systems Management,* forthcoming.

Biggiero L. (2001b), Sources of complexity in human systems, *Nonlinear Dynamics, Psychology, and Life Sciences*, vol. 5, no.1, p. 3-19.

Biggiero L. (2002), The location of multinationals in industrial districts: knowledge transfer in biomedicals, *Journal of Technology Transfer*, vol. 27, p. 111-22.

Biggiero L. and Laise D. (1998), Economia politica ed economia aziendale: un confronto inter-teorico, *Economia e Politica Industriale,* vol. 97, p. 79-126.

Biggiero L. and Sammarra A. (2001), Similarity and Complementarity in Inter-organizational Networks, paper accepted at APROS Conference, December 3-5, Hong Kong.

Boari C. and Lipparini A. (1999), Networks within industrial districts: organising knowledge creation and transfer by means of moderate hierarchies, *Journal of Governance and Management,* vol. 4, p. 339-360.

Borroi M., Minoja M. and Sinatra A. (1998), The relationship between cognitive maps, industry complexity and strategies implemented: the case of the Carpi textile-clothing industrial system, *Journal of Management and Governance,* vol. 2, no.3, p. 233-266.

Braczyk H., Cooke P. and Heidenreich M. (1997), (eds.), *Regional Innovation Systems*. UCL, London.

Breschi S. and Malerba F. (1997), Sectoral innovation systems: technological regimes, schumpeterian dynamics, and spatial boundaries, in, C. Edquist (ed.) *Systems of Innovation. Technologies, Institutions and Organizations,* Pinter, London.

Brusco S. and Solinas G. (1997), *Competitività e partecipazione*, Il Mulino, Bologna.

Brusco S. and Paba S. (1997), Per una storia dei distretti industriali italiani dal secondo dopoguerra agli anni novanta, in F. Barca (ed.), *Storia del capitalismo italiano*, Donzelli, Rome.

Brusco S. (1982), The Emilian Model: productive decentralization and social integration, *Cambridge Journal of Economics,* vol. 6, p. 167-184.

Brusco S. (1986), Small firms and industrial districts: the experience of Italy, in, D. Keeble and E. Wever (eds.), *New Firms and Regional Development*, Croom Helm, London.

Camagni R. (ed.) (1991), *Innovation Networks: Spatial Perspectives*, Belhaven Press, London.

Casti J.L. (1994), *Complexification* Harper Collins, New York.

Coleman J.S. (1990), *Foundations of Social Theory*, The Belknap Press, Cambridge (MA).

Cooke P., Uranga M.G. and Etxebarria G. (1997), Regional innovation systems: institutional and organizational dimensions, *Research Policy*, vol. 26, p. 475-491.

Corò G. and Grandinetti R. (1999), Evolutionary patterns of Italian industrial districts, *Human Systems Management,* vol. 18, p. 117-129.

Dei Ottati G. (1994), Cooperation and competition in the industrial district as an organization model, *European Planning Studies,* vol. 2, p. 463-483.

DiMaggio P.J. and Powell W.W. (1983), The iron cage revisited: institutional isomorphism and collective rationality in organizational fields, *American Sociological Review,* vol. 48, p. 147-160.

Donzelli F. (1986), *Il concetto di equilibrio nella teoria economica neoclassica*. NIS, Rome.

Dutton J.E., Dukerich J.M. and Harquail C.V. (1994), Organizational images and members identification, *Administrative Science Quarterly,* vol. 39, p. 239-263.

Ellemers N. (1993), The influence of socio-structural variables on identity enhancement strategies, *European Review of Social Psychology,* vol. 4, p. 27-57.

Eve R.A., Horsfall S. and Lee M.E. (1997), (eds.), *Chaos, Complexity, and Sociology. Myths, Models, and Theories*, Sage, Thousand Oaks (CA).

Fama E.F. (1980), Agency problems and the theory of the firm, *Journal of Political Economy,* vol. 88, no. 2, p. 288-307.

Ferrucci L. and Varaldo R. (1993), La natura e la dinamica dell'impresa distrettuale. *Economia e Politica Industriale,* vol. 80, p. 73-97.

Flood R.L. and Carson E.R. (1988), *Dealing with Complexity: An Introduction to the Theory and Application of Systems Science*, Plenum Press, New York.

Foerster K.H. (1982), *Observing Systems*, Intersystems, Seaside (CA).

Gagliardi P. (1990), (ed.), *Symbols and Artefacts. Views of the Corporate Landscape*, De Gruyter, Berlin.

Gambetta D. (1988), (ed.), *Trust: Making and Breaking of Cooperative Relations,* Blackwell, Oxford.

Gioia D.A and Sims H.P. (1986), (eds.), *The Thinking Organization*, Jossey-Bass, San Francisco.

Goodman E. and Bamford J. (eds.) (1989), *Small Firms and Industrial Districts in Italy,* Routledge, London.

Grabher G. (ed.) (1993), *The Embedded Firm*, Routledge, New York.

Hayek F.A. (1976), *Individualism and Economic Order*, Routledge & Kegan Paul, London.

Hogg M.A., Terry D.J. and White K.M. (1995), A tale of two theories: a critical comparison of identity theory and social identity theory, *Social Psychology Quarterly,* vol. 58, p. 255-269.

Howells J. (1999), Regional systems of innovation? In D. Archibugi, J. Howells and J. Michie (eds.), *Innovation Policy in a Global Economy*, Cambridge University Press, Cambridge.

Jensen M. and Meckling W. (1976), Theory of the firm: managerial behavior, agency costs and ownership structure, *Journal of Financial Economics*, vol. 3, p. 305-360.

Jones E.E. (1998), Major developments in five decades of social psychology, in, D.T. Gilbert, S.T. Fiske and G. Lindzey, (eds.), *The Handbook of Social Psychology*, McGraw-Hill, Boston (MA).

Keeble D. and Wilkinson F. (1999), Collective learning and knowledge development in the evolution of regional clusters of high technology SMEs in Europe, *Regional Studies*, vol. 33, p. 295-303.

Klir G.J. (1991), *Facets of Systems Science*, Plenum Press, New York.

Kogut B. and Zander U. (1996), What firms do? Coordination, identity, and learning, *Organization Science,* vol. 7, no. 5, p. 502-518.

Kramer R.M. (1993), Cooperation and organizational identification, in, Murnighan J.K. (ed.) *Social Psychology in Organizations. Advances in Theory and Research,* Prentice Hall, Englewood Cliffs (NJ).

Kramer R.M. and Tyler T.R. (1996), (eds.), *Trust in Organizations. Frontiers of Theory and Research*, Sage, Thousand Oaks (CA).

Lane C. and Bachmann R. (eds.) (1998), *Trust Within and Between Organizations,* Oxford University Press, Oxford.

Lange O. (1965), *Wholes and Parts*, Pergamon Press, Oxford.

Lave J. and Wenger E. (1991), *Situated Learning: Legitimate Peripheral Participation,* Cambridge Univ Press, Cambridge.

Lazerson M.H. and Lorenzoni G. (1999a), Resisting organizational inertia: the evolution of industrial districts, *Journal of Management and Governance,* vol. 4, p. 361-377.

Lazerson M.H. and Lorenzoni G. (1999b), The firms that feed industrial districts: a return to the Italian source, *Industrial and Corporate Change,* vol. 8, p. 235-266.

Levinthal D. and March J.C. (1993), The myopia of learning, *Strategic Management Journal,* vol. 14, p. 95-112.

Lipparini A. (1995), *Imprese, relazioni fra imprese e posizionamento competitivo*, Etaslibri, Milan.

Lorenzoni G. (1997), Trasformazioni organizzative nell'impresa distrettuale: alcune premesse sulla definizione della unità di analisi, in, R. Varaldo and L. Ferrucci (eds.) *Il distretto industriale tra logiche di impresa e logiche di sistema,* Franco Angeli, Milan.

Luhmann N. (1984), *Sistemi sociali,* Il Mulino. Bologna.

Luhmann N. (1990), *Essays on Self-reference.* Columbia U.P., New York.

Mael F. and Ashforth B.E. (1992), Alumni and their alma mater: A partial test of the reformulated model of organizational identification, *Journal of Organizational Behavior,* vol. 13, p. 103-123.

March J.C. and Simon H.A. (1958), *Organizations*, John Wiley & Sons, New York.

Marshall A. (1919), *Industry and Trade,* Macmillan, London.

Maturana H.R. and Varela F. (1980), *Autopoiesis and Cognition: The Realization of the Living,* Reidel, Boston.

Maturana H.R. and Varela F. (1987), *The Tree of Knowledge: The Biological Roots of Human Understanding,* New Science Library, Shambala.

Monge P.R. and Contractor N. (2000), Emergence of Communication Networks, in, F.M. Jablin and L.L. Putnam (eds.), *The New Handbook of Organizational Communication,* Sage, Thousand Oaks (CA).

Organization Science (1998), *Special Issue on Complexity.*

Organization (1998), *Special Issue on Complexity,* vol. 5, n. 2.

Paniccia I. (1998), One, a hundred, thousands of industrial districts. Organizational variety in local networks of small and medium-sized enterprises, *Organization Studies,* vol. 19, no. 4, p. 667-699.

Peitgen H.O., Jürgen H. and Saupe D. (1992), *Chaos and Fractals. New Frontiers of Science,* Springer, New York.

Perry M. (1999), *Small Firms and Network Economies,* Routledge, London.

Peteraf M. and Shanley M. (1997), Getting to know you: a theory of strategic group identity, *Strategic Management Journal,* vol. 18 (Summer Special Issue), p. 165-186.

Piore M. and Sabel C. (1984), *The Second Industrial Divide,* Basic Books, New York.

Porac J.F. and Thomas H. (1990), Taxonomic mental models in competitor definition, *Academy of Management Review,* vol. 15, p. 224-240.

Porac J.F. and Thomas H. (1994), Cognitive categorization and subjective rivalry among retailers in a small city, *Journal of Applied Psychology,* vol. 79, p. 54-66.

Porac J.F., Thomas H. and Baden Fuller C. (1989), Competitive groups as cognitive communities: the case of Scottish knitwear manufacturers, *Journal of Management Studies,* vol. 26, p. 397-416.

Porac J.F., Thomas H., Wilson F., Paton D. and Kanfer A. (1995), Rivalry and the Industry Model of Scottish Knitwear Producers, *Administrative Science Quarterly,* vol. 40, p. 203-227.

Pyke F., Becattini G. and Sengenberger W. (1990), (eds.), *Industrial Districts and Interfirm Co-operation in Italy,* International Institute for Labour Studies, Geneva.

Pyke F. and Sengenberger W. (1992), (eds.), *Industrial Districts and Local Economic Regeneration,* ILO, Geneva.

Rabellotti R. (1997), *External Economies and Cooperation in Industrial Districts: a Comparison of Italy and Mexico*, Macmillan, New York.

Reger R.K. and Huff A.S. (1993), Strategic groups: a cognitive perspective, *Strategic Management Journal,* vol. 14, p. 103-123.

Rosch E.H. (1978), Principles of categorization, in, Rosch E. and Lloyd B.B. (eds.), *Cognition, Categorization,* Erlbaum, Hillsdale (NJ).

Rothbard M..N. (1979), *Individualism and the Philosophy of Social Sciences,* Cato Institute, San Francisco.

Rullani E. (1997), L'evoluzione dei distretti industriali: un percorso tra decostruzione e internazionalizzazione in, Varaldo R. and Ferrucci L. (eds.) *Il distretto industriale tra logiche di impresa e logiche di sistema*, Franco Angeli, Milan.

Sammarra A. (2000), *L'identificazione sociale nei distretti industriali: antecedenti e conseguenze. Un'analisi applicata al distretto biomedicale di Mirandola.* Unpublished Doctoral Dissertation. University of Molise, Campobasso (IT).

Sammarra A. and Biggiero L. (2001a), Identity and Identification in Industrial Districts, *Journal of Management and Governance,* vol. 4, p. 61-82.

Sammarra A. and Biggiero L. (2001b), Identity, Cooperation and Learning in Local Production Networks, paper presented at the EGOS 17th Colloquium, July 5-7 2001, Lyon, France.

Schmalensee R. and Willig R.D. (1989), *Handbook of Industrial Organization,* North-Holland, Amsterdam.

Simon H.A. (1983), *Reason in Human Affairs,* Stanford UP, Stanford (CA).

Storper M. (1993), Regional worlds of production: learning and innovation in the technology districts of France, Italy and the USA, *Regional Studies,* vol. 27, p. 433-455.

Tajfel H. (1978), The achievement of group differentiation, in, Tajfel H. (ed.) *Differentiation between Social Groups: Studies in the Social Psychology of Intergroup Relations*, Academic Press, London.

Tajfel H. (1981), *Human Groups and Social Categories: Studies in Social Psychology*, Cambridge University Press, Cambridge (UK).

Tajfel H. and Turner J.C. (1979), An integrative theory of intergroup conflict, in, Austin W.G. and Worchel S. (eds.), *The Social Psychology of Intergroup Relations*, Brooks-Cole, Monterey (CA).

Tenbrunsel A.E., Galvin T.L., Neale M.A. and Bazerman M.H. (1996), Cognitions in organizations, in, Clegg S.R., Hardy C. and W.R. Nord (eds.), *Handbook of Organization Studies*, Sage, London.

Turner J.C. (1982), Towards a cognitive redefinition of the social group, in, Tajfel H. (ed.) *Social Identity and Intergroup Relations,* Cambridge University Press, Cambridge (UK).

Turner J.C. (1984), Social identification and psychological group formation, in, Tajfel H. (ed.), *The Social Dimensions of European Developments in Social Psychology*, Cambridge University Press, Cambridge (UK).

Turner J.C. (1985), Social categorization and the self-concept: a social cognitive theory of group behavior, in, Lawer E.J. (ed.), *Advances in Group Processes: Theory and Research*, JAI Press, Greenwich (CT).

Turner J.C., Hogg, M.A., Oakes P.J., Reichner S.D. and Wetherell M.S. (1987), *Rediscovering the Social Group: A Self-categorization Theory,* Blackwell, Oxford.

Unctad (1994), *Technological dynamism in industrial districts: An alternative approach to industrialisation in developing countries?* Unctad, United Nations, New York and Geneva.

Weick K.E. and Roberts K.H. (1993), Collective mind in organizations: heedful interrelating on flight decks, *Administrative Science Quarterly*, vol. 38, p. 357-381.

Weick K.E. and Bougon M.G. (1986), Organizations as cognitive maps: charting ways to success and failure, in, Gioia D.A. and Sims H.P. (eds.), *The Thinking Organization*, Jossey-Bass, San Francisco.

Weick, K.E. (1995), *Sensemaking in Organizations,* Sage, Thousand Oaks (CA).

Whetten D.A. and Godfrey P.C. (1998), (eds.), *Identity in Organizations. Building Theory Through Conversations*. Sage, Thousand Oaks (CA).

Williamson O.E. (1985), *The Economic Institutions of Capitalism,* Free Press, New York.

Williamson O.E. (1993), Calculativeness, trust and economic organization, *Journal of Law and Economics*, vol. 36, p. 453-486.

Zucker L.G. (1986), Production of trust: Institutional sources of economic structure, 1840-1920, *Research in Organizational Behavior*, vol. 8, p. 53-111.

Chapter 11

The industrial district and the "new" Italian economic geography

Fabio Sforzi

11.1 Introduction

The industrial district is one of the theoretical concepts by which Italian economic geography has redefined a large part of its scientific and methodological status. The industrial district has asserted itself in Italian economic geography following rediscovery of *Marshallian external economies* and revaluation of the *territory* as territorial society, and of *place* as the unit of analysis, and classification, of the economy and society. This has come about through recovery of a tradition of geographical analysis long neglected by geographers themselves since the time when economic geography was recast as spatial economics.

The resurgence of the industrial district has undoubtedly been due to the debate on the role of the small firm in the Italian economy. This was already an important problem at the time of the "economic miracle" (in the post-war period until the early 1970s), but for which the theory then current – bound to the notion that only economies of scale yield production efficiency – could not find an explanation and did not deem one necessary. The interpretative difficulties raised for social scientists by Italian economic change in the mid-1970s led to rediscovery of the industrial district as a possible conceptual framework for "empirical observations in search of a theory", namely the flourishing of the Italian small firm, the efficiency and competitiveness of which could only be explained in terms of Marshallian external economies.

Further impetus was given to use of the concept of industrial district by reflection on the crisis of Fordism and by the onset of a new model of industrialization – "post-Fordist", for the sake of brevity – where a production process based mainly on “internal” (corporate) integration and co-ordination was replaced by another and very different one based mainly on the “external” (territorial) integration and co-ordination of individual production units of small and medium size by a set of local institutions centred on a system of phase markets.

These theoretical reflections on the nature of Italian industrial capitalism have been accompanied by the growth of a robust body of empirical research which has identified industrial districts in Italian economic reality by using the local system as its unit of analysis (Tessieri, 2000).

Legal recognition of the industrial district (law 317/91) and the economic policy provisions consequent upon it have helped popularize the term “industrial district”, to the extent that it is now part of current language.

In this article I shall address the theme of the industrial district from the point of view of economic geography, focusing on some of the aspects mentioned above.

11.2 Marshallian external economies and the industrial district

The empirical finding of cumulative processes and increasing returns in a geographically concentrated production process – as compared to an isolated one – has found theoretical explanation in Marshall’s treatment of the external economies connected with the concentration of specialized industries in particular localities. Hence derives the expression *external economies of localization*, to which the neo-Marshallian geographical literature refers as “local external economies” or “economies of organization”. The external economies provide the basis for the theoretical construct that goes by the name of *industrial district* (Becattini, 2000).

The fact that Marshall analysed local external economies within the theoretical framework of industrial organization may probably account for their neglect by economic geographers and regional economists, and not only in Italy. The main scholarly concern has been the problem of industrial localization – that is, the universal factors responsible for the decision to locate a plant in a particular geographical site – rather than explanation of the process of industrialization to yield understanding of the variety of patterns assumed by local development. Influencing this attitude has been the fact that the *firm* has been taken to be the fundamental unit of analysis, while the *industry* has been pushed into the background. Conversely, in the Marshallian system the two notions – of firm and industry – are inseparable, to the point that some authors have declared that in Marshall «the industry is conceived as the *milieu* in which individual firms live and is the very condition for their life» (Bruguier Pacini, 1953; p. 75; quoted in Becattini, 1962). This is because a firm, like an individual, is never considered in isolation from the socio-economic relations of either the sector (the set of entrepreneurs) or of the place (the local society) to which it belongs. The key to understanding the origins of the notion of local external economies in Marshall's thought lies in his inclusion of organization among the agents of production, and in his emphasis on knowledge (cognitions) as a constitutive element:

> The agents of production are commonly classed as Land, Labour and Capital. [...].
>
> Capital consists in a great part of knowledge and organization: and of this some part is private property and other part is not. Knowledge is our most powerful engine of production; it enables us to subdue Nature and force her to satisfy our wants. Organization aids knowledge; it has many forms, e.g. that of a single business, that of various businesses in the same trade, that of various trades relatively to one another, and that of the State providing security for all and help for many. The distinction between public and private property in knowledge and organization is of great and growing importance: in some respects of more importance than that between public and private property in material things; and partly for that it seems best sometimes to reckon Organization apart as a distinct agent of production (Marshall, 1920; p. 115).

The social, and particularly industrial, organization examined by Marshall is subject to processes whereby

> [...] the development of the organism, whether social or physical, involves an increasing subdivision of functions between its separate parts on the one hand, and on the other a more intimate connection between them. Each part gets to be less and less self-sufficient, to depend for its wellbeing more and more on the other parts, so that any disorder in any part of a highly-developed organism will effect other parts also.
>
> This increased subdivision of functions, or «differentiation», as it is called, manifests itself regard to industry in such form as the division of labour, and the development of specialized skill, knowledge and machinery; while

> «integration», that is, a growing intimacy and firmness of the connections between the separate parts of the industrial organism, shows itself in such forms as the increase of security of commercial credit, and of the means and habits of communication by sea and road, by railway and telegraph, by post and printing-process (Marshall, 1920; p. 200-201).

Described here is a twofold process of differentiation/integration which involves both the vertical and horizontal relations of the industrial organism while it also influences the local sphere (rootedness in the local territory) and the global sphere (the overall competitive arena in which it operates).

In the light of this conception of a reality comprising processes and the relative linkages among the elementary units and entities of which it is made up – what today would be unhesitatingly called a biological conception – it is clear that achieving economies in production depends less on the size of the firm than on how production is organized locally and interacts with the social and productive environment in which it takes place. Moreover, the Marshallian industry is marked by a commonality of technology and of social conventions and customs. All of which is reflected in Marshall's endeavour to explain

> what are the causes which make different forms of business management the fittest to profit by their environment, and most likely to prevail over others; but it is well that meanwhile we should have in our minds the question, how far they are severally fitted to benefit their environment (Marshall, 1920; p. 220).

At the basis of Marshall's concept of local external economies, therefore, lie the advantages of territorial (local) concentration and (sectoral) specialization. These advantages foster (a) the reproduction of skills; (b) the circulation of knowledge; (c) the development of subsidiary activities in both manufacturing and services (through variety in production); (d) the use of specialized machinery; (e) the formation of a specialized labour market; and (f) the development of complementary industries (through variety in employment).

These advantages are evident, and of maximum importance, especially in the case of industrial local systems consisting of small production units. This is because their industrial specialization springs from the above-described mechanism of differentiation/integration, which operates through the de-composition of firms' production processes into individual phases, and then their re-composition into a dynamic interweaving of competition and co-operation at the level of the local system as a whole. This is not to imply, however, that local external economies are not equally important in the case of industrial local systems focused on (or constituted by) large plants, with the exception of the extreme case of a single, vertically integrated plant orienting local society.

The expression *industrial atmosphere* in its turn comprises the set of social and productive features that constitute the Marshallian local system, fashioning it into a single whole and qualifying it as a cognitive social system.

In the 1960s, these reflections, backed by theoretical analysis of the concept of "industry" (Becattini, 1962), led to the rediscovery in Italy of the Marshallian industrial district: as the *unit for analysis* of the industrial economy as well as the sector; as a *model of light industrialization* as well as classic (or Ford-Taylorist) industrialization; and as a *theoretical paradigm* for the interpretation of economic change because it reintegrated the territory and society – considered always in its concrete and "local" nature – in economic analysis.

This was the interpretative key that served the endeavour to situate every economic process in the territory in which it occurred, and in relation to local society.

It is natural, therefore, that the industrial district should represent the principal theoretical-practical locus for the local – i.e. geographically based – interpretation of development, given that the linkages between economic-productive relations and socio-cultural relations are inseparable in the industrial district. At the same time, the dynamic congruence of these linkages gives external competitiveness to the firms operating in it, and satisfies the need for social integration of the people who live in it.

Since this analytical approach arose in Italian economic geography – and more in general in the Italian scientific community – economic activities have no longer been studied exclusively in terms of their geographical distribution, as the local projection of an abstract industry defined on a national scale, or in terms of location processes. Analysed instead has been the industry as a factor constitutive of the individual local economies which make up the country's economic system. This new conception of industry assumes the local territorial dimension as its defining component, so that the firm – hitherto detached from the industry and local society – finally is once again embedded in them.

11.3 A definition of industrial district

In contrast to the theoretical construct of "growth pole" (Perroux, 1955), the industrial district is a local system characterized by the active co-presence of a human community and a dominant industry constituted by a set of small independent firms specialized in different phases of the same production process.

This "active co-presence" consists of the fact that the local society exerts an influence on the organization of production, which springs from its social culture. A system of values and norms – dominated by a spirit of initiative and largely reflected in the principal aspects of life, like work, consumption, saving, attitudes to uncertainty – produces a cultural environment favourable to economic enterprise, influencing industrial relations and the activities of local government and administration. The marked propensity of individuals and households for self-employment – in its various forms of home work, crafts and small entrepreneurship – favours the formation and spread of organizational skills, creativity, pragmatism, and individual and group abilities.

On the other hand, the organization of production based on small and medium-sized firms, largely coincident with individual production units, connected by specialized transaction networks and co-ordinated by more or less explicit forms of co-operation, is made possible by the technical divisibility of the production process, and it benefits from local external economies and internal ones of scale.

This division of labour among firms derives from an expansion in demand for non-standardized goods. This demand is characterized by marked qualitative fragmentation and temporal variability, in reaction to which the dominant industry of the district displays a dynamic adaptability which meets the needs of both elasticity (quantitative change in demand) and flexibility (qualitative change in production) and arises from the functional skills of its workforce and the specificity of its production formula.

Organizational intelligence, practical abilities, creative talent, craftsmanship, technical skills, and innovativeness confer dynamism on the industrial district, as a whole, and give it competitive advantage in the international market.

The dominant industry of the district – together with its auxiliary industries and the numerous services functional to it – pervades the local environment, providing employment for almost all sections of the population: young people, adults and the elderly, men and women. The result is a local society dominated by small entrepreneurs and self-employed workers, besides employees in industry, and by a high level of labour-market participation by young people and women, both single and married, while families tend to be extended rather than nuclear. It is therefore easy to understand why local society in the district identifies with the *territory*, rather than with the *company*, as happens in industrial poles dominated by large-size firms.

An industrial district, therefore, enables a geographical concentration of numerous small specialized firms to organize production efficiently, in a similar manner to what happens inside a single large factory. This is made possible by the flows of external economies generated locally among firms and arising from the knowledge, values, typical behaviours, and institutions through which local society acts on the industrial organization.

11.4 The industrial district and the "new" Italian economic geography

The rise of the industrial district as a theoretical paradigm has compelled Italian economic geography to reflect on the soundness of the interpretation of its origin – established in spatial economics – and to re-assess its geographical identity as the "science of places". This has been accompanied by reorientation of empirical research which has led to definition of a *territorial classification grid* (local systems) as an instrument of analysis which matches the theoretical construct of the industrial district (Istat, 1997),

The traditional view of economic geography, and not only in Italy, ties it to the conceptual framework of location theory (and of economic space) centred on the optimal allocation of economic activities in space, where distance plays a decisive role in terms of transport costs. In Italy, this view has been gradually replaced by a conceptual framework of different theoretical inspiration. This framework centres on industrial organization (and geographical concentration), and focuses on the increasing returns deriving to a firm from the way in which it organizes production by integrating with other firms to create a single productive process, and also with the local society in which such production takes place. This conceptual framework, too, was present in economic geography at its origins, but it failed to receive the development that it deserved and was soon abandoned under the predominance of Isard's neo-Weberian interpretation of the spatial nature of economic phenomena (Isard, 1956).

Belatedly reinstated in the early 1980s by *Regional Science*, Isard's theory was welcomed by economists in search of a way to integrate space into economic analysis, and also by economic geographers, who saw it as a theory able to emancipate them from the role assigned to them by mainstream economics of investigating the influences of the physical environment on economic facts, and who were therefore happy to accommodate regional economists in a geographical tradition stretching back to the Weberian location theory developed between the two wars (Toschi, 1941). This was a tradition that gave priority to theoretical analysis of the firm/economic space relationship, while neglecting the theoretical-empirical strand that had in the meantime consolidated with study of industrial localization in Italy, drawing on Marshallian theory to explain the geographical concentration of certain manufacturing sectors in particular areas of the country (Milone, 1930).

Under the impetus of the notion of "industrial district", which in opposition to *space* reinstated the *territory* as the versatile integrator of firms, sectors and society, all of them localized in the same place, and therefore belonging to a network of places, which approximates the global to the local dimension, over the last twenty years a "new" economic geography has arisen in Italy. This has achieved the reversal of perspective urged by numerous geographers, and by some economists, whereby the territory becomes the unit of analysis and classification of economic and social facts. That is to say, the territory is viewed as a *territorial society* identified by common residence and formed by demographic features and the operation of an indeterminate number of interests (Nice, 1987). Thus the territory denotes the geographical area pertaining to a population that has settled therein with all its legal, economic and social structures. This sense of *territoriality*, which on the dialectical plane comes almost to symbolize the man/environment relationship (social and natural together), is also apparent when one passes from the territory, as a universal concept, to consideration of its concrete components – however these may be identified and in whatever order of magnitude – constituted by places.

For the "new" Italian economic geography, a "place" is a portion of territory to which a human group attributes an individuality, which derives from the functions and overall role that it performs in the society's system of spatial structures. And since the subject, besides the purpose of these functions is the population, whether it lives and acts in the place, or whether it visits the place occasionally, the place becomes a social construct.

The most evident function of a place is its habitative one. However, alongside the purely habitative or residential function, inhabited places obviously perform a variety of productive functions, by furnishing goods and services to both the inhabitants of the place and those in other places lying a greater or lesser distance away from it. The differing weights of each of these economic, and more broadly social, functions help determine the individuality of individual places, and their degree of importance, giving rise to *hierarchies of places* and *networks among places*.

A place, therefore, is multifunctional in nature because it results from a diversified set of residential and productive settlements, each of which contributes to its individuality and importance, not simply by adding itself to the others, but by exerting a multiplier effect on the whole. It is for this reason that the term *place* can be straightforwardly replaced by that of *local system*.

The economic, social, and political organization of the territory takes the form of systems of places, so that it is possible to interpret a country's society and economy in terms of a pattern of local systems: that is, a multiplicity of local systems characterized by different types of productive activity and, especially, by different levels of socio-economic development.

The definition of the local system as a "territorially-based socio-economic unit" enables its use to explain the structure and evolution of society and the economy. Not surprisingly, therefore, some scholars have claimed that local systems constitute the basic category for division of the Globe (Wolch and Dear, 1989). Moreover, with regard to the competitiveness of countries on a global scale, there are those, like Porter (1990), who declare that the economy of a nation is the set of its multiple local economies – or in other words, its concentrated local systems of industrial firms which gain competitive advantages from their geographical proximity – and stress that it is a world-wide phenomenon in that it is to be found in all the most competitive countries.

The fact that a firm's local external environment influences its competitive success over time explains why the most widespread feature of industrial production is geographical concentration, while dispersion (or the relative isolation of a firm) is exceptional.

What comes to be geographically concentrated, however, is not simply a set of firms operating in the same sector, but rather *a local society* and *systems of specialized firms*. The latter may consist of either populations of small firms or one (or a few) large firm(s) and the relative industrial tissue. Operating in both cases is the mechanism that generates economies in production by virtue of geographical concentration, although only in the former case can one say that these economies are decisive for the organization of the production process; in the latter case, they are a supplementary factor, albeit one of increasing importance. Indeed, as large firms shift from mass production to mass differentiation – where markets are relatively fragmented and volatile, rather than homogeneous – the role of external economies of organization becomes more and more crucial (Storper and Salais, 1997).

In local systems, differentiated by the socio-economic features of the population and the economic-productive ones of firms, there arise localized networks of traded (economic exchanges) and untraded interdependencies (exchange and acquisition of knowledge) which lie at the basis of competitive advantage. Effective achievement of the latter depends, besides the individual efficiency of firms, on the degree of socio-cultural cohesion, but above all on the presence in the local system, i.e. among the members of local society, of a system of values and of institutions, which represent their interests and regulate their everyday lives. These are conditions that determine the greater or lesser ability of local systems to develop attitudes, and to mediate creatively among different forms of knowledge – the knowledge that derives from technical progress (codified knowledge) and that arises from practical experience (tacit knowledge) – thereby fostering the formation of the technical skills required to participate in that particular form of collective action constituted by production. It is through this process of "versatile integration" that it becomes possible to associate the production of goods and services with the re-production of the human and material factors of production itself, and thus ensure the permanence and vitality of a local system.

The complex organization of the contemporary economy and society displays a relational logic increasingly based on technical, organizational, and communicative interactions, involving the productive systems of both small and large firms and their places of geographical concentration. Disappearing as a consequence is the duality between the two traditional paths of industrialization – the "classical" one, based on internal economies and vertical integration, and the "district" one, based on external economies and without vertical integration – which instead tend to converge. Because these non-material local economies are intangible and specific to particular places, they are difficult to transfer from one place to another. This determines the inevitability of *local* as the pivot for revision of the criteria of economic-social analysis and, consequently, for integrated analysis of the territory as the theoretical and empirical place where the relationships among population, productive sectors, and territory meet and merge (Conti and Sforzi, 1997).

Finally, whilst it is true that, in the current phase of capitalism, relational networks tend to contract in time and expand in space (so that they are virtually global in their extent), the relations essential for production and the creation of knowledge are still territorially circumscribed, that is geographically concentrated.

11.5 Final remarks

Without the discovery of the industrial district, and its affirmation in the social sciences as a theoretical paradigm, the renewal that in Italy today enables us to talk, without rhetoric, of a "new" economic geography would not have been possible. Rather than a manifestation of actual reality – a concrete case of development or an organizational formula for production – the industrial district is a theoretical paradigm, which allows us to focus on the place and re-connect technical-productive and social-cultural relations, as inseparable components of economic behaviour; or, in other words, as the sources of productiveness and innovativeness (Becattini, 2000).

Initially proposed as the unit of analysis of the industrial economy, the industrial district has thus helped to consolidate interpretation of the local system as an integrated production unit, highlighting that production is a geographically concentrated process, because its genesis occurs in a place, not simply in a firm.

This emphasis on the industrial district has produced a change of approach in both theoretical reflection and empirical study. As regards the latter, in particular, this different approach to economic-social phenomena has sought to define *a territorial classification grid*, represented by local systems, which would enable empirical research that matched the theoretical construct of the district.

Crucial to this end has been the availability of census data with which to overcome the classificatory rigidity of traditional territorial grids (administrative or statistical) – which are entirely inadequate as well as misleading – and devise one suited to the analysis of geographical concentrations and productive specializations: that is a grid comprising the main causes of the phenomenon to be studied.

Intense methodological and empirical research has led to identification of a now well-known classificatory grid: that of *local systems* (Istat, 1997). The use of the latter for the territorial analysis of economic change – including identification of industrial districts as concrete cases of local development – has yielded positive results, not only in socio-economic research, but also as regards industrial and labour policies, and, more generally, the planning of local development (Sforzi, 1989; Istat, 1996; Ministero del Tesoro, 1999).

One may therefore say that the industrial district, framed by a neo-Marshallian reading of the organization of production and of local society, has engendered a knowledge circuit where theory and empirical research blend and evolve together, producing that change of perspective – frequently urged, but never effectively accomplished – whereby it is today the *territory* which provides the key to interpretation of society and economy, with the *local system* as its unit of analysis. This theoretical framework of reference naturally provides the basis for the formulation and implementation of local policies, which recast traditional economic, social, and infrastructural ones as specialist policies aimed at the creation of the institutional and environmental conditions for the competitiveness of places.

References

Becattini G. (1962), *Il concetto d'industria e la teoria del valore*, Boringhieri, Turin.

Becattini G. (2000), *Il distretto industriale. Un nuovo modo di interpretare il cambiamento economico*, Rosenberg & Sellier, Turin.

Bruguier Pacini G. (1953), 'Intorno alla nozione di industria. Per la storia di un termine economico', *Studi in memoria di Gino Borgatta*, vol. I, Bologna.

Conti S. and Sforzi F. (1997), 'Il sistema produttivo italiano', in, Coppola P. (ed.), *Geografia politica delle regioni italiane*, Einaudi, Turin.

Isard W. (1956), *Location and Space-Economy. The Technology Press of Massachusets Institute of Technology*, Cambrige (Mass.).

Istat (1996), *Rapporto annuale. La situazione del Paese nel 1995*, Istituto Poligrafico e Zecca dello Stato, Rome.

Istat (1997), *I sistemi locali del lavoro 1991*, Sforzi F. (ed.), Istituto Poligrafico e Zecca dello Stato, Rome.

Marshall A. (1920), *Principles of Economics,* Eighth edition, Macmillan, London.

Milone F. (1930), 'La localizzazione delle industrie in Italia', in, *Atti dell'XI Congresso geografico italiano*, vol. III, Naples.

Ministero del Tesoro (1999), Orientamenti per la programmazione degli investimenti nel periodo 2000-2006 per lo sviluppo del Mezzogiorno, Ministero del Tesoro, del Bilancio e della Programmazione Economica, Dipartimento per le Politiche di Sviluppo e Coesione, Rome.

Nice B. (1987), *Introduzione alla geografia economica*, CUSL, Florence.

Perroux F. (1955), 'Note sur la notion de pôle de croissance', *Economie Appliquée*, vol. 7, p. 307-320.

Porter M. (1990), *The Competitive Advantage of Nations*, Macmillan, London.

Sforzi F. (1989), 'The geography of industrial districts in Italy', in, Goodman E., Bamford J. and Saynor P. (eds.), *Small Firms and Industrial Districts in Italy*, Routledge, London.

Storper M. and Salais, R. (1997), *Worlds of Production: the Action Frameworks of the Economy,* Harvard University Press, Cambridge (Mass.).

Tessieri N. (2000), 'Una rassegna bibliografica sui sistemi locali di piccola impresa', in, Becattini G., Bellandi M., Dei Ottati G.and Sforzi F. (eds.), *Il caleidoscopio dello sviluppo locale. Trasformazioni economiche nell'Italia contemporanea*, Rosenberg & Sellier, Turin.

Toschi U. (1941), *La teoria economica della localizzazione delle industrie secondo Alfred Weber*, Istituto di Geografia della R. Università di Bari, Memorie, no. 9, Arti Grafiche Macrì, Bari.

Wolch J. and Dear M. (1989) (eds.), *The Power of Geography: How Territory Shapes Social Life*, Hyman, Boston.

Chapter 12

Behavioural rules in industrial districts: loyalty, trust, and reputation[1]

Maurizio Mistri and Stefano Solari

12.1 Introduction

The already wide literature on local systems and, in particular, that on industrial districts, has highlighted the role of reputation and trust in the competitiveness of these organizational forms compared to Fordist organization. The work of Becattini (1990), Bellandi (1987), Dei Ottati (1994a, 1994b) can be interpreted in this sense and a similar point is made by Sako (1998; p. 23). Reputation and trust are important parts of what Storper (1997) calls *untraded interdependencies,* and of the relational factors of local competitive advantage considered by Amin (1998). Many sociologists and economists utilize the term "trust" in an intuitive way. The common-sense semantics of this term let us conceive a positive relational attitude opposed to distrust. We believe that the concept of trust should be defined in an operational way, as an expectation concerning the convergence between the promised and effective behaviour of an agent. In this paper, the interaction of behavioural rules, reputation, loyalty, and trust will be examined as being framed by a system of norms in the context of industrial districts. Once we have outlined the features of the Marshallian industrial district, we will define the main interacting concepts showing their interdependency. It will be argued how the real central theoretical point is the existence and the influence on contracting of a moral order understood as a

[1] Stefano Solari would like to thank Bart Nooteboom for his advice and help in finding materials. Mistri was responsible for Sections 12.2, 12.3, 12.7, 12.9; Solari for Sections 12. 4, 12.5, 12.6; a joint effort was made in Section 12.8.

system of rules. Finally, it will be shown how, in a district, a principal–agent interaction is affected to achieve the best dynamic efficiency.

12.2 An outline of the district form

Concrete industrial districts take several forms, depending on local contexts and history. However, what we are discussing is not a specific industrial district, but the ideal type, the defining class of a set of concrete organizational realities, which share certain relational features and some modes of agents' behaviour. This logic stylisation is theoretically consistent with the economic methodology based on abstraction of factual reality. According to the Becattini's (1989) reinterpretation, the district is «a socio-territorial entity characterized by the active coexistence, in a circumscribed territorial area, naturalistically and historically determined, of a community of people and a population of industrial enterprises» (Becattini, 1989; p.112).

Becattini's definition is sufficiently wide to highlight the district's dual nature as an industrial and a social organizational form. This definition is substantially compatible with some statements made by Porter concerning the application of the "value chain" to "clusters", which is a good definition of the districts' value-creation mechanisms.

There are two basic issues concerning the district which should be recalled. First, firm size is assumed to be small, but such enterprises are as efficient as bigger ones. The second issue is the emergence of loyal behaviour inspired by a principle of respect of reciprocal trust.

District firms are smaller, compared to non-district firms, and in order to achieve levels of efficiency comparable to non-district firms, we should suppose that the production function of a sector can be subdivided into some basic producing units. Therefore, the production function should be linear and homogeneous of the first order, so that production can be made indifferently by a large or by many small firms in a "network" form of organization.

The substitution of a single enterprise with many complementary and, at the same time, competing firms, corresponds to a precise implicit project, or self-organization, since it allows the maximum of flexibility to be achieved and the reduction of the overall sector's risk. Each firm, given its competition relationship, is stimulated to implement both technological and product innovations and cost reduction. Competition among subcontracting firms allows buyers to achieve some cost reduction. At the same time, the search for innovative spaces is obtained, minimizing the overall system's risk, since risk is assumed by innovative firms, while followers benefit from first-comer's experiences. However, the efficiency of industrial districts cannot be simply explained by specific hypotheses on the technology side, since external economies and some relational factors provide a competitive advantage compared to other organizational forms.

Our rapid glance on organizational forms aims only to highlight the importance of the relationships among many suppliers and some buyers and between entrepreneurs and workers.

12.3 Behavioural rules and trust relations

We have intentionally stressed the socio-cultural aspect of the district organizational form, while, as is well known, in the analysis of the "Fordist" enterprise, this element is completely omitted, minimizing the relevant aspect of personal relationships, as if such relationships were univocally defined by the nature of the production process. Becattini himself (1990) considers the district mainly as a community of people sharing a relatively homogeneous system of values concerning the ethic of work, entrepreneurship, family, reciprocity, and modes of change.[2] These values pertain to a sufficiently restricted geographical area, in this space the district's culture is practised and constitutes the discriminatory element dividing the district from other zones. Culture is defined here in the anthropological sense (Mistri, 2000). Behavioural rules are part of the culture, and they are at the same time learned and the result of a process of learning. On the other hand, the analysis of learning processes in the district systems of small firms allows us to highlight how the small dimension is functional for the transmission of fundamental information on production and the market, without the appearance of information asymmetries. In other words, the space of "private" information appears as extremely limited and, given the nature of social relations taking place in the district, it is difficult for the individual entrepreneur to keep specific information secret. Dei Ottati (1994b), following Becattini, has emphasized the *communitarian* nature of districtual markets. Therefore, district actors do not only share a system of values or a local culture, but they "perform" a system of values and rules. They grant and

[2] In a society sharing a system of values there is always some norm fixing what has to remain stable and where and how improvements must be concentrated.

preserve them for future generations.[3] It follows that institutions are permeated and inseparable from district culture.

The way social norms are transmitted is coherent with the mode of reproduction of the firms' population, because the same birth of enterprises is favoured by the sharing of values and by family structure. At the same time, given the dominant social features of the district, the birth of new firms reinforces the sharing of communication protocols (or systems of meaning) (Mistri, 1997). Therefore, the mode of creation of new firms appears relevant to reinforcing behavioural rules inspired by trust and loyalty. In fact, the birth of new enterprises often takes the form of spin-off, that is to say through the creation of a firm by a former worker of a mother-firm, or the form of the fragmentation of family enterprises. It is a widespread practice that the workers of an enterprise leave it to found their own firm, often imitating the former firm's production process or improving it. The fact of working in a small factory makes all the workers conscious of the production techniques, the overall organization, the commercial channels and their specificity. There is no aspect of a firm's life that can be kept secret from workers intent on learning. This peculiar dimension of knowledge and cognition of small firms conceives a market situation where all agents – worker or entrepreneur, often subcontractor – have access to the informational stock concerning the local market. In other words, as stated before, "private" information is reduced. In the efficiency equivalence of small and large firms, the differential factor determining small firms' competitive advantage is to be found in relational immaterial factors as trust structured by behavioural rules defined in the system.

One of the basic relationships takes place between suppliers and buyers, the other between entrepreneurs and workers. In the first class of interactions, trust is tied to the system's possibility of defining loyal behaviours, which leads to positive expectations. The second class of relationships can be seen through the common framework of the principal-agent model, but a specific modification is needed as regards the local labour market, which includes family work.

[3] Actually Becattini (1989; p.114) says that they give birth to "a system of values and rules which these values spread in the district, guaranteeing them and handing them down from one generation to the next".

12.4 Theorizing the connecting structure of a small community economy

Trust has been seen as a constitutive feature of districts compared to other economic systems. However, as pointed out by Nooteboom (1999a), trust is typical of relational contracting and cannot be directly referred to a system. Districts are economic systems where, because of the resistance of some socio-cultural factors, some kind of dynamically efficient relational contracting is possible. Therefore, in order to understand the relative competitive advantage of districts in many productions, we should study what is the *connecting structure*[4] of a small community economy.

An appropriate description of the district implies abandoning the perfect rationality approach for an evolutionary and relational view of choices, or at least a "rationality perspective of rule following" (Vanberg, 1994). The fact is that the social features of these socio-economic systems allow us to describe them as models historically opposed to individualistic urban societies (Gambarotto and Solari, 2000). As a consequence, a number of concepts help explain the peculiar mode of economic interaction by pointing out some synthetic cognitive attitudes which play a basic role in the actor's decisions. We have already proposed the concepts of trust, loyalty, reputation, culture, system of values, system of meanings, and system of rules. These slippery concepts need to be precisely defined and epistemologically framed to use them correctly in economic theories, without superposition, semantic ambiguities, and excessive concept stretching. In particular, we must stress the appropriateness of such concepts with relation to the levels of analysis, whether individual or collective.

In economics, the best way to cope with the concepts of values, norms, culture, or any factor which structures the economic environment, is to understand them as a system of rules (institutions). That is to say a number of interrelated self-enforcing rules, which limit human action by inducing specific behaviour in well-defined situations. The relationship between a system of norms, and the organizational arrangement of an economic system, has been theorized by Aoki (2000) in the perspective of *multiple equilibria*. In his view, organizational forms match best some – formal and informal – institutions, and give rise to some self-reinforcing co-ordination schemes. Also in Brian Loasby (1999), productivity increase processes are framed by some schemes of division of labour where institutions play a major role. Following that perspective we

[4] This term is explicitly taken from Bateson (1979).

may consider the industrial district as one of the organizational equilibria found by good matching between some organizational scheme and a shared system of norms. The scheme of division of labour in the district, based on a highly interconnected network of small entrepreneurs, fits best a communitarian kind of culture and institutions. Here we conceive culture as a system of social norms and, therefore, of mainly informal institutions.

The concept of culture has been analysed with relation to economic theory by the German historical school, where the concept of *ethical system* and *moral order* were elaborated and conceived as the basic structuring element of the economy. More recently, Kapp (1985) conceived culture as a structure connecting symbols, people, goods, values, and institutions in a unitary whole. Culture includes shared systems of beliefs and it materializes in artefacts. It cannot be interpreted as a simple externality. In the substantialism of K. Polanyi (1944), culture is a process continuously constructed and reconstructed through interaction by a network of individuals.[5] The economy itself is "culture". Local institutions, which frame reciprocal behavioural expectations by making them predictable, can be taken as the backbone of the local culture, which constitutes a basic factor in the organizational equilibrium of the economy. In that way trust, reputation, and loyalty are some relational elements, which emerge from the good matching of the local culture, expressed by its institutions, and the district's scheme of division of labour (fig. 1).

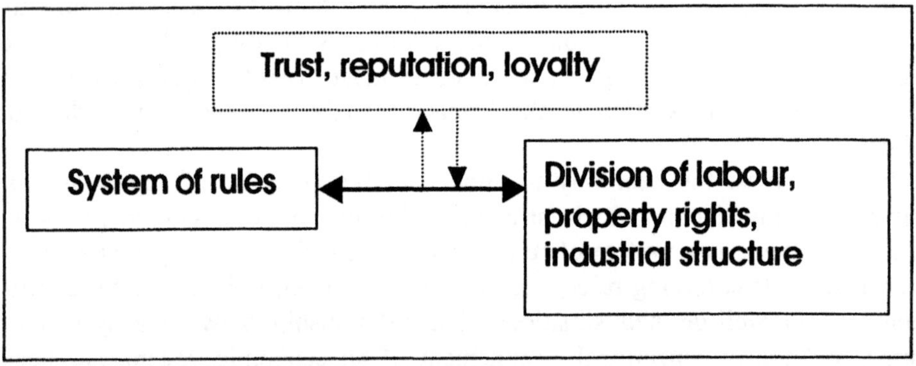

Figure 1: Multiple equilibria and relational variables

Many kinds of institutions exist. Voigt and Kiwit (1998) ordered them by kind of enforcement and by origin (tab.1). The peculiarity of Italian industrial districts is the prevailing role of "internal institutions", that is to say of

[5] That view coincides with the symbolic interactionism of Mead.

spontaneous rules of civil society. Ethical rules and customs play a major role, while also some kind of private rules are discernible in some systems (Dei Ottati, 1994a).

Table 1: Types of institutions (Voigt and Kiwit, 1998; p. 86)

Kind of Rule/Norm	Kind of Enforcement	Type of institution
1. Convention	Self-enforcing	Type 1, internal
2. Ethical rule	Self-commitment of the actor; possibly amplified by religious convictions	Type 2, internal
3. Custom	Via non-organized societal control	Type 3, internal
4. Private rules	Organized private enforcement	Type 4, internal
5. State law	Organized State enforcement	External

Norms and institutions do not constitute an additive set, since they show strong complementary effects resulting in *systems of rules* (Hayek, 1973). Some norms reinforce the power of other norms, and the overall system tendentially displays a coherence, which implies the absence of conflicting rules of behaviour.[6] A system of norms produces a structuration effect of economic space and emerges as interrelated regularities in the repeated interaction of individuals. It defines an actor's available behaviours in given situations, a viability domain of action in the community, the latter being the space where these norms are implicitly accepted.

A system of norms is defined at a collective level or, more precisely, at a mesoeconomic level, because it comprehends the space of interaction of a set of individuals. We can study the behaviour of an actor in a system of norms, but in this case the system is understood as the actor's context. When we take the system of rules as the object of study, we situate ourselves at an intermediate level between micro and macro, which is mainly relational in nature, and characterized by some emergent phenomena (spontaneous order). Another example of an emergent phenomenon is "reputation". It refers to an actor, but it emerges at the community level by communicative interaction. It can be considered as an actor's immaterial capital, but of a vanishing kind, due to past experience. The most important fact is that, when we study reputation, we analyse the link between an actor at the individual level and an emerging

[6] We are obviously not talking about formal rules which can be conflictual, but of the effective rules of a community.

phenomenon at the collective level. Therefore reputation is a collective or "context phenomenon", which falls down to the individual level by assigning an attribute to a precise actor.

Table 2: Concepts and levels of description

Appropriate level of analysis	**Concept**	**Features**
Collective level	System of **NORMS** (culture, values)	Emergence of a structuration effect of economic space by repeated interaction of actors
Collective →individual	**REPUTATION**	Immaterial capital proper to an actor due to the spread opinion on its propensity of action due to past experience
Collective ←individual	**LOYALTY**	Maintenance of a commitment with relation to a context of shared rules and values
Individual-relational level	**TRUST**	Relational cognitive attitude of an actor with relation to a precise partner or institution
Collective ←individual	**CONFIDENCE**	Cognitive attitude of an actor relative to a viable way of acting in a structured, but unpredictable environment

The same is true for the concept of "loyalty". In this case, the appropriateness of an actor's behaviour is evaluated with relation to a system of norms. In any case his action is evaluated, comparing it with some expected behaviour which is to fall within the viability domain of a system of rules.[7] Therefore, expectations on his behaviour are not precisely defined, but are defined in a fuzzy set. The reference norms can be general – the norms prevailing in the community, and specific – a contractual agreement. The interaction of the two gives form to the expected domain of actor "loyal behaviour". No loyalty can be defined independently from a system of shared values and norms, which at least represents the interpretative code of personal agreements.

At the opposite side of a transaction relation we can find "trust", which is totally defined at the individual level. An individual puts his trust in another actor and in his reputation, and this fact concerns his communicative interaction and his cognitive processes, therefore the evaluation of an actor's reliability is part of the individual decision-making process. Following this line of reasoning,

[7] Often, one or more rules also indicate the expected direction taken inside the viability domain.

trust is directly linked to procedural rationality and is therefore an essential part of the deliberative process, on which it is based. Trust is not a collective good or an external factor – a positive externality – but a constitutive factor of the decision-making process at the individual level. Therefore, it is not correct to claim that trust is a feature of a local system. There, we can find shared values and a system of norms which constitute a frame to individual reciprocal trust, which can also disappear for many reasons in one relationship while persisting in another. Trust is typical of relational contracting between two subjects – or a subject and an institution – in a procedural rationality framework of an individual decision process. In conclusion, trust refers to an expected behaviour based on reputation. The concept of *confidence* will be discussed in the next paragraph.

The kind of reasoning relevant to our study is that *an actor trusts another individual who has a reputation for behaving loyally according to a system of norms*. The process displays a positive feedback between these concepts, which reinforces community ties and the relative system of norms. What is crucial in this process is the system of values and rules of the local culture, which allows the emergence of relational factors leading to the best economic coordination. According to Lyons and Metha (1997; p. 244), trust is norm-based in orientation, and social relations are experienced in certain normative, or mutually understood ways. However, it is doubtful that a distinction between a *self-interested trust* and a *socially oriented trust* can be meaningful to the actual analysis of given economic systems, as well as the distinction of economic and sociological man. Since self-interested behaviour takes place in a structured social environment, the middle way of a purposeful action of the individual in a structured space is the basic working hypothesis.

Therefore, the peculiar system of norms is the distinctive feature of an economic system and not trust or loyalty, which are contingent and emerging phenomena in specific interactions. However, many kinds of trust can be found in the economic literature, and some of them can be useful theoretical devices in economic explication.

12.5 The many kinds of trust

Nooteboom (1999b) defines trust in the following way:

"To have trust in the narrow sense, or "real" intentional trust, is to accept or neglect the possibility that a partner will utilize opportunities for opportunism even if it is in his interest to do so."

When we hypothesize an imperfect rationality and a non ergodic environment, actors find themselves in a situation of uncertainty which they have to reduce in order to make their transactions.[8] This problem becomes even more radical in the case of long-term engagements, specific investments, or common technology development. Therefore, if in a perfect rationality framework, trust can simply help overcome prisoner dilemma situations, in procedural rationality trust is understood as an essential pre- and post-contractual element constitutive of the "rationality". It is also a "cheap" element for the governance of systems of transactions. This fact has been pointed out in the analysis of organizations by Arrow (1974) and Leibenstein (1987), and in district studies by Dei Ottati (1994a), and by You and Wilkinson (1994). Arrow, in particular, argues that trust is an economic good with a real value, and that it enhances exchange efficiency, while it is not a tradable good since its exchange has no meaning (Arrow, 1974; p.23). Leibenstein goes even further by saying that trust is fragile in comparison with its substitutes (1987; p. 600). That means that if to a relationship based on trust – producing a certain efficiency – we add some control mechanisms – devices for the reduction of informational asymmetries – this fact would in general make trust disappear, leading to a minor exchange efficiency: trust substitutes are distrust signals (Leibenstein, 1987). Therefore trust cannot be set by a rational evaluation of actors, otherwise it would be contradictory, but it relies on collective beliefs and institutions. At the same time it is characterized by an explicit intentional element.

Niklas Luhmann (1988) has proposed a definition of trust by relating it to the concepts of *familiarity* and *confidence. Familiarity* is an unavoidable fact, which depends on experience and on the ability to cope symbolically with the universe of our perceptions. Familiarity is obviously a precondition to trust, which in turn is a solution for specific problems concerning risk.[9] The difference between *trust* and *confidence* is particularly interesting, since there is not a comparable distinction in other languages such as French and Italian. Both terms refer to expectations and their distinction is based on the different perception and evaluation of possible happenings.

Generally, confidence involves a reliable way of acting, which cautiously disregards some possible negative events or outcomes. It means to neglect some

[8] In fact, Bénédicte Reynaud (1998; p.1458) has defined trust as "*a relationship of delegation which is founded on an anticipation of the behaviour of the delegatee. The asymmetry and the reciprocity of the commitment* are the two essential and inseparable characteristics of the trust relationship ".

[9] Luhmann argues that "risk" is a modern concept since once the concept of chance – fortuna – was used in the same situation.

events which are possible, but impossible to predict, and which cannot be considered without making action slow and uncertain. This happens for the reason that there is no alternative: taking into account all possible events prevents us from taking decisions and co-ordinating with other people. So some "nonchalance" is necessary to act.

Trust, on the contrary, implies a specific commitment and a risk, where the latter is relevant compared to benefits. Risk is a reintroduction of an uncontrollable element in a controllable situation, and trust fills the lack of control by a simple decision yes/no instead of making an insurance.

Both the concepts of trust and confidence, are relevant in the study of industrial districts and small firms systems. Besides Luhmann's definition of the two concepts, what we would like to stress is the different projection of individual cognition in the two cases. While trust refers to specific relations with specific people, confidence implies a wider perspective of a larger context of interaction and not necessarily relational. Therefore confidence is particularly useful when describing an actor taking decisions or acting in a structured context without referring to relational contracting. In this context, confidence is connected to a reliable way of behaving in an environment characterized by a given system of norms.

The basic distinction has been made by Luhmann (1988) himself, when he argues that we can talk of a "*confidence in the system* and a *trust in partners*". For this purpose also the concept of *institutional trust* proposed by Lynne Zucker (1986, 1988) is quite relevant.[10] Institutional trust in an industrial district is best defined as a *"confidence in a system of rules"* linked to informal local institutions (Solari, 2000). Bart Nooteboom (1999b) underlined the relevance of a "systemic trust" in organizations, and argued that it depends on the distribution of roles to individuals. The same can be said of a larger system such as a district.

[10] The classification proposed by Lynne Zucker includes: trust based on experience of past exchanges and acquired reputation; trust based on people's characteristics, mainly based on their belonging to a group or to a given category; lastly she proposed an *institutional trust*, connected to formal social structures, based on codified signals, requiring a specific learning. Williamson (1993) has proposed another taxonomy more oriented to economics: *calculative trust* of a pure microeconomic kind; *personal trust*, which is not based on calculation but refers to a context of discrete structures; *institutional trust*, which refers to the social contexts in which contracts are embedded. In the last case, if institutions are theorized as exogenous, then trust is non calculative. Williamson (1985, 1986) proposes different supports of institutional "discrete" trust: social culture, politics, regulation, professionalization in the sense of specialization of roles, networks – and here he cites Northern Italy small firms systems – and corporate culture.

In this case we do not have any corporate culture, but a communitarian culture, which frames social interactions and the specific economic roles of agents.

12.6 The interaction of loyalty, reputation and trust

An issue, which is worth mentioning, is why some communitarian ties resist in most Italian (and also non-Italian) districts. The process of economic development, which, as described by Weber (1957), consists of an extension of the market form over traditional communitarian economies, can also be understood as a process of extension of trust from immediate relationships toward wider systems of interaction. Personal and communitarian trust is progressively substituted by institutional trust tied to modern institutions (brands, stock exchange, antitrust, etc.). However, traditional local relationships are still relevant when dealing with fictitious goods in areas where, for historical reasons, some continuity of institutions exists. This process of conservative modernization is characterized by a progressive opening up of communities to market transactions, without losing completely local systems of norms. In these areas, people insist on using conservative architecture of transactions for contracts concerning intermediate production, employment relations, etc.. In this way they show a preference for some kind of "conservative" governance structure, in which they are more confident. Therefore, the concept of "confidence in a governance structure" is best suited to explain the historical resistance of communitarian rooted exchange systems, while the notion of "relational trust" is best suited to describe the functioning and the good performance, at low transaction costs, of relational contracting between buyers and subcontractors in a synchronic theorizing setting. Other kinds of trust, such as "irrational proximity trust", do not exist, or are irrelevant for the economic theorization of districts. Therefore, we will concentrate mainly on a synchronic perspective of district interactions, without renouncing an evolutionary perspective.[11]

Nooteboom (1999a) distinguishes an *ex ante* trust, based on norms, shared values, kinship or friendship, and trust based on a process which builds up as a "capitalization" of direct personal interactions. The former is more relevant for local systems, the latter is a feature of all relational contracting. What is peculiar to industrial districts is the existence of common values and culture, which act as

[11] That is the great antinomy underlined by Walter Eucken. Evolutionary theorizing can obviously be non-historical and at the same time be synchronic, since it explains interactive relationships between variables, in a context of non equilibrium.

a reference point when taking decisions. Therefore, the relevant fact in the cognitive process of the districtual agent, who decides upon a transaction internal to the district, is the framing role of the accepted system of norms.

The problem concerned here is not the alternative between self-interested behaviour or altruism, but the kind of cognitive processes which frame self-interested decision-making of dispersed agents in a small community economy. Therefore, what is under analysis here is the model of rationality, the way self-interest is understood by actors acting in a structured space. This kind of rationality, which refers to a collective level by taking into account reciprocity, norms, and shared values, is called "Kantian".[12] Ethics and values become catalysers of actors' behaviour at the macroscopic level, while at the micro level the object of analysis is represented by interpersonal and communitarian ties. Concise concepts at the macro level become context factors at the micro level, showing the different space structuring needed for theorization.

A similar approach to the understanding of the functioning of an economic system was that of some authors of the German historical school. In particular, Wilhelm Roscher and Karl Knies with their concepts of history and culture contributed to an economic theorization of the *sense of community,* which could have fruitful application in an industrial district. As Pirker *et al.* (1998) point out, Roscher in his late writings

> "inserts conscience between self-interest and *sense of community*. The latter sense then not only represents one aspect of human economy, but its synthesis – it is the *sense of community* that grounds the institutions (...) The sense of community thus claims to be one part of the economic principle, the one that corrects self-interest and refers to the whole. On the other hand, this sense represents the synthesis of self-interest and conscience" (Pirker and Rauchenschwandtner, 1998; p. 417).

Therefore, institutions and localised culture are ontologically interconnected and not only overlaying, and the district economy is strictly embedded in this structured space. We will not develop further this connection with the historical school. What is important here, is to find a correspondence between this theoretical approach to the economic system and present-day theories of

[12] Kant, in the *Groundwork of the Metaphysic of Morals*, explained what he calls the unique *a priori* categorical imperative "Act only on that maxim through which you can at the same time will that it should become a universal law". This statement can be reconducted to a form of rationality, which is self-interested, but relational in nature, given its intrinsic reference to the social system.

industrial districts, which are much more localised and circumscribed systems, but also for this reason a better suited object of analysis compared to the national economy. A more compatible approach can be that of Hayek (1973), when theorizing about cultural evolution, and that of Vanberg (1994). Both authors showed how rule- following can be the best strategy to achieve economic coordination when rules constitute a *moral order* (or, as in Schmoller's *Grundriss*, an *ethical system*). From this perspective, a procedural rationality becomes rational at the level of rule choice (Vanberg, 1994). Here, we would also like to show that a moral order can also lead to a dynamic efficiency of the system.

Often, the birth of a new small firm is possible thanks to the trust that some local actors (family, friends, banks, and contractors) place in the competencies of the new entrepreneur to put in practice his design and to respect commitments. In this case, we experiment a specific form of trust, called by Mongeon and Edmondson (1998; p. 259) "competence trust". Competence trust refers to the expectations that the partner will be able to do a certain thing in a certain way. Its nature is rather different from "intentional trust", which simply indicates the will of the partner to maintain its commitment (Nooteboom, 1999b). Competence trust derives from the fact that transactions concern goods and services featuring many qualitative aspects that are difficult to measure.

The many kinds of trust can be synthesised in a single index measuring the degree of trust of an actor in the competence and reliability of the partner. Nooteboom *et al.* (1997) show that this trust can also be estimated. This degree of trust can be considered to depend on the *reputation* of the partner in the community of producers. Reputation cannot be identified in trust, since it is one of the conditions for its emergence, but it can be seen as a kind of immaterial capital accumulated in repeated interactions. It takes the form of a spread opinion, and in fact, it is not an intrinsic property of an individual, but a relational property of the interaction between buyer and seller, principal and agent, which involves the reference to a collective emergent phenomenon as a general opinion.

We could even imagine the existence of a subjective reputation index of subcontractors or agents, elaborated by each buyer or principal from spread rumours, similarly to what is commonly done by ordering preferences of goods, comparing them two by two. Also subcontractors and agents could determine analogous indexes, conscious of the fact that the number of individuals concerned by this ranking is limited. Reputation can be considered a kind of quality index, and this fact could explain the interest market actors normally have in this kind of rankings.

12.7 Trust in strategy games between buyers and subcontractors

Two of the economic characters of the district are the buyer and the subcontractor. We can assume that, in a general district, *D*, we can find a set of buyers, $\{c_i\}$, with i=1,2,...m, and a set of subcontractors, $\{s_i\}$, with i=1,2,...n, and m<n; buyers are all involved in the same kind of production, while subcontractors are all performing the same kind of service. Given the different number of the two sets, ($n \neq m$), we can assume matching between each buyer and one or more subcontractors. The temporal horizon *T* can be subdivided into a number of periods $<\Delta t_i>$, with i=1,2,...,n. At the beginning of the period Δt_1, buyers and subcontractors meet casually, in a way that each buyer can make contracts with at least one subcontractor. In the contract the features of the service will be specified (which can be a good) with the commitment of determining it in Δt_2. Let c_j and s_j be two general actors, the formulation of the contract depends on their respective degree of reputation, the continuation of contractual relations depends on the coherence between reputation and effective behaviour. In fact, both of them know that disloyalty would lead to the impossibility to go on with their relationship and that they would find difficulties to sign contracts with other district actors, even with the actors who in Δt_1 exhibited non loyal behaviours.

It is obvious that a loyal c_j, is not obliged to make contractual agreements with a s_j who manifested disloyal behaviour, as well as a loyal s_j would not make agreements with a previously disloyal c_j. This scheme substantially reproduces a strategy game repeated for the sequence of periods composing *T*, and every choice made by c_j and s_j is immediately known by other actors. This is a problem analysed by Kreps (1990a) in the field of game theory. Kreps' solution is that in a series of transactions between a firm – with a pluriperiodal time horizon – and a certain number of clients, the firm will act in a way to keep its reputation level high, since it assigns a higher value to expected future income compared to present gains. In fact, in the case of disloyal behaviour, we can talk of a singular present gain, since such behaviour would become immediately known by all the clients.

Since we are dealing with a repeated game, it is easy to demonstrate that it will converge towards a co-operative behaviour equilibrium, and this strategic game can be seen as a *tit for tat* variant, able to solve the *prisoner dilemma* in a finite horizon. The dominant strategy will be that of choosing a loyal behaviour in Δt_1, and in the continuity of loyal behaviours as far as the counterpart will be loyal. Otherwise, some "sanctions" will be adopted as an answer to defections by the counterpart.

Kreps' model, as all models deriving from *tit for tat* schemes, relies on the assumption that disloyal behaviour would be immediately known by clients, which is possible in our case thanks to the "physical" proximity of agents which characterizes a district. This assumption should be further specified with relation to the way in which actors could become aware of disloyal behaviours. In an industrial district, the common knowledge of other actors' behaviour is based on two intertwined elements. The first is the spatial contiguity of actors, given by the limited physical space in which economic activities take place. The second is the common communicative protocol, which characterizes the district *communitarian market*. As stated by Lazaric and Lorenz (1998; p. 3), *"complex relations marked by unanticipated contingencies encourage trading partners to develop a private language"*, with which the autopoietic closure characterizing industrial districts is expressed (Mistri, 1997), and which tends to exclude from the system of transactions those who do not share this language. Language sharing is possible thanks to the sharing of life experiences and to the existence of common business.

From our point of view, both the spatial dimension and the closure of district society[13] are important, because of the micro-size of firms and the nature of their relationships. It has been demonstrated (Bicchieri, 1993) that sequential transactions in a "small" community manifest the outcome of loyal and co-operative behaviours much more easily than in bigger size societies. In "small" societies loyal behaviour is favoured by the closure of a system of meaning related to a common language which also constitutes the boundary of socio-cultural relationships. That is also what Lazaric and Lorenz (1998; p. 3-4) say when they point out that trust behaviour is founded on the non articulated nature of knowledge, and this means that trust behaviour cannot spread in large populations. Therefore, as the number of actors increases, the possibility of interaction between individuals (e.g. clients) becomes more and more difficult, and the communication link becomes disconnected.

If trust is, in a certain way, inversely proportional to the interpretation ambiguity of other actors' behaviour, it is evident that the better the direct knowledge of such behaviour, the greater the reliability of the estimation of counterparts' expected behaviours. This aspect explains the role of proximity and of direct knowledge in industrial districts.

[13] The closure is obviously selective and refers to the system of meanings.

Going back to Kreps' scheme, we stated that a loyal behaviour becomes rationally obliged for entrepreneurs who have long-term income flow expectations. However, this scheme risks being affected by a contradiction, when an entrepreneur reaches the last time period of *T*, when he could be induced to adopt disloyal behaviour.

This scheme is not sufficient to understand what is happening in small systems of firms, therefore some reference to systems of norms is needed to improve the theorization, and Kreps himself (1990) takes the same direction, analysing corporate culture.

12.8 The parable of the three talents

Eggertsson (1998; p. 347) affirms that "Rational choice social science is reluctant to recognise personal ethics and internalised norms as exogenous variables, and, moreover, social science does not offer any generally accepted theory of preference formation". Therefore it is even more difficult to endogenize relational variables. However, some studies which recognize that shared beliefs, culture, ethnic networks constitute a focal point for equilibria in repeated games exist (Weingast, 1994; Kreps, 1990; Grief, 1994). On the other hand, Dopfer (1991) and Witt (1985, 1989) propose some frequency dependency models of the evolution of institutions, which somehow endogenize rules to economic evolution.[14]

Our theoretical strategy is to consider the system of norms as a structuring of economic space, which constitutes a focal point for the determination of the form of economic equilibria. Therefore, we will put the principal–agent model in a context of norms which do not only restrict, but also shape decision making. The result is that institutions do not emerge from individuals' interaction, as in the rational approach to institutions, but we embrace the Weberian view of economic activity shaped by cultural factors. In our interpretation, the game takes place in a structured environment, and the rules which structure the context are reinforced, or destabilized, by the result of the economic interaction as in frequency- dependency models (Witt, 1989).

In order to include the structured environment in which economic interaction takes place in the theorization of buyer-subcontractors interactions, let $N(n_1, ..., n_n)$ be the set of norms constituting the system of values of the district, which

[14] However, in these models preferences, are not endogenized. We can imagine that preferences change if some basic norms change.

leaves a free domain of action *N*. When a principal and an agent make a contract, the latter is constituted by a set of rules, let $C(c_1, \ldots, c_m)$ be the set of *m* rules of the contract. Contracts are always incomplete, and often very reticent of the main features of the expected behaviour of the parties. They are fuzzy and weak compared to social norms. Therefore, we can assume that social norms complete the contract, and the set of rules governing the transaction then becomes $V(v_1, \ldots, v_n)$, where $(v_1, \ldots, v_n)$ are simply $(n_1, \ldots n_n; c_1, \ldots c_n)$ with the viability domain $N<V$, where *V* is the new restricted viability domain. *V* can be limited by conflictual norms, and in this case some uncertainty is left. However, *V* is not only space left free for action, but is also an oriented space, since norms are not only negative, but also prescriptive and constitute expectations of certain behaviours. Therefore, *V* is:

1. a viability domain intended as a subset of possible actions, that is to say a restriction of the possible behaviours;
2. an oriented set, where some outcomes produce more reputation then others, and this effect is given by the existence of some incitation rules.

The second point is particularly important, since a system of norms, as emphasized in the *German ethical economy*, is also a moral system. In Koslowski (1995; p. 11) interpretation of the ethical economy, economic and social order is understood as a self-interpreting and self-interpreted order. The importance of systems of social norms for the economic processes of the industrial district is a brilliant empirical example of this conception.

The subcontractor is expected to show a positive engagement in his buyer's activity, which goes further than a simple respect of the formal terms of the contract. We can mention metaphorically the biblical parable of the talents, where God asked to justify the quality of the commitment, rewarding the partner who made the best effort to obtain a result superior to the one implicitly demanded by his master. In the same way, the buyer leaves free space to the subcontractor's actions, and uses the fuzziness of formal commitments as an instrument to stimulate the subcontractors' engagement beyond minimal requirements. It is on this fuzziness and on the moral orientation of society that incremental innovation of small districts firms is based.

In our case, sanctions on disloyal behaviour are also social sanctions (Posner, 1996), as information spreads rapidly, reputation is crucial, and social control is tight.

Competitive pressure and personal trust framed in social norms are therefore the best incentive to innovation and efficacy of production, to the best efficiency of critical resources, such as fictitious goods, and to the reduction of transaction costs in small firms systems. So, the expected orientation of actions in a viability domain puts some incentives towards improving production, collaborative relations, technical innovation, personal engagement, etc.. Paradoxically, the rigidity of contracts, the specification of tasks or the use of insurance mechanisms could cause a diminishing of efficacy of the whole system by reducing relational trust and personal engagement in other subjects' results.

12.9 Effort and rewards

The peculiarity of the theoretical scheme of industrial districts, compared to perfect competition, resides in the interaction of the conditions favourable to contractual trust and competition among subcontractors. In a perfect market, competition is based on prices until equilibrium is reached and prices become fixed. In the district form, we can assume that price is given as in market competition, but we also assume that competition exists among subcontractors on quality, and on effort and engagement. The consequences in terms of the settlement of contractual relations are straightforward. In standard competition, contracts are written in order to clearly fix the quality-price relation so that the demanded performance is clearly defined at a couple of quality and price values. Any subcontractor will therefore adopt a level of effort compatible with the respect of the minimum expected qualitaty standard. His interest will be to avoid exceeding the engagement which satisfies market equilibrium.

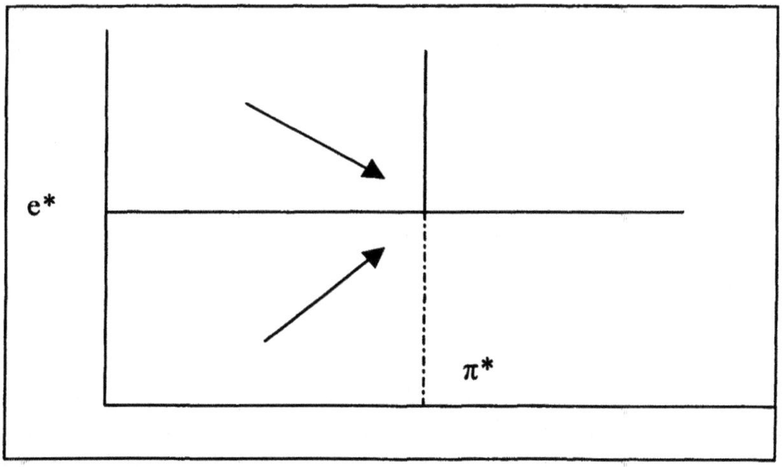

Figure 2: Market situation; e is the effort, π is the reward.

In Figure 2, equilibrium is defined by a couple of values related to the engagement of subcontractors e and the assigned reward π. Below e level (which is the standard effort level demanded in the contract) no contract is stipulated. Subcontractors are stimulated to converge their effort to e^*. Since efforts below equilibrium level are not accepted, the reward is represented by a dotted line.

In the district case, the guarantee that the minimum effort level will always be fulfilled lies in the features of social relations already outlined. In particular, the oriented viability domain of action determines a different tendency compared to the e^* equilibrium level.

The buyer is therefore interested in leaving the scheme of the contract fuzzy, since he knows that the joint effect of the competition forces and local norms will induce firms to engage in a supplementary effort to improve production.

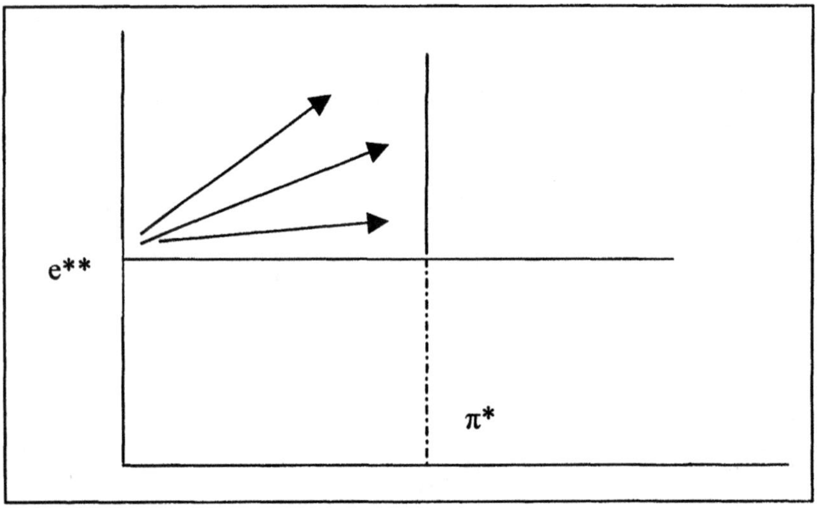

Figure 3: The district; e is the effort, π is the reward.

In Figure 2, subcontractors, even if rewarded at π^*, can engage in various efforts, all higher than e^*. Therefore contracts are left open, social norms grant a positive direction of engagement, and competition mainly takes place on the effort level. The effort is obviously directed to improve and grant quality, to suggest improvements and innovations, to speed-up delivery terms, etc..

From the comparison of Figures 2 and 3, we understand how the subcontracting firms in the district are likely to perform better by assuring that the effort can be spontaneously driven to a level equal to, or better than, that prevailing in markets where $e^{**} \geq e^*$.

12.10 Conclusion

We have tried to develop a framework in which the main social phenomena happening in a district can be introduced in economic analysis. The basic concept is that of the embeddedness of economic interaction in local cultural rules and values. Systems of rules, which constitute a "moral order", have the effect of structuring relational factors in such a way that incomplete contracts are framed in an oriented framework, which induces a dynamic efficiency of relational contracting. From the assumption of competitive pressure prevailing both in markets and districts, we concluded that, given its social features, the district is characterized by a pressure, which also takes the form of spontaneous qualitative improvements. This competitive pressure shows some adaptive innovation effects on both the technological level (process innovation) and the product level. What remains to be explained is the extent to which this form of competition urges firms to invest in internationalisation, technological research, and the establishment of consociative services.

References

Amin A. (1998), "Globalization and regional development: a relational perspective", *Competition and Change,* vol. 3, p.154-165.

Aoki M. (2000), *Information, Corporate Governance, and Institutional Diversity,* Oxford University Press, Oxford.

Arrow K. J. (1974), *The Limits of Organization,* Norton, New York.

Bateson G. (1979), *Steps to an Ecology of Mind,* Dutton, New York.

Becattini G. (1989), "Riflessioni sul distretto industriale marshalliano come concetto socio-economico", *Stato e Mercato,* no. 25, p. 111-128.

Becattini G. (1990), "The Marshallian industrial district as a socio-economic notion", in, Pyke F., Becattini G. and Sengenberger W. (eds.), *Industrial Districts and Inter-firm Cooperation in Italy,* ILO, Geneva.

Bellandi M. (1987), "La formulazione originaria", in, Becattini G. (ed.) *Mercato e Forze Locali: il Distretto Industriale,* Il Mulino, Bologna.

Bicchieri C. (1993), *Rationality and Coordination,* Cambridge University Press, New York.

Dei Ottati G. (1994a), "Trust, interlinking transactions and credit in the industrial district", *Cambridge Journal of Economics,* vol. 18, p. 529-546.

Dei Ottati G. (1994b), "Cooperation and competition in the industrial district as an organization model", *European Planning Studies,* vol. 2, p. 463-483.

Dopfer K. (1991), "Toward a theory of economic institutions: synergy and path-dependency", *Journal of Economic Issues,* vol. 25, p. 535-550.

Eggertsson T. (1998), "Limits to Institutional Reforms", *Scandinavian Journal of Economics,* vol. 1, p. 335-357.

Gambarotto F. and Solari S. (2000), " Social ties, rent creation and regulation: territorial patterns of local development" Working paper presented at the annual EAEPE conference, Berlin.

Grief, A. (1994), "Cultural beliefs and the organisation of society: A historical and theoretical reflection on collectivist and individualist societies", *Journal of Political Economy,* vol. 102, p. 912-950.

von Hayek F.A. (1973), *Law, Legislation and Liberty,* Routledge, London.

Kapp K. W. (1985), *The Humanization of Social Science,* (posthumous, ed. by J. E. Ullmann) University Press of America, Lanhamm.

Koslowski P. (1995), "Economics as ethical economy in the tradition of the historical school. Introduction", in, Koslowski P. (ed.), *The Theory of Ethical Economy in the Historical School,* Springer, Berlin.

Kreps D. (1990), "Corporate culture and economic theory", in, Alt J. and Shepsle K. (eds.), *Perspectives on Positive Political Theory,* Cambridge University Press, Cambridge.

Lazaric N. and Lorenz E. (1998), Introduction: the learning dynamics of trust, reputation, and confidence, in, Lazaric N. and Lorenz E. (eds.), *Trust and Economic Learning,* Edward Elgar, Cheltenham.

Leibenstein H. (1987), "On some economic aspects of a fragile input: trust", in, Feiwel G. (ed.), *Arrow and the Foundations of the Theory of Economic Policy,* Macmillan, London.

Luhmann N. (1988), "Familiarity, confidence, trust: problems and alternatives", in, Gambetta D. (ed.), *Trust: Making and Breaking Cooperative Relations,* Basil Blackwell, Oxford.

Lyons B. and Metha J. (1997), "Contracts, opportunism and trust: self-interest and social orientation", *Cambridge Journal of Economics,* vol. 21, p. 239-257.

Mistri M. (1997), "Distretti industriali e 'competenza comunicativa' come processo autopoietico", in, Benedetti E., Mistri M. and Solari S. (eds.), *Teorie Evolutive e Trasformazioni Economiche,* CEDAM, Padua.

Mistri M. (2000), *Economia Cognitiva*, Cedam, Padua.

Mongeon B. and Edmondson A. (1998), Trust and organisational learning, in, Lazaric N. and Lorenz E. (eds.), *Trust and Economic Learning,* Edward Elgar, Cheltenham.

Nooteboom B. (1999a), "Innovation, learning and industrial organisation", *Cambridge Journal of Economics"* no. 23, p. 127-150.

Nooteboom B. (1999b), "Trust as a governance device", in, Casson M.C. and Godley A. (eds.), *Cultural Factors in Economic Growth",* Springer, Berlin.

Nooteboom B., Berger H.and Noorderhaven N.G. (1997), "Effects of trust and governance on relational risk", *Academy of Management Journal,* vol. 40, no. 2, p. 308-338.

Pirker R. and Rauchenschwandtner H. (1998), "Sense of community: a fundamental concept of institutional economics", *Journal of Institutional and Theoretical Economics (JITE),* vol.154, no. 2, p. 406-421.

Polanyi K. (1944), *The Great Transformation,* Holt, Rinehart & Winston, New York.

Posner E. A. (1996), "The regulation of groups: The influence of legal and non-legal sanctions on collective action", *University of Chicago Law Review,* vol. 63, p. 133-197.

Reynaud B. (1998), "Les conditions de la confiance", *Revue Economique,* vol. 49, no. 6, p. 1455-1471.

Sako M. (1998), "The information requirements of trust in supplier relations: evidence from Japan, Europe and the United States", in, Lazaric N. and Lorenz E. (eds.), *Trust and Economic Learning,* Edward Elgar, Cheltenham.

Solari S. (2002), Fiducia e capitale sociale nell'analisi economica della regolazione locale, in, Camagni R., Fiorentini R. and Mistri M. (eds.), *Scritti in memoria di Eugenio Benedetti*, Cedam, Padua.

Storper M. (1997), The Regional World, The Guilford Press, New York.

Vanberg V. (1994), *Rules and Choices in Economics*, Routledge, London.

Voigt S. and Kiwit D. (1998), "The role and evolution of beliefs, habits, moral norms, and institutions", in, Giersch H. (ed.), *Merits and Limits of Markets*, Springer, Berlin.

Weber M. (1957), *Wirtschaft und Gesellschaft*, vol. II Mohr, Tübingen (Italian translation 1995, *Economia e Società*, vol. II Comunità, Milan).

Weingast B.R. (1995), "A rational choice perspective on the role of ideas: Shared beliefs systems and State sovereignty in international cooperation", *Politics and Society*, vol. 23, p. 449-464.

Williamson O.E. (1985), *The Economic Institutions of Capitalism. Firms, Markets, Relational Contracting*, The Free Press, New York.

Williamson O.E. (1993), "Calculativeness, trust, and economic organization", *Journal of Law & Economics*, vol. 36, p. 453-502.

Williamson O.E. (1996), *The Mechanisms of Governance*, Oxford University Press, New York.

Witt U. (1985), "Coordination of individual economic activities as an evolving process of self-organization", *Economie Appliquée*, vol. 38, p. 569-595.

Witt U. (1989), "The evolution of economic institutions as a propagation process", *Public Choice*, vol. 62, p. 155-172.

Zucker L.G. (1986), "Production of trust: institutional sources of economic structure, 1840-1920", *Research in Organizational Behavior*, vol. 8, p. 53-111.

Zucker L.G. (1988), "Where do institutional patterns come from? Organizations as actors in social systems", in, Zucker L.G. (ed.), *Institutional Patterns and Organizations*, Ballinger, Cambridge.

Part III

The new design of evolutionary industrial districts: some case studies

Chapter 13

Italian industrial districts: performance and evolution

Ivana Paniccia

13.1 Introduction

13.1.1 The semantic ambiguity of the term industrial district

The literature on industrial districts (henceforth: IDs) counts such a wide number of contributions that it has become difficult to denote with only one term a large variety of phenomena. In fact, under the umbrella term of ID, extensively used in various disciplinary fields, different models of (local) development and inter-firm arrangements can be recognised.

Most of the literature follows a case study approach, which is very useful for shedding light on how IDs function and particularly on the rules governing the relationships among firms and people. What is wrong in case studies is their use, when taken as illustrating general characteristics or the evolution of industrial districts, which however are not applicable to other areas. On one hand, specific and not-replicable features of specific industrial districts are attached to a general, though sometimes unspecified model; on the other hand, general features of the model - as delivered by the literature - are automatically applied, without any empirical assessment, to areas showing simply a high concentration of (small) firms.

The semantic ambiguity of the term ID may distort industrial policy interventions, since 'blanket' solutions are applied to all alleged IDs, with effects that may not be similar as they impact on genuinely different local phenomena.

Such ambiguity also affects theories on IDs. On a theoretical ground, a critical question is to understand the relationship between the empirical forms of alleged IDs and a 'normative' model of ID. This requires the isolation of the ID's rationale, whether it is rooted in economic or socio-economic theory.

The economic rationale of IDs may be found in the notion of external economies, as provided by Marshall and then further refined by other scholars in economic geography and economics. Geographers conceive the ID as a model of economic development based on external economies. Among geographers, some emphasise the role of agglomeration or proximity of firms (in economic studies) and people (in urban studies) as the sole factor able - by itself - to produce positive effects, in terms of efficiency and growth.

Becattini (1990), his collaborators and other authors (e.g. Piore and Sabel, 1984) would claim that the mere agglomeration of firms, even of a small size, is not a sufficient condition to denote an ID, but other features attaining to the *attitudes to co-operation* of local population are important as well to explain the positive economic and social performance of IDs, while Brusco (1982, 1990) would require an *extended division of labour* to recognise an ID and its effects on social integration.

The ID is therefore conceived in Becattini as a *form of capitalism joining together society (a community with a local identity) and economy (market)*. The contributions of Becattini, Piore and Sabel as well as their collaborators and, to some extent, Brusco[1], are referred to throughout this work as the 'canonical' approach, while the 'canonical' ID corresponds to the definition provided by Becattini (1990), and to his distinction between *population of firms* and *community of people.*[2]

A specific research question therefore arises concerning whether a successful ID requires the existence of additional conditions, such as co-operation or an extended division of labour for its higher performance.

1 There are some differences between the approach of Becattini and that of Brusco, as far as the Marshallian background has not played so crucial a role in the latter who discovered the concept of ID simply by studying inter-firm relationships. These differences are disregarded here and the two authors are considered as linked to a 'canonical' view of IDs, in so far as both include social welfare in their analysis of IDs.

2 Becattini defines the ID as follows: "a socio-territorial entity which is characterised by the active presence of both a community of people and a population of firms in one naturally and historically bounded area" (Becattini, 1990; p. 39).

13.1.2 Research questions

This Chapter contains some of the results of a wider work aimed at addressing such a general research question. A practical notion of ID that is comprehensive enough to include areas possibly showing different organisational arrangements is assumed. The ID is used throughout this work as any agglomeration of small to medium-sized firms specialised in one or a few industries in a limited area.

This step avoids qualifying the ID with precise socio-economic features (e.g. horizontal and vertical networking, innovativeness, co-operation, trust, etc.), which instead should come out from empirical scrutiny. Along the lines set by Marshall, the invariant traits of IDs are agglomeration of firms, wide diffusion of firms of relatively small size, specialisation in one industry and self-containment. As a consequence, our research takes as a unit of observation the notion of local labour market area (henceforth: LLMA), which is a relatively self-contained area as far as labour demand and supply are concerned (ISTAT-IRPET, 1994; Sforzi, 1989, 1990).

These areas are specialised in one or few complementary industries and show a prevalence of small and medium sized enterprises (henceforth: SMEs). Performance is then related to the structural factors highlighted by literature. An assessment of to what extent these areas comply with the conceptual framework developed by Becattini (1987, 1989, 1990) and other scholars (Bellandi, 1987; Brusco, 1982, 1990; Capecchi, 1990; Piore, 1990; Pyke and Sengenberger, 1990; Sforzi, 1990), who paved the way to research on IDs, is also offered as a first attempt to test the normative value of that framework. This also implies looking at the real diffusion in Italy of the 'canonical' model of ID in different periods of time.

The discussion is based on a review of the few works focused on IDs' performance and in a first attempt to measure the performance, as well as the socio-economic structure of 39 Italian local/sub-provincial areas.

Here we succinctly review the conceptualisation of Becattini as a canvas to discuss other contributions with the specific aim of highlighting causal relationships between the structural elements of IDs and their performance. Becattini offered a stylisation or a framework for analysis rather than a theory, as he himself declared; however, as we shall shortly review, very different theoretical approaches can be applied to IDs. They range from industrial organisation (new institutional economics: Baumol *et al.*, 1982; North, 1990; Williamson, 1975; 1985) to organisation science (network approaches: Nohria and Eccles, 1992) and sociology (Granovetter, 1985, 1992).

A second object of analysis of this paper is to assess the contribution of IDs to employment creation and to analyse, in an exploratory way, the impact of external economies on growth. The indicators calculated permit us to observe the pattern of firms' and employment growth between 1961 and 1996 in the 39 areas of the sample.

The following section discusses the main characterising features of IDs, as stressed in economic literature. Section 13.3 illustrates the strategy of "operationalisation" and methodology used for the empirical research: Results of multivariate analysis are analysed in section 13.4, while section 13.5 describes the relationship between clusters and performance, and section 13.6 analyses the relationship between external economies and economies of agglomeration, on one hand, and growth, on the other hand.

13.2 The contribution of economic literature on industrial districts

13.2.1 External economies and economies of agglomeration

The analytical framework provided by Becattini has a clear and declared ancestry with Marshall's idea of external economies.[3] According to Marshall, these are advantages deriving from the concentration of a given industry in a territory. As examples of external economies, Marshall mentions the advantages of splitting the production process into specialised phases, the increasing knowledge of markets accompanying the expansion of industrial output, the creation of a market for skilled labour, for specialised services and for subsidiary industries and, finally, the improvement of physical infrastructures such as roads and railways (Marshall, 1919, 1950). The economies so far described consist of the localisation advantages considered from the point of view of the economy of production (Marshall, 1950), but the author suggests considering the convenience for the customer too. Almost a century later, these proximity advantages were called transaction costs (Williamson, 1975, 1985). The approach of Williamson has been used to describe the economies of IDs and to show how agglomeration of firms causes savings on transaction costs (Dei Ottati, 1987, 1994). Actually, transaction economies are not only related to

[3]The purpose of the following remarks on external economies is far from being exhaustive, given the controversial nature of the concept, notwithstanding its popularity. They only intend to draw some implications useful for the empirical section of this work.

agglomeration but to a flat division of labour as well. A large number of firms having a similar contractual power sweeps out problems of opportunism and asymmetry.

Marshall himself did not omit to warn of the diseconomies of industrial concentration as well, these consisting of the higher cost of labour for one kind of work, or of the high cost of land. Recent literature on agglomeration in big conurbations also stressed the high social costs in terms of air and water pollution or traffic congestion. A decreasing function of external economies, in the long run, is also coherent with Marshall's approach, and may suggest a life-cycle model of development of IDs.

The concept of external economies has a static and a dynamic dimension as the aggregation of firms reduces costs, enhancing allocative efficiency, while at the same time forcing them continuously to innovate and ameliorate their performance, thereby increasing their capacity to survive. For this reason, the idea of external economies is used as a crucial explanatory variable in the context of regional and country growth theories (Glaeser, Kallal, Scheinkman and Shleifer, 1992; Henderson, Kuncoro and Turner, 1995; Henderson, 1996). Since spillovers of knowledge are localised, as Marshall noticed, firms and/or households tend to cluster together and therefore the growth of that industry, or city, is faster. Knowledge spillovers between firms in an industry or between households in a city are recently renamed as the Marshall-Arrow-Romer (MAR) externalities. After Marshall, Arrow (1962) presented an early formalisation, and the paper by Romer (1986) gave another contribution for an approach predicting that regionally specialised industries grow faster, because neighbouring firms can learn from each other better than geographically isolated firms. In contrast, Jacob's theory (1969) predicts that industries located in areas that are highly industrially diversified should grow faster. The MAR theory also predicts, as Schumpeter does, that local monopoly is better for growth than local competition, because local monopoly restricts the flows of ideas to others and so allows externalities to be internalised by the innovator. When externalities are internalised, innovation and growth speed up (Glaeser *et al.*, 1992). This view is in contrast with the Italian literature on IDs, which sustains that local competition, as opposed to local monopoly, fosters the pursuit and rapid adoption of innovation (Bellandi, 1987).

The logical pairing of external economies, and of economies of agglomeration enables us to define IDs better. Asheim (1994) defines external economies of scale as a necessary condition for the existence of IDs, and economies of agglomeration as the sufficient condition. In fact, in order to have external economies of scale, proximity of firms is not necessary. We are thus obliged to resort to the notion of economies of agglomeration, as an additional one, to cast the economies deriving from the localisation of an industry in a given territory. In this view, economies of agglomeration are seen as a specification of external economies of scale, as depending on the concomitant decisions of different entrepreneurs to concentrate in a certain area.

The discussion of the specification of external economies is useful in distinguishing different models of local development. In economic geography handbooks, at least four different kinds of models of local development are described: the localisation model of Weber, the pole of development of Perroux (1970), the Japanese just-in-time production systems or the networks of subcontractors that can be observed in Italy (the Benetton nation-wide system) and the ID model.

However, the juxtaposition of different categories of models of development risks not acknowledging how they evolve and may mutate from one form to another. As long as the local industry evolves, different types of economies may appear, depending on technical and historical circumstances. In fact, the concept of external economies as well as economies of agglomeration have a static and a dynamic dimension.

Another source of dynamic advantage of agglomeration has been singled out by theorists of technological competence in local systems of firms (Malmberg and Maskell, 1996). In modern innovation theory it is clearly stressed how territorial agglomeration is fundamental for innovative processes. In the theory of technological competence (Cantwell, 1989; Cantwell and Iammarino, 1998) technology is defined as partially tacit, specific to the context in which it has been created or adapted, and tied to the skills and routines of those who have developed and operate it. The relations with information sources external to the firm, as for example with scientific infrastructures, or between producers and users at inter-firm level, are strongly influenced by spatial proximity mechanisms that favour processes of polarisation and cumulativeness (Lundvall, 1988, 1992). Furthermore, the employment of informal channels for knowledge diffusion (the so-called tacit or uncodified knowledge) provides another argument for the tendency of innovation to be geographically confined (Lundvall, 1992). Based on this theoretical background, a view of IDs as innovatory systems may be placed. Bellandi (1987) clearly shows how incremental rather than radical innovations, as associated with a pattern of learning by doing, are typical in IDs.

The cumulative creation of professional know-how implied in agglomeration processes and in specialisation allows skilled workers to leave the factory and found a new firm, for which activity they are better rewarded. Social mobility within the areas characterised by agglomeration realises an efficient allocation of resources.

Division of labour and IDs

The way a production process is organised has obvious implications on the degree of efficiency and effectiveness of the industry concerned. Efficiency derives from specialisation that may regard human or physical assets, and from competition conditions.

The kind of specialisation of firms may differ, according to whether a vertical or horizontal division of labour is applied. Very roughly, the former coincides with specialisation of firms in different tasks, while the latter concerns specialisation in the same task (Leijonhufvud, 1986).[4]

A horizontal division of labour may be considered to have the same efficiency effects of a vertical one, as the specialisation principle holds in both cases.

In the vertical division of labour there are some further implications for effectiveness which the horizontal one does not have. Firstly, vertical division of labour realises a flexible system of production, because each task can be re-organised with a different mix of specialised producers. Flexibility, in turn, has two other effects: it enables quick response to variation in degree and quantity of final demand, and gives a spurt to innovative processes. Secondly, it has effects on the welfare of some of the *stakeholders* in the industry concerned. According to many authors, IDs as a form of agglomeration are flexibly specialised, and therefore enjoy all the benefits of such a form of organisation (Scott 1988a, 1998b).

Concerning welfare effects, an Italian author has illustrated the outcomes of a demand reduction on an extended division of labour, which is vertical in its

[4]Supporters of this distinction hold that the former leads to a reduction in the human-capital requirements, that is, in the craft skills or capabilities of those involved in the production process (*à la* Penrose). This view may contrast with certain conceptualisations of the ID as a model of organisation alternative to mass production, where a social, rather than a detailed (*à la* Smith) labour division, is achieved (Piore, 1990, 1992). But, a vertical division of labour may also be compatible with the integration of conception and execution at each task stage. What differs is that the governance, or the 'intelligence' of the de-verticalised structure, is the principal function of coordinators. Provided the latter are included in the same environment as the subcontracting firms, the local territory is undoubtedly the recipient of capabilities. In other words, even if single firms do not own the overall knowledge and capabilities to govern the entire process, the inclusion of co-ordinators and local social mechanisms of sanctions and rewards ensures that these capabilities are produced and transmitted on a strictly local level.

nature: "the impact of a fall in demand for the products of a particular firm depends on its level of vertical integration: where this is high, such a fall in demand will produce unemployment; where it is low, the workers employed in subcontracting firms will simply receive their orders from more successful competitors" (Brusco, 1982; p. 175). According to Piore, the allocation of external demand (positive or negative variation) among different firms is ensured by a set of rules and standards of behaviour, rather than (only) by the organisational structure of the industry. An attitude to co-operation of local protagonists would ensure the full employment of the communities' resources or the sharing of the burden of unemployment, and the like (Brusco, 1990; p. 58). Co-operation fosters innovation and entrepreneurship also in the discussion presented by Dei Ottati (1987), a collaborator of Becattini.

In addition, the extended presence of a myriad of small firms realises an even model of society. Two radical academicians, Piore and Sabel (1984), saw in a flexible system of production, and in IDs, the realisation of the promise of a democratic world, the community of equals. As Perrow stressed, speaking of small-firm networks:

> “The heads of 1,000 firms related to furniture production will receive a great deal less in salary and benefits than the two heads of the two large firms. (…) Furthermore, one of the problems of uneven development and uneven economies associated with multidivisional and giant firms is that locally generated wealth is spent or invested non-locally” (Perrow, 1992; p. 462).

The resulting hypothesis is that the more extended the vertical division of labour, the lower the extent of the unemployment rate, and the reduction of employees, when facing a negative external factor.

In the ID everyone finds his/her own place in society. This means that the district manages to allocate each individual to his/her optimal place, at least to a certain extent. As a result, we observe in IDs high activity rates, very low rates of unemployment, and a strong trend towards entrepreneurship and self-employment.

If the area is characterised by traditional economies of localisation (e.g. low labour costs), it is likely that firms will generally exhibit a cost advantage; on the contrary, if the area stands for a well-established tradition in the manufacturing of a given product, the competitive advantage will rest in the quality or uniqueness of the product with respect to other competitors. This also authorises us to identify the sources of the competitive advantage of IDs in local knowledge, *savoir faire* and competencies. This is also coherent with the theory of the resource-based firm (Langlois, 1988; Wernerfeld, 1984) and of technological competence (Cantwell, 1989). The competitiveness of an area may also rest in the way production is organised. The so much celebrated IDs flexibility is a product of their organisation structure.

Competing theories of IDs

As we mentioned earlier, the ID – as conceptualised by Becattini - is a socio-economic complex having efficiency and welfare effects. Because of its nature, it cannot be paired to a pure market representation as competition conditions affect either the competitive advantage of the industry or the welfare of the local community. Similarly, the transaction cost theory, owing to its static nature, may explain why the sourcing of raw materials, or intermediate inputs, or the hiring of human resources, occur at lower cost, but it may not give full account of the dynamic effects of agglomeration in IDs. The emphasis given to opportunism also impairs appreciation of the real nature of co-operation in those IDs where a manufacturing tradition, and, therefore, competencies and knowledge are locally rooted (Paniccia, 1998). In these situations, *confidence* in other people's competencies, rather than *trust,* explains why parts and components or services are internalised or externalised/put out.

It is our claim that all the apparent conflicting approaches on IDs may be considered compatible, once the empirical variety and evolving nature of IDs are acknowledged.

13.3 Testing the concept of ID

13.3.1 The measurement of IDs' performance. A short review of literature

The literature on IDs in the last 20 years has been mostly of a qualitative nature. Few attempts to offer a sound and fine measurement of IDs' economic rationale on a comparative basis have been made. However, the contribution of IDs to Italian economy, employment, and export, is largely acknowledged (ISTAT, 1996), and it is also one of the easiest ways to look at the IDs' performance.

In addition, an integrated approach looking at both economic and social performance based on quantitative measurements is lacking.

An initial attempt to measure the performance of firms in IDs has been provided by Signorini (1994). This research is based on the balance sheets of firms specialised in the woollen cloth industry located in the province of

Florence (which includes the district of Prato) and compares their financial and economic ratios to the average of woollen cloth manufacturers located outside the province. Recently, the Bank of Italy Research Department, has devoted most of its financial and human resources to the measurement of the performance of firms belonging to an ID[5], the so called "ID effect". An initial result of this research is available in a recent work, where the profitability and productivity ratios of firms belonging to those areas defined as IDs by ISTAT (1996) and Sforzi (1990) are analysed in comparison with a control sample of firms having the same characteristics in terms of size and specialisation (Fabiani and Pellegrini, 1998).[6]

The analysis confirms the existence of positive externalities in belonging to IDs, such as a higher profitability, measured by ROE, ROI, and gross operating margin over sales, and superior technical efficiency, as measured by using a parametric function. Profitability always appears higher, even when different industries and average size of firms are taken into account. The area-specific factors, such as proximity to outlet markets, are not significant. The authors conclude that the ID is an efficient organisation mode. These external advantages are higher for firms belonging to specialised industry, rather than for those belonging to different industries in the same ID. Another result of agglomeration is a lower per capita cost of labour.

In such a work, IDs are conceived as an alternative model of local development to those characterised by large firms, but there is no structural analysis of the areas selected. We are, therefore, left with the doubt whether such results are due to agglomeration factors only, or to other additional conditions, such as, for instance, a pervasive co-operation or a flat division of labour. Secondly, this work looks at the economic performance only, neglecting the effects on society's welfare. In addition, a doubt may also arise about the representative nature of a sample deriving from a balance sheet database including data on limited companies only.

Our claim is that there is such a wide variety within IDs that it may not be correct to take them as a homogeneous category.

In any case, this new avenue of research goes in the direction of overcoming the intangible nature of Marshallian "industrial atmosphere", and demonstrates

[5] The identification of IDs is based on the notion of LLMA. Those LLMAs which show the following characteristics are considered IDs: rate of industrialisation higher than the national average; rate of specialisation in one dominant manufacturing industry higher than the corresponding national average; and a presence of SMEs higher than the national average for manufacturing industries.

[6] The analysis is based on the whole universe of the IDs, as defined in the previous note.

how an analytical framework of hypothesis testing may be applied to IDs. However, it does not address the research question concerning whether agglomeration of SMEs is, by itself, the sole reason of efficiency and competitiveness of IDs. A further direction of research should be not to test the superiority of clustering with respect to other ideal types, whether organisation forms or models of development, but to look at the multifaceted world of IDs. In the present phase of development of Italian IDs, an urgent question is to understand, first, to which degree of agglomeration some effects accrue, or which is the critical threshold after which IDs devolute and, secondly, which type of internal restructuring of the labour division model ensures the survival or success of the local systems from which IDs originate. From a policy-maker perspective, it is important to understand how, and to what extent, agglomeration may be fostered, and why some agglomerations of firms are more successful than others. It may be the case that agglomeration *per se* is not a sufficient explanation of the ID's performance, or that not all agglomerated systems are flexible- specialised, provided that flexibility is a factor of competitiveness.

A study on the relationship between external economies at a local level and growth has recently appeared in Italy (Cainelli and Leoncini, 1999), though it still considers an aggregate object of analysis (e.g. the Italian 'provincia'), instead of the LLMA, as we do.

According to Becattini's approach, a superior performance is associated with the 'canonical' ID. A normative although implicit view of the ID, as a paradigm of flexible-specialised systems, is also evident not only in Piore and Sabel (1984), but also in Best (1990), Saxenian (1994), and Scott (1988a, 1988b). However, these authors do not yet provide a model, or a theory, as Becattini had done with the conceptualisation of the ID, even though he intended only to provide a 'framework for analysis'. Indeed, Becattini is reluctant to use the term model. In fact, in canonical literature, a formal treatment of the relationship between the distinctive characters of the model and the performance is not offered, neither is there a clear (hierarchical and structured) logic arrangement of concepts with linkages of inclusion/exclusion, opposition/consistency. In particular, it is not clear whether all the listed characteristics are necessary or just sufficient conditions for ID success. The distinctive features of IDs, as listed by Becattini (1990), all appear to be equally indispensable for achieving the ID's effect. A reasonable distinction between 'necessary' and 'secondary', or sufficient conditions is not present. Indeed, confusion between 'structural' and 'behavioural' or performance conditions of IDs does persist.[7]

13.3.2 Methodology

We are rather convinced that operationalisation may help provide a theoretical grounding for the concept of ID. Comparison on the basis of controllable criteria may allow the emergence of 'functional equivalents' (Pyke and Sengenberger, 1990) that identify the mechanisms, which ensure some results, without their being linked to given static, historical and contingent forms.

Operationalisation may also enable us to distinguish between those conditions which are responsible for the specific effects of the ID, such as competitiveness, as well as some positive welfare effects, or which instead do more and ensure their stability, survival, or other characters still to be determined. In addition, an attempt to measure and to operationalise the list of criteria helps us recognise that some requisites are implied in others, and thus their listing may be considered valid on a narrative rather than explanatory level.

For this purpose, operationalisation, as well as a measurement of concepts, is necessary. Once data are available, a comparison of different LLMAs of SMEs, and an assessment of the explanatory factors of their performance will be possible.

By operationalising the ID concept, we are concerned in the process of linking abstract concepts to empirical indicators. However, the task is quite complex, as abstract concepts do not have a one-to-one correspondence with empirical indicators; they can be operationalised and measured in an almost infinite variety of ways. The selection may, in any case, be guided by previous choices in the same or germane disciplines. In this attempt we are also aware that translation may involve misunderstandings or the use of inappropriate indicators (the 'words'), which may for example have a semantic content which goes beyond that meant by the 'translator'. However, the choice of variables has often been the result of a compromise between what is desirable and which data are available. In this sense, our choice is a proposal and not a definite statement.

[7] This is particularly evident when Becattini claims that IDs are characterised by a continuous tendency to change, which ensures their survival. It is hard to accept this claim on an objective basis, as we expect the structural characters of the ID to ensure the conditions for survival.

The strategy starts from propositions, then testable relations among phenomena are derived and, finally, indicators and parameters are associated with these phenomena.

The methodological plan in our research may be summarised as follows:

- Territorial identification of units of analysis
- Selection of the sample
- Operationalisation of the ID concept;
- Collection of data and surveys on a local level;
- Interviews and mail questionnaires to selected 'qualified' observers;
- Calculation of indicators;
- Multivariate and econometric/correlation analysis.

13.3.3 Object of inquiry and data source

As already mentioned, the notion of LLMA is assumed as our object of inquiry. Because of the limited area in which people tend to live and work, we are close to what Becattini describes as 'the merging of people and economy in a self-contained area'.

For the purpose of objectivity and comparability, it is very crucial to choose a data source that is reliable and homogeneous for all the units of observations. Census data fit the requisites of reliability, homogeneity in space and time and subdivision at municipal level. However, the number of indicators that can be built from census data is limited. No data on firms' linkages or on overall or firms' profitability are given in the census. We had therefore to integrate it with other more qualitative sources (interviews of qualified observers, secondary sources). For the purposes of this paper, only the 1991 (however, all the dynamic variables are referred to the inter-census variation between 1961 and 1996) census year has been considered.

13.3.4 The sample

As useful and general criteria of discrimination between LLMAs of SMEs, we chose two critical factors: industry and space. The idea is that the industrial sector sets the technical constraints (and therefore the range of variability) of different forms of organisation of labour within firms, while the local space is the geographical *locus,* where particular relationships and systems of values develop, influencing the linkages between firms, people, and institutions.

We thereby obtained a sample of industrial systems located in different regions of Italy and specialised in different industries. A third of our sample consists of LLMAs where medium-sized rather than small firms are prevalent.

We divided Italy into three different macro-areas according to a classical distinction: North, Centre-North, and Centre-South. As regards industry, we distinguished between fashion-led (or traditional) and innovative (but not science-based) industries (Table 1). Relying on Sforzi's work (1990), some of them correspond to the model of the Marshall-style ID. We prefer to term these IDs as 'canonical', as they correspond to the Becattini model, while Marshall observed a slightly different phenomenon in the past century. This procedure of selection enables us to take into account technology and national or international demand, as implied in industrial sector, and culture - as related to geographical location - as external conditions that may affect the structure of performance of the selected areas.

The sample represents about 20% of total employment in IDs in the industries considered. The degree of representativeness rises if ancillary industries are considered. In particular, tanning IDs cover 77.4% of the total employment in the tanning industry, footwear 30%, textiles 19.5%, knitwear 13.9%, ceramics 25.1% (ceramic tiles: 76.5%), agriculture machines 6.4% and furniture 8.2%.

13.3.5 The indicators

Here we present a short list of the indicators used as explanatory variables in the empirical testing.[8] The data set compiled is a quite comprehensive one, including a large variety of indicators, mainly built from census data at a communal level. For the purpose of this paper only some indicators are illustrated.

[8] A wider description of the indicators deriving from the data set is in Paniccia (2002).

Table 1: Final sample of 39 LLMAs
Key: the letter C denotes "canonical" IDs.

	Fashion-led industries							**Innovative industries**	
	Upstream industry: Tannery	**Downstream industry: Furniture**	**Downstream industry: leather footwear**	**Downstream industry: sport & leather footwear**	**Upstream industry: textiles**	**Downstream industry: knitwear**	**Downstream industry: clothing**	**Non-specialised industry: ceramics**	**Specialised industry: mechanical engineering**
North	Arzignano (Veneto) **C**	Desio (Lombardy) Bassano del Grappa (Veneto) Oderzo (Veneto)	Vigevano (Lombardy) San Giovanni Ilarione (Veneto)	Montebelluna (Veneto)	Biella-Cossato (Piedmont) Schio (Veneto)	Ostiglia (Lombardy)	Palazzolo sull'Oglio (Lombardy)	Marostica (Veneto) Sant'Ambrogio in Valpolicella (Veneto)	Suzzara (Lombardy) **C**
Centre-North	Santa Croce dell'Arno (Tuscany) **C**	Pesaro (Marche) **C** Poggibonsi (Tuscany) **C**	Fermo (Marche) **C** Porto Sant'Elpidio (Marches) **C**	Civitanova (Marche) **C** Montecatini Terme. (Tuscany)	Prato (Tuscany) **C**	Carpi (Emilia Romagna) **C**	San Benedetto del Tronto (Marches) Senigallia (Marches)	Sassuolo (Emilia Romagna) Castellarano (Emilia Romagna) **C**	Guastalla (Emilia Romagna) **C** Cento (Emilia Romagna) **C**
Centre-South	Solofra (Campania)	Matera (Basilicata)	Guardiagrele (Abruzzo) Casarano (Puglia)	Barletta (Puglia)	Ascoli Piceno-Val Vibrata (Abruzzo)	Giulianova (Abruzzo)	San Marco dei Cavoti (Campania) Putignano (Puglia)	Civitacastellana (Lazio)	non existent

The main source of data is ISTAT and data are calculated at communal level, unless otherwise specified. The features of 'extended labour division' and of economies of agglomeration have been 'proxied' as follows (name of the variable in Italics):

- firms' density: local units or establishments[9] in leading industry by inhabitants (*finh*);
- absolute number of local units and employees (*firm*, *empl*);
- rate of specialisation (by number of employees) in the leading industry (*spce*);
- rate of specialisation (by number of employees) in machinery ancillary industry (*meche*);
- rate of specialisation (by number of employees) in ancillary manufacturing industries (*ancye*) as defined in footnote[10];
- the rate of diversification in manufacturing industries (first ten industries) calculated as a Herfindhal index (*HH*);
- the percentages of micro-firms with fewer than 10 employees (C1), small firms with between 20 and 50 employees (C2), the percentage of medium-sized firms with more than 50 employees (*C3*);
- average size of local units (*size*);
- the proportion of artisan local units over the total in the leading industry (*arti*);
- distribution of population in industry sector by employment status: self-employed workers (*isem*), entrepreneurs - professionals, owners of

[9] A local unit or an establishment coincides with a single-unit firm. Several local units may belong to one company, which is the legal entity comprising various local units owned by the same person or group of persons.

[10] For each of the 5 industries considered in the sample, ancillary manufacturing industries (excluding mechanical engineering) have been defined as follows:

Specialisation or leading industry	Upstream ancillary industry	Downstream ancillary industry
Textiles	-	Clothing and upholstery
Clothing	Textiles	-
Tanning	Chemical products (chrome)	Footwear and leather clothing industry
Footwear	Tanning and plastics	-
Agricultural machines & tractors	Metals and motors	Other machinery
Ceramics	Other non-ferrous products	Paper

companies and members of cooperatives (*enti*), dependent workers (*iwor*), clerical workers and managers in industry (*icle*), ancillary workers (*ancy*) [11];
- degree of concentration of local production in the leading industry (% of local production controlled by the first 5 larger firms: *C5*);
- proportion of local turnover accounted by independent firms versus firms dependent on external (non-district) commissioning firms (*auto*).

A spatial concentration of firms in a bounded territory and in a given industry is a pre-requisite for external economies. What is required to grasp this feature is a proxy of the density of firms in a local context, rather than indexes of localisation or specialisation, which are instead better suited to measuring the territorial distribution of one industry in a given territory, for example a region or a province. In order to measure the concentration of one industry in an area, the absolute values of firms and employees in that industry were also included. In fact, in an area poor in firms, an index of specialisation may yield high values relatively to the presence of other industries in the area, but the number of firms may in any case be too small to generate significant external economies. *Firm* and *empl* are further indicators of agglomeration, given that it is to be expected that a greater industry size, in a given location, will permit the pursuit of greater specialisation.

The concept of an agglomeration of firms, or people, seems to come closer to the idea of a large number of small firms, rather than to the densest concentration in a given region. For this reason, two indicators of density are calculated, as the ratios between inhabitants and, respectively, firms (*finh*) and employees (*einh*).

The proportion of autonomous workers (self-employed people and entrepreneurs) is closely correlated with the absolute size of the leading industry; this suggests that an extended degree of labour division may be found in areas where the industry has grown to a considerable size.

Another hypothesis prompting the choice of the indicators for the number of firms and employees (*firm* and *empl*) is that the larger the industry, the more the local industry is fragmented; and the higher the specialisation rate, the more extended the labour division. Furthermore, the social weight of entrepreneurs

[11] Entrepreneurs: people who run their own business, where they do not employ their own or family manual work but those of employees only. Professionals: people who practice a liberal profession (engineers, technicians, etc.). Ancillary workers: people who collaborate with a member of the family who owns his own business, without having a regular labour contract. Managers: people having a supervisory position. Clerical workers: white-collar workers, e.g. secretaries or book-keepers. Workers: foremen, skilled workers, unskilled workers, auxiliaries, apprentices, home workers.

and self-employed people is a variable of crucial importance: in Piore's view of the ID as a 'community of equals' especially, since the role, income, status, and competencies of employers, *vis-à-vis* workers do not differ significantly. The more an industry is characterised by self-employed and ancillary workers, the closer it approaches an artisan, pairwise and family-based model of labour division. The precondition for an extended class of small entrepreneurs is, in fact, high work mobility. This in turn explains, according to Becattini, how internal mobility in districts creates job opportunities, high activity rates, very low rates of unemployment, and a strong propensity for independent work.

The rate of specialisation calculated both for the leading industry and for the other main industries of the *filière* grasps the dominant presence of one or a few complementary industries in an ID. Since external economies also extend to the diffusion of complementary or ancillary industries, the rate of specialisation was calculated in ancillary industries as well. In particular, the rate of specialisation in machinery manufacturing (*meche*) was calculated in order to test to what extent the leading industry is able to foster the development of complementary industries.[12] Complementary industries (*ancye*) were defined *ex ante* as related to the leading industry technological constraints. A condensed indicator of the degree of specialisation /diversification is the index of Herfindhal (built on the first ten industries). Owing to the classification of data available, it was not possible to assess the complementarity between industries pertaining to different processes or products but classified under the same code.

The proportion of artisan firms (*arti*) in the leading industry gives further information on the nature of the production process, since one may expect to find that the larger this proportion is, the less mechanised the latter will be.

The indicator of firms' average size (*size*) and the indicators of size distribution (*C1*, *C2*, and *C3*) capture a distinctive feature of IDs and of their market structure: their high degree of competition and labour division. In fact, the larger the number of firms and the smaller their size, or the larger the percentage of small firms in the total, the more extended the division of labour will be. However, caution is required when analysing these data, because a

[12] Industry data at municipal level between 1951 and 1961 are provided by ISTAT at a two-digit level classification. This means that mechanical industry comprises the manufacturing of all machinery, not only that used by the specialisation industry. Even considering that in 1996 a four-digit industry classification became available, it is very likely that the case of a strong mechanical industry and a scant presence of 'specialised' manufacturers is not possible, given the high degree of interrelatedness among different industries and therefore the ease with which the knowledge accumulated in the production of one machinery can be transferred to the manufacturing of a different one, provided there is demand for it.

similar distribution of firms by size may conceal different inter-firm relationships. Consequently, other information was used in the empirical research.

The market structure also affects the rate of growth of an industry. Authors who stress the value of non-pecuniary economies maintain that spreading the same employment over several firms increases the diffusion and dissemination of knowledge, which promotes growth. This result is also compatible with the view that small firms grow faster, although this does not perfectly fit the Marshallian model or other evidence.

The variable C5 was introduced to test whether processes of supply/enterprise concentration were occurring in IDs. Unlike the other indicators selected, the degree of concentration of production was extracted from secondary sources or estimated.

Complementary information analysed within the framework of the other indicators on human capital allowed construction of a qualitative variable (*auto*) assuming value 0 when the majority of local turnover was accounted for by subcontracting firms dependent on one or two commissioning firms, value 1 when it was covered by mostly independent firms, and value 0.5 when the contribution to local turnover by both subcontractors and independent firms was balanced. Since mostly qualitative information was used for this indicator, there was an obvious risk of inaccuracy for areas where the proportion is close to 50 per cent.

Apart from external economies indicators, other indicators depicting the economic and social features of the areas have been calculated:

- density of immigrants (*immi*): number of immigrants (resident and present foreign population) by inhabitants (*1000);
- rate of specialisation to the services sector: quota of population employed in services activities (Public Administration) over total employees (*serve*);
- density of social services (amenities): quota of firms in social services (sports, cinemas, gyms, etc.) over inhabitants (*sosf*);
- rate of specialisation in banking and financial services: quota of employees in banking and financial services over total employment (*banke*);
- rate of inactivity: population economically inactive over population in working age (*inac*);
- distribution of population (aged six and over) by education: higher education (*high*) (diploma, bachelor degree), compulsory education (*comp*) (primary and secondary school), illiteracy (*illi*); these indicators are relevant to grasp cultural views of local population and, at the same time, they give clues on the diffusion of skills and abilities among population;

- schooling drop-out rate in the 19-to-24 age group (*drop*);
- rate of unionisation: members of Italian main trade unions (Cgil and Cisl) over total employees (*unio*);
- plurality of votes: quota of votes of the larger party or coalition in national polls in 1992 over total votes; it measures the presence of a dominant political subculture, as in Trigilia (1986, 1990) (*poli*);
- rate of housekeeping: percentage of housekeepers over female population aged 14 and over (*hous*);
- proportion of young people (aged 14-29) involved in industrial activities (variable *apyo*);
- quota of 'extended families' (as those enclosing grandparents, nephews/nieces, and aunts) over total family classes/typologies (*famy*).

The aforementioned indicators do not exhaust the cultural dimension of a local area, for which we collected further evidence from secondary sources which cannot, however, be easily quantified.

13.3.6 Measuring performance

Performance is a multifaceted concept, which can be measured at a firm or system level. While company performance has its standardised indicators, it is more difficult to select performance indicators of a system of firms and people, as in our case. Indicators of social performance are very difficult to build. In addition, empirical testing of Italian IDs' competitiveness finds a substantial obstacle in the absence of National Account data on investments, added value (and sales) at a local level.

However we decided not to look at data at corporate level for two reasons. First and foremost, because the nature of the concept of ID, and of local system of SMEs in general, requires us to look at collective, that is social performance. Secondly, as already mentioned, available data on limited companies are not very representative. We have, thus, restricted our definition of performance to turnover per capita, competitiveness on export markets, social welfare and growth indicators - for the whole of the productive system included in the LLMA. We thus look at the following features for each area:

- competitiveness: variation of export in the leading industry *i* in the period 1986-1991 (on the corresponding variation in the same industry at the national level) (exp1) and 1991-1996 (exp6) - on the corresponding variation in the same industry at the national level[13];

[13] Source of data: ISTAT, 1997.

- industry global productivity: per capita turnover of the industry *i* (*pcsale*). The average value of turnover for each district (and industry) was divided by the corresponding average indicator at national level, in order to take into account the different structural features of each industry. Data on sales were collected by asking local trade associations or local institutions, or by conducting estimates on the basis of the available information (chambers of commerce, local trade unions, 'Club Distretti').[14] All the data refer to 1996;
- employment growth: percent variation in the number of employees and local units in the leading industry *i* between the four census years 1961, 1971, 1981 and 1991 (*vale, vafi*)[15];
- inclusiveness of local production system measured positively by the rate of economic activity of local population (*inac*) and negatively by the rate of unemployment (*unem*);
- welfare: per capita income in 1987 and 1996 (*inco87* and *inco96*) [16] and housing facilities (*M2in*: average square metres per building occupants; *M2av*: average square metres by dwelling; *room*: number of rooms by dwellings);
- infrastructure provision (*infr*) – ISTAT (1996).

Each of these indicators has its content of ambiguity, as they have a meaning which overcomes what they have to denote. This suggests that they should be used with caution, and their results critically appraised, rather than relying with confidence on only some of them.

The concept of competitiveness is generally linked to export performance, but this feature particularly suits the case of IDs, as they are deemed to be responsible for Italy's leadership position on foreign markets. Regarding export, we have to recall that data are in this case referred at the provincial level as data at municipal or LLMA level are not yet provided by ISTAT. Whereas most of the provincial employment of one industry is located in one LLMA, data on export referred to a province may be taken as a good approximation of the LLMA's export.

With regard to income, as well as production and sales, few objections can be made on a theoretical basis, as they are typical economic performance

[14] In order to estimate local turnover, due consideration was also given to the estimations by Moussanet and Paolazzi (1992), who carried out a survey between 1990 and 1991 in most of the IDs examined here.

[15] The calculation of variation in the number of firms and employees at 3-digit level and at communal level since 1961 has been made possible thanks to a new data-set provided by Istat (1998) where the figures for the various census years are made comparable.

[16] Source of data: Banco di Santo Spirito (1987).

indicators. Objections may be valid, according to the way they are calculated. The income indicator is, in fact, an average value (million lire per capita), which may conceal major inequalities in the distribution of wealth within local population. Concerning the variable on the average square metres per building occupants, one objection is that this variable may well reflect certain cultural attitudes, rather than the actual welfare of population.

Infrastructure provision may, to some extent, be conceived as an indicator of context-related performance. It is also a measurement of economies of urbanisation. The more a population is culturally attached to its own territory, the more it should devise political instruments to equip the local area of the necessary infrastructures.

Unemployment is an indicator of both structure and performance[17].

One of the main and distinctive characteristics of the 'canonical' ID is cooperation, and many researchers in the field assume a definition of them as networks of firms based on trust and cooperation. As we illustrated in a previous paper (Paniccia, 1998), cooperation is a dynamic phenomenon affected by structural and contingent facts, which must not be taken for granted in any place where SMEs are agglomerated. For this reason, either because it is a structural dependent variable, or because its analysis is not specifically addressed in this paper, we did not calculate any cooperation indicator.

To the extent that our measurements of external economies of agglomeration, efficiency, and effectiveness are the most appropriate, the second step was the assessment of their causal relationships.

13.3.7 The adequate technique for analysing IDs' performance

For hypothesis or for testing causal relationships, at least three instruments are available: multivariate analysis, correlation analysis and econometrics. From what has been said before, it should be quite clear that performance results from a complex interaction of different factors. Hence, to proxy these explanatory causes with only one variable may be misleading or disappointing. In this respect, multivariate techniques may be appropriate. However, they are

[17] Performance may also be analysed in terms of innovative capacity; indicators of innovative activity at municipal level may prove very difficult to find. In addition, R&D expenditure indicators are entirely inappropriate in areas dominated by SMEs, which tend not to formalise or create dedicated departments for their innovative activities. A low propensity to patent is also to be expected, since innovative processes in areas dominated by SMEs take place within an unformalised structure, and generally concern incremental advances that are unlikely to be patented (Cantwell, 1989).

conceived mainly as descriptive tools of analysis by economists, more suited to social or natural sciences, who prefer 'harder' and more robust tools of analysis, such as econometrics, which are believed to have an explanatory value.

However, multivariate analysis may also have a predictive value, provided that certain conditions are satisfied. In our case, cluster analysis certainly has a taxonomic aim, but it is also intended to verify if the conceptual category of ID, which is indeed a theoretical classification, corresponds to an empirical unit, or group, and then if a similar and superior performance is associated with this cluster. For this reason, the cluster and factor exercise will consider only 'independent' variables. The factors thereby obtained can be regressed on the indicators of performance.

For the purposes of this paper we illustrate only the results of multivariate analysis (omitting here the regression analysis) for all that concerns the relationship between static indicator of performance and socio-economic factors. As well, the determinants of employment growth are here presented in an exploratory way.

13.4 Multivariate analysis

Multivariate analysis has been carried out on the 'independent', i.e. 'structural' variables used to operationalise the distinctive features of 'canonical' IDs.

As a preliminary process, we excluded dynamic variables, such as the variation in the rates of industrialisation, activity, and so on, as they may be considered a performance result. Moreover, we dropped closely correlated variables and those indicators whose sum amounted to 100 (e.g. all the composition indexes), retaining only those with the highest statistical explanatory value, the purpose being to avoid statistical redundancy. A total number of 31 active or independent variables were considered by the multivariate analysis, and eight of them were deemed to be active in 1991. According to the resulting correlation values, the independent variables considered were: *hous, apyo, clei, isem, arti, HH, C3, auto*.

13.4.1 Factor analysis

The analysis was carried out using a French software program: SPAD (*Système Pour l'Analyse de Donnée*). The procedure followed the principal component method for factor analysis, and then a hierarchical method of clustering. The number of relevant factors was set at three, according to a common criterion, since they account for more than 61 per cent of the total sample variance.

The resulting factors have been interpreted as a coherent combination of economic and social features, each roughly describing a typology of local economy. Factor analysis partitions the variables between a positive and a negative quadrant of the axis (factors). A summary description of the positive and negative variables combined in each factor is provided below. Factor loading (that is, the correlation between factors and variables) is given in parentheses.

An ideal factor analysis would have produced only one single 'pure' significant factor, and cluster analysis would have distinguished two groups: canonical and non-canonical IDs. On the contrary, our results show the existence of a multidimensional space of analysis in local economies, which probably requires a richer lexicon than the distinction between Marshall-style, canonical or non-Marshall-style, non-canonical IDs.

Factor 1: according to the loading values and their signs (in parentheses the variables placed on the negative quadrant) [18], the first factor, the one with the highest explanatory value, clearly denotes an industry based on firms employing a relatively lower (higher) proportion of clerical workers, generally of a smaller (larger) size and of an artisan nature, generally non unionised, mainly dependent on external clients (mainly independent); while the specialisation industry appears not to be integrated (integrated) with mechanical engineering activities. The low (high) proportion of young workers and immigrants (among the illustrative variables) and the high (low) rate of housekeeping, together with high rates of unemployment and illiteracy, describe the features of a non-inclusive (inclusive), and marginal economy, mainly based on traditional services rather than industry (Table 2). This factor may be termed the *marginality factor*, when the positive segment (quadrant) of the axis is considered, and the *firms' tertiarisation, concentration* and *integration factor* when the variables negatively correlated are considered (henceforth, the first name of the factors' label is related to the variables on the positive quadrant, the second to the negative ones).

The second factor, which explains 23 per cent of the total variance, summarises on the negative quadrant the features of an economy based on family-owned, artisan and often independent enterprises run by self-employed workers, employing 20-50 units on average, and typical of a society where the

[18] In multivariate analysis each factor identifies the axis of a four-dimensional Cartesian space.

relative weight of extended families is still substantial. It also combines most of the (both active or illustrative) indicators that describe the features of an extended labour division, based on a densely concentrated and sizeable leading industry, where the degree of market concentration is low and the leading industry is highly specialised.

Table 2 Coordinate values of explanatory factors – 1991
Source: author's calculations.

Variables	**Factor 1**	**Factor 2**	**Factor 3**
ACTIVE VARIABLES			
arti - proportion of artisan firms	0.49	-0.17	0.35
HH - diversification index	0.22	-0.52	-0.69
C3 - medium firms weight	-0.61	0.44	-0.08
clei - clerical workers	-0.74	-0.21	0.36
isem - self-employed workers	0.09	-0.84	-0.02
hous - rate of housekeeping	0.30	0.40	-0.40
apyo - young workers	-0.39	0.29	-0.51
auto - independent firms	-0.67	-0.57	-0.18
ILLUSTRATIVE VARIABLES (first 20)			
einh - employees by inhabitants	0.36	0.31	0.45
emp - lead industry employment	-0.21	-0.57	-0.03
serv - rate of tertiarisation	0.32	0.02	0.31
C1 - micro firms weight	0.32	-0.69	0.07
C2 - small firms weight	0.63	-0.03	0.06
C5 - degree of market concentration	-0.61	0.23	0.03
enti - entrepreneurial rate	0.37	-0.47	-0.16
iwor - dependent workers	0.32	0.79	-0.15
ianc - ancillary workers	0.13	0.72	-0.21
meche - mechanical special.	-0.35	-0.20	0.31
immi - rate of immigration	-0.40	0.18	-0.15
famy - extended families	`0-35	-0.38	-0.01
isef - business services	-0.30	-0.23	-0.04
bank - financial services	-0.32	-0.32	-0.12
unio - rate of unionisation	-0.23	-0.42	0.06
high - higher educated people	-0.14	-0.20	0.44
unem - unemployment rate	0.48	0.10	-0.13
asso - associations rate	-0.22	-0.17	0.01
illi - illiteracy rate	0.69	0.15	0.02
drop - schooling drop-out	0.06	0.27	-0.02

On the same negative quadrant, the rate of unionisation also appears as highly correlated (Table 2). Since all the indicators of agglomeration and specialisation, and of an extended labour division based on self-employed and ancillary workers are closely (with a negative sign) correlated with this axis, we call it the *non canonical/canonical* factor.

Factor 3, which explains a lower amount of variance, is a specification of factor 1 that stresses on the negative quadrant diversification (specialisation), low employment of young workers, but a low rate of housekeeping. Looking at the illustrative variables, it appears that the specialisation industry is not homogeneously extended or geographically concentrated, while the rates of specialisation in the service sector and the proportion of highly educated people (on the positive quadrant) are highly correlated. All together these variables point to still immature processes of industrial agglomeration and specialisation. It is therefore apparent that a fragmented business structure may arise even when the size of the industry is not large (Table 2). This factor may be termed the *diversification/specialisation* factor.

13.4.2 Cluster analysis

Hierarchical clustering techniques are very common in multivariate analysis and with respect to optimisation techniques they present less arbitrary steps. According to this technique, data are not partitioned into classes in one step. Rather they are first separated into a few broad classes, each of which is further divided into smaller classes, and each of these further partitioned, and so on until terminal classes are generated and not further subdivided. By using an agglomeration method[19] a distance matrix between the entities is accordingly selected. According to the method of Ward, the units of observations are clustered in a manner which results in the minimum increase in the error sum of squares (ESS), as quoted by Everitt (1981).

The obtained clusters are the following:
Cluster 1: Solofra, Vigevano, Marostica, Bassano, Carpi, Civitanova, Pesaro, Guastalla, Desio, Poggibonsi, San Benedetto del Tronto, Fermo, Montecatini Terme, Sant'Ambrogio in Valpolicella.
Cluster 2: Santa Croce dell'Arno, Porto Sant'Elpidio, Prato.

[19] Essentially hierarchical techniques may be subdivided into *agglomeration* methods, which proceed by a series of successive fusions of the N entities into groups, and *divisive* methods, which partition the set of N entities successively into finer partitions (Everitt, 1981). As the two methods are equivalent, we chose the agglomeration one.

Cluster 3: Arzignano, Montebelluna, Sassuolo, Civitacastellana, Cossato, Suzzara, Cento, Schio, Oderzo, Palazzolo sull'Oglio, Castellarano.
Cluster 4: Giulianova, Ostiglia, Ascoli Piceno (Val Vibrata), Barletta, San Marco dei Cavoti, Senigallia, Putignano, San Giovanni Ilarione, Guardiagrele.

The fourth cluster groups together areas belonging to the same geographical partition, while areas in northern and central Italian areas are split in all four clusters.

Cluster 1: Craft-based or urbanised/tertiarised industrial districts

The first cluster is correlated very positively with the *non-canonical/canonical* factor and with the *diversification/specialisation* factor (the values of correlation are respectively -2.3 and 3.0). It is therefore explained by an extended division of labour based on family-owned and run firms that have few employees and rely on family members or relatives. The structure of the industry is highly fragmented, in that it consists of independently-owned firms scattered among small units. The distinctive features of this cluster are the relatively large proportion of entrepreneurs and self-employed workers, a low rate of schooling drop out, extended families, high rates of associationism and unionisation and political subcultures. These areas are however not homogeneous as concerns, for example, a shared political subculture, their unionisation rates, infrastructures, and specialisation in the mechanical engineering industry (Table 2). When the *auto* indicator is analysed, together with indicators related to the quality of human capital and unionisation rates, differences emerge internally in the areas with strong agglomeration economies.

This typology comprises IDs like Civitanova, Pesaro, Poggibonsi and San Benedetto del Tronto, in which a productive structure based on an extended labour division in a leading industry is embedded in a typically urban context: this demonstrates that an ID may be compatible with a non-manufacturing environment. The large weight of the variables *micro*, *high*, *arti*, *serve* suggests calling these IDs 'craft-based' or 'urbanised/tertiarised' IDs.

Cluster 2: Canonical (or quasi-canonical) industrial districts

The second cluster, which consists of a small number of IDs, exhibits the largest number of 'canonical' features. Therefore they coincide, on average, with the canonical IDs. The group is in fact correlated with the second and third factors with values of respectively – 3.6 and – 3.5. Indeed, this cluster groups together the best known and most frequently studied IDs, with Prato often taken as a paradigmatic example - although even a glance at its descriptive statistics shows that it is hardly imitable. The degree of fragmentation of production in these IDs is even more extreme than in the previous case (the quota of employment in local units with fewer than 50 employees is over 80 per cent). Self-employment is one of the main distinctive social features of these IDs, as in the case of Sforzi's Marshallian IDs (1990). Among the most distinctive features there are a high specialisation, high proportion of self-employed people and ancillary workers, a wide extension of the leading industry and entrepreneurial density.

The social cohesion of these areas may be deduced from their relatively pronounced political subcultures, and from a high level of social associationism and trade-unionisation (with the exception of Porto Sant'Elpidio), although unionisation is a distinctive feature of typology 3 as well. A weakness of this cluster of IDs is their relatively high rate of schooling drop-out, which reveal a distinct attitude towards education among the local population (Table 3). Education is perceived as a means to achieve social and economic emancipation; but this concerns only compulsory schooling, while higher education is not regarded as providing easier entry to the local labour market. Another distinctive feature of this typology is the high proportion of business services, which confirms that while firms are lean and focused on manufacturing, they are able to rely on an abundant supply of services provided by public and private organisations in the neighbourhood. This is the case, in particular, of Santa Croce and Prato (Table 3). For craft IDs the evidence is instead inconclusive in this regard, since firms specialised in business services are more abundant in the urban and generally tertiarised IDs only.

Cluster 3: Concentrated and organisationally upgraded industrial districts

This cluster is characterised by a high proportion of clerical workers matched by a relatively low proportion of ancillary workers. Firms within these IDs tend to be larger, and most of them have more than 50 employees (Table 3). The predominant specialisation is integrated with a strongly-developed mechanical industry which makes the tools and machines used in the production process of the leading industry. Society and industry comprise high proportions of immigrants and young working, while housekeepers are only a minority in the female population. Considering that the most significant variables for this cluster are *medium*, *C5*, *apyo*, *auto*, *immi* and *clei*, we may call these IDs 'concentrated', 'inclusive', 'organisationally upgraded' and 'integrated' (with the mechanical engineering industry).

Table 3: Distinctive features of the obtained clusters - average values
(*) manufacturing industry
Source: author's calculation on ISTAT (1995a, 1995b).

	immi	enti	clei	isem	ianc	apy	high	drop	illi	sosf	serve	hous	unio
Canonical IDs	7.4	6.2	16.5	16.4	2.0	16.4	20.8	20.2	15.1	506.9	35.4	30.2	50.0
Craft-based IDs	7.9	7.9	11	22.4	3.5	18.2	16.3	26.5	14.2	461.4	44.8	28.6	41.1
Concentrated IDs	10.4	5.1	16.6	13.1	1.3	21.2	17.5	20.8	12.1	517.4	35.8	30.3	39.5
Embyronic IDs	5.5	5.8	9.8	13.1	1.4	15.7	17.1	23.1	20.8	636.4	45.7	32.9	26.7
Italy average	*10.1*	*5,1*	*13.9*	*13.9*	*17.5*	*10.7*	*3.8*	*15.4*	*14.3*	*90.0*	*50.3*	*31.7*	*18.0*

	poli	Firm	Empl	finh	HH	C3	size	C5	meche	auto	arti	bankf
Canonical IDs	36.8	3483	18820	172.1	247.4	22.3	9.2	22.5	3.8	0.9	72.8	1.4
Craft-based IDs	36.3	793	5720	47.1	5592.0	9.6	6.7	8.0	1.9	1.0	72.1	0.9
Concentrated IDs	35.9	311	6252	290.8	112.5	56.0	23.5	29.7	3.3	1.0	55.1	1.1
Embyronic IDs	34.1	172	2140	359.7	48.6	26.4	12.1	19.2	0.3	0.1	59.5	0.7
Italy average	*38.0*	-	-	*61.6**	-	*57.9**	*8.8**	-	*9.2*	-	*66.1*	*1.8*

The proportion of clerical workers is greater than that of self-employed. As discussed later, a less 'extended' division of labour has formed in this group. Clerical workers include those occupying positions of responsibility in product development, sales or quality assurance departments. These workers are required to satisfy the increasing demand for innovation, internationalisation and quality control. This indicator therefore grasps the inclusion of new professional figures in the production cycle.

The integration of the leading industry with their ancillary industries describes clusters of activity *à la* Porter (1990). Garofoli would describe these districts as 'system areas' (1991).

Cluster 4: Peripheral/embryonic industrial districts

The fourth typology is significantly explained by all the three factors. This suggests a relatively higher internal heterogeneity that is in fact mirrored in the higher degree of inner inertia. The correlation values are –2.0, –3.4 and –2.6

respectively for the first three factors. This typology includes not only all the southern IDs, with the significant exception of Solofra, but also areas located in the so-called 'Third Italy' regions (those with the highest density of districts), notably Lombardy and Veneto. The districts of San Giovanni Ilarione and Ostiglia may in fact be viewed as offshoots of larger and stronger IDs in contiguous LLMAs. The Southern areas do not all display the same pattern of labour division. Ascoli Piceno and Giulianova are slightly more upgraded than San Marco dei Cavoti or Guardiagrele, while Matera and Casarano, on the other hand, display a model of labour division based on small firms and more vertically integrated industries coexisting side by side (Table 3).

This cluster includes IDs whose industrial vocation is still weak, given the small extent of the specialisation and the manufacturing industry and this impacts on the local labour market, which is characterised by high levels of unemployment. A metalworking or mechanical engineering tradition and, more generally, the development of pecuniary external economies are absent from southern areas. With regard to social features, the most significant and distinctive variables in cluster 4 are rate of unemployment, rate of inactivity and rate of illiteracy. As shown below, a negative social performance, a low level of infrastructure and social services (*sosf*) qualify these areas.

13.5 The performance of industrial districts' typologies

We now look at the performance of the different typologies. Table 4 shows that concentrated IDs exhibit the best performance on average, in that they score low on unemployment, high in global productivity and per capita income (mainly in 1991). They also achieve good export performance, although the level is lower than in the case of embryonic IDs. As regards individual indicators, the results are again varied. Canonical IDs outperform the other with respect to the index of infrastructure, while craft-based or diversified IDs have a higher average size of dwellings. Embryonic IDs are the most dynamic and overall they exhibit a variation in export which is much higher than the national average. These results point to the fact that good performance may be achieved even if the area does not comply with all the requisites of an ideal-typical ID.

Concerning infrastructures, all southern areas show very bad performance (the range of variation of the indicator is between –100 and +100), while in the other areas they are positive. According to some recent works, physical provision is always considered inadequate by local operators, especially in the

Veneto region (Perulli, 1998). This fact clearly shows how economies of urbanisation are not always a result of agglomeration processes.

Table 4: The performance of IDs 1991 – average values
Source: author's calculation on various sources (see Section 13.3.6).

	unem	Inco 87	Inco 96	M2	M2in	room	infr	Saled	Expo 91	Expo 96	vafi 81-91	vale 81-91	vafi 91-96	vale 91-96
Canonical IDs	10.1	13.4	25.4	108.5	34.5	4.9	19.8	0.6	1.2	0.8	-17.0	-12.5	-1.6	4.9
Craft-based IDs	9.4	13.1	25.9	108.7	36.2	4.5	8.5	0.7	0.7	1.2	13.1	12.7	-23.0	-15.7
Concentrated IDs	7.1	13.4	25.6	110.7	38.3	4.8	14.7	0.9	1.4	1.7	21.1	22.4	-11.7	-20.2
Embryonic IDs	13.5	1.8	21.3	107.5	36.2	4.9	-48.4	0.5	2.8	1.1	65.8	99.6	-25.4	-18.8
Italy average	17.8	13.3	28.8	94.1	33.0	4.3	100.0	-	-	-	-4.1	-6.2	-0.1	-6.8

From a dynamic point of view, the areas, which share the minor characters of the 'canonical' model, are those where employees (and also export) grow faster. The variation in exports between 1991 and 1996 – period that benefit from the Lira's devaluation - occur above all in the areas where density of firms is lower and where the average size of firms is higher. Looking at the average values inside each cluster, we can observe that the most dynamic cluster in the same period is cluster 4, which includes the less mature areas.

But the relationship between agglomeration and growth is better addressed in the following section.

13.6 The relationship between economies of agglomeration, competition and growth

In this section we test the relationship between proximity and industry structure (degree of specialisation and vertical integration), on the one hand, and growth on the other, in the period 1951-1996. The growth of the areas in the sample is compared with the national average in order to find evidence that firms in IDs grow faster than those located in non agglomerated areas.

In other words, here we test whether the ID, as a local agglomeration of SMEs, ensures average or better-than-average growth over time and whether agglomeration insulates the area from the job loss and firm failures of short-to-intermediate-term business cycles.

The influence of specialisation on growth is also examined. A critical question concerns whether agglomerative specialisation and success in one industry can actually impede the development of other sectors, whose presence may diversify the economy, and counteract maturation or instability in the original sector. The impact of specialisation may also vary, according to whether it is associated with some degree of market power and/or dominance over local factor markets.

In Table 5 the rates of employment growth (in percentage) of the specialisation industry in each ID are calculated as index numbers (1961 = 100), and they are divided by the indexed rate of growth of the corresponding industry at the national level. The data offer evidence that, on average, the sample of IDs grew more than the corresponding (specialisation) national industries between 1961 and 1996. The various clusters, on the contrary, do not appear to identify clearly distinctive patterns of growth. The obtained typologies may have a predictive value as regards short-term performance, but they are unable to discriminate clearly among different patterns of development. Apart from the cluster of embryonic IDs, which groups together all the areas in which development started in the 1970s, the more mature IDs are scattered among the three typologies, which do not distinguish between old and more recent IDs. At the same way, it does not emerge that IDs belonging to the same industry have a similar pattern of growth. This evidences that even different points of departure may have converged on a similar pattern of labour division in the 1990s, or the reverse, that starting from common initial conditions or facing similar technological or market conditions, the various IDs then follow idiosyncratic processes of growth. For this reason, the sample was divided into three groups according to their average values in 1951. On inspection of the size of the leading industry in 1951, just after World War Two, three groups can be distinguished according to average values of employment. The first group, the smallest, includes the oldest IDs, those whose origins date back to the beginning of the last century, or which were the sites of the Italian industrial revolution in the 1800s. The group of 'post-war' IDs comprises those whose specialisation industries already averaged 3000 employees in 1951, while the younger IDs grew in the 1970s. Figure 1 shows the different patterns of development of these three groups very clearly. The group of historical districts grew at a steady rate, while the younger ones expanded dramatically. The group of post-war IDs display intermediate rates.[20]

Table 5: Growth of employment (index number deflated by the corresponding national index) between 1951 and 1996. Intercensus variations – average values
Source: author's calculation.

	51-61	61-71	71-81	81-91	61-91	61-96	91-96
Canonical IDs	2.1	1.4	1.3	0.9	1.7	1.8	1.1
Craft-based IDs	1.8	1.8	1.5	1.0	3.0	2.8	0.9
Concentrated IDs	1.4	2.2	1.3	1.2	3.3	3.2	1.0
Embryonic IDs	0.9	2.7	2.8	2.3	9.3	9.6	1.1
Sample average	*1.1*	*1.2*	*1.1*	*1.0*	*1.3*	*1.3*	*1.0*

As regards the relationship between employment growth and economies of agglomeration, growth is therefore taken as the dependent variable, while rate of specialisation, firms' density, absolute size, and average size of firms in the leading industry, are taken as explanatory variables. The first three variables are considered as a specification of external economies at a local level, while the variable size is considered as a proxy of the degree of competition in that industry and as an indicator of the extent of labour division. Variables C1, C2, C3 and C5 have been discarded because they showed low values of correlation with growth indicator.

The correlation values are illustrated in Tables 6. Growth has been calculated for each of the ten-year inter-census periods, between 1951 and 1991 and between 1991 and 1996, and for the longer period 1961-1991. For each decade, the relative growth indicators have been correlated to the explanatory variables at the beginning of that decade. Similarly, the rate of variation in the period

[20] *Young IDs*: Solofra, Casarano, Bassano, Giulianova, Val Vibrata, Montebelluna, Civitanova, Barletta, Matera, Oderzo, Poggibonsi, San Benedetto del Tronto, San Marco dei Cavoti, Senigallia, Putignano, San Giovanni Ilarione, Fermo, Guardiagrele, Sant'Ambrogio di Valpolicella.
Post-war IDs: Arzignano, Santa Croce, Sant'Elpidio, Marostica, Sassuolo, Civitacastellana, Carpi, Suzzara, Guastalla, Cento, Palazzolo, Montecatini Terme, Castellarano.
Historical IDs: Vigevano, Prato, Cossato, Desio, Schio.

1961-1996 has been correlated to 1961 variables. In order to avoid a 'size' effect - that is, a high percentage variation linked with very small absolute variations - the growth indicator *vale* has been calculated in logarithms. Also the variables with the highest standard deviation, such as *finh*, *firm* and *empl*, have been calculated in logarithms in order to compress a scale effect.

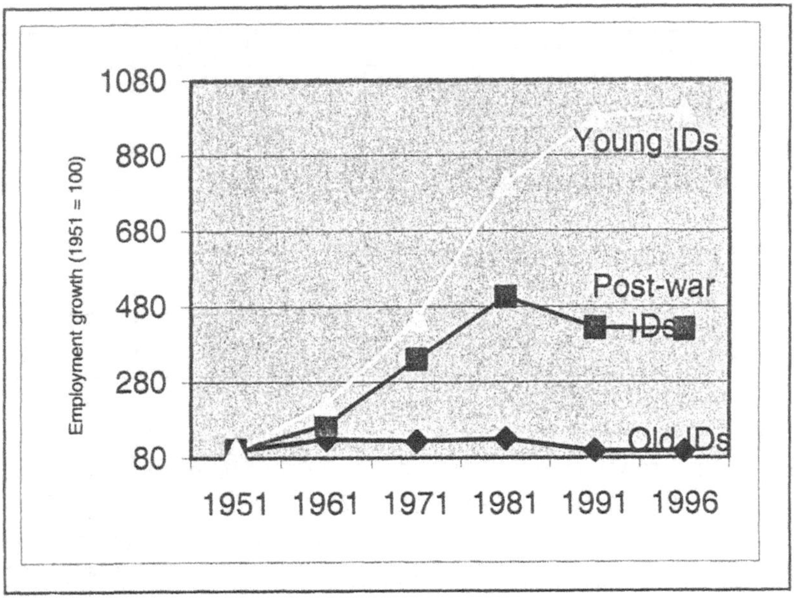

Figure 1: Patterns of growth of selected industrial districts. 1951 – 1996
Source: author's calculations.

The first thing to be pointed out is that growth reacts differently to the same factors in different periods of time, since the sign and the value of correlation are not the same for each of the four ten-year periods considered. 1981 appears to be a turning point in the relationship between growth and the agglomeration and competition indicators, given that the signs of correlation – though not always significant - are the same for all the variables between 1961-71 and 1981-91, but not between 1951-61 and 1991-96. The variable size is weakly significant owing to its lower variance.

The results also show that specialisation, density of firms, stock of firms, and employees are negatively (or insignificantly) related to growth over the period 1951 - 1996. Positive feedback in agglomeration processes may therefore not be expected to persist. Indeed, beyond a certain point, congestion reduces the attractiveness of an existing agglomeration for new entrants – particularly if the specialisation industry is entering a maturity phase in which technological opportunities are almost exploited.

Table 6: Correlation values among employment growth, agglomeration, specialisation and competition. 1951 – 1996

Source: author's calculations.

	Agglomeration and competition indicators					
Growth indicators	Ln(finh)	Ln(einh)	HH	Ln(firm)	Ln(empl)	Size
Lnvale 51-61	-0.2	-0.3	-0.2	0.3	0.3	0.0
Lnvale 61-71	0.2	0.2	-0.2	-0.1	0.1	0.1
Lnvale 71-81	0.4	0.5	-0.4	-0.2	-0.5	-0.3
Lnvale 81-91	0.5	0.6	-0.4	-0.4	-0.5	-0.2
Lnvale 91-96	0.0	0.0	0.2	-0.2	-0.1	0.2
Lnvale 61-91	0.4	0.4	-0.5	-0.2	-0.3	-0.2

Growth in the period between 1971 and 1991 is also negatively correlated with the variables measuring the degree of competition (*firm*) and average size of firms (*size*). Overall these results highlight the marked dynamism of the less industrialised areas in the sample. When the correlation values are calculated for each of the three sets of Ids, divided according to their age of formation, the results are slightly different. According to these results, the correlation values between growth and agglomeration are high for the first years until 1981, for both old and post-war IDs, and negative for young IDs. A specular situation characterises the decade between 1981 and 1991. The correlation signs of the variable related to employment (*empl*) seem to suggest that the role of agglomeration and specialisation is stronger, or positive, only in an ID's (and industry's) first stage of development, while it loses in importance when the industry reaches maturity. Similarly, in the period of an ID's maturity, a high rate of specialisation or a large scale of industry is not a necessary condition for growth. The maturity of an ID coincides with a stage at which fewer new firms are born and the growth externality for incumbent firms perhaps no longer operates. But even if employment decreases, this in itself does not necessarily imply decline; rather, it may simply imply stabilisation at a particular size (Baptista and Swann, 1998). Since a decrease in employment may derive from a rise in productivity rates, it can also be paired with the growth of more structured firms, thus signalling that processes of organisational restructuring are at work.

We may therefore conclude that neither agglomeration nor specialisation factors induce one industry to grow faster than others in any case: the history of the area has to be taken into account. In our sample, a distinction has been drawn among areas whose origins date back to the 1970s, others, which arose just after World War Two, and a further few, which began two centuries ago. The correlation values may therefore be appraised with respect to each of these groups separately. Our results suggests that any test of the MAR versus the Jacob hypothesis would require preliminary verification that the sample selected is homogeneous with respect to the age of formation of the ID under investigation.

13.7 Concluding remarks

13.7.1 Industrial districts' performance

Our results show that external economies of agglomeration have a positive effect on the performance. However, in a dynamic perspective, the growth of export, employment, firms, or production, does not happen where they are strongest. A life-cycle scheme does apply.

The results show that a superior performance is not strictly linked to a specific socio-economic model, specifically characterised in its organisational or social features. The same results also allow for an evolution of the model of 'canonical' ID towards more services-oriented or more integrated organisations, where the role of the external context, the active institutions, may have been different from those underlined in the literature on 'canonical' IDs (extended families, political parties, local councils or provinces). The conditions which Becattini requires for the existence of the ID, such as an extended division of labour, diffusion of local banks, the presence of a political subculture, etc, do not appear able to ensure the stability of the 'canonical' ID. Indeed, the latter does not stand out as a normative model. As we found that ID performance is not fully explained by socio-economic factors, we have to admit that a high performance can be associated to different classes of local systems, that is 'canonical' IDs as well as less canonical ones. We did not find strong evidence supporting the idea that one model is superior to all others, as each system evolves according to its internal resources and capabilities. However, there are signs that among the first three clusters, embracing the areas located in north-central Italy, the LLMAs, where a more structured model of labour division has been achieved, are those showing the best indicators of productivity and export growth.

The higher concentration of production which can be observed in all the clusters with the exception of the 'canonical' one, and the replacement of spontaneous or informal relationships with formal subcontracting relations, the creation networks and constellations of firms, leading or 'interconnecting' firms, does not necessarily enlarge the gap between social and economic performance. Concentrated IDs are an example of how a restriction of firm population involved in industrial activities does not impair the social welfare of local population. However, these results should be evaluated in the long term, while our research provides only first accounts. The new organisational forms, such as the emergence of large firms, may also generate external economies for the local environment. In this evolution, the role of new and single protagonists is very crucial (Ferrucci and Varaldo, 1993; Corò and Grandinetti, 1999). They may introduce radical innovation thereby stimulating smaller firms to innovate their product/process, which they would not be able to reconceptualise, without an external pressure. Large firms can produce public goods which generate externalities for smaller firms (logistic infrastructures, training centres). This is, for example, the case of Matera, where the leader firm, Natuzzi, has promoted several initiatives for the diffusion of information and technological capabilities (Baculo, 1996; Belussi, 1999).

Hence, the formation of protagonists out of the scale of the district - larger firms or groups[21], or firms resorting to delocalisation processes – have not hindered the performance of the area, at least in the period observed in our research.

13.7.2 Policy implications

The large variety of real IDs resulting from our work suggests an application of differentiated industrial policies. Blanket solutions must be avoided, since IDs are very dissimilar with respect to the structure of inter-firm relationships, the nature of the competitive advantage, weakness and strength points, and the propensity growth and innovation. Also the variety within IDs, as related to the

[21] In this respect, Becattini excludes that a structure of relationships between firms of an asymmetric nature, where some firms may have a major role than others, can be considered as an ID: "Coalitions and agreements of various types and importance between firms in the district also occur, but whenever the weight of those which take a financial form increases too much, or the growth of some firms sends it "out of scale", as it were, we are already out of the canonical form of the Marshall district" (1990; p. 41).

emergence of groups, or of integrated firms, may suggest selective policies aimed to pick up and foster the more dynamic actors. To the extent that the function of "out of the scale" actors in IDs does not damage the collective performance or the welfare of IDs - as our first results show - they should be sustained by policy measures. The presence of a leading firm as orchestrator of flows of goods and information in local areas may, in fact, be a point of strength rather than of weakness. Our results also show how IDs having a more integrated industrial structure, with a bunch of firms playing an interconnecting role, providing new ideas, techniques, best practices, technologies, have better performance. Therefore these firms or networks should be sustained and encouraged to disseminate their knowledge locally. It is of paramount importance that such policies be devised locally by partnerships between leading firms, local associations of employers, trade unions, and councils, though within a national framework.

Networking of IDs, in order to extend their sources of knowledge and stimuli to change, is essential. If the agglomeration of many SMEs in bounded places has historically represented in Italy a protection against the diseconomies of small scale and, at the same, time a propellant for its virtue, in the present scenario, SMEs might suffer from isolation, even if they are 'integrated' in an industrial district.

The industrial district as an organisational model helped in the recent past to overcome the diseconomies associated with small size in the access to critical resources. But the new competitive challenge will penalise the industrial districts, which are isolated and not connected to the external centres of production of new ideas and knowledge, remaining locked-in in their industrial trajectory.

The entry of new firms, whether domestic or multinational, may represent a path-breaking event, capable of introducing a new rationality of action, and producing a new pattern of division of labour among local firms (see also Amin, 1997).

Given that change may originate from outside, improving co-operative relationships and building networks that reach outside agents or institutions, may prove more productive for some localities than concentrating on indigenous resources.

The enacting of these policies requires an upgrading and a deep transformation of economic and social structures: a higher degree of education of workers and entrepreneurs, a better quality of the urban and natural environments, more effective infrastructures, and a more co-ordinated and 'interconnected' division of labour. Indeed new collective entrepreneurs are necessary.

References

Asheim B. T. (1994), Industrial districts, inter-firm co-operation and endogenous technological development: the experience of developed countries, in, Van Dijk M. P. *et al.* (eds.), *Technological dynamism in IDs: An alternative approach to industrialisation in developing countries?*, Unctad, United Nations, New York and Geneva.

Amin A. (1997), Institutional foundations and problems of adaptation in entrepreneurial regions: illustrations from Emilia-Romagna, *mimeo*, Durham, University of Durham.

Arrow K. J (1962), The economic implications of learning by doing, *Review of Economic Studies*, vol. 29, p. 155-173.

Baculo L. (1996), (ed.), *Impresa forte politica debole,* Edizioni Scientifiche italiane, Naples.

Banco di S. Spirito (1987), *Il reddito dei comuni italiani*, Rome.

Baptista R. and Swann P. (1998), 'Do firms in clusters innovate more? *Research Policy,* vol. 27, p. 525-540.

Baumol W. J., Panzar and Willig R. (1982), *Contestable Market and the Theory of Industrial Structure*, Harcourt J. Brace Jovanovich, New York.

Becattini G. (1987), (ed.), *Mercato e forze locali: il distretto industriale*, Il Mulino, Bologna.

Becattini G. (1989), Sectors and/or districts: Some remarks on the conceptual foundations of industrial economics, in, Goodman E. and Bamford J. (eds.), *Small Firms and Industrial Districts in Italy*, Routledge, London.

Becattini G. (1990), The Marshallian ID as a socio-economic notion', in, Pyke F., Becattini G. and Sengenberger (eds.), *IDs and Inter-firm Co-operation in Italy*, International Institute for Labour Studies, Geneva.

Bellandi M. (1987), The industrial district in Marshall, in, Goodman E. and Bamford J. (eds.), *Small Firms and Industrial Districts in Italy*, Routledge, London.

Bellandi M. (1992), The incentives to decentralized industrial creativity in local systems of small firms, *Revue d'Economie Industrielle*, vol. 59, p. 99-110.

Belussi F. (1999), Path-dependency vs. industrial dynamics: the analysis of two heterogeneous Italian districts specialised in leather upholstery, *Human System Management,* vol. 18, no. 2, p. 161-174.

Best M. (1990), *The New competition: Institutions of Industrial Restructuring*, Harvard University Press, Cambridge.

Brusco S. (1982), The Emilian model: productive decentralisation and social integration, *Cambridge Journal of Economics* vol. 6, p. 167-184.

Brusco S. (1990), The idea of the ID. Its genesis, in, Pyke F., Becattini G. and Sengenberger W. (eds.), *IDs and inter-firm co-operation in Italy*, International Institute for Labour Studies, Geneva.

Cainelli G.and Leoncini R. (1999), Esternalità e sviluppo industriale di lungo periodo in Italia. Un'analisi a livello provinciale, *L'Industria*, n. 1, p. 147-166.

Cantwell J. A. (1989), *Technological Innovation and Multinational Corporation*, Basil Blackwell, Oxford.

Cantwell J. A. and Iammarino S. (1998), MNCs, technological innovation and regional systems in the EU: some evidence in the Italian case, Discussion Papers in International Investment & Management, Reading.

Capecchi V. (1990), Flexible specialisation and industrial districts in Emilia-Romagna, in: *IDs and inter-firm co-operation in Italy*, Pyke F., Becattini G. and Sengenberger W. (eds.), International Institute for Labour Studies, Geneva.

Corò G. and Grandinetti R. (1999), SMEs and districts: hybrid governance forms, knowledge creation & technology transfer, *Human System Management,* vol. 18, n. 2, p. 117-129.

Dei Ottati G. (1987), Il mercato comunitario, in: *Mercato e forze locali: il distretto industriale*, Becattini G. (ed.), cit..

Dei Ottati G. (1994), Trust, interlinking transactions and credit in the ID, *Cambridge Journal of Economics*, vol. 18, p. 529-546.

Everitt B. (1981), *Cluster Analysis*, 2nd edition, Halsted Press, New York.

Fabiani S. and Pellegrini G. (1998), Un'analisi quantitativa delle imprese nei distretti industriali italiani: redditività, produttività e costo del lavoro, *L'Industria*, vol XIX, no. 4, p. 811-831.

Ferrucci L. and Varaldo R. (1993), La natura e la dinamica dell'impresa distrettuale, *Economia e Politica Industriale*, vol. 80, p. 73-91.

Granovetter M. (1985), Economic action and social structure: the problem of embeddedness, *American Journal of Sociology,* vol. 91, p. 481-510.

Garofoli G. (1991), *Modelli locali di sviluppo*, Progetto finalizzato CNR, Franco Angeli, Milan.

Glaeser E.L., Kallal H.D., Scheinkman J.A. and Shleifer A. (1992), Growth in cities, *Journal of Political Economy*, vol. 100, p. 1126-1152.

Granovetter, M. (1985), 'Economic Action and social structure: the problem of Embeddedness', *American Journal of Sociology,* vol. 91, p. 481-510.

Granovetter M. (1992), Problems of explanation in economic sociology, in, *Networks and organisations: structure, form, and action*, Nohria N. and Eccles R. G. (eds.), Harvard Business School Press, Boston, Massachusetts.

Henderson V., Kuncoro A. and Turner M. (1995), Industrial development in cities, *Journal of Political Economy*, vol. 103, p. 1067-1090.

Henderson V. (1996), *Externalities and Industrial development*, mimeo, Brown University, Providence, USA.

ISTAT (1995a), Census of population: *7° Censimento generale della popolazione e delle abitazioni*, 25 October 1991, Rome.

ISTAT (1995b), Census of industry: *7° Censimento generale dell'Industria, del Commercio e dell'Artigianato*, 26 October, 1991, Rome, 1995.

ISTAT (1996), *Rapporto annuale 1995*, Rome.

ISTAT (1997), *Il commercio internazionale italiano per province*, Rome.

ISTAT (1998), *I censimenti delle attività produttive dal 1951 al 1991. Dati comunali*, CDRom, Rome.

ISTAT (1999), *Censimento intermedio dell'industria*. www-istat.it.

ISTAT – IRPET (1994), *I mercati locali del lavoro*, mimeo, Rome, 1994.

Istituto Tagliacarne (2000), *Il reddito delle province italiane*, Franco Angeli, Milan.

Jacobs, J. (1969), *The Economy of Cities*, Random House, New York.

Langlois R. N. (1988), Economic change and the boundaries of the firm, *Journal of Institutions and Theoretical Economics*, vol. 144, no. 4, p. 635-657.

Leijonhufvud A. (1986), Capitalism and the Factor System in Economics as a Process, in, Langlois R. N. (ed.), *Essays on the New Institutional Economics*, Cambridge University Press, New York.

Lorenzoni G. (1990), *L'architettura di sviluppo delle imprese minori. Costellazioni e gruppi*, Il Mulino, Bologna.

Lundvall B. Å. (1988), Innovation as an interactive process: from user-producer interaction to the national system of innovation, in, Dosi G. *et al.* (eds.), *Technical Change and Economic Theory*, Pinter Publishers, London.

Lundvall B. Å. (1992), (ed.), *National Systems of Innovation*, Pinter Publishers, London.

Malmerg A. and Maskell P. (1996), Proximity, institutions and learning – towards an explanation of regional specialization and industry agglomeration, paper presented at the EMOT Workshop on "Learning and embeddedness", Durham, 28th-29th June.

Marshall A. (1919), *Industry and Trade*, Macmillan, London.

Marshall A. (1950), *Principles of Economics*, 8th edition, Macmillan, New York.

Moussanet M. and Paolazzi L. (1992), (eds.), *Gioielli, bambole, coltelli*, Il Sole 24Ore, Milan.

Nelson R. R. and. Winter S. C (1982), *An Evolutionary Theory of Economic Change*, Harvard University Press, Cambridge.

Nohria N. and Eccles R. G. (1992), (eds.), *Networks and Organisations: Structure, Form, and Action*, Boston, Massachusetts, Harvard Business School Press.

North D. (1990), *Institutions, Institutional Change and Economic Performance*, Cambridge University Press, Cambridge.

Paniccia I. (1998), One, a hundred, thousands of industrial districts. Organisational variety in local networks of small and medium-sized enterprises, *Organisation Studies*, vol. 19, no.4, p. 667-699.

Paniccia I. (2002), *Industrial Districts: Evolution and Competitiveness in Italian Firms*, Edward Elgar, Cheltenham.

Pavan M. (1992), 'Il distretto ceramico di Sassuolo-Scandiano', in, Fabrizio Onida *et al.* (eds.), *I distretti industriali: crisi o evoluzione?*, EGEA, Milan.

Perroux F. (1970), Note on the concept of 'growth poles", in, McKee D., Dean A. and Leahy W. (eds), *Regional Economics: Theory and Practice*, New York.

Perrow C. (1992), Small-firms networks, in, Nohria N. and Eccles R. G. (eds.), *Networks and Organisations: Structure, Form, and Action*, Harvard Business School Press, Boston, Ma..

Perulli P. (1998), (ed.), *Neoregionalismo. L'economia-arcipelago*, Bollati Boringhieri, Turin.

Piore M. J. (1990), Work, labour and action: Work experience in a system of flexible production, in, Pyke F., Becattini G. and Sengenberger W. (eds.), *IDs and Inter-firm Co-operation in Italy*, International Institute for Labour Studies, Geneva.

Piore M. J. (1992), Fragments of a cognitive theory of Technological Change and Organisational Structure, in, Nohria N. and Eccles R. G. (eds.), *Networks and Organisations: Structure, Form, and Action*, Harvard Business School Press, Boston, Ma..

Piore M. J. and Sabel C. F. (1984), *The Second Industrial Divide*, Basic Books, New York.

Pyke F. and Sengenberger W. (1990), Introduction, in, Pyke F. *et al.*, (eds.), *IDs and Inter-firm Co-operation in Italy*, International Institute for Labour Studies, Geneva.

Porter M. E. (1990), *The Competitive Advantage of Nations*, Free Press, New York.

Romer P. M. (1986), Increasing returns and long-run growth, *Journal of Political Economy*, vol. 94, p. 1002-1037.

Saxenian A. (1994), *Regional Advantage: Culture and Competition in Silicon Valley and Route 128*. Harvard University Press, Cambridge, Ma..

Scott A. (1998a), Flexible production systems and regional development: The rise of new industrial space in North America and Western Europe, *International Journal of Urban and Regional Research*, vol. 12, p. 171-86.

Scott A. (1998b), *New industrial space,* Pion, London.

Sforzi F. (1989), "The geography of industrial districts in Italy", in, Goodman E. and Bamford S. (eds.), *Small Firms and Industrial Districts in Italy*", Routledge, London.

Sforzi F. (1990), The quantitative importance of Marshallian IDs in the Italian economy, in, Pyke F., Becattini G. and Sengenberger W. (eds.), *IDs and Inter-firm Co-operation in Italy*, International Institute for Labour Studies, Geneva.

Signorini L. F. (1994), The price of Prato, or measuring the ID effect, *Papers in Regional Science*, vol. 73, p. 369-92.

Trigilia C. (1986), *Grandi partiti e piccole imprese*, Il Mulino, Bologna.

Trigilia C. (1990), Work and politics in the Third Italy's IDs, in, Pyke F., Becattini G. and Sengenberger W. (eds.), *IDs and Inter-firm Co-operation in Italy*, International Institute for Labour Studies, Geneva.

Wernerfeld B. (1984), A resource-based view of the firm, *Strategic Management Journal*, vol. 5, p. 171-180.

Williamson O. E. (1975), *Market and Hierarchies: Analysis and Anti-trust implications*, Free Press, New York.

Williamson O. E. (1985), *The Economic Institutions of Capitalism*, Free Press, New York.

Chapter 14

Is a district possible in the car industry? The case of the Turin area[1]

Ronny Bianchi and Aldo Enrietti

14.1 Introduction

This paper sets out to examine how changes to production methods have affected the component manufacturers operating in the area. In other words, we have tried to see what implications the reorganisation of Fiat had on the suppliers and sub-contractors and what the consequences were for the firms which continued to supply Fiat and firms which were excluded. We also wanted to see whether those firms, which were excluded have been able to survive without Fiat. Another purpose was to verify whether the region possessed an innovative/competitive capacity independently from Fiat and, if this did exist, how it had developed and what its decisive elements were. Further, considering these factors, we wanted to show how, in the Piedmont industrial system, from the end of the Second World War to the 1990s, an industrial area had evolved into an automotive technological district.

14.2 The consequences of the changes in the production model on the component manufacturers in Piedmont

Technically, the car, as a product, is the result of people joining together to assemble finished items (specific parts, components, sub-systems, and systems,

[1] This research benefitted from the financial support of the European TSER/INLOCO project and of the Italian Ministry of the University. The authors are grateful to Cristiano Antonelli for the stimulating discussions.

numbering about 20,000) which are produced separately, but are complementary and depend on each other in order to work properly when they are assembled.

Notwithstanding the changes which have occurred in the last 20 years, the Piedmont Region is still heavily dependent on the car industry, with more than 40% of the Italian employees working in that sector. The car system (car and body production, components, tyres and related activities such as mechanics, plastics, chemistry and glass) accounts for 27% of Piedmont's industrial employment, while the core business activities (car and body production, components) account for 14%. However, the importance of Piedmont is better emphasised when the Italian car industry is considered as a whole for:

- 40% of car production
- 46% of components production
- 32% of the value of national car exports and 39% of components exports are located in that region.

Fiat Auto has three plants in the region: Turin Mirafiori, Rivalta, and Verrone with 25,000 employees out of 60,000 employees in Italy and 93,000 in the world (in 1999).

Notwithstanding the presence of Fiat Auto, a fundamental role is played by SMEs: if the production of cars and trucks is totally concentrated in firms with more than 1,000 employees, the SMEs prevail in the production of components and accessories (in fact, in 1996, 94% of firms had has less than 250 employees and 71% less than 50).

The most important strengths are: the car manufacturer and its suppliers are geographically close to each other, which is especially important during the design and development stage, ensuring that a wide range of components are available; there is a close-knit network of subcontractors (engineering, moulding, die-casting); a highly developed and well-qualified styling and design sector; and a highly skilled labour force.

14.2.1 Particular features of the Italian industrial model

In order to understand the new dynamics that component suppliers have developed in Piedmont, it is necessary first of all to consider the relationship between the new model of production and the historical and production conditions particular to Italy. In the 1980s, many researchers judged the Toyota model to be much better than Fordism, and they believed it was certain to replace the latter (Womack *et al.*, 1990), because the Japanese ways of organising production were clearly superior or, at least, there was a kind of one best way towards which every firm's management strategy should be directed

(Porter, 1989). Experience during the 1990s has shown that there are various production models, which, while having the same aims, tend to interact with the domestic institutional context (Boyer, 1991; Boyer and Freyssenet, 1995) and with the national system of innovation (Lundvall, 1992). This is because firms must work with a much more open production structure. In Fordism, a firm based its production strategy entirely on schemes decided internally, independent from the external situation, while the new model of production made necessary instead the creation of external relations built up at various levels - suppliers, markets, financial system, labour market and institutions.

Regarding the Fiat case, the switch from Fordism to the new manufacturing principles is a special case compared to what happened in other industrialised countries, where there are many more large industrial firms. When it was necessary to reorganise production relationships through a more open system and, above all, through a greater decentralisation of the decision-making process, Fiat inevitably had to consider the Italian industrial tradition in which production centred on SMEs plays a leading role (Bianchi, 1996). These indications suggest that it is possible to develop a trend of industrial innovation, which is no longer separated between large companies and SMEs, as was the case up to the end of the 1980s. In other words, new co-ordination technologies have allowed SMEs to grow to the size of some large firms in the form of the group in a very short time, while maintaining the original competitive advantages of a small firm. However, this organisational model is also a characteristic of the new production methods adopted by large firms.

The car industry, in general, has not led to the establishment of relations between firms, which are typical of industrial districts. The case presented here shows an industrial area that had been transformed into a specialised technological district. To understand how it was possible, it is necessary to turn to the macro institutional level.

14.2.2 The macro institutional level

If we consider the model from the point of view of the institutional rules, which govern production, there are various tools we can use. At a macroeconomic level we can examine the overall dynamics of economic systems, drawing attention to the conditions for reproduction through the evolution of employment contracts, kinds of competition, monetary systems, institutional structures and the international framework. The interaction of these five structural factors determines the rate of accumulation, which characterises a country's growth rate (Boyer, 1986; Boyer and Saillard, 1995) or a given geographical area (Benko and Lipietz, 1992). Instead, at a microeconomic level, we have the Theory of Conventions (Orlean, 1994), which defines the rules accepted by the members of an organisation together and how they affect the behaviour of the individual players, emphasising dynamics which orthodox theory does not consider (fig. 1). The connection between the two approaches becomes difficult, however, without the help of a link. It is thus necessary to introduce an intermediate stage to link the two approaches: intermediate spaces, which have their own *raison d'être* so that it is possible to interpret the relationships between strategies of the organisations (especially firms) and their structural logic.

In this way the mesoeconomic system (production filière, industrial group, technological district, inter-firm co-operation network etc.) brings together:

- the organisational strategies, the general conditions of accumulation, the innovation processes, and the development of the global technical and industrial paradigm;
- the institutions which operate in the global economic system and in the local institutional framework (Gilly, 1994, p. 296 and 298)

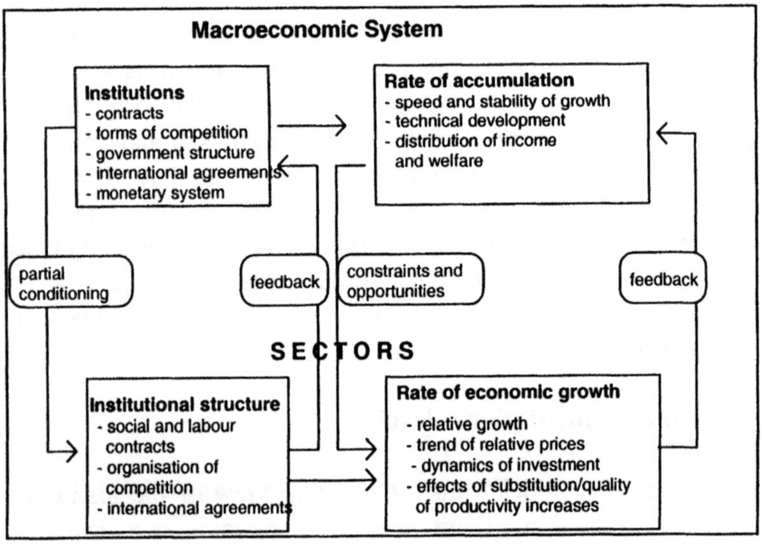

Figure 1: The Structure of Macroeconomic and Sectorial Systems

The relations within the mesoeconomic system, despite being based on stability over time, are affected by changes caused by the evolution of the role and the power of the organisations which belong to it. Mesoeconomic systems differ. They are not always able to develop an endogenous economic surplus, in contrast with production structures, where individual abilities are simply grouped together in a one way modality, and do not benefit from positive externalities, deriving from the co-operation and interaction processes found in the mesoeconomic system itself.

The advantage of the mesoeconomic analysis is that it does not only consider the production dimension, as the orthodox model does, but it also considers the institutional side, highlighting the relations between them. The mesoeconomic system is thus able to determine its own institutional dynamic, hierarchically subordinate to national institutions, but independent from them with regard to production. It is, therefore, an open system, which must be able to react to external changes, for example a new production model, and renew itself from within. In fact, a dynamic mesoeconomic system tends to generate a local system of innovation (Lundvall, 1988), where there is a strong element of "co-operative culture" which can develop competitive industrial dynamics even when external changes are present (Carlsson, 1992).

The ability to adapt continuously to the evolution of the market within the new model of production is the decisive element in determining whether a firm's or industry's competitive level will be maintained or increased. Its success depends on the ability to reconcile the theoretical or ideal objectives which have been set, for example lean production, with the institutional structure within which the firm must operate[2] (the meso structure). The positive interaction between the two forces is fundamental for the growth of the system, although it can differ from country to country or from region to region, as it is dependent on the historical and socio-economic framework (Boyer *et al.* 1998).

14.2.2.1 The technological district

Within the mesoeconomic system, industrial districts become one of the possible forms of the organisation of production (Gilly, 1994). However, in the case of the Turin area, and specifically the car industry, it is useful to adopt the concept of "technological district" (Antonelli, 1986):

"The technological district is created from the coexistence in an urban area of:

[2] Obviously a production network can be created from nothing (ex novo). See the cases of new car factories being set up by Fiat in Melfi, Italy, by Ford and Volkswagen in Portugal, or in South Carolina by BMW etc., but, again, in these cases there were always institutional incentives to the creation of a system from a "green field" (taxation, labour costs, etc.).

a) the industrial district, and therefore a system of inter and intraindustrial relations including the tertiary and financial sectors which will accelerate the diffusion of innovations and the introduction of incremental innovation;
b) a science park, and therefore the concentration of sufficient research and development activities to offer scientific/ technological economies of scale
c) an industrialising pole, and thus the area's ability to absorb and possibly generate leading firms, to act as an incubator for the innovation process and be a stimulus for diffusion throughout the area. In fact, the innovative process in a technological district appears to be partly exogenous for the single firm, but endogenous for the local economic system" (Antonelli, 1986; p. 22-23).

With this in mind, it is useful to remember the characteristics of the canonical industrial district, as it has been defined by Becattini (Becattini 1987, 1992): a socio-territorial organisation characterised by a lot of independent SMEs specialised in one phase (or in few phases) of a specific industrial sector, with strong co-operation links among them.

Even if the technological district contains one of the most important characteristics of the canonical industrial district, that is to say, it is a network of different relationships linking the firms, it is different in other three aspects: 1) the role of the large firm - instead of the network of SMEs - both in the economic development of the area and in the spread of innovations; 2) the central position of formal institutions (firms, research centres, universities) that carry out research and development activities and that are able to spread technological knowledge[3]; 3) the localisation in a metropolitan area based on the city of Turin.

14.2.2.2 The economic and institutional system of car production in Piedmont.

The internal reorganisation undertaken by Fiat Auto in the 1990s was later extended to the whole car manufacturing filière (Enrietti, 1995 and 1997). The crisis during the second half of the 1980s made the Turin firm consider the new model of production and eventually adopt it; without it, further market share would have been lost. The first step was to reorganise the internal structure - decentralisation, deverticalisation, reduction of the hierarchy , and the introduction of the integrated factory (Bonazzi, 1993) - so that it would be possible to introduce the necessary process innovations (just in time, diversified

[3] In other words, external economies have a strong technological character.

batch production, improvements in quality), and product innovation - new models. The process and product innovations were thus induced exogenously by the evolution of the world markets, but achieved thanks to the know-how (or path-dependence) existing within the firm and the region of Piedmont[4].

At a mesoeconomic level, the innovations introduced by component manufacturers were certainly endogenous, determined (or imposed) by the evolution of Fiat's strategies. However, if single firms are considered, the roles can be different. Firms chosen as prime suppliers by Fiat were restructured according to their relationship with Fiat, therefore it was exogenous, but these firms were chosen according to their endogenous innovative ability (Enrietti, 1995). It is more difficult to analyse the behaviour of the firms, which were excluded from dealing directly with Fiat, but it is possible to imagine that the firms which managed to survive possessed important innovative abilities due, in part, to the skills and know-how gained over the years (path-dependence). They also survived because they belong to a regional production structure in which there is a technological district, which stimulates learning by interacting, "where external tacit knowledge is acquired through informal and social exchanges that make the innovation systems in which each firm operates possible" (Antonelli, 1999; p. 6).

There are, in reality, three levels at which to analyse the car manufacturing filière in the mesoeconomic system of Piedmont (see Figure 2). The first, in order of importance, is the technological district, which can be considered as a partially closed local (or meso) system. It is characterised by a high number of firms operating in the sector, vertically integrated and horizontally concentrated. Within this area, the multinational Fiat is the leading firm in the sector and in the region. As such, it plays a central role in determining the growth of the whole system. It acts as a filter for the input which comes from international markets (macro economic relations, the second level) and emits the necessary impulses so that the firms in the region will adopt suitable production strategies to maximise their own strategies. However, Fiat must act within an institutional structure, which includes players who can by no means be considered as secondary, such as, for example the trade unions, which have always had a counterbalancing role, often to such an extent as to determine the strategic

[4] "Interrelatedness, complementarity, cumulability and more generally technical distance among technological artefacts are not exogenous specifications of technology, but on the contrary should be regarded as the outcome of a path-dependent process of the introduction and adoption of technological changes under the constraints of the environment and the topological characteristics of the multidimensional space into which the economic and technological action of firms is embedded" (Antonelli, 1999; p. 5).

choices made by the multinational. Analysis at the last level (micro) covers all the component manufacturers in the technological district. They operate in conditions historically determined by Fiat in that they are linked directly or indirectly to the company. The degree of dependency is, however, less than for firms operating in similar conditions, as for example in the case of Toyota City, or with Fiat at Melfi. This demonstrates the fact that, even within the new post-Fordist methods of production, there is not only one best way, but various solutions depending on the different historical and economic conditions[5].

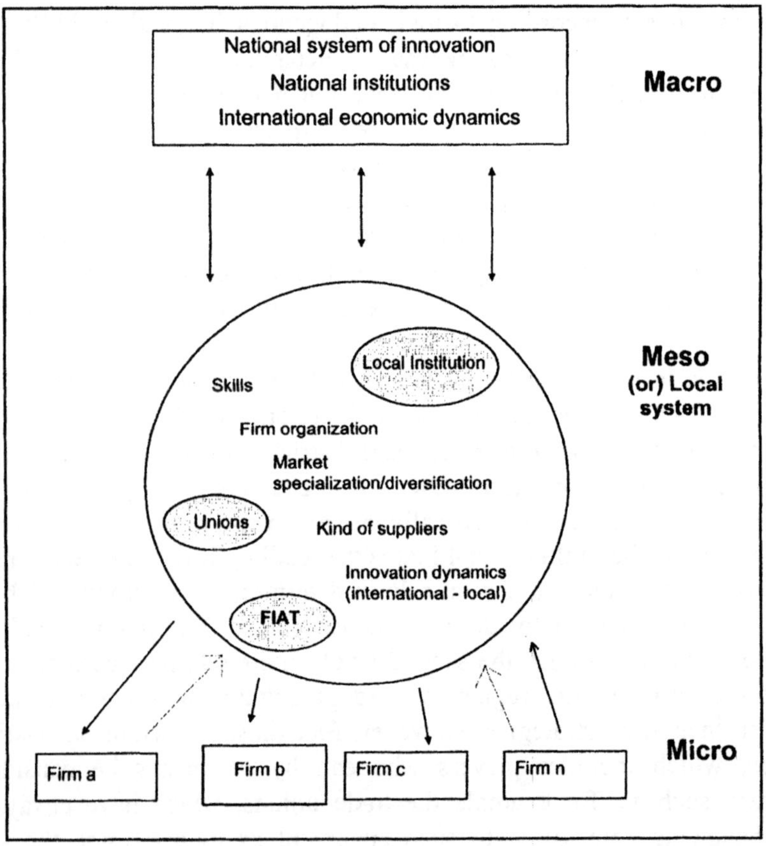

Figure 2: Relations between the macro and micro levels

[5] This is one of the main results that emerged from GERPISA (Groupe d'Etudes et de Recherche Permanente sur l'Industrie et les Salariés de l'Automobile): see Freyssenet, Mair, Shimuzu, and Volpato (1998).

The whole range of information and technological know-how flows can be briefly summarised as follows. Up until the 1990s, the companies active in the region operated within a system relatively impermeable to the outside world. Most of the information from the market was collected by Fiat, which sifted and often distributed it to the other firms (similar to what a service centre does in an industrial district). The reorganisation during the 1990s made it more permeable. The adoption of the new method of production has led Fiat to build up privileged relations with a small number of first tier suppliers, whilst at the same time allowing the components sector in Piedmont to establish more direct contacts with the external market, underlined by the arrival in Piedmont of foreign purchasers (GM, Ford), but also, and more generally, there was a big increase in the purchase in Italy of components by Fiat's competitors[6]. It should be noted that these components in many cases incorporate advanced technology[7], produced by Italian firms, which goes to show how over time the process of innovation has become more autonomous from Fiat through the development of the process of learning by doing and by interacting with international car manufacturers.

14.2.2.3 The local structures for the diffusion of know-how and innovation

Fiat has clearly always had a central role in the diffusion (and/or imposition) of innovation[8] including the contributions made by its own companies involved in this field.

For example, Fiat's Research Centre (FRC), which is the largest privately owned research centre in Italy (850 employees) has aimed to increase the competitiveness of its clients, which are often Fiat's own suppliers, by transferring research results. In fact, in 1992, FRC set up a new department called External Diffusion of Innovation (EDI), whose task is to identify market niches in which to spread "in-house" innovations. This department has also

[6] Some examples: BMW increased its purchases from 286 billion lire in 1993 to nearly the 700 billion forecast for 1998. Volkswagen purchases have increased from 300 billion in 1994 to 1,000 billion in 1997.

[7] For example BMW has said that its purchases in Italy include passenger airbags, brake pumps, gear boxes, ABS systems.

[8] "Information, the competence of the firm and the technological externalities from the technological and regional innovative system represent three inputs in the process of the creation of new knowledge. This knowledge is the result of a complex process of creation of new information which builds upon the competence mix acquired by means of the learning process, the socialisation of experience, the rearrangement of available information and formal R&D activities" (Antonelli, 1999; p. 6).

developed new products based on the technology and skills which were already available. Moreover, FRC has drawn up a project with the Chamber of Commerce in Turin called "Technology transfer to the SMEs" with the aim of creating a regional system of innovation in the area around Turin. The needs of the 100 SMEs interviewed can be classified as follows:

- 43% of the SMEs expressed the need for experiments and tests to be carried out on their product
- 31% thought it was urgent to develop new products and processes
- 30% needed specialised tests and analyses to be carried on materials and
- 29% wanted help with structural calculations.

It should be noted that FRC also collaborates with state bodies, universities, as well as public and private research centres, and has developed many EU research projects (BRITE, ESPRIT, Telematic Application, and JOULE). In this way, although FRC's main activity is applied research, the centre also builds up its links with pure research, which will enable it to choose the most promising and potentially most useful innovations for its clients. In short, although FRC is part of an industrial firm, a microeconomic unit, it acts in the diffusion of know-how at a mesoeconomic level as a link between international, domestic and regional institutions and firms.

Fiat also controls a training institute (ISVOR), one of Europe's most important full-line training companies in terms of turnover, size, quantity and quality of staff. ISVOR was founded in the 1970s through the merger of all Fiat's training bodies, thus inheriting a history and experience that dates back to the establishment of the "Scuola Allievi Fiat" in 1922.

However, other regional institutions have also set up interesting structures for the diffusion of know-how. Firstly, the Polytechnic of Turin carries out various activities ranging from research to consultancy and to teaching for firms operating in the region. As from Autumn 1999 it has offered a degree course in automobile engineering. Professional training in design in the Turin area is offered by three important institutions: the Faculty of Architecture, which has offered a course in industrial design since 1969; the School of Applied Art and Design, which was founded in 1978; and the European Institute of Design, which was established in 1989 expressly to satisfy the need for specialised skills in the car industry.

14.2.3 The transition towards a "technological district"

Since the end of the Second World War, there have been radical changes in the relationship between firms involved in the car manufacturing filière and the external environment of the Turin/Piedmont area. Changes can thus be summarised as the move from an industrial area, dominated by Fiat, to a car technological district (Figure 3).

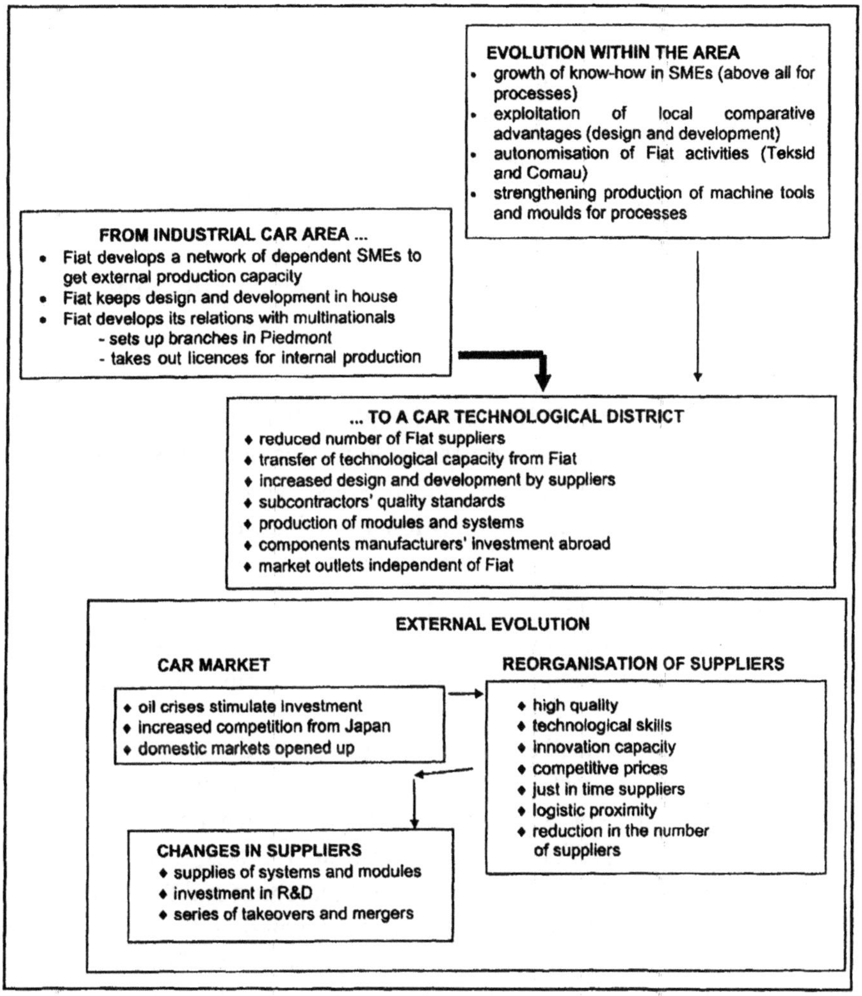

Figure 3: From an industrial area to a car technological district

The Turin industrial area was characterised in the first two decades after the war by a profound hierarchy of relations within the filière, with Fiat playing an absolutely central and dominant role. However, these elements were the basis for the development of the next steps.

During the 1960s and 1970s Fiat surrounded itself with a group of small and medium-sized firms, which had been helped and encouraged to such a extent that it was as if they formed a separate department of Fiat. These firms were expected to supply Fiat with production capacity at a cost lower than Fiat's own departments.

This network of SMEs relied on Fiat not only for market outlets, but also for design, which was developed by Fiat itself or by firms, which depended on Fiat[9].

Fiat built up a close relationship with multinational component manufacturers at two levels: setting up new local branches in Piedmont and taking out licences to manufacture components internally[10] (Enrietti, 1982).

The 1980s and, above all, the 1990s were periods of radical reorganisation of suppliers at Fiat, due both to "natural" evolution (supply effects) and to external shocks (demand effects).

Natural evolution

The local network of SMEs mentioned earlier developed a degree of specialisation, above all in process technology (including the manufacture of measuring machines, machine tools and moulds) which enabled them to take part in a process of partial autonomisation from Fiat, both in terms of market outlets and the development of their own know-how.

There was a rapid development, in this period, of design and development - crucial stages in car manufacturing. In fact, the Turin area has absolute international leadership with its designers, (Pininfarina, Bertone, Giugiaro, IDEA Institute, etc.).

The Fiat Group, a vertically integrated firm up to the middle of the 1970s, introduced a number of drastic changes to its organisational structure, creating independent companies for a series of activities which had previously been either in its internal divisions (cars, industrial vehicles, and earth moving equipment), or stages in the production process (smelting and the manufacture of machine tools). The latter two cases were firms, situated in the Turin area, which have since become leaders in their field. These two firms, Teksid and Comau, have become internationally famous, taking advantage of the presence of local suppliers of components of moulds and machine tools.

[9] Remember that, even at the beginning of the 1990s, the degree of vertical integration in Fiat Auto in components design was 76% (Enrietti, 1995; p. 141).

[10] For example, radiators, assisted steering, cardanic transmission, brake components, valves.

External shocks

The two oil crises in the 1970s introduced a series of innovations in car manufacturers' strategies: more product innovations, reduction in pollution and weight, greater comfort, and, in general, more applications of new technologies in the car.

The increased competition from Japanese companies during the 1980s was associated with the greater efficiency of lean production compared to Western methods. These two factors forced Fiat (and its European competitors) to reorganise their networks of suppliers in the following ways (Lamming, 1994; Sinclair and Mercer, 1993):

- *High quality*: this affected all aspects of the company, including design, product, manufacturing processes, logistics and service;
- *Technological Know-how and Innovative Capacity*. Car manufacturers expect their suppliers to be able to supply technologically advanced components, i.e. complete systems, using the most advanced manufacturing processes. The crucial factor for the component manufacturing sector is investment in research and development;
- *Competitive Prices*. Fiat needed this, whilst at the same quality and innovation had to be ensured. Car manufacturers asked for regular price reductions during the 1990s[11], as they expected to share the benefits from productivity gains and increased overall efficiency with the component manufacturers;
- *Just in time*. The car manufacturer[12] asked their suppliers to located close by, in order to maximise exchanges between the two parties, reduce costs and make logistics less complex.

All these factors together led to radical changes in the kind and number of suppliers.

Kind of Supplier.

Suppliers to car manufacturers were subjected to radical restructuring.
As a car is no longer simply a collection of different parts but a group of systems[13], the suppliers' role is not just to supply parts designed by the

[11] Some examples of price reductions might be mentioned: Renault, -18% between 1997 and 2000; PSA, -25% from 1997 to 2000; Ford, -5% per annum between 1996 and 2000; Honda UK, -1.5% per annum and Fiat 3% per annum.

[12] There are even examples of suppliers having their production line inside the customer's factory, as is the case of Recaro, which manufactures seats in the Mercedes factory at Bremen, of Continental, which manufactures tyres in the Volkswagen factory at Salzgitter, and of Magneti Marelli, which assembles suspension units in Fiat Auto plants.

customer, but to supply complex units. The supplier is now responsible for development, design, planning, quality and reliability of the systems it produces and, in this way, it must show its specific ability and know-how. This can be done in two ways:

1. The supplier takes on the role of systems integrator, being able to integrate the design and development of a complete working system for a car (ignition, brakes, exhaust, air conditioning). This kind of supplier must choose and deal with specialist manufacturers of single parts and organise the logistic chain for deliveries;
2. The supplier only provides modules which are pre-assembled and delivered to the car manufacturer.

They also need to invest in R&D together with large investments in production processes, and setting up factories close to the assembly lines, even abroad. This means that firms have to be able to generate a constant flow of finance over a long period.

Number of Suppliers.

At the centre of the reorganisation of the suppliers' network we could observe a drastic reduction in the number of firms. At the end of the 1980s the average number of suppliers to a European car manufacturer fluctuated between 900 and 2500 compared to 160 to 260 for Japanese manufacturers[14]. Thus, a process began which led European car manufacturers to reduce the number of suppliers by 40% to 80%[15]. The result of this selection process was the creation of a pyramid-shaped hierarchical structure, in which there are first, second and third level suppliers. First level suppliers are mostly large firms, often multinationals. The process of concentrating first level suppliers into a small number of leading firms also brings a further effect: greater internationalisation of supplies (global sourcing) than in the past, when purchases were made mostly

---

[13] In 1993, suppliers of modules accounted for 22.2% of the purchases of European car manufacturers while it is forecast to reach 42.8% in 2000 and a strong growth of the request of modules (+ 166%) is forecast from 2000 to 2010.

[14] In 1990 Mercedes had 2500 suppliers, GM Europe 2000, Ford Europe and VW 1700, Audi and PSI 1250, BMW and Renault 1100, and Fiat 720, while Honda had 270, Toyota 260, Mazda 180 and Nissan 162.

[15] The aim of many European car manufacturers for the middle of the 1990s was: 500 suppliers for Mercedes and Fiat, 450 for BMW, 400 for Audi, 600 for Ford and PSA, 700 for Renault, and 1000 for Volkswagen. In fact, as shown in Table 1, in 1994 Fiat had already reduced the number of its suppliers to 410.

from domestic suppliers, (even in 1993/1994, 95% of Volkswagen supplies came from Germany, and Fiat bought 86% of its components in Italy).

It is worth considering what happened to Fiat's supply network during the years under review. At the beginning of the 1980s, direct suppliers to Fiat numbered about 1200 and in 1980 a selection process began (table 1) based on the firm's ability to ensure innovation (especially through development contracts), competitive prices (Fiat agreed to price increases which were below the rate of inflation) and reliability. The effect was that, between 1980 and 1983, about 350 firms lost their business with Fiat. However, the number of suppliers rose again to 1200 in 1987 with Fiat's takeover of Alfa Romeo. The number of suppliers starts to fall again in 1987, but more rapidly; so much so that today, 11 years later, it is less than one third of the 1987 figure. There has therefore been a drastic concentration of supplies in a relatively small number of manufacturers and within this number an even smaller number (130 in 1994, and 110 in 1995) account for 90% of Fiat supplies.

Table 1: Trend of the number of Fiat Auto suppliers

1987	*1988*	*1989*	*1990*	*1991*	*1992*	*1993*	*1994*	*1995*	*1996*	*1997*	*1998*
1200 (100)	1050 (87.5)	990 (82.5)	723 (60.27)	670 (55.8)	560 (46.6)	520 (43.3)	410 (34.2)	380 (31.7)	370 (30.8)	350 (28.3)	364 (33.3)

During the same period Fiat Auto reduced the degree of vertical integration, from 38% to 30%, and therefore there was a corresponding increase in outsourcing[16]. It should be noted that, at the beginning of the 1990s, a significant number of components manufactured by suppliers were still wholly designed internally by Fiat. If we consider the break-down of the value of the parts designed and developed by Fiat, it can be seen that the degree of vertical integration was even higher in 1991 (76%), and it was forecast to fall to 30% in 1997.

In brief, why do the processes described above reveal changes which lead to the creation of a technological district ?

The reduction in the number of suppliers, and the switch from suppliers of single parts to producers of modules and systems, has obviously caused component manufacturers to change (especially small and medium-sized firms). They are now involved in design and development. This has been possible at a

[16] According to VDA (2000), in the last few years the level of vertical integration of eastern producers has on average decreased from 48% to 36%, with rates as low as 20% in the case of Porsche.

local level because there was already a network of design firms working on Fiat contracts, which thus transferred their skills to suppliers. In addition to that some designers left Fiat to work with the suppliers. It is therefore possible to argue that for these two different reasons, design capabilities and know-how, which had previously been internal to Fiat, or which had depended on Fiat, came onto the market, thus becoming available for the local system.

There were other consequences for the local structure as well as those mentioned in the above points. As the number of prime suppliers was reduced, and they concentrated on supplying Fiat with modules and systems, the quality of the subcontractors increased. In this way, there was a sort of top-down effect on the second level suppliers, which were able to reduce there production costs (Fiat expected an annual fall in costs of between 1% and 2% from 1996 onwards).

The local supply system became less dependent on Fiat, through the development of new market outlets, supplying Fiat's competitors, (especially abroad), other sectors of the automotive industry, and being able to diversify into other sectors.

These processes are linked to the other dynamics, considered earlier, which have facilitated the growth of design work and machine tool production which are increasingly independent from Fiat

Furthermore, Fiat itself has played an important role in the spread of innovation both through the Fiat Research Centre's programme of transferring innovation to SMEs and, indirectly, by setting increasingly rigorous standards for its suppliers.

It is interesting to note that, with the structural changes described above, a change has also occurred in the terminology defining the group of companies composing the automotive filière situated in Piedmont: in the 1960s and 1970s it was called "induced car activities" or, even better, "induced Fiat activities" to indicate a situation in which suppliers were dependent on Fiat. Later on, in the 1980s and at the beginning of the 1990s, it was first defined as a "system"[17], meaning the fact that the final producers' performance depends heavily on the behaviour of all the companies in the industrial filière and, in particular, that it was necessary to switch from a dependent relationship to a more cooperative one between the car company and its suppliers.

[17] See the 1991 volume of IRES Piemonte, the Research Centre of the Piedmont Region, entitled "From induced activities to system".

During the last three years, following both the political debate on the future of Turin and Piedmont, and more theoretical contributions[18], this localised system of companies has been described as a 'technological car district'. During the last 30 years, the production activities related to the car industry in Piedmont have evolved from being a car-based industrial area dependent on Fiat into a technological district. This structure is characterised by the intensification of relations between firms to such an extent as to generate localized knowledge, which becomes an innovative input and a new source of growth. The dynamics of this transition is shown in Figure 4.

The transition from industrial area to technological district (Figure 4) has led to changes in the roles of the two main players (Fiat and suppliers), with the reinforcement of other players internal to the district.

Fiat Auto reduced the level of vertical integration not only in manufacturing (components manufacture, car assembly), but also in a series of tertiary activities which had previously been considered to be crucial (design and development, administration and logistics) so that know-how strongly related to the car industry came onto the market and was available to suppliers and later to all the firms in the technological district.

On the other hand, suppliers began to play an increasingly important role both in manufacturing (no longer supplying only simple components but systems/modules) and, above all, in advanced tertiary activities such as design and development. The process of learning (focussed on tacit knowledge) passed by interacting with Fiat to the appropriation of external knowledge (linked to the absorption of codified knowledge).

There was an increase in the number of players who helped to build up localized knowledge in the area:

- The growing importance and number of designers led to the growth of a series of schools and institutes supporting design and engineering activity.
- Specific help was available in design and development, and, more generally, the local universities provided a more clear-cut orientation towards the needs of the car sector.

A decisive step was taken when the Fiat Research Centre became independent and opened its doors to the outside world.

[18] See Bianchi, R. and Enrietti, A. (1999), 'La dinamica innovativa nel distretto tecnologico dell'auto in Piemonte'.

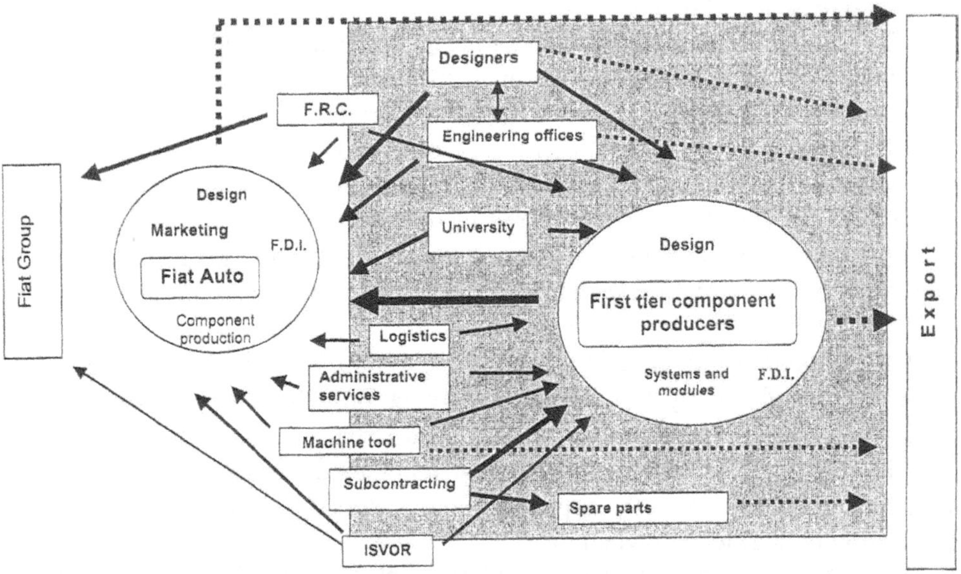

Figure 4: Relations between Fiat, suppliers and other players within the car technological district in the 90s

Further innovation came in 1998 when Fiat, Unione Industriale of Turin, and a number of first-tier suppliers, took the initiative to involve the second- tier suppliers directly by organising a programme of "guided growth for second level suppliers" (Follis and Enrietti, 2001).

14.2.4 Company Strategies, and the development of "Localized Knowledge"

The transformations briefly described above can be interpreted as typical of a number of significant component manufacturers in Piedmont, as shown in the data collected for research carried out during the period 1990 to 1997 (Enrietti; 1997; Bianchi and Enrietti, 1999). The most interesting results are in Table 2.

Table 2: The Evolution of the components manufacturing sector (% changes between 1990-1997)

	1990	1997	%Variation 97/90
Total sales (in billions of Lira)	2,067.9	4,332.97	+109.53
Employees	12,398	14,559	+17.42
- blue collars	9,655	11,240	+16.42
- white collars and managers	2,743	3,319	+21.00
Market shares:			
Fiat	40.74	47.47	+16.54
Car market	64.40	65.09	+1.07
OE market	50.31	59.69	+18.64
Industrial vehicle market	9.60	7.97	-16.94
AM market	14.09	5.34	-62.10
Subcontracting	9.61	9.40	-2.27
Total export	22.39	26.87	+20.00
Car: OE export	7.07	10.38	+46.90

OE: Original Equipment; AM: After Market

What we argue here is that the strategy adopted by some firms was strongly influenced by the localized knowledge developed in the technological district (Figure 4).

Three kinds of firm, which represent "extreme" features, are considered:

- *Firms which never supplied Fiat during the period 1990 to 1997* are able to use localized knowledge in the district to achieve significant results, in some ways unexpected: very high export shares (65%) and, above all, entry to the car market directly through the OE exports. Thus, it is not their relationship with Fiat that gives them access to the foreign market, but the specific density of localized knowledge (relationships, technology, know-how, learning by doing and by interacting).

- *Firms which stopped supplying Fiat Auto directly* could be expected to adopt a passive strategy when they left a customer like Fiat Auto. Instead, their reaction was aggressive, shown especially by the significant increase in OE exports. Therefore, their previous experience of learning by interacting and the access to the area's localized knowledge enabled them to pass from being subcontractor to becoming a direct supplier of Fiat's competitors.
- *Firms which increased their share of sales to Fiat Auto by more than 100%* by moving from being subcontractors to the direct OE market adopt another kind of innovative strategy based as much on building up tacit knowledge as a consequence of learning by doing and interacting.

Confirmation of the relatively high level of technology is seen by analysing the data, which the Turin Chamber of Commerce collected on the technology used and what kind of links there are between suppliers and car manufacturers (Table 3) (Camera di Commercio di Torino, 1988).

Table 3: ICT adoption by component manufacturers in Piedmont
Source: Camera di Commercio di Torino (1988)

Kind of technology	**Piedmont firms using this technology**	**% of Piedmont firms using this technology against all Piedmont firms**	**Italian firms with this technology**	**% Piedmont firms with this technology out of all Italian firms**
Columns	*1*	*2*	*3*	*4*
Codesign	93	*45.15*	135	*68.89*
ISO	88	*42.72*	136	*64.71*
CAD	98	*47.57*	141	*69.50*
CAD + Cod	64	*31.07*	91	*70.33*
ISO + CAD	53	*25.73*	85	*62.35*
ISO + Cod	54	*26.21*	82	*65.85*
ISO + CAD + Cod	38	*18.45*	58	*65.52*

The figures in column 4 reflect the intensity of technology and relations of Piedmont firms. Taking into account the fact that they represent 72.5% of the total national firms producing car components, the situation does not appear to be particularly rosy. However, it should be remembered that subcontractors are over-represented in Piedmont, i.e. ones which have a lower level of the technologies in question. If we also consider that most of the other Italian component firms not located in Piedmont are in the regions of Lombardy, Emilia Romagna and Veneto, which have a long tradition of incremental technological innovation, the picture that emerges is of a Piedmont production system that has shown itself capable of developing technological and relational structures to respond to the needs of auto manufacturers, whether independently or stimulated by Fiat Auto.

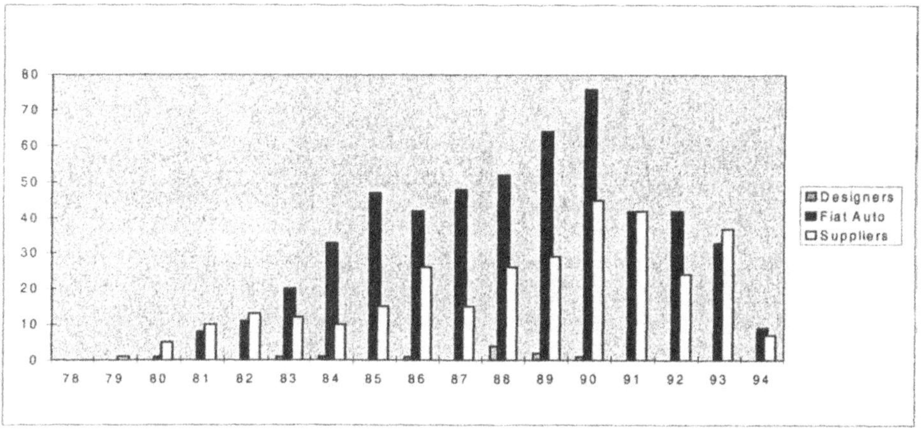

Figure 5: Dynamic of patents registered by Fiat Auto, component manufacturers and bodywork firms

Further confirmation of the growth of the innovative capacity of the technology district for cars in Piedmont can be seen in the data from the European Patent Office in Munich (Beaudry, Breschi, and Swann, 1998) (Figure 5 and 6). Figure 5 shows the dynamics of patents registered by Fiat and by its Piedmont suppliers: from 1978 to 1982 the number of patents registered by the suppliers was higher than those by Fiat, albeit in small absolute numbers (the highest being 13 in 1982). After this date, there is a sharp rise in the number of patents registered by Fiat, reaching a maximum of 76 in 1990[19]. After that we

[19] Figures for the years after 1990, however, are not very significant because of the procedures at the European Patent Office in Munich. Data were collected in 1995, and at least five years have to pass before there is the certainty that the patents have effectively been registered.

observe a notable decline of Fiat patenting activity, while patents registered by producers of components and bodywork remain at a high level (see fig. 5 and 6).

Component manufacturers showed a similar trend, but did not reach the levels of Fiat. In fact, they reached a maximum of 45 in 1990 too. To conclude, it would appear that the cyclical trends of patents registered by component manufacturers is similar to that of Fiat, showing a significant capacity to adapt to increasingly high standards. Figure 6 highlights the importance of patents registered in Piedmont by companies belonging to foreign multinationals, which has increased over time. This would seem to indicate that the technology of the car district in Piedmont is an area that has attracted foreign firms because of its specific localised know-how.

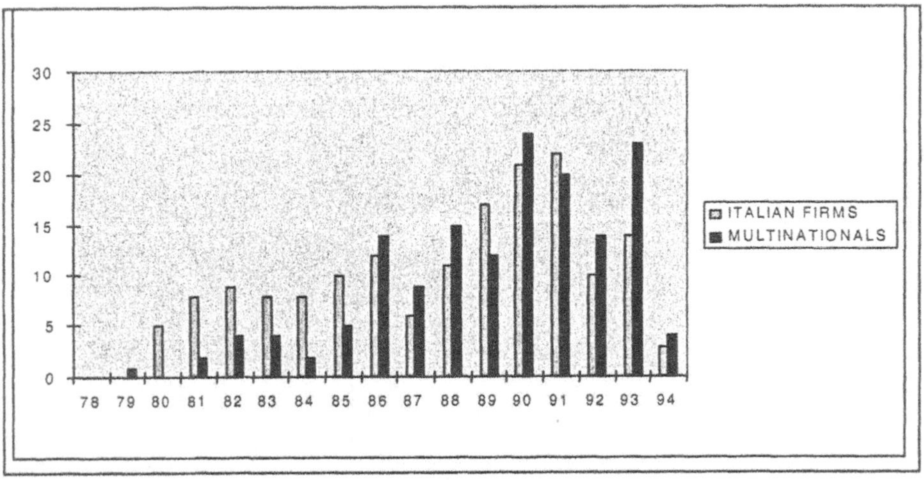

Figure 6: Patents registered by component manufacturers, by company nationality

Figure 7 shows, extremely briefly, the changes in the relationship between Fiat and the component manufacturers during the 80s and the 90s through the analysis of changes in the technological knowledge.

In general, the technological knowledge can be classified in four types[20] [14] along two axes, tacit or codified and internal and external: i) internal tacit knowledge is generated by means of processes of learning by doing and learning by using; ii) external tacit knowledge is acquired through informal exchanges and socialisation; iii) internal codified knowledge is the result of formal activities of R&D; iv) finally, external codified knowledge consists of structured

[20] For a recent discussion on this concept see the special issue of *Industrial and Corporate Change*, no. 2, 2000.

information available in a generic form and reorganised and applied in different contexts from the original ones (Nonaka and Takeuchi, 1995; Antonelli, 1999). The car technological district contains the different types of knowledge listed above: the internal tacit knowledge is the consequence of a century of specialisation by the firms involved in the car filière; tacit external knowledge is generated by the continuous interchange of information flowing among the firms in the district, also thanks to Fiat Auto.

Internal codified knowledge is mainly represented by the R&D activity of Fiat and its most important suppliers, in particular the multinational firms with laboratories located in Piedmont (for instance Valeo); external codified knowledge is generated by the co-operation both between research firm laboratories (for example the co-design activities between Fiat Auto and its suppliers) and between firms and University.

As regards what was discussed previously, the situation in the 1970s and 1980s was polarised, with FIAT Auto situated between internal codified knowledge and external codified knowledge, while for the suppliers the knowledge was mainly tacit and internal. The changes which came about in the 1990s completely modified the suppliers' position in relation to knowledge: the transition to the technological district ensured that knowledge moved mostly towards a dimension between tacit and external, with a shift also towards the codified.

14.3 The technological district as an institutional subject.

From the political and institutional point of view, the existence of the technological car district was officially acknowledged in the "Pact for the development of Piedmont"[21]. This Pact, in fact, included a precise undertaking in favour of developing a specific policy supporting those local production systems which possess district characteristics: the car technological district – with top level productions in mechanics, planning, design and research – as well as robotics, machine tools and air-space.

[21] Signed on 5th May 1999 by the Minister of Employment, the President of the Piedmont Region, the President of the Turin Province, the Mayor of Turin and the representatives of economic and social groups.

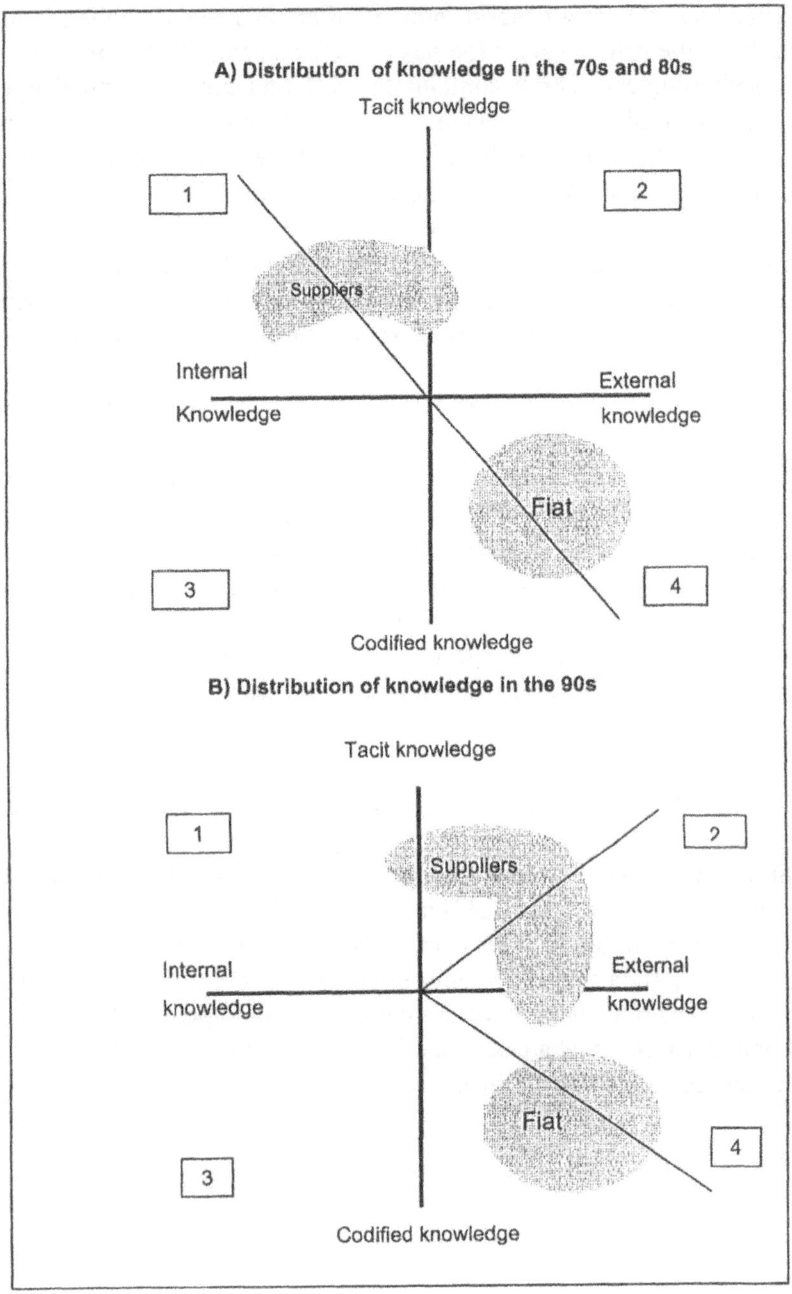

Figure 7: Distribution of knowledge in the meso system

The application of this policy, however, requires a set of conditions:

1. The definition of the district from a geographical and industrial point of view:
2. The definition of the objectives and of the beneficiaries:
3. The initiatives to be taken in favour of the district:
4. Who is responsible for implementing this policy.

1. As a ready-made classification does not exist, it is quite a complex problem to establish a definition of the district. The use of official statistical classifications, for instance, can create problems of under or over estimation of the size of the car system. This problem is due to the fact that only some of the companies, which form the filière, are classified in specific car sectors and sub-sectors[22], while other parts, which are not irrelevant when their size is considered, are included in non-car specific manufacturing activities[23]. There is a further problem of delimitation if a wider definition of filière is adopted, for example, including non-manufacturing activities[24].

The question of the geographical identification is obviously linked to the previous question and can be addressed together with the characteristics ofd the level of specialisation required in the local industrial structure.

By using a loose classification, the car technological district can be reasonably considered to be different from traditional districts, which are characterised by a high degree of territorial agglomeration (i.e. Biella or Valenza). The car district, on the other hand, because of its geographical dispersion, is not exclusively located in the Province of Turin, since there are important activities in other provinces of Piedmont. So, the borders of the car district are significantly larger than those of the other districts in Piedmont defined by the Regional Law n° 24/1997.

2. The existence of the car technological district is not sufficient to legitimate the use of government funds (local or national) made available by public bodies. In fact, the European Community limits the public aid given to companies. A main question, therefore, is the definition of the goals which are to be reached by the car district, and the definition of who should benefit from public initiatives.

Referring to the first question, there is the general goal of increasing the competitiveness of all the companies of the car system, perhaps paying special

[22] For example: motor vehicles, bodywork, parts and accessories, electrical parts for motor vehicles, and tyres.

[23] It is possible to indicate the case of metal melting, forging, pressing, products in plastic and rubber.

[24] It can be the case of R&D centres, of engineering and product certification offices. Repair workshops, sales rooms and insurance can also be taken into consideration.

attention to SMEs or specifying main themes and priority areas of action, such as, for example, internationalisation, technological transfer, access to net-technologies and services, professional training, logistic modernisation, quality and environmental certification.

Considering the beneficiaries, it is possible to think of interventions aimed at supporting a single company, and/or a limited number of companies. We must also think of financing district policies, which include actions to improve the infrastructure and services, and so to ensure a high degree of territorial externalities.

14.4 Conclusions

In our opinion, if we want to understand the evolution of the car technological district in Piedmont, it is necessary to interpret its special aspects and its uniqueness, with relation to the peculiarity of the Italian industrial model.

The Italian industrial model can be described as a Fordism hybrid (Wolleb, 1986). As the first industrial development began late, after Great Britain and Germany, in Italy productive capacity, even during the twentieth century, still relied very much on a large number of traditional SMEs.

The car area in Piedmont clearly developed around Fiat's needs. The firm was created in the early twentieth century, and especially after the second world war, from the 1950s to the 1970s, Fordism was extended to include relations with its suppliers.

In the 1990s, Fiat progressively abandoned Fordism, and introduced the new principles of production, which fitted in with the general restructuring based on flexible specialisation of the Italian industrial district.

During the 1990s, the osmosis between the two models was the result of two elements. The first is the loss of centrality of Fiat, which allowed the creation of a mesoeconomic system, that brought about the diffusion of an external codified knowledge to Fiat and non Fiat suppliers and subcontractors, thanks to the creation of institutions which are independent from the multinational and which reflect a district character. This is unlike the case of Toyota City (or to a lesser extent Melfi), where the diffusion of knowledge and creativity in production was and is programmed, and rewarded, by the leading enterprise (characterising a model of a verticalised firm).

The second element is the reduction of the number of direct suppliers creating a network of firms operating independently on the market, where Fiat can be a client (more or less important) among others. The need to survive of the companies excluded by the Fiat selection process has created the need for new production strategies, and obliged firms to seek new markets: the know-how existing within the district helped this evolution, as shown by the positive performances of the firms excluded by Fiat, and by those that had never been part of the system. Here we find important similarities with the canonical district model.

It is probably too early to come to the conclusion that the peculiarities of the Italian industrial model allowed the development of a specific local production model, based on the interaction between large companies and the industrial structures formed by SMEs.

However, the evolution of the Piedmont technological car district shows an interesting and peculiar situation, which requires further investigation.

References

Antonelli C. (1986), *L'attività innovativa in un distretto tecnologico*, Fondazione Giovanni Agnelli, Turin.

Antonelli C. (1999), *The Microdynamics of the Technological Change*, Routledge, London.

Beaudry C., Breschi S. and Swann P. (1988), Statistical and Econometric Analysis of Clustering and Innovation, EC TSER INLOCO, mimeo.

Becattini G. (1987), (ed.), *Mercato e forze locali: il distretto industriale*, Il Mulino, Bologna.

Becattini G. (1992) "Le district marshallien: une notion socio-économique", in Benko G. and Lipietz A., op. cit.

Benko G. and Lipietz A. (1992), *Les régions qui gagnent*, PUF, Paris.

Bianchi R. (1996), "Pour une lecture politico-institutionelle du modèle industriel italien et en particulier de la forte présence des PME", *Revue internationale P.M.E.*, vol. 9, no. 2, p. 104-123

Bianchi R. and Enrietti A. (1999), "La dinamica innovativa nel distretto tecnologico dell'auto in Piemonte", Quaderni di Ricerca 01/99, Dipartimento di Economia, Turin.

Bonazzi G. (1993), *Il tubo di cristallo*, Il Mulino, Bologna.

Boyer R. (1986), (ed.), *La théorie de la régulation: une analyse critique*, La Découverte, Paris.

Boyer R. (1991), "New direction in management practices and work organisation: General principles and national trajectories", CEPREMAP, Cahiers Oranges, Paris.

Boyer R. and Freyssenet M. (1995), *L'émergence de nouveaux modèles industriels*, GERPISA, Paris.

Boyer R. and Saillard Y. (1995), *Théorie de la régulation. L'état des savoirs*, La Découverte, Paris.

Boyer R. (1996), (ed.), *La flexibilité du travail en Europe*, La Découverte, Paris.

Boyer R., Charron E., Jurgens U. and Tolliday S. (1998), *Between Imitation and Innovation*, Oxford University Press, London.

Camera di Commercio di Torino (1998), Osservatorio sulla componentistica autoveicolare italiana, Turin.

Carlsson B. (1992), "Industrial dynamics: a framework for analysis of industrial transformation", *Revue d'économie industrielle*, no. 61, p. 7-32
Dosi *et al.* (eds.) (1988), *Technical Change and Economic Theory*, Pinter Publishers, London.
Du Tertre C. (1995), La dimension sectorielle de la régulation in: Boyer R. and Saillard Y. (eds.), *Théorie de la régulation. L'état des savoirs*, La Découverte, Paris.
Enrietti A. (1982), "Le imprese italiane a partecipazione estera nel settore della componentistica auto", in Fornengo G. and Rullani E. (eds.), *L'industria dell'auto e dei componenti: integrazione e internazionalizzazione produttiva*, Franco Angeli, Milan.
Enrietti A. (1995), "Il settore dei componenti auto: struttura e dinamica", *Economia e Politica Industriale*, n° 88, p. 131-151.
Enrietti A. (1997), "Il processo di selezione nella componentistica auto piemontese", IRES Lucia Morosini, Quaderni di ricerca, no. 26, Turin.
Follis M. and Enrietti A. (2001), "Improving performances at the second tier of the automotive supply chain. A draft case study of an innovative initiative in the Italian car industry", *International Journal of Automotive Technology and Management*, vol. 1, no 4.
Fornengo G. and Rullani E. (1982), (eds.), *L'industria dell'auto e dei componenti: integrazione e internazionalizzazione produttiva*, Franco Angeli, Milan.
Freyssenet M., Mair A., Shimuzu K. and Volpato G. (1988), *One Best Way? Trajectories and Industrial Models of the World's Automobile Producers*, Oxford University Press, London.
Gilly J.P. (1994), "Dinamiche industriali e meso-analisi: Il caso dei sistemi di innovazione", *L'industria*, no. 2, p. 295-309.
Industrial and Corporate Change (2000), Special Issue, no. 2.
Lamming R. (1994), *Beyond Partnership*, Prentice Hall International, London.
Lundvall B.A. (1988), "Innovation as an interactive process: from user-producer interaction to the national system of innovation", in Dosi *et al.* (eds.), *Technical Change and Economic Theory*, Pinter Publishers, London.
Lundvall B.A. (1992), (ed.), *National Systems of Innovations*, London, Pinter Publishers
Nonaka I. and Takeuchi H. (1995), *The Knowledge Creating Company*, Oxford University Press, New York.
Orlean A. (1994), (ed.), *Analyse économique des conventions*, PUF, Paris.
Porter M. (1989), The Competitive Advantage of Nations, Free Press, New York.
Pyke F., Becattini G. and Sengenberger W. (1990), (eds.), *Industrial Districts and Inter-firm Cooperation in Italy*, International Labour Office, Geneva.
Sinclair J. and Mercer G. (1993), "Successful Automotive Suppliers in the 1990s: Lessons from the 1980s", *Europe's Automotive Components Business*, 2nd trim., p. 67-91.
VDA (2000), "*Report 2000*", Frankfurt.
Wolleb E. (1986), "Mutamenti nei rapporti di produzione nella crisi: ascesa e tracollo del fordismo in Italia", in Boyer R. (ed.), cit.
Womack J.P., Jones D.T. and Roos D. (1990), *The Machine that Changed the World*, Macmillan, New York.

Chapter 15

The generation of contextual knowledge through communication processes. The case of the packaging machinery industry in the Bologna district

Fiorenza Belussi

15.1 Introduction

This chapter presents the results of an empirical research[1] on the Bologna district, specialised in automatic packaging machinery. The Bolognese packaging machinery industry has been studied for the excellence of some companies for a long time (Lipparini, 1995; Lipparini and Lorenzoni, 1994). However, the positive performance of some dynamic firms may be related to the local context, where firms have developed a network of historically path-dependent, locally based, specific interactions: a nexus of productive relationships, and a wide net of social relationships conducive to a fast innovation activity. Within this technological district a clustering of innovations has occurred; in turn, this has led to a self-reinforcing mechanism of industrial growth. The generation and diffusion of knowledge have been promoted throughout specific communication processes which have crossed large firm nets, with suppliers, subcontractors, and local institutions. The empirical study presented aims to show the close relationship existing between the dynamics of localised technological change and the role of local communication processes.

[1] A first analysis of the Bologna packaging machinery district was presented at the "Inloco project" workshop, organised in Nice, at the Latapses Institute (Sophia Antipolis, Nice University), October 1998. This international project, sponsored by the UE (Tser) was organised by the Fondazione Rosselli (Turin) under the scientific supervision of Cristiano Antonelli. See Belussi (1998).

The Bologna district represents an urban industrial sub-system which concentrates skills and a complex structure of locally defined interactions, quite similar to the notion of "the concentration of specialised industries in particular localities" (e.g. the Marshallian industrial district). However, many (more or less stable) large production networks form this district. The Bologna district is made up of a few leader firms, and about 150 smaller niche producers. In addition, numerous intermediate firms, that are not classifiable as packaging producers, are involved in sub-contracting. The packaging district, in the Bologna area, is formed by about 1,000 mechanical firms, altogether employing nearly 20,000 workers. The networks are co-ordinated by final firms, on the basis of orders, which may be very complex and may correspond not just to the purchase of a single machine but to an entire packaging line (which includes the operations of preparation, making up, and packaging). This type of specialised machinery is settled downstream from the production cycle of various manufacturing firms, and large multinational firms, serving mass production markets. The industry is characterised by a wide variety of products, technologies and systems[2]. The extreme customisation of demand, and the low number of machines sold annually (packaging systems must be very reliable, and as a consequence they have a long life cycle: an average of 10-20 years) in the past allowed Italian firms entry as specialised firms, or niche producers[3].

[2] The main users of packaging machinery are large mass producers of consumer goods, often large multinational firms that include packaging lines at the end of the production cycles in the food, pharmaceutical, printing, beauty treatments, secondary chemicals, detergents, and tobacco industries. Italian firms operate in specific sectors, e. g. in the food industry (packaging for bouillon cubes, tea, chocolate, or beverages: such as milk, water, and wine); the pharmaceutical industry (machines for blistering), the paper industry (machines for paper wrapping), the cosmetic industry (machines for boxing, dispensing, and bottling).The packaging systems in manufacturing are non-homogeneous (e. g., in the ceramic industry the main business is storage on pallets, while in pharmaceuticals they require thermoforming of individual doses). The basic materials utilised also vary: plastic film, boxes, iron strips, cases, tin foil, cardboard, and so on. Besides firms have very different needs which depend on many factors, e. g., the evolution of production and processing technologies, or firm marketing strategies, that may require particularly attractive new forms.

[3] At international level, two countries are the leading producers and exporters of the packaging industry: Germany and Italy. The Italian industry, which at present is in a very strong competitive position, being the second worldwide exporter after Germany, has exploited over the last twenty years an impressive catching up mechanism, being rewarded on the international level for the quality of the machinery manufactured and the originality of their ideas. Over the period 1980-1992, the competitive position of Italian firms in international trade (UN source) has improved noticeably, rising from 14.8 to 23.0. In the mid 1990s

Each production chain is concentrated on customised products and is dedicated only to the time necessary to fill in-house orders. While the local production chains are continually re-adjusted and changed, the district is characterised by stable, long-term, and co-operative relationships among the firms. In the district, the final firms are the principal agents that receive orders and know the market outlets. They possess not only all the technical skills of engineering and building the machinery, but also the relational skills for governing the production network (who in the district is able to perform a specific task, how, at what cost and when).

What is interesting to observe, in this context, is the density of the production relationships built around the most significant leader firms. In this respect, this district is quite different from the typical morphological structure of the Italian industrial "districts"[4], characterised by the presence of small-size firms competing in the same market. The Bolognese packaging industry illustrates, on the contrary, the case of a local production system where the structure of firms has evolved into more visible organisation systems (Lazerson and Lorenzoni, 1999). It is, indeed, similar in its evolutionary complexity (Witt, 1997) to some advanced complex technological systems, based on a growing hierarchisation of the activities, where the model of "network firm" plays a central role[5]. Complex relational webs, with mobile and changing boundaries, characterise this local system. They represent the communicative channels where technological knowledge is transferred, disseminated, and elaborated.

The paper is organised as follows: Section 15.2 presents the specific characteristics of the Bolognese packaging district. Section 15.3 discusses the historical formation of the Bologna district. The model of localised technological learning is discussed in Section 15.4, where a list of communicative channels for knowledge creation and diffusion is presented. Local interactions among firms are analysed in Section 15.5, where the evolution of production networks is described. Some preliminary conclusions and thoughts are presented in the final section.

(UCIMA source) sales were estimated around 4,000 billion Italian liras, of which 3,160 exported. The share exported has risen from 66.6% in 1992 to 83.2% in 1995.

[4] For a review see Becattini (1987), and Harrison (1992).

[5] Regarding the post-Fordist development of firm organisation cf. Freeman (1991); Belussi (1992); Di Bernardo and Benedetti (1997); Callon (1991 and 1993); Belussi and Arcangeli (1998).

15.2 A profile of the Bologna district of the packaging industry

The Bologna packaging district [6] does not have very clear-cut boundaries. It has developed around the city of Bologna, and now covers more or less the entire province, extending in the direction of Modena and Reggio Emilia, along the old Roman road: the "Via Emilia". The quantification of the Bologna district may be obtained by the elaboration of the 1991 ISTAT census, available in a disaggregated form (with data referred to individual municipalities). As we have mentioned, the Bolognese "macro-district of machinery", mainly specialised in packaging, which gravitates around the province of Bologna is made up, once we include the related mechanical sector (cogs, electrical-mechanical components, and service firms), of 1,000 mechanical firms and about 20.000 employees[7].

Using the INPS data base (see Table 1) we can analyse in detail the most crucial segment of this district: the population of final producers localised in the Bologna area. In the mid 1990s they were 153 (this source, unlike ISTAT data, refers to the firm as a legal entity, and it does not disaggregate information at the establishment level). Let us describe the main characteristics as they appear from Table 1. Two-thirds (105 units) of the firms appear to be very small, having less than 19 employees, and many of them, under Italian law, are defined as artisan firms. Five large firms, having more than 500 employees, dominate the district, with about 6,000 employees. From the analysis of the INPS data base, there emerges a strong model of firm specialisation: only 119 firms out of the total are strictly speaking final producers, 7 firms perform only design work, 9 are assemblers, 8 produce spare parts and are involved in machinery repair, 8 firms manufacture only bearings/conveyor belts, and 5 make labelling machines. By the year of their foundation, it emerges that 9 firms were founded before 1970, 27 during the 1970s, and 85 during the 1980s. 32 firms in the district were established after 1990; on the whole, new start-ups are of small size.

The largest firms are among the oldest established firms in the district. They are part of the historical nucleus that formed the district. This confirms the

[6] In this section we have adopted a wide definition of the sector, including all producers of automatic machinery (and complete production lines) for wrapping and packaging, and involving the phases of forming, splitting, metering, and dose dividing.

[7] Note that in the province of Bologna the entire mechanical sector in 1991 employed nearly 100,000 workers in 9,500 establishments. Using ISTAT and UCIMU data, we can observe that the Bologna packaging machinery district represents about one third of the employment of the Italian industry and about 40-45% of the total sales.

existence of a slow mechanism of creation of internal capabilities and specific skills and a mechanism of firm growth related to it.

Table 1: Firms belonging to the Bologna packaging district: year of establishment
Source: our calculation on Inps (1996).

Year	1-19 emp.	20-49 emp.	50-99 emp.	100-199 emp.	200-499 emp.	500-999 emp.	More than 999 emp.	Total Prov. Bologna
Before 1970s			2	1	1	4	1	9
1970s	18	4	5					27
1980s	60	17	3	3	2			85
1990-6	27	3	1	1				32
Total	105	24	11	5	3	4	1	153

15.3 The formation of an innovative district as an engine of growth for the Italian packaging industry[8]

Numerous works, some of which are extremely well documented historically (Biggiero and Forni, 1987; Lipparini, 1995; Capecchi, 1997) have analysed the genesis of the packaging industry in the Bologna district.

The expansion of this industry occurred primarily in the post-war period. The main impulse was the transformation of distribution systems, with the increase of over-the-counter products, and the utilisation of promotional packaging as a means of brand differentiation. The growth of the packaging sector is correlated to the rise of mass markets and the shift towards large scale distribution. This required, as a consequence, the rapid substitution and/or structural modification of the old packaging systems, even if the machinery had not yet become technologically obsolete or worn out.

The sector blossomed in the city in the 1930s with the foundation of the most important Italian firms on national level: ACMA, GD, and SASIB.

The evolutionary path started with a restricted number of firms which we may call "district founder firms". In these firms, knowledge and technical skills become consolidated. In the 1960s and 1970s, several technicians left the founder firms to create dozens of new firms.

[8] The following sections are based on the result of our empirical research, conducted through an open questionnaire and many in-depth interviews organised in a representative sample of firms localised in the Bologna packaging district.

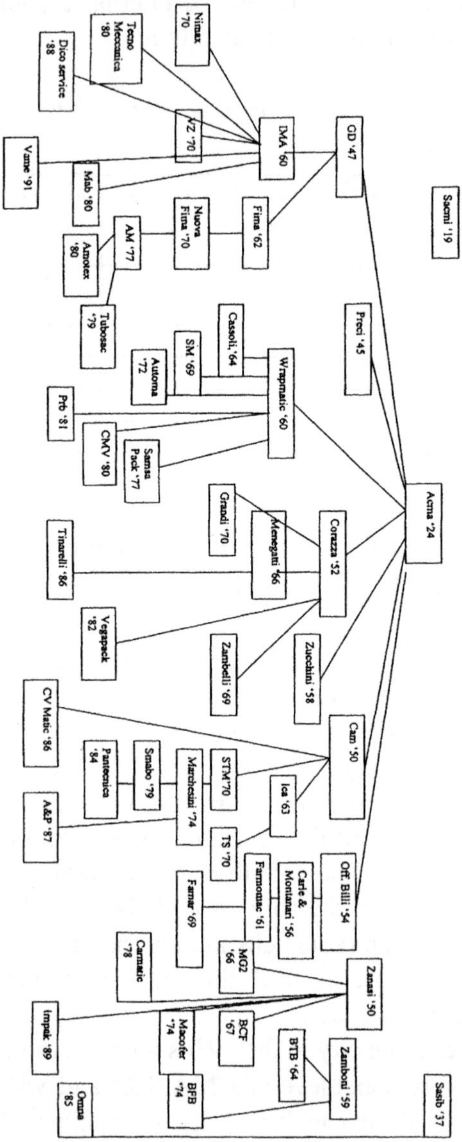

Figure 1: The historical formation of the Bologna packaging district by firm scission. Source: our interviews and Lipparini (1995). The year in brackets refers to the year of foundation.

The wave of spin-offs which occurred populated the district with small innovative producers. This process, thus, was highly path-dependent and built upon a nucleus of original technical skills and internally created tacit knowledge. Network externalities (Witt, 1996) were significant and the critical mass of the district allowed a rapid adoption of innovation.

This intense process of firm creation was centred on new technical entrepreneurs, or technical designers, and cadres (sales persons, machinery maintenance personnel). The latter, having accumulated technical skills and managerial experience, in order to exploit new ideas, left their factory for new entrepreneurial positions. But the process of firm creation was also sustained by skilled blue-collar workers, for whom Emilia-Romagna represents a highly stimulating social environment, and a favourable location for becoming small entrepreneurs (Amin, 1998).

The formation of the industrial structure, as it now exists, was thus a process of firms scission, not specifically directed by the profitability of markets, as it is simplistically explained by the standard theory (see Fig. 1). This district owes its success to the mechanisms typical of the Italian industrial districts. Some common elements are:

a) High levels of new firm start-ups (to absorb the subcontracted work and to launch new products in final markets);
b) The existence of a solid substratum of specialised small mechanical firms, serving a vast array of specialised subcontracting (Bonaccorsi and Lipparini, 1994);
c) A highly regulated labour market, based on skilled workers, mixed with forms of quasi-participation and social consensus between workers and the firms' management;
d) The high innovativeness of firms (see Table 2) and the localised generation of technical knowledge (see Antonelli, 2000; and Belussi and Pilotti, 2000).

The continuous start-up of firms has fed the expansion, replication, and renewal of knowledge. Former employees, both workers and technicians, establishing new entrepreneurial initiatives, transferred to others (their new employees) the historical firm-specific skills of the founders.

Social and spatial proximity (Grabher, 1993; Gilly and Torre, 1998) has facilitated the imitation process. Many newly created small firms tended to stick to the products already manufactured in the district. With time, the most dynamic firms shifted their production towards more advanced models. So, many staple products, manufactured by the former, were bought by the larger ones, just to complete their packaging lines (complementary firms).

Table 2: The formation of the entrepreneurial paths in the Bologna packaging district
Source: our interviews, and Capecchi 's (1997) historical reconstruction.

	Firm main characteristic at the start-up	Origin of the entrepreneur	Task performed before the start-up
Carle Montanari	Original inventor	Technician in Milan	Reviser of machinery
Corazza	Started as a machine repairman, in 1956 original inventor	Acma	Reviser of machinery
Tetra Brick	Original inventor	Multinational	Came to make use of local skills
Sasib T	Innovator (US pat. Using)	Mechanical industry of the area	Ex-blue collar worker
Sasib B	Innovator (US pat. Using)	Mechanical industry of the area	Ex-blue collar worker
Baumer	Imitator	Mechanical industry of the area	Technical designer
Acma	Original inventor (Z wheel, that divides a rotatory movement into phases)	Bologna entrepreneur, entered the sector with the idea of automating the packaging of Idrolitina powder	Gazzoni, a local entrepreneur, used the technical competence of Carpigiani, a student of Aldini Valeriani
Macofar	Imitator	Zanasi	Ex-blue collar worker
MG 2	Improver	Firm of the same sector, based in the area	Machine designer\
Sacmi	At the beginning imitator, then inventor	9 unemployed blue collar workers from Imola who founded a co-operative in 1919	Mechanics
Casmatic	Original inventor	Wrapmatic	Machine designer
Ima	Started as imitator, then original inventor	Acma	Machine designer
GD	Original inventor (innovation related to machine performance)	ACMA (Ariosto Seragnoli) and a firm of motorcycles and bicycles called GD	Machine designer

Some new firms having limited internal engineering skills were using all the information and knowledge embedded in the district to compete with larger firms in the less profitable niches (cost-driven firms).

Some small firms entered the district with new ideas and unique skills for the customisation of their products. They were competing directly with the largest firms in the district (technology-driven small firms).

Looking back at the historical development of the packaging machinery industry, it is important to stress the social institution that established mechanical skills in the territory of Bologna. The diffusion of technical skills was realised by the Aldini Valeriani Institute, a local technical training centre founded by the local authorities and by the local associations of entrepreneurs. This technical institute was founded at the end of the nineteenth century, and, from then on, many generations of local entrepreneurs have passed through it, as well as many professionally skilled workers. Thus, the origin of mechanical skills in the district goes back to a collective social investment.

15.4 The innovation model

15.4.1 The sources of innovation

The development of technical change found among the firms in the district does not follow the traditional or standard model discussed in the orthodox literature (Stoneman, 1995), where technological innovation derives primarily from science, in which one-shot innovations are developed within R&D departments, strategic innovations are protected by patents, and firms are rewarded for their innovative efforts by the complete appropriability of the results of their inventions, and not even the so called interactive chain-link model, proposed by Kline and Rosenberg (1986) few years ago, and soon became popular among scientists.

On the contrary, in the district we found very few firms that conducted R&D, a very low level of patents, and an intense circulation of information and knowledge. Thus, the creation of localised technological change has been neither only internal to firms nor simply external.

15.4.2 The accumulation of knowledge as a collective mechanism stimulating the learning process

The investment of firms is not the only element to be considered within this ample framework. A crucial role was played by the Aldini Valeriani centre. Since its establishment, it has developed the necessary skills and the mechanical know-how for the district by building a technical school for workers and technicians. The institute has provided a kind of overall supervision for the mechanical and engineering activities developed in the district, observing, ordering, and testing new technologies applicable to the field (innovation validating). The activities of Aldini Valeriani may be considered the largest investment in the district for the reproduction of codified knowledge within the packaging system (about 70% of the alumni find jobs in packaging firms). We have calculated an estimate of the annual expenditures for training (the data refer to mid 1990s). Each year, collective investment for the development and transfer of codified knowledge equals about 2% of the total sales of the district. The spread of codified (new and old) knowledge partially explains the sources of the innovative behaviour exhibited by local firms. But within the district the most important innovations were introduced by local new entrepreneurs, coming very often from the oldest firms in the district.

Their innovative efforts, directed towards the exploration of new solutions to specific problems, were based on, firstly, the re-combination of the highest levels of accumulated and diffused knowledge existing in the district, and, secondly, on some firm-specific idiosyncratic knowledge possessed and developed by them through practical experience (Table 2). The development of tacit[9] and codified knowledge was essential for the process of building up a local innovative capability in these specific technologies.

15.4.3 The origin of tacit knowledge

The localised process of knowledge creation in firms was heavily dependent on the types of multiplicative relationships described by Antonelli (1999), mainly based on several forms of <u>internal</u> organisational learning (Lundvall, 1993) plus the absorption of <u>external</u> knowledge (originating in the other firms in the district but also outside). This absorption was facilitated by the presence

[9] On this point see the fundamental work of Polanyi (1967). Some recent works have discussed this issue in depth; see: Senker, 1995; Foray and Lundvall, 1996; Cowan and Foray, 1997; and Belussi and Pilotti, 2000.

of multiple informal and social communication channels. In Table 3 we have tried to estimate the "innovative" efforts conducted by firms in their engineering departments.

Table 3: The investment of firms in engineering activities for new products (average of the last three years)
*Source: our interviews * estimated*

	Expenditures in engineering as a percentage of sales
Carle Montanari	4.0
Corazza	2.0
Tetra Brick	47.0
Sasib T.	12.0
Sasib B.	12.0
Baumer	1.0
Acma	5.0
Macofar	5.0
MG 2	3.0
Sacmi	4.0
Casmatic	3.0
Ima	4.0
GD	7.0
Total	5.0*

It is significant to observe the high figure of Tetrabrick, a firm which delegates all manufacturing activities to external subcontractors. During our research we organised many in-depth discussions related to how firms were able to introduce original innovations in their products.

Many innovations were described as the result of a collective effort among: a) the designers and engineers of the factory who at the same time perform indistinguishable routine tasks in designing new machinery and develop conceptionally new projects, on the basis of the specific suggestions received from the marketing department and after-sales assistance, b) some professional skilled workers (and, in particular, those assigned to the final assembling of packaging machinery and the outside inspectors and testers of machinery who have the responsibility of checking the assembly and installation of machinery by the firm's clients).

Table 4: "Innovation inducing" communicative channels in the Bologna packaging district. Source: our interviews (The percentage refers to the relative importance of each source for the innovative activity of the firm)

	Evaluation of the importance of the following channels Internal Engineering %	Clients %	Universities %	Technical exhibitions %	Total %
Carle Montanari	90	10	0	0	100
Corazza	80	10	10	0	100
Tetra Brick	100	0	0	0	100
Sasib T.	30	50	10	10	100
Sasib B.	30	50	10	10	100
Baumer	100	0	0	0	100
Acma	100	0	0	0	100
Macofar	0	100	0	0	100
MG 2	0	100	0	0	100
Sacmi	70	0	10	20	100
Casmatic	100	0	0	0	100
Ima	80	10	0	10	100
GD	100	0	0	0	100
Total	**67.7**	**25.4**	**3.1**	**3.8**	**100**

The model of innovation found was neither based on costly R&D investment in a traditional firm's laboratory, nor dependent on public research sponsored and carried in universities. In this perspective, the accumulation of technological knowledge in the district may be described as a collective process that combines pieces of information and knowledge that are dispersed among different agents (Maskell and Malmberg, 1999). Our empirical work has shown that innovative activity in firms is not only sustained by internal resources, but by a variety of sources. A significant part of the innovative process lies in the interaction of the informational circuits (Becattini and Rullani, 1996; Belussi and Gottardi, 2000) created among the technicians and expert entrepreneurs, clients, universities, and external agents (see Table 4).

As argued above, in the industrial district analysed, we found a model of innovation without elevated levels of intramural R&D (Gottardi, 1996), where new knowledge was created in an interactive process among different technological channels and communication mechanisms. This model is clearly focussed on organisational learning (Nonaka, and Takeuchi, 1995).

The principal actors were suppliers, innovative subcontractors, knowledge business services and institutional infrastructures. Within the district, tacit knowledge has been developed in firms where an accumulation of know-how has occurred (learning by doing activities). Tacit knowledge must be considered mainly a type of practical knowledge, which in the district was possessed by technicians and qualified blue collar workers. Tacit knowledge spilled over: a) when knowledgeable individuals found new entrepreneurial initiatives (new firm start-ups by scission), b) with the inter-firm mobility of technical personnel (the Bologna district, like all Italian districts, is characterised by elevated levels of intra-district mobility of labour-forces), and c) through the various contacts among professional people. Tacit knowledge with time become embodied and embedded in firm routines, giving rise to a growing expertise in building the packaging machinery.

For this reason this type of knowledge was very difficult to copy from outside.

15.4.4 The origin of codified knowledge

As already mentioned, the Aldini Valeriani, a technical school founded by the local government for the technical education of Bologna workers, has been the main source of diffusion of codified knowledge in the Bologna district. For instance, in the complex matter of metallurgy, the resistance of metals, the new materials applied in sophisticated treatments, new applications of the laws of rotating movement (crucial for this type of industrial machinery, in which the movements of boxing or canning must be repeated thousands of times every day), and, more recently, in the process of integration of electronic controls into machine tool controls. New external scientific knowledge produced by Italian and foreign universities has been tested here in advance, simplified and adapted to the linguistic codes used by local firms, and socialised through the organisation of technical courses, addressed to the re-training of the personnel of firms located in the Bologna district. In these years, the Valeriani institute has played the role of innovation "watchdog". Many teachers and alumni come from the best firms in the district, so this facilitates the absorption of external knowledge and the upgrading of knowledge among local firms.

15.4.5 The accumulation of contextual knowledge

While traditional economics emphasises one-shot innovation, the district has experienced a slow but constant construction of technological skills over time, and the nature of the innovative process has been one of sequential inter-related innovations in mechanical technologies (Patel and Pavitt, 1994) based on a social accumulation of knowledge. Among the interviewed firms, we observed a fair balance of development between tacit and codified knowledge. Codified knowledge has been developed by training schools and by universities, while local firms have represented the main social laboratories and repositories, where tacit know-how has been developed and transferred to new people, and amalgamated with the available information and knowledge.

Our empirical analysis allows us to introduce a new category within the classical dichotomy of codified and tacit knowledge. Some of the knowledge shared by the principal actors in the district can be defined as "contextual knowledge", in the sense that it is the collective result (Amin and Cohendet, 1999) of a slow process of knowledge building, experimentation, tacit know-how, interpretation and transposition of abstract knowledge (Arora and Gambardella, 1994). This knowledge is transmitted orally, and is learned on the shop floor or in the design department, after years of experience of working side by side. This contextual knowledge is therefore partially tacit and partially codified. Within the district, it is the simultaneous evolution of tacit and codified knowledge in firms that has created a significant barrier to imitation. While knowledge and information were at the same time widely shared among agents operating in the district (firms co-operate in large networks, their personnel is highly mobile, and the "district effect" produces an inevitable spill over of localised knowledge within the area), contextual knowledge *per se*, on the whole, never goes beyond the invisible walls of the district. This reflected a process of historical accumulation of localised capabilities, which any economic agent cannot possess individually, and which cannot, as a whole, be transformed, encoded, and stored in transmissible files.

So, as a whole, this type of knowledge can never undergo a process of conversion into a more transmissible knowledge. This type of knowledge is "situated" in human actions, and it tends to remain "sticky" (Hippel, 1994) within the firms and subcontractors belonging to the district. It is shared by the people living in the district, in their memory of industry, in the success stories told about the entrepreneurial adventures of the pioneers, in the memory of how technical problems were resolved, and in the now "natural" ability in the area through which every artisan knows how the pieces must be made (the working, re-working and treatments of metals, until they reach the perfect shape and the required specific actions in terms of precision, often calculated to the thousandth of a millimetre, i.e. in microns).

This body of knowledge that firms make use of, produce, or transform, cannot as a rule be considered as a free public good, available and usable by all economic agents, as is frequently assumed by the neo-classical tradition. Contextual knowledge is a collective good within the district but it has a quasi private nature. It is incorporated in specific contexts, in communicative channels, networks, and social relationships.

Generally speaking, firms tend to minimise the disclosure to their rivals of their technological skills and the know-how they have acquired by direct experience. But within specific local production systems, spill over of knowledge occurs, throughout the various communicative channels. So, a continuous process of inventing, catching up, copying, imitating, and improvement is promoted. Network externalities, on the supply, side related to technological production externalities (e.g. use of innovative suppliers and subcontractors) may render this process even more efficient from a welfare point of view. Therefore, a growth mechanism with increasing returns is endogenously activated at spatial level with an acceleration of the rate of technological change.

The accumulation of knowledge in the past gave rise to a path-dependent process. The accumulation of knowledge reinforced the ability of firms to recognise the potential value of several innovations introduced by other sectors (thus, these systems display a high capacity for absorption[10]). Moreover, the accumulation of localised knowledge increased the probability that the distribution of innovative efforts will end successfully. This has occurred in the district during the 1980s, when most of the machinery incorporated new electronic devices, produced elsewhere, but adapted by the firms in the district to packaging machinery.

[10] This point has been developed with accuracy by the literature on learning organisation (Gilbert and Cordey-Hayes, 1996; Cohen and Levinthal, 1990, 1991).

15.5 The model of local interactions: production networks as channels of technological communication

Although the economic literature on industrial districts has been quite vague in the operationalisation of the concept of the "industrial district"[11], the case of the Bologna packaging district forms part of the class of phenomena that, starting with Marshall, have been used in order to explain the spatial agglomeration of industries. Here we find a local agglomeration of small and large producers of packaging machinery, and a network of subcontracting firms, specialised in producing specific parts or special processing. The expansion of the system has allowed, over time, a larger division of labour, organised on a cognitive basis shared among different "productive units". In the district, some firms focus mainly on productive tasks (they are only able to replicate simple routines), while others are creative agents, endowed with specific skills and dynamic capabilities.

Routine work and simple structural steel work are generally decentralised by the district firms to "systematic" or "occasional" subcontractors. Firms with better skills are included in the former type. The latter are buffer firms used to fulfil peaks in demand. The former provide and enlarge the knowledge of final firms, the second allow for flexibility. Systematic long-term subcontracting may be distinguished as:

1. "volume-subcontracting" (firms which make parts "on order", or sub-systems with engineering skills for the final firm with which they have a stable relationship), and
2. "specialised subcontracting" (firms to which the final firms go for specific processes or specialised parts, e. g. camshafts, which are a strategic component in packaging machinery, manufactured only by few firms in the Bologna area (also many foreign firms come to Bologna to purchase their unique, specialised products).

Volume subcontracting firms generally have a lower level of knowledge on the technological complexity of packaging machinery, but they may be considered very efficient firms, because they can focus on a small segment of the production cycle, and their internal technological levels are general very

[11] For the orthodox approach (see Becattini's school), an industrial district can be defined only when there is an exclusive identification between productive specialisation in an industry (or branch of an industry) and a place, as in the case of Prato (characterised by the presence of the textile industry). In the case of the Bologna district, which is a vast area made up of many industrial and tertiary activities, this exclusiveness is lost.

high. Specialised subcontractors may be very competent in their specialist work, or they may own very specialised machinery that no one "core firm" or final firm could individually afford to buy.

The most creative agents are the cycle co-ordinators, the larger final firms mentioned above, able to design and direct all phases, from the conception/project of the machinery up to the phase of after-sales assistance. They take the order, design and engineer the machinery, and delegate to subcontractors a large part of the manufacturing of parts and sub-components. In some cases, the design of the parts is realised in partnership with some subcontractors (see also Lipparini and Lorenzoni, 1994). These "core firms" perform the final assembly, the processing of particular strategic components, and the critical phases of format adjustment, testing, and delivery (being responsible for the final installation of the packaging line in the client's factory). In their structure, they have accumulated the deeper levels of codified and tacit knowledge. The core firms are immersed in a dense, locally embedded network of relationships and communicative channels, through which technology is developed. The crucial channels are represented by the commercial outlets, the suppliers of routine components, and the specialised part producers. They know how to realise the entire machine, because they possess both the design and technical skills in manufacturing. In practice, they know the entire packaging production cycle. However, very often their engineering departments concentrate on more innovative and strategic activities, as well as those that can add value to the machinery.

In comparison with foreign firms, firms located in the Bologna district are quite small. While foreign firms are often part of multinational conglomerates, the Bologna packaging machinery producers are small firms: they occupy their international niches together with other 3 or 4 large international producers. Italian firms concentrate more on differentiated products, and they subcontract much more than their foreign competitors. The district "effect" allows a higher decentralisation of production activities. Because many routine activities are externalised, firms may lighten their internal organisation.

In this chapter we have discussed a conceptualisation of the innovation process as an interactive social phenomenon, where the dominant feature is learning by networking.

Table 5: Analysis of production networks developed by final firms in the Bologna packaging district
Source: our interviews

	Number of subcontractors		Types of subcontractors		Productive characteristics of subcontractors				
	Commercial suppliers A	External outsourcing B	Long-term B’	Occasional B’’	Single pieces B’1	Also assembled B’2	A+B	B’/B	B’2/B
Carle Montanari	500	100	80	20	60	20	600	80.0	25.0
Tetra Brick	300	30	20	10	5	15	330	66.0	75.0
Sasib T	600	70	50	20	43	7	670	71.0	14.0
Baumer	200	20	10	10	6	4	220	50.0	40.0
Acma	520	190	133	57	113	20	710	70.0	15.0
Macofar	15	25	10	15	5	5	40	40.0	50.0
Sacmi	300	200	150	50	40	10	500	25.0	6.0
Casmatic	250	80	50	30	30	20	330	62.5	40.0
Ima	650	350	100	250	80	20	1000	71.4	20.0
GD	175	75	75	0	50	25	250	100.0	33.0

The strong relational capabilities of packaging firms located in the district emerge from Table 5. Large organisations typically involve in their production cycles 500-1000 suppliers of parts, single pieces, assembled sub-groups, or machines. In fact, many packaging machines often involve the assembly of 8,000-10,000 components. Commercial suppliers provide raw materials, or half-finished (tailor made) components included in a catalogue.

External subcontractors participate in the construction of packaging machinery. The interviewed firms were using a mix of long-term and short-term subcontracting arrangements. However, the involvement of subcontractors in systematic and co-operative relationships was the norm. In the firms analysed the share of subcontracting as a percentage of output varied between a minimum of 13% and a maximum of 80% (see Table 6). Subcontracting involves essentially the manufacture of mechanical parts, and sometimes the co-design of the machinery or partial assembly (see Table 7). Final firms never delegate commercial functions, product innovation activities and the conception of novelties, the final testing, and the final assembly of the packaging line: in other words the core of their firm-specific patrimony.

During the 1990s the organisation of subcontracting has shifted towards a more pyramidal post-Fordist hierarchy. External firms have been tied to final firms, in relation with their production capacity.

Subcontractors have been selected and stratified and the number of links reduced, but reinforced in terms of connectivity.

The first layer of subcontracting is now represented by relatively large size firm-system suppliers, which work in a strategic partnership with the final firm. The further organisation of the subcontracting chain (subcontractors for parts, sub-systems, or highly specialised components) is delegated to them. These integrated systems now tend to work together for the manufacture of a complete machine, which is co-designed with the final firm. Tacit knowledge and specific technical skill are in this way shared by final firms with their first layer subcontractors. The topology of subcontracting networks has moved during the 1990s to less betweenness centrality nets (from star models to overlapping pyramids).

Table 6: Analysis of final firms in the Bologna packaging district: some economic indicators
*Source: our interviews *our estimation + positive but unquantifiable*

	Employees	Exports as a percentage of total sales	Sales per employees	Variation of sales 94/97	**Share of subcontracting work as a percentage of total sales**
Carle Montanari	350	95.0	257.1	+25.0	**30.0**
Corazza	170	93.0	251.8	+13.2	**24.0**
Tetra Brick	375	90.0	489.0	+39.8	**75.0**
Sasib T.	391	100.0	See GD	+22.5	**20.0**
Sasib B.	381	95.0	See GD	+60.1	**13.0**
Baumer	36	65.0	330.0	+	**37.0**
Acma	499	85.0	260.0	+	**40.0***
Macofar	55	95.0	327.0	+	**35.0***
MG 2	130	70.0	250.0*	+	**43.0***
Sacmi	808	90.0	1299.0	+32.2	**15.0**
Casmatic	250	85.0	425.0	+	**80.0**
Ima	917	92.0	412.2	+37.7	**55.0**
GD	1653	97.0	459.8	+22.5	**70.0**
Total	**5,965**	**95.0***			

In turn, subcontractors have increased the number of clients (final firms normally never overcome the saturation of 30-40% of the productive capacity of their subcontractors). Prices for subcontracting work are not spontaneously set by markets: they are normally bargained between the subcontractor and the final firm, very often they are "fixed" by the final firm.

A price cap strategy (with a planned time-based diminishing of prices) is used by final firms to recover competitiveness, but it obliges subcontractors to invest heavily in technology and in the improvement of their organisation in order to maximise the efficiency of their production cycle. We found that subcontractors located in the district were quite technologically advanced. They were controlled obsessively in the performance of their tasks by final firms which exerted a strong influence in forcing the adoption of modern organisational methods, such as just-in-time, total quality, ISO 9000 etc.

Table 7: Analysis of outsourcing by final firms activated in the Bologna packaging district
Source: our interviews

	Share of externalisation of specific functions calculated on the value of each function			Commercial functions
	Design Machinery	Mechanical parts	Assembling	
Carle Montanari	0	80.0	0	0
Corazza	10.0	80.0	0	0
Tetra Brick	50.0	100.0	100.0	0
Sasib T.	0	50.0	0	0
Sasib B.	0	30.0	30.0	0
Baumer	0	100.0	0	0
Acma	10.0	50.0	20.0	0
Macofar	0	50.0	0	0
MG 2	0	50.0	0	0
Sacmi	0	50.0	0	0
Casmatic	30.0	100.0	50.0	0
Ima	0	80.0	50.0	0
GD	0	80.0	0	0

15.6 Conclusions: the formation of a district based on the exploitation and accumulation of contextual knowledge

This paper has analysed the packaging district located in and around the province of Bologna, specialised in packaging machinery.

The composition/form of the local industrial structure does not resemble that typically described by Marshallian Italian industrial districts: hundreds of small undifferentiated and flexible firms competing in the same large mass market. On the contrary, a few leading firms form the heart of the packaging district. They account for about 40% of the total output produced in the province. Firms are integrated in large overlapping production networks, which constitute the main technological channels where knowledge is recombined and developed. It has been documented that innovative activity, within traditional Italian districts, built around established products, has been quite modest. They compete internationally by using their internal flexibility to adjust production volumes to demand needs and devote their creativity to product re-styling. In contrast with this, our study demonstrates how, in the local context of Bologna, a continuous process of innovation has given rise to progressive spatial clustering of very innovative producers, internationally renowned, and on the cutting edge of the technological frontier of the sector.

The success of the district is based on an evolutionary institutional mechanism of knowledge building. The formation of firms in the district has supported the process of socialisation of tacit (practical) knowledge, but this has also activated a powerful market selection mechanism (under the pressure of competition expressed by new entries, vis-à-vis aggressive neighbours, and competitive spin-off, etc.).

Within the district we have found, as expected, a dense substratum of competitive and co-operative relationships among firms[12].

Despite the impressive social bounds found, horizontal competition among firms was quite severe. Horizontal competition among firms played an important role: it activated a Darwinian selection, pushing firms to make incessant improvements in their products. Each firm was also threatened by the potential entry of other competitors located in the district working in different markets (for instance, firms may move from food industry packaging to pharmaceuticals, etc.). Thus, this district, as some would argue, was not protected from the stresses of market mechanism, because of its climate of profound social integration (the value of the community). The district, having internalised market pressures, was shaped by the concrete operation of competitive forces (high firm turnover, frequent entries and exits, rapid adoption of innovations, strong rivalry). The fierce competition which exists is one of the ingredients of the district success, because over time it has provided the proper incentives to innovation. Not to mention the geographic visibility of firms' behaviour, which has contributed to rendering market competition more transparent, fluid, and selective.

The continuous improvements, combining innovation and incremental modifications, introduced by firms has raised over time their competitive position. No one dominant firm in the Bologna district could rest on the laurels of a comfortable temporary oligopoly.

This path-dependent, historical agglomeration of packaging firms has been a process sustained by the existence of high levels of local/contextual knowledge. Practical and tacit knowledge have been transmitted through apprenticeship and

[12] In the idealised popular view, firms in industrial districts are immersed in a climate of trust, reciprocity and kinship bounds (Dei Ottati, 1996). Studying the concrete operation of firms in the district, it is important to stress that the district is characterised by a melange of co-operation and competition: co-operation is mainly vertical (governed by final firms with their suppliers) while not much horizontal co-operation was found.

personal contacts. The diffusion of codified knowledge has been related to the activity of local training institutions. From the perspective discussed in the paper, the stock of contextual knowledge embedded in the district may be viewed as an existing strategic (but immaterial) resource. We must reflect on the fact that this cognitive form of social capital, historically accumulated in the area, cannot be thought of as a mobile factor or input, simply transferable outside the district walls. Only agents operating in the district have access to it through the various channels described in this article: labour market, knowledge business services, user-producers relationship, after-sales assistance, use of specialised suppliers and subcontractors, and institutional infrastructures. They may further enlarge and exploit it through strategies of entrepreneurial growth.

References

Amin A. (1998), Il modello emiliano, sfide istituzionali, *IPL Working paper,* Bologna.

Amin A. and Cohendet P. (1999), Organisational learning and governance through embedded practices, paper presented at the workshop "Information processes and path-dependent evolution: local systems' response to changes in context, Padua 27th November.

Antonelli C. (1999), *The Microdynamics of Technological Change*, Routledge, London.

Arora A. and Gambardella A. (1994), The changing technology of technological change: general and abstract knowledge and the division of innovative labour, *Research Policy*, 23, p. 523-532.

Becattini G. and Rullani E. (1996), Local systems and global connections: the role of knowledge, in F. Cossentino, F. Pyke, and W. Sengenberger (eds.), Local and regional response to global pressure: the case of Italy and its industrial districts, Ilo, Geneva, 159-174.

Becattini G. (1987), L'unità di indagine, in Becattini (ed.), *Mercato e forze locali: il distretto industriale*, Il Mulino, Bologna.

Belussi F. (1992), (ed.), *Nuovi modelli di impresa, gerarchie organizzative ed imprese rete*, Franco Angeli, Milan.

Belussi F. (1998), Industrial districts and the generation of localised technological change. The case of the packaging machinery industry in the Bologna district, Inloco, mimeo, Turin.

Belussi F. and Arcangeli F. (1998), A typology of networks: flexible and evolutionary firms, *Research Policy*, vol. 27, p. 415-428.

Belussi F. and Gottardi G. (2000), (eds.), *Evolutionary Patterns of Local Industrial Systems*, Ashgate, Aldershot.

Belussi F. and Pilotti L. (2000), Knowledge creation and collective learning in the Italian local production systems, Working Paper of the "Marco Fanno" Department of Economic, *WP no. 21*, Padua University.

Biggiero L. and. Forni F. (1987), *L'industria delle macchine automatiche per dosatura, confezionamento e imballaggio*, Ervet, Editrice Compositori, Bologna.

Bonaccorsi F. and Lipparini A. (1994), Strategic partnership in new product development: an Italian case story, *Journal of Production Innovation Managing*, no. 11, p. 135-156.

Callon M. (1991), Réseaux technico-économiques et irréversibilité, in Boyer R. *et al.* (eds.), *Figures de l'irréversibilité en économie*, Editions EHESS, Paris, p. 107-127.
Callon M. (1993), Variety and irreversibility in networks of technique conception and adoption, in Foray D. and Freeman C., (eds.), *Technology and the Wealth of the Nations*, Pinter, London, p. 232-268.
Capecchi V. (1997), La ricerca della flessibilità: l'industria meccanica bolognese dal 1900 al 1992, *Sviluppo Locale,* IV, no. 4, p. 80-130.
Cohen W. and Levinthal D. (1989), Innovation and learning: the two faces of R&D, *Economic Journal*, vol. 99, p. 569-596.
Cohen W. and Levinthal D. (1990), Absorptive capacity: a new perspective on learning and innovation, *Administrative Science Quarterly*, vol. 35, p. 128-152.
Cowan R. and Foray D. (1997), The economics of codification and the diffusion of knowledge, *Industrial and Corporate Change*, vol. 6, no. 3, p. 595-622.
Dei Ottati G. (1996), Trust, interlinking transactions and credit in industrial districts, *Cambridge Journal of Economics*, vol. 18, p. 529-546.
Di Bernardo B. and Benedetti E. (1997), Reti: un nuovo paradigma?, in Benedetti E., Mistri M.and Solari S. (eds.), *Teorie evolutive e trasformazioni economiche*, Cedam, Padua.
Foray D. and Lundvall B. (1996), From the economics of knowledge to the learning economy, in, Lundvall B. and Foray D. (eds.), *Employment and Growth in the Knowledge-based economy*, Oecd, Paris.
Freeman C. (1991), Networks of innovators: a synthesis of research issues, *Research Policy*, vol. 20, p. 499-519.
Gilbert M. and Cordey-Hayes M. (1996), Understanding the process of knowledge transfer to achieve successful technological innovation, *Technovation*, vol.16, no. 6, p. 301-312.
Gilly J.P. and Torre A. (1998), Introduzione, Dinamica di prossimità e reti, *L'industria,* vol. 19, no. 3, p. 433-461.
Gottardi G. (1996), Technology strategies, innovation without R&D, and the creation of knowledge within industrial districts, *Journal of Industry Studies*, vol.3, no. 2, p. 119-134.
Grabher A. (1993), (ed.), Rediscovering the social in the economics of interfirm relations, in Grabher A., 1993, (ed.), *The embedded firm*, Routledge, London.
Harrison B. (1992), Industrial districts: old wine in new bottles, *Regional Studies*, vol. 26, p. 469-483.
von Hippel E. (1994), "Sticky Information" and the Locus of Problem Solving: Implication for Innovation, *Management Science*, vol. 40p. 429-439.
Kline S. and Rosenberg N. (1986), An overview of innovation, in Landau R. and Rosenberg N. (eds.), *The Positive Sum Strategy*, National Academic Press, Washington.
Lazerson M. and Lorenzoni G. (1999a), The firms that feed industrial districts: a return to the Italian source, *Industrial and Corporate Change*, vol. 8 no. 2, p. 235-266.
Lazerson M. and Lorenzoni G. (1999b), Resisting organisational inertia: the evolution of industrial districts, *Journal of Management and Governance,* vol. 3, no. 4, p. 361-377.
Lipparini A. (1995), *Imprese, relazioni, tra imprese e posizionamento competitivo,* EtasLibri, Milan.
Lipparini A. and Lorenzoni G. (1994), Neither strengthening nor dismantling firm's wall. External ties and architectures promoting unfolded innovations, paper presented at the workshop "The changing boundaries of the firm", Como, 21-23rd October.

Lundvall B. (1993), Explaining interfirm cooperation and innovation, in, Grabher A. (ed.), *The embedded firm*, Routledge, London.

Maskell P. and Malmberg A. (1999), Localised learning and industrial competitiveness, *Cambridge Journal of Economics*, vol. 23, p. 167-185.

Nonaka I. and Takeuki H. (1996), *The Knowledge Creating Company*, Oxford Univ. Press, Oxford.

Patel P. and Pavitt K. (1994), The continuing, widespread (and neglected) importance of improvements in mechanical technologies, *Research Policy*, vol. 23, p.533-545.

Polanyi M. (1967), *The Tacit Dimension*, Anchor Books, New York.

Senker J. (1995), Tacit knowledge and models of innovation, *Industrial and Corporate Change*, vol. 4, no. 2, p. 424-447.

Stoneman P. (1995), (ed.), *Handbook of the Economics of Innovation and Technological Change*, Basil Blackwell, Oxford.

Witt U. (1996), Lock-in vs. critical masses: industrial change under network externalities, *International Journal of Industrial Organisation*, vol. 15, p. 753-773.

Chapter 16

The Biomedical Valley: structural, relational and cognitive aspects

Lucio Biggiero and Alessia Sammarra

16.1 Introduction

This study furnishes empirical evidence from the biomedical district of Mirandola (Italy) with the aim of offering two distinct contributions to the literature on industrial districts. The first objective is to challenge the stereotyped model of Italian industrial districts whose claimed distinctive features are to be specialized in traditional sectors of the Italian economy, to be relatively self-contained and close productive systems, and to be mainly composed of small and medium size enterprises. The Biomedical Valley contradicts this model, being a successful example of a recent, high-tech, high-wages, and open district led by some medium sized companies, some of which have been acquired by large European and American multinationals. The district emerged in 1963, triggered by a true Schumpeterian entrepreneur, who created the "first firm" of the district, from which many others (the most important) generated through spin-off processes.

The second objective of this study is to present an integrated analysis which complements the traditional approaches, mainly focused on the investigation of the structural characteristics (production cycle, inter-firm division of labour, firms' size, export) and the historical traits of industrial districts, with some recent perspectives more concerned with the relational (Ebers, 1997; Grandori, 1999) and cognitive interpretation of the district model (Borroi *et al.*, 1998; Sammarra and Biggiero, 2001a, 2001b). Indeed we argue that cooperation, coordination, knowledge transfer and innovation depend not only on structural and relational features, but also on how firms perceive and evaluate their space and how they evaluate their belonging to the local territory. Specifically, we claim that the firms' identification with the district (as a social category), by supporting the emergence of group processes among the district members, involves qualitative changes in the way firms interact that contribute to enhance the reservoir of social capital available through the generation of intangible resources such as depersonalized trust and cooperative attitudes (see Biggiero and Sammarra in the first part of this volume). Furthermore, we attempt to link the analysis of social identification processes with the main structural and relational variables. We believe that, in order to explain district performance and evolution, all these aspects must be taken into account. In this view, we propose an integrated analysis of the biomedical district localized in the province of Modena (Italy), which compares and complements the structural description of the district with some empirical findings concerning both the relational texture and the processes of social identification enacted by the district members.

In order to develop a full integrated approach, we first highlight the district's structure and history, illustrating the productive filière and the role that different typologies of firms play within the productive system. Next, we analyse the relational factors of the biomedical district describing the district as hyper-network. Finally, we focus on the cognitive aspects illustrating the role and the mechanisms of formation of a district identity among the biomedical firms.

16.2 The Biomedical district: an overview

The Biomedical Valley, located in a small group of municipalities centred in the city of Mirandola, in the province of Modena (fig. 1), is the most important European centre for the biomedical industry, representing one of the few examples of innovative and high-tech Italian industrial districts. Worldwide, the biomedical industry developed in the 1950s, with the application of electronic instruments and accessory technologies for health care, and developed during the subsequent decades. The firms that came into being in the biomedical field tended to group together in a few geographical locations, which is evidenced by the presence of production agglomerates in the United States and Italy (Lipparini and Lomi, 1999).

Since the district's birth in 1963, the area of Mirandola has changed from being an agricultural area to an industrial one, thanks to the rapid development of the mechanical and biomedical sector. The Biomedical Valley has in a few years achieved international status with about 74 manufacturing firms and 3,200 employees (R&I, 1999).

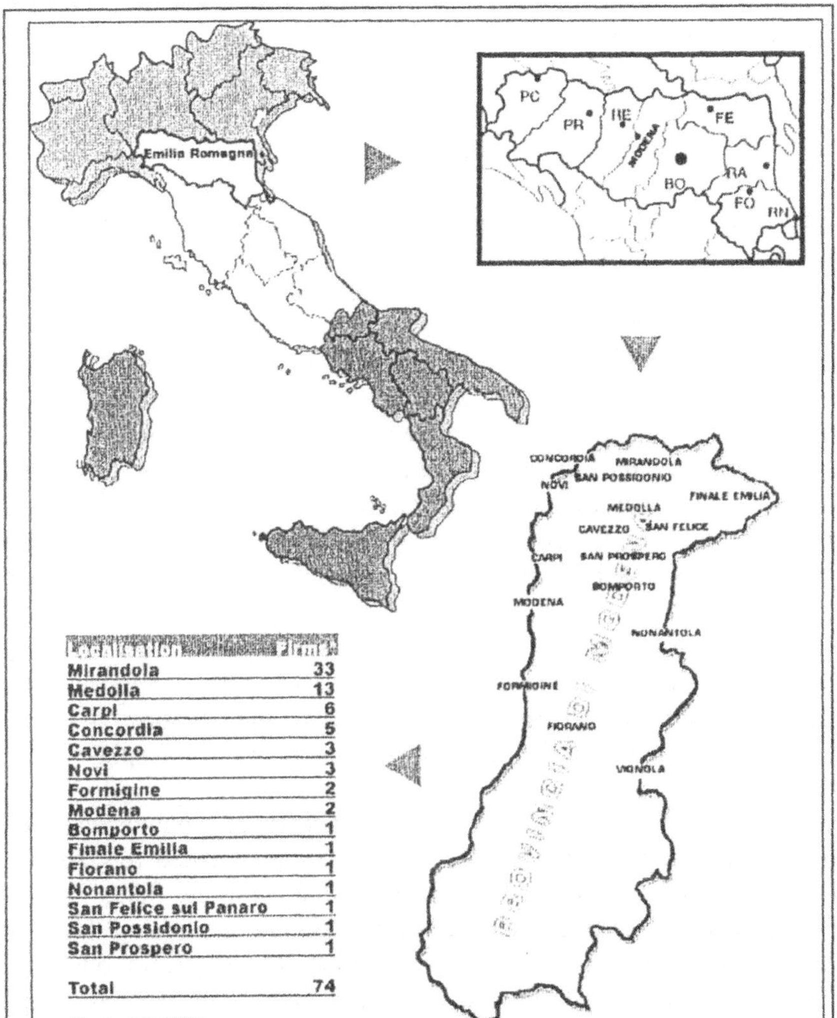

Figure 1: The geographical location of the Biomedical Valley

The most distinctive feature is the significant role played by multinational companies, which entered the district through the acquisition of firms founded by local entrepreneurs. In the district, most firms (77%) are independent, while 10 are controlled by external groups, of which 6 are non-Italian companies. The most important national and foreign external groups in the district are: Gambro,

Sorin-Snia, B. Braun, Mallinckrodt, Baxter, and Fresenius. District firms controlled by these external groups are also the largest, employing nearly 60% of workers and covering 79% of sales.

16.3 The District's History

The origins and development of the biomedical district are quite original compared to other Italian industrial districts. This district grew in an innovative sector characterized by advanced and constantly evolving technological and scientific features. While the formation of most of the Italian industrial districts dates back to the 19th century and is based on the presence of craft tradition and/or particular resources in the area, the biomedical district has a very recent history, directly linked to the entrepreneurial initiative of the pharmacist Dr. Mario Veronesi, whom everyone considers to be the founder of the district.

In 1963 Veronesi founded the first biomedical enterprise in the district, Mirastet and, in just over 30 years, from 1963 to 1998, founded, developed and then sold four of the most important enterprises in the district: Dasco, Bellco, Dideco and Dar. His business history has become the development process typical of the biomedical district: the creation of new enterprises by local entrepreneurs, size expansion, total or partial sale of these companies to major Italian and multinational groups and, finally, the creation of new small enterprises, above all in the role of subcontractors, as a spin-off of larger enterprises.

In 1963, when Veronesi created his first company, the biomedical sector was wholly absent in the Mirandola area even though it could rely on the skills developed in the near area of Bologna and Modena in the fields of mechanics and plastics, and in particular on the special skills to be found in the production cycle and later used in starting up the production of biomedical products.

This district owes its birth to the identification of the needs of hospitals in terms of single-use products for the circulation of liquid solutions (e.g. blood): traditional products (latex tubes) used by hospitals in fact demanded continual sterilization, and hence the need to produce disposables. For his first idea, that of disposable products, developed in 1963, he put together workers in the mechanical trades and workmen who processed plastic. Furthermore, his management approach was oriented towards the workers' participation in company risk.

There are many examples of the creation of firms by former Veronesi workers. Many left Dasco, Bellco or Dideco to create either a firm that would be a subcontractor of the firm in which they had formerly worked or an independent firm concentrating on competitive or lateral niches. The process of emulation of the Veronesi-style creation of firms came together with spin-off processes, even if the two things are, of course, separate and distinct. The former show up as a process of becoming independent and of taking on entrepreneurial risk. The latter are processes stimulated by the leading firm, which seeks to farm out the activities that are more economical to outsource than to produce in-house.

The spin-off process went ahead, as did acquisitions of local firms by foreign multinationals. There are 6 multinationals and 4 large companies that are owned by non-indigenous firms. During the last 20 years, both entered the district through acquisitions of incumbent firms. District firms did not succeed in competing with large firms because of their economic, financial and technological constraints. In fact, during the growth phase of their life cycle they did not manage to go beyond a merely local dimension and they were forced to sell the developing firms to foreign companies.

The value of the multinationals (Biggiero, 2002a) became manifest with the arrival of foreign capital and investment, which meant the transfer of ownership of the local firms: companies such as Gambro, Baxter, Pfizer and Sorin acquired firms of solid local tradition, thus feeding the process of the internationalization and globalization of the biomedical district. In particular, two eras of big acquisitions can be identified: the period 1987-88 in which Gambro, Pfizer and Sorin entered the district (Gambro buying Hospal Dasco, Pfizer buying Dideco, and Sorin starting up its joint venture with ENI), and the period 1994-95 when Sorin and Gambro strengthened their positions (the former purchasing Dideco and Cortex in 1995 and the latter Cobe in the same year); Mallinckrodt, Fresenius and B. Braun also entered the district through the acquisition of Dar in 1994, and of Biofil and Carex in 1995. From the viewpoint of district incumbent firms, the entry of multinationals and their establishment there were motivated, as has already been underscored, by the financial difficulties encountered by the small firms in taking on the transition from the launching of the product to its development. In particular, the management of the small firms succeeded in identifying the entrepreneurial idea and in producing new products. However, it encountered hardships tied to the production cycle and to finance and marketing, which limited the development of the product they had identified.

Biggiero (2002a) shows that the multinationals' entry benefited the district with new financial resources and with direct access to the world market demand. At the same time multinational headquarters did not change strategies, capabilities and human resources of acquired firms. They contributed to change the district structure by pushing and improving a selection process of subcontractors. The entry rationale is based on three reasons. The first is to neutralize the potential competition of the other non-district, worldwide competitors. After recognizing the effectiveness of many products supplied by district firms, multinationals entered in a fast succession, in order to prevent or hinder domination over the district by competitors. The second reason is related to the exploitation of agglomeration economies and of knowledge transfer within the district. Both are linked to the diffusion of tacit and explicit knowledge between district members (firms and other organizations, like trade associations, specialized schools, etc.). Both aspects reveal the presence of lock-in effects (Arthur, 1988). Currently, there is no evidence whether, and to what extent, some kind of technology transfer between multinational headquarters (or its subsidiaries) and its acquired district firms takes place. What is sure is that until now multinationals have played the role of bridges, opening the district to the external world. Linking local and global, they did not destroy the Biomedical Valley.

If the entry of multinational companies allowed the local firms to overcome the constraints of their local dimension, the widespread entrepreneurial spirit of the local population and its marked tendency towards innovation have led to the emergence of a constellation of specialized suppliers and firms offering complementary products. Many small enterprises of the district did not content themselves with the role of simple sub-contractors of the major enterprises. The search for new market niches inaugurated by the founder of the district has become the typical behaviour of the smaller enterprises, which have adopted strategic paths and objectives which are completely independent and original with respect to the major enterprises.

Many of the small entrepreneurs of the district proudly stress their role as innovators with respect to the major enterprises of the district, and also, in some cases, with respect to the Veronesi example. Indeed, there is no lack of traditionally-minded small entrepreneurs who have a strong attachment to their own firms and do not consider selling them to foreign investors as a viable although profitable strategic option. In this regard, they blame the entrepreneurial model inaugurated by Veronesi, proudly stressing the close link with their companies founded "by the creation of completely new products", and their conscious "search for paths which are original compared to what others have already done".

The continuous search for new market niches by product innovation undertaken by the small enterprises represents another specific feature of the biomedical district, which has made cohabitation between small and large front-end enterprises both possible and economically profitable, a situation which is hard to find in other areas of localization of the biomedical industry, and equally anomalous in most of the Italian industrial districts.

16.4 Products and production cycle

The biomedical industry includes technologies and products for health-care applications, excepting those for pharmaceutical use (CNR, 1987). The ISTAT (Italian National Institute of Statistics) classification includes the biomedical industry in those categories that refer to the manufacture of electro-medical equipment (33.10.1) and the manufacture of medical and surgical devices (33.10.2). In the US, the revised SIC classification of 1987 includes the biomedical industry in the categories SIC 3844 and SIC 3845.

The biomedical district is specialized in the manufacture of disposable products, which represent 80% of the district's sales, and in the manufacture of electro-medical equipment (13% of total sales). Other types of biomedical products, such as sterilized solutions and concentrates have a marginal role on the district's sales (tab. 1).

Table 1: Product specialization of the biomedical district
Source: R&I, 1999

Type of products	Sales (billions of lira)	Sales (%)
Electro-medical equipment	103	13.2
Disposables	643	83.0
Other medical devices	13	1.7
Non medical products	9	1.2
Others	7	0.9
Total	775	100.0

Disposable products have many applications in the health care sector. The most important are: 1) disposables for the circulation of the blood outside the body (dialysis, oxygenation, transfusion, plasmapheresis, filtering/purification); 2) disposables for the administration of injectable sterilized solutions (flow aids and nutritional membrane sacs); 3) disposables for respiratory use in reanimation and in anaesthesia. These three sectors of applications are also the most traditional, the one that led the district's birth and development as a consequence of the product specialization of the oldest firms, such as Bellco, Hospal Dasco, Dideco and Dar, which are at present the largest of the district.

In the latest years, the growing commitment towards product innovation of many small and medium size firms led to the development of new niche markets and to the increasing diversification of the district's production in new sector of applications such as nutrition and gynaecology (tab. 2). The organization of the production cycle is quite different according to the type of product. In the production of electro-medical equipment, all the firms in the district adopt a highly decentralized manufacturing system. With respect to the main phases of the production cycle (design, mechanical processing, moulding, painting, assembly, testing, packaging, consignment), only the final phases are performed in-house, while the others are outsourced through the use of subcontracting and alliances. The suppliers' network for the production of equipment is mostly localized outside the district area, especially in the province of Bologna (about 40%), and the suppliers do not work exclusively for the biomedical sector.

Table 2: Applications in the health care sector of the district's production
Source: R&I, 1999

Applications in the health care sector	Percentage value on total sales
Haemodialysis	48.1
Cardio-surgery	17.9
Anaesthesia and intensive care	11.2
Blood transfusion	9.5
Aphaeresis	3.3
Infusion/nutrition	1.8
Urology	1.1
Surgery	0.8
Orthopaedic	0.6
Gynaecology	0.6
Other medical devices	4,2
Non-medical products	1.0
Total	100.0

In the production of disposable products, the biomedical firms in the district adopt a quite different organization of the production cycle according to their size. The largest firms (more than 200 employees on average), mostly owned by multinationals, have a highly vertical integration. Among the 6 largest firms of the district none of the production phases (design, extrusion, moulding, cutting, assembly, testing, sterilization, packaging, consignment) is totally outsourced.

Only some of them, such as assembly, are partially outsourced to subcontractors. On the contrary, the smaller firms adopt a highly decentralized manufacturing system: they completely rule out the initial phases of extrusion and moulding. These firms extensively rely on the local suppliers for both standardized and customized components in order to produce customized products in small batches for market niches.

The supply network for the production of disposables is mainly localized within the Biomedical Valley, which in terms of supply relationships represents a highly self-contained local production network. Only 12% of the production is decentralized in the nearest Italian regions, 4.4% in South Italy, and 5.2% abroad (R&I, 1999). These figures are very meaningful if we take into account that: (i) the largest manufacturing firms are all controlled by multinationals, (ii) in the Mirandola area the labour cost is not competitive because of the high local employment rate. Considering these two factors, the absence of substantial de-localization processes is an expression of a deliberate choice of the district manufacturing firms to keep most of the supply chain within the district's boundaries in order to exploit the distinctive skills and reliability of local suppliers.

The decision made recently by the major enterprises of the district to reorganize production seems to have reduced the risk of location transfer for the near future. In recent years, the major enterprises of the district have invested in automatic production lines for the pre-assembly phase of disposables. This implies that at least in the medium term, the production of disposables will not be transferred outside the Mirandola area. One of the most significant examples in this respect is Gambro, the most important multinational group of the district, which has started to build a computerized warehouse in the district, designed to become the centre of distribution and handling for the entire production of the European market.

Among the suppliers involved in the production of disposables, most are specialized in the assembling phase, while a few (less than 10) concentrate on the production of plastic components. This latter group of suppliers have a more highly evolved relationship with respect to traditional subcontracting enterprises specialized in moulding and assembly for third parties, which are generally characterized by dependence on a major customer.

16.5 The District's Structure

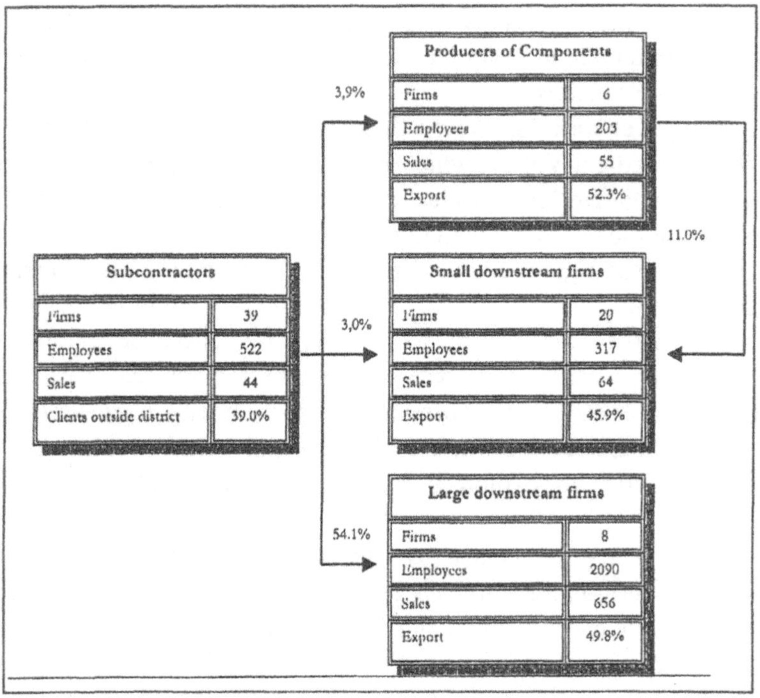

Figure 2: The Mirandola district filière
Source: R&I, 1999

The biomedical district has been defined as an organizational community (Lipparini and Lomi, 1999) because of its internal structure of relationships among differentiated but highly interdependent actors with complementary resources, competencies and products.

With regard to the different role played in the district's filière, four types of firms emerge which are characterized by different structural features and, at the same time, strong organizational interdependencies (Fig. 2).

16.5.1 Large downstream firms

This first type includes Gambro, B. Braun-Carex, Bellco, Dideco, Mallinckrodt Dar, Baxter, and Haemotronic. Those are the largest firms in the district in terms of employees and sales: 4 have more than 250 employees, the others belong to the category 50-249 (tab. 3).

Table 3: Front-end firms in the Biomedical Valley
Source: R&I, 1999

Employees	Firms	Employees (%)	Sales (%)	Independent Firms	Controlled by Italian multinational	Controlled by foreign multinational
Less than 50	25	17.0	13.5	24	-	1
50-249	6	20.2	16.2	3	-	1
More than 250	4	62.8	70.2	-	2	1
Total	35	100.0	100.0	27	2	3

With the exception of Haemotronic, all the largest firms are controlled by external multinationals and produce both disposables and electro-medical equipment. For the manufacturing of electro-medical equipment they rely extensively on subcontractors, while for the manufacturing of disposables they adopt a highly vertical integrated structure, usually focusing on one sector of application for which they offer a complete line of products. The two largest firms are specialized in the manufacture of disposables for dialysis, the second in disposables for anaesthesia and intensive care, and the last in disposables for cardio-surgery. Consequently these are the three most important sectors of application in terms of total sales. The strategic orientation of the largest firms is aimed at improving efficiency, mainly through automatization of the production cycle, and innovation in the specific sector of application they are specialized in. These firms are directly linked to the market through their own sales networks. In the domestic market, mostly represented by public and private hospitals, sales to end customers represent about 65% of their production, while 28% is absorbed by internal exchanges with the other subsidiaries of the group. In the foreign market the quota absorbed by internal exchanges raises up to ¾ of the largest firms' production.

16.5.2 Small downstream firms

Most of the downstream firms in the district are independent, small firms with less than 50 employees (tab. 3). They focus only on the production of disposables and, unlike the larger enterprises, they show a significant differentiation of the sectors where their products are used, including above all the less traditional ones. In fact, these small firms are the most committed towards product innovation. In recent years they have discovered new niche markets for which they offer customized products in small batches. These firms are very careful to keep a direct contact, also of an informal nature, with physicians as the end users of their products, in order to acquire a strategic orientation toward product innovation in relation to the fast and continuous evolution of health care needs and medical techniques. Trade intermediaries are the major customers for this type of firms, accounting for about 45% of sales, even if most of the production is placed on the domestic market with the manufacturer's trademark. The rest is sold directly to the health care system or other producers. Dependence on trade intermediaries is particularly important for exports, and over half of the production is placed on the foreign market with the trademark of the distributor or the customer.

In general the commercial function is the weak point of the small downstream firms. This depends on cultural and economic aspects. With regard to the former, it should be pointed out that most small entrepreneurs in the district have acquired previous experience in the production and technical departments of the major companies before starting their own business. This has led to an in-depth knowledge of the sector and the technical aspects of production, but few have managerial training and experience, which is reflected above all in the commercial field. Together with this cultural aspect, there are the difficulties that all the small enterprises encounter when developing an effective commercial activity due to their size constraints and the lack of financial resources. Up to now, these difficulties in the commercial sector have limited the growth potential of these enterprises, above all on the foreign market.

With regard to the organization of production, the small enterprises are much more decentralized compared to the large biomedical companies in the district. In the production of disposables, the smaller companies totally exclude the two main phases upstream from the production cycle: extrusion and moulding. The enterprises producing the finished product purchase the moulded components and extruded tubes directly on the market, and these purchases can involve both standard production and articles customized on the basis of particular requirements.

The formation inside the district of a small group of firms specialized in the production of components for disposable products has enormously enhanced the growth potential of the small firms offering finished products. The latter, thanks to the relationships with the component producers, can assemble the components acquired in the area in many different combinations, and can access the market with products customized on the basis of a competitive quality/price ratio.

16.5.3 Component producers

This group focuses on the production of plastic components. These companies cannot be classified among the downstream firms because they do not produce a finished product, but at the same time they present a number of significantly different characteristics with respect to the traditional role of subcontractors.

Indeed they play a very important role within the district (tab. 4): all except one have less than 50 employees and, on the whole, account for 7-8% of the sales of the district. In percentage values, component manufacturers export more compared to the firms specialized in finished products, with an export quota of 52.3% compared to 45.9% recorded by the smaller firms and 49.8% recorded by the larger ones (R&I, 1999).

Component manufacturers work above all for the small downstream firms, since the larger tend to produce all the components they require to produce their own specific products and materials. Together with the main activity on behalf of third parties, some component manufacturers also sell their own line of products. This direct relationship with the market and the end users further strengthens the competitive position of the component manufacturers with respect to the downstream firms. Through contact with the end users, the component manufacturers acquire a more direct knowledge of the new requirements of the final market and are able to create new components with original designs to propose to the downstream firms of the district, according to the development of demand.

Differently from traditional subcontractors and assemblers, the component manufacturers have a high degree of customer portfolio diversification, and there are no cases of dependence (even prevalent) on a single customer. Relationships with the customer enterprises are generally long-lasting. The aspects governed by written contracts above all regard prices and terms of delivery and payment, while the products to be processed are never accurately defined, so that in practice production is organized with the customer.

The technology in the companies is generally proprietary, and the specialized skills, above all the mechanical ones, developed by these enterprises not only ensure compliance with the rigid quality standards required by customers, but also the design and production of completely innovative components on the basis of the specifications defined by the final firms.

16.5.4 Assemblers

Most of the subcontracting firms of the biomedical district are specialized in the assembly of disposables (tab. 4). Assemblers are typically small firms with less than 10 employees. They employ about 364 employees, and in 1997 they undertook work amounting to 22 billion lira (R&I, 1999). These smaller units mainly work for the major downstream firms of the district. They receive the design, technical specifications, and all the necessary materials and components from the customer. Their degree of dependence on the main contracting company is quite high: all the assemblers have a maximum of 3 clients, and many of them have only one. In most cases, the relationship with the customers is long term.

Table 4: Suppliers of the Biomedical Valley
Source: R&I, 1999

Employees	**Firms**	**Employees (%)**	**Sales (%)**
Less than 10	13	13.6	12.2
10-19	15	42.8	42.3
20-49	7	43.6	45.5
Total	35	100.0	100.0
Type of supplier	**Firms**	**Employees (%)**	**Sales (%)**
Electro-medical equipment	2	2.1	2.8
Components producers	7	22.1	44.47
Assemblers	26	77.9	55.3
Total	35	100.0	100.0

The enterprises working for third parties and specialized in assembly are the weakest link of the district production chain. This group of firms, whose activity is highly manual and with a very low technological content, are subject to strong pressure by the larger enterprises, both in the determination of prices and in delivery times. They have the lowest profit performance and are exposed to the highest risks in relation to possible outsourcing of some production phases by the larger companies.

16.6 The relational view: the district as a hyper-network

The structural analysis presented in the previous sections underlines the strong heterogeneity which characterizes the Biomedical Valley. This peculiar feature raises two fundamental questions: (i) is the increasing structural diversity weakening the relational texture of the district?; (ii) has the entry of multinational firms relegated the younger and smaller firms of the district towards marginal and secondary positions within the inter-organizational network?

Lipparini and Lomi (1999) have analysed the relational structure of the Biomedical Valley in a sample of thirty-nine firms, with respect to the exchanges of: raw materials, subcontracting, components and parts, finished products or services. The overall picture emerging from their study suggests a negative answer to both questions.

The relational data referring to the totality of exchanges show that the number of direct ties among the biomedical firms is particularly high and suggests the existence of an intense relational activity within the district. Lipparini and Lomi (1999) found that there are very few disconnected firms and they argue that the development of close working relationships with proximate organizations benefited the largest and oldest firms of the district, but also the youngest and smallest ones. The former play the role of "orchestrators of different relationships in the community which they have helped to develop" (Lipparini and Lomi, 1999; p. 136). The latter use external relationships to reduce liability of newness and smallness, and avoid lack of legitimacy and resources.

The analysis of the relational data referring to the use of subcontracting and exchange of components and parts indicates that the smallest and youngest firms in the district have not been relegated to the role of marginal actors. In the subcontracting network, smaller firms have the highest outdegree scores, confirming their fundamental role in supplying working capacity. However, also the relational data confirm that the smaller firms in the district do not depend generally on one of the large companies. In the subcontracting network, even the smaller firms work for more than a single client, and sometimes act as a bridge among the older and larger firms in the district. It is especially in the exchange network of components and parts that the smaller firms emerge as prominent players. Indeed they are able to interact with actors of different size and age, acting as strategic interfaces between the large biomedical firms which offer complete lines and the small firms specialized in customized products for niche markets.

Lipparini and Lomi (1999) argue that the formation of different types of organizations was a key factor in the development of the Biomedical Valley. The larger and older firms play a crucial role within the district, since they are visible actors in the global market and they contribute to transfer resources and "legitimacy" to the smaller and younger firms in the district. The authors (1999) refer to this type of firms as the generalists. Their involvement in the relational texture of the district indicated by their high centrality scores also confirms that the entry of the multinational companies has not weakened the relationships among the acquired firms and their local network; on the contrary, the entry of multinationals has reinforced the role of the largest and oldest firms in the district as facilitators for knowledge transfer and access to global markets (Biggiero, 2002a). Empirical evidence from the relational study also highlights the role played by the smaller and younger firms, the specialists, which include the suppliers, especially the producers of components, and the small manufacturers producing customized disposables for niche markets.

From a relational perspective, the overall picture of the biomedical district is well expressed by the concept of hyper-network (Biggiero, 1999), that is a network of networks. Indeed, the district's web is composed of single firms and institutions, vertical networks formed by each focal firm and their own set of suppliers and service providers, and horizontal networks represented by the consortia and/or associations between manufacturers. The hyper-network refers to the structure of the overall district, in which we can identify second-level networks linked by vertical and/or horizontal ties.

Vertical second-level networks (entrepreneurial networks) are formed on the basis of the stable, recurring relationships that each focal firm has with its own network of subcontractors, component producers, trade intermediaries and customers. Relationships are often based on personal links of friendship or kinship, on verbal agreements or long-standing contracts. There is a strong mutual interdependence generated by the division of labour among the network members, although this does not prevent an actor from forming part of several entrepreneurial networks at the same time, as, for example, in the case of the component producers who, given the diversification of their customer portfolio, act as bridges among several entrepreneurial networks, built around different focal firms.

In particular, we can distinguish two different types of entrepreneurial networks according to whether the focal company is a large downstream enterprise or one of the small downstream firms operating in market niches. In the first case, the focal firm is linked above all to the other enterprises of the group localized outside the district and to its set of local subcontractors (mostly assemblers). In the second case, the focal firm is linked to its own network of component producers, trade intermediaries and possibly to other enterprises of the district through consortiums and/or associations.

Horizontal second-level networks are formed by associations and/or consortia, as in the case of the Consobiomed consortium, currently linking 22 companies in the biomedical district. The associated companies are small downstream firms, component producers and assemblers. The consortium was set up in 1990 with the aim of helping small and medium size enterprises in the district to achieve quality certification and to deal with all the problems related to European regulations and the EC trademark. In 1999 it was converted into a consortium for foreign trade with the intention of sustaining the associated companies in commercial and promotional activity for foreign markets.

16.7 The cognitive view: the formation of a district identity

The Biomedical Valley is an example of an innovative industrial district that emerged through the mobilization and generation of intangible resources rather than through the exploitation of pre-existing tangible assets. Many authors maintain that this mechanism of formation of localized production networks lies in the opportunity of reducing inter-firm coordination costs, and of enhancing cooperation, knowledge transfer, and innovation through identity creation and participation at the local level (Biggiero, 1999, 2002b; Storper 1993, Rullani, 1997).

In the literature on industrial districts, many authors have emphasized the catalytic role played by a shared collective identity in fostering social and economic relationships and in sustaining local development. In this regard, Rullani states that 'enterprises that cohabit in the same place and have experience in the division of labour and communication, produce a district only when they tune their cognitive and decision-making processes by the formation of a collective identity' (1997; p. 65). Marshall has coined the ambiguous term of 'industrial atmosphere', referring to something intangible that is somehow present in the district and makes it something more than a simple aggregation of firms.

In order to affirm that an industrial district is not merely a conceptual category, used by researchers to identify a particular cluster of firms, but also a "real" socio-economic entity, associated with a particular competitive model, it is necessary to show that the firms are aware of the district's existence, and that they recognize their belonging to the district as an important element of their organizational identity which, in turn, contributes to shape their strategic sense-making and behaviour. This claim follows many recent arguments which state that theoretical explanations of industry models cannot disregard the constitutive role of the managerial mind in making markets, and that collective beliefs shape inter-organizational relationships (Porac *et al.*, 1995).

Following this view, it becomes central: 1) to analyse how the actual members perceive and evaluate the district they belong to, and 2) to verify to what extent they feel they are part of the district. Sammarra and Biggiero (2001b) raised these questions in their empirical study of the biomedical district of Mirandola, which focuses on cognitive aspects that explain how firms perceive and evaluate their belonging to the district and whereby they come to view themselves and other firms in the district as members of a particular social entity, and adopt pro-social behaviours towards co-members as a consequence of their identification with the group. Drawing upon previous studies on social and organizational identification, the study conceptualizes social identification with the district as an instance of self-categorization (Tajfel, 1978). Consequently, identification is defined as the firm's awareness of its own belonging to the district as an inter-organizational group, which can assume different degrees of intensity (Dutton *et al.*, 1994). The research findings support this definition showing that, in the 40 firms of the sample, identification with the district is not an on/off condition, but assumes different degrees of intensity among respondents. Specifically, the authors found that the existence of the district is explicitly recognized by the interviewed firms and that, for some of them, their belonging to the district is perceived as a defining element of their organizational identity.

In the Biomedical Valley, firms that declared they had a stronger identification with the district were also more likely to have a positive evaluation of the district (perceived attractiveness of the district), based on the awareness of the existence of unique tangible and intangible resources within the Biomedical Valley, such as specialized skills, diffused know-how, unique competencies or high status firms. The firms' evaluation of the district was also found to be positively related with the perception of a positive external image. This means that, in the Biomedical Valley, the firms that declared that they identified strongly with the district were more likely to express a positive evaluation of the district and also to believe that the district has a good reputation in the national and international markets and that the local population and institutions attribute an important role to the district for the local development and well-being.

Interestingly, the entry of multinational firms was found to be one of the facilitating factors in the process of identity formation. Traditionally, industrial districts are described as closed socio-economic systems pooled together by the strength of a common cultural and territorial bond. The district's firms and entrepreneurs are supposed to have strong ties with the local environment and these ties are supposed to be the catalytic mechanism of the district model. In this view, the entry of external actors is likely to appear as a serious threat for the district's texture and identity. However, evidence from the study presents a rather different picture. Indeed, the entry of the multinationals through the acquisitions of the larger and older firms of the district has contributed to enhance the reputation and legitimization of the district, providing an international status to the entire local business community. In the Biomedical Valley, this status-enhancement effect seems to increase the willingness of local actors to identify with the district as a prestigious social group. From the theoretical point of view, this finding is coherent with the social identity theory assumption that social identification is motivated by self-enhancement needs, and with Peteraf and Shanley's model of strategic group identity (1997). Specifically, the two authors argue that the presence of high-status firms within a strategic group affects the perceived prestige of the group, and thus reinforces the adoption of mimetic activity, which facilitates the process of identification with the strategic group based on self-enhancement needs. From the empirical point of view, the positive role that the entry of multinationals has played in reinforcing social identification processes with the district is also coherent with some of the implications emerged from the relational study, which emphasized the important role played by the larger and older firms in reinforcing the relational texture of the district.

Furthermore, Sammarra and Biggiero (2001b) found that, in categorization processes, firms distinguish organizations that belong to the biomedical district (the in-group) from the external firms (the out-group) on the basis of those technical, economic, and cultural attributes which highlight strong and mutual interdependencies among district members, disregarding their degree of similarity in terms of structural characteristics, such as size, ownership structure and age. This result is coherent with the relational analysis conducted by Lipparini and Lomi (1999), which showed that the formation and development of the biomedical district was favoured by the generation of different, but highly complementary, organizational forms, allowing for the constitution of an organizational community. The cognitive analysis conducted by Sammarra and Biggiero (2001b) confirms this view, highlighting that the perception of conditions of interdependence and complementarity with other firms of the district is one of the main factors which can lead to the construction of a system identity, especially in those complex inter-organizational contexts, such as in the case of the industrial districts, in which inter-firm relationships involve both competition and cooperation.

Furthermore, this finding emphasizes that the strong heterogeneity in terms of the firms' structure and typology which emerged from the structural analysis of the Biomedical Valley does not prevent the firms' believing that they are part of a "real industrial district", although from a different perspective, this result supports recent studies questioning the empirical foundation of the canonical model of industrial district. Indeed, the model of canonical district defined as an internal homogenous community of firms, largely composed by small and medium enterprises that are typically family-owned and locally controlled, appears more and more as an exception rather than the rule among the many different industrial districts existing in Italy (Paniccia, 1998), suggesting that researchers should update and modify their belief systems of what an industrial district is. The cognitive analysis carried on in the Biomedical Valley emphasizes this claim because it suggests that the canonical model of industrial district does not correspond to the collective beliefs that the same firms hold.

In their study of the Biomedical Valley, Sammarra and Biggiero (2001b) found that the firms' identification with the district positively affects also their behavioural attitudes towards other district members, such as trust and cooperation. Specifically, they found that firms identifying more with the district were also more likely to adopt trustful and cooperative attitudes towards other firms in the district. Likewise they declared that they attributed more importance to the district for their own success and that they were more inclined to undertake collective actions and promote projects aimed to strengthen the district competitiveness.

These findings suggest that identification with the district, by supporting the emergence of group processes among the other biomedical firms in the district, involves qualitative changes in the way firms interact that contribute to enhance the reservoir of social capital available through the generation of intangible resources such as trust and cooperative attitudes. Moreover, by influencing inter-firm behaviours, identification with the district may contribute, at a collective level, to enhance the competitiveness of the whole district. Indeed, by increasing inter-firm trust based on group membership and cooperative attitudes towards co-members, identification facilitates the exchange of information within the group, reduces transaction costs, facilitates the adoption of collective actions and cooperative strategies, thus favouring also inter-organizational learning among group members (Sammarra and Biggiero, 2001a). However, in this regard a word of caution is in order. There are potential negative effects of identification on group performance which have not been included in this study, such as the tendency to adopt mimetic attitudes towards co-members that reduces the firms' diversity with respect to viewpoints, cultural values and behaviours that can result in a loss of strategic flexibility and variety within the network (Peteraf and Shanley, 1997; Sammarra and Biggiero, 2001a).

16.8 Conclusions

The Biomedical Valley is a hi-tech recent industrial district, and thus it shows that industrial districts (especially Italian ones) are not necessarily a low-tech and old economic phenomena (Amin and Robins, 1991).

The history of the biomedical district shows that the triggering factor can be a true Schumpeterian entrepreneur, and that multinationals can enter a district through acquisitions, play a catalyst role through subcontracting, and open the district to the world market. They did it without destroying the hyper-network structure of the district. Actually, the relational and cognitive analyses carried on in the Biomedical Valley show that the formation of different typologies of firms, mainly caused by the entry of multinationals, favoured the development of an organizational community characterized by strong inter-organizational ties, and supported the emergence of a collective identity, based on the firms' perceptions of strong interdependencies with other district members.

Furthermore, the analysis of identification processes occurring in the biomedical district suggests that industrial districts are not only in the minds of researchers, but also in those of its members. Moreover, it implies that the stronger the identification, the stronger district competitiveness will be, confirming the model proposed by Biggiero and Sammarra (2001a), and developed by the same authors in the first part of this volume. However, this conclusion must be qualified in several ways. Indeed, the empirical analysis of the identification processes was conducted in one single industrial district with many peculiar features which results in a serious limitation to the generalizability of the findings briefly presented in this study.

References

Amin A. and Robins K. (1991), 'These are not Marshallian times', in, R. Camagni (ed.), *Innovation Networks*, Bellhaven Press, London.

Arthur B. (1988), 'Competing technologies: an overview', in Dosi G., Freeman C., Nelson R., Silverberg G. and Soete L. (eds.), *Technical Change and Economic Theory*, Pinter, London.

Biggiero L. (1999), 'Markets, hierarchies, networks, districts: a cybernetic approach', *Human Systems Management,* vol. 18, p. 71-86.

Biggiero L. (2002a), 'The location of multinationals in industrial districts: knowledge transfer in biomedicals', *Journal of Technology Transfer*, vol. 27, p. 111-122.

Biggiero L. (2002b), 'Self-organizing processes in building entrepreneurial networks: a theoretical and empirical investigation', *Human Systems Management*, vol. 20, p. 209-222.

Borroi M., Minoja M. and Sinatra A. (1998), 'The relationships between cognitive maps, industry complexity and strategies implemented: the case of the Carpi textile-clothing industrial system', *Journal of Management and Governance,* vol, 2, p. 233-266.

CNR (1987), Analisi di mercato delle tecnologie biomediche, vol. 1, sez. 1, progetto finalizzato tecnologie biomediche.

Ebers M. (1997), 'Explaining inter-organizational network formation' in Ebers M. (ed.), *The formation of inter-organizational networks*, Oxford University Press, Oxford.

Grandori A. (1999), (ed.), *Interfirm networks,* Routledge, London.

Lipparini A. and Lomi A. (1999), 'Interorganizational relations in the Modena biomedical industry: a case study in local economic development' in Grandori A. (ed.), *Interfirm Networks,* Routledge, London.

Paniccia I. (1998), 'One, a hundred, thousands of industrial districts. organizational variety in local networks of small and medium-sized enterprises', *Organization Studies* vol. 4, p. 667-699.

Peteraf M. and Shanley M. (1997), 'Getting to know you: a theory of strategic group identity', *Strategic Management Journal*, vol. 18, p. 165-186.

Porac J.F., Thomas H., Wilson F., Paton D. and Kanfer A. (1995), 'Rivalry and the industrial model of Scottish knitwear producers' *Administrative Science Quarterly* vol. 48, p. 203-227.

R&I (1999), Osservatorio sul settore biomedicale nel distretto mirandolese. Prima rilevazione.

Rullani E. (1997), 'L'evoluzione dei distretti industriali: un percorso tra decostruzione e internazionalizzazione', in, Varaldo R. and Ferrucci L. (eds.), *Il distretto industrial fra logiche di impresa e logiche di sistema*, Franco Angeli, Milan.

Sammarra A. and Biggiero L. (2001a), 'Identity and identification in industrial districts', *Journal of Management and Governance,* vol. 5, p. 61-82.

Sammarra, A. and Biggiero L. (2001b), 'Identity, trust and cooperation in localized production networks', paper presented at the Egos 17th Colloquium July 2001, Lyon, France.

Storper M. (1993), 'Regional worlds of production: learning and innovation in the technology districts of France, Italy and the USA', *Regional Studies,* vol. 27, p. 433-455.

Tajfel H. (1978), 'The Achievement of Group Differentiation', in, Tajfel H. (ed.), *Differentiation between Social Groups: Studies in the Social Psychology of Intergroup Relations*, Academic Press, London.

Chapter 17

Sophia-Antipolis as a technopolis phenomenon: is myth becoming reality?

Michel Quéré

17.1 Introduction

This contribution discusses the importance of the spatial dimension of localized knowledge by focussing the analysis on a specific territorial area: the Sophia-Antipolis park, located in the south of France. This chapter establishes how the production of knowledge is closely related to the process of accumulation that has moulded the economic working of companies located in the Sophia-Antipolis park. Economic working refers partly to quantitative aspects, but mainly to the nature of inter-firms relationships, to the physiology of local science-industry partnerships, and, more largely, to the characteristics of the local labour market (Quéré, 1999). In that respect, Sophia-Antipolis is very interesting. Obviously, it starts from scratch in the seventies and progressively appears for some sectoral patterns as one of the most suited locations in Europe where "high-tech" activities have concentrated (mainly information and communication technologies [ICTs]). As this project is only 30 years old, it is possible to have a complete vision of its development and to analyse its transformation toward an actual technopolis-type of economic working. Section 17.2 analyses the role of history in the emergence of localized knowledge for some of the dominant activities. Section 17.3 deals in more detail with the transition toward the technopolis-type of economic functioning. Section 17.4 discusses the necessary conditions to promote the development of localized knowledge by focussing more especially on major requirements for the local institutional infrastructure. Concluding remarks address some related policy implications.

17.2 Origins and development: Sophia-Antipolis as an "international park of industrial activities"

Sophia-Antipolis has recently celebrated its thirtieth birthday. The experiment is known today as an important centre of high-tech activities in Europe, and often presented as a success of local development. Different indicators are currently used to support this image of a successful experiment. The number of firms and employment on site has reached respectively 1,193 and 21,535 in the year 2000, and the number of students and researchers is around 5,000, when the initial numbers were zero (see Figure 1). However, this obvious quantitative success does not guarantee either that the local accumulation of economic activities in that place is stable, nor that the viability of the experiment in the long run can be ensured.

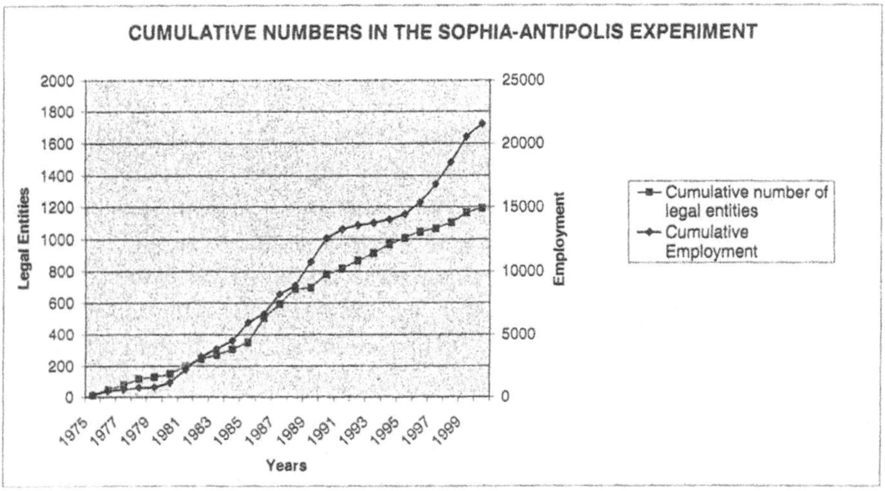

Figure 1: Cumulative numbers in the Sophia-Antinopolis experiment
Source: SAEM-SACA, Sophia-Antipolis

This section will go through the history of the project in order to understand the economic process that has structured the experiment, the main constraints faced by the park in its development and the still existing fragility in the process of setting-up localized knowledge for some of its activities.

The Sophia-Antipolis experiment developed through the accumulation on site of diversified resources that decided to locate in the project mostly because of the general advantages exhibited by the so-called French Riviera. Among the latter, the Nice international airport, the pleasant climate, the quality of accommodation and local infrastructures, and a cosmopolitan cultural tradition appear central to the argument. During the 1960s, a diversification process of the French Riviera had already started and resulted in the location of activities related to private R&D and "high"-technologies. During that period, an IBM research centre and a R&D plant by Texas Instruments located in the neighbourhood of the park. Those locations gave credibility to a new mode of development based on a "sun belt effect". It is in this context that Pierre Laffitte started the Sophia-Antipolis project in 1972 and decided to experiment a "city of science, culture, and wisdom". This utopian vision of Laffitte resulted in a private experiment that rapidly faced infrastructural constraints and collapsed financially. As the project initiated in a complete desert, all the usual physical infrastructures were needed and this was actually very costly. To be continued, the experiment required substantial participation and direct financial support from local and national public authorities. The management of the experiment was therefore transferred from private initiative to local public authorities in 1977 and this induced a significant change in its content. It became an international park for selected industrial activities, where the location of innovative, non polluting, high value added, research and development activities was centrally favoured. This change in the management of the project went hand in hand with a change in its scale: if the first stage was supported by an individual initiative and resulted in the local accumulation of activities based on the activation of the personal network of Pierre Laffitte, the public character of the experiment allowed a strategy of international marketing to be developed in order to attract external investments, which resulted in a numerous location of R&D plants from US companies in Sophia-Antipolis. As a consequence, the scale of the experiment changed significantly but the resulting accumulation of activities was somewhat casual. This characteristic is in fact a major feature in the development of Sophia-Antipolis, which can be thought of as an exogenously driven process characterized by the location of R&D units from large external companies.

This first stage in the process of local economic development resulted in a model characterized by:

- the location of R&D plants from large international companies with the aim of adapting their products to the European market;
- a dominance of exogenous decision-making on the internal working of the project;
- a high distribution in the range of industrial activities located on site;
- a local growth characterized by a random accumulation process;

– a low level of local interactions among the components of the project.

However, this first stage ensured the quantitative success of the Sophia-Antipolis experiment in that it resulted in the location of 12,000 jobs at the end of the 1980s in a project that started from nothing. According to local public policy, two dominant characteristics explain the success of the experiment at this stage: one is the active marketing strategy towards the international environment and the US market more especially; the other is the accumulation on site of technical facilities that contribute to give the experiment an image of high modernity (in that respect, the establishing on site of a telecommunications network based on fibre optic technology had a considerable impact on the site attractiveness).

As a result, the main factor explaining the success of the experiment lies in the general advantages exhibited by the French Riviera that were well suited to large multinational companies for locating in Sophia-Antipolis either European administrative centres or R&D units. At that stage of its development, Sophia-Antipolis was exogenously-driven and the experiment exhibited a significant comparative advantage, as regards other potential locations in Europe. However, the economic working of Sophia-Antipolis was very insecure because innovative activities were not based on local relationships, market contracts or co-operative relations (Gaffard and Quéré, 1996; Longhi and Quéré, 1997).

From this first stage of the experiment, the random accumulation of companies on site results in two dominant sets of activities. One of these is related to computer science, telecommunications and microelectronic industries. This first set of activities has been the essential growth engine for Sophia-Antipolis and has structured the development of the local industrial context. Those activities still represent around 75% of the technological jobs, and their accumulation balances large French (Air France, Thalès, Organic, Schneider/Télémécanique, etc.) and international firms (Digital Equipment Corporation/Compaq, Amadeus, Accenture, etc.). Most of the public research institutions delocalized on site have hopefully been complementary to those activities, as will be stressed later on (Ecole Nationale des Mines de Paris à Sophia-Antipolis, INRIA, CNRS, University of Nice-Sophia-Antipolis, etc.). Because of that complementarity, those activities have progressively reached a critical mass with local positive feedback effects.

The second set is made up of life and health sciences activities, which gradually exhibited a significant volume on site, but it is still insufficient to obtain a comparable mass-effect that can set up similar local positive feedback effects. One reason is that large international companies have located in Sophia-Antipolis R&D units that are exclusively participating in an internal coordination for their European activities (this was the case for Dow Chemical, Dow Corning, Welcome, Allergan, Rohm & Haas, etc.). Those companies' units are not using Sophia-Antipolis as an operational platform open to capabilities existing in its local environment. However, some of them have nevertheless engaged in research activities (Cird/Galderma, Cordis/Zeneca) including a few local partnerships. But the latter do not constitute a sufficient critical mass that allows for a more endogenous growth process, despite a total of 50 companies and 2,000 jobs on site.

Apart from those two dominant sets of industrial activities, a third specific sectoral area deserves attention. Earth sciences, that appeared as the initial set of activities accumulated in Sophia-Antipolis, as regards the historical development of the project, represent a set of resources that have never taken off, and still remain constant since the beginning. The location of those activities results from the personal networking effect of Laffitte and those activities have never reached a sufficient mass-effect on site to be identified as a structuring sectoral pattern of the experiment.

During the 1980s, the nature of the experiment changed progressively, with a relative slowdown in the location of large international companies. The assimilation of the experiment to a technopolis-type of local development appears in that period, due to an increased accumulation of training and research institutions, and of high-qualified resources that better match the emerging labour market and the growth of local services activities. However, the accumulation of those scientific and technological activities on site has contributed to a mass-effect inducing further endogenous development, mainly in ICTs.

17.3 Towards a "Technopolis-type" of economic development

What characterizes this second stage is the transition from an external-oriented growth process toward a more endogeneous-oriented development allowed by the mass effect resulting from the first stage of the Sophia-Antipolis development. It is mainly during this second stage that a significant increase in the number of academics and students has been made possible and risen to an amount of 5,000 students and 2,000 academia on site.

From the start of the experiment, a few training and research institutions have located in Sophia-Antipolis. The siting of some public research institutes in Sophia Antipolis was the result of either Laffitte's influential role (l'Ecole Nationale Supérieure des Mines de Paris), or of a decentralisation policy followed by some major French research institutes: Centre National de la Recherche Scientifique (CNRS), l'Institut National de Recherche en Informatique et Automatique (INRIA). Their role in the development of Sophia-Antipolis has been very important. In particular, INRIA had a fundamental role in structuring local technological development and played a large part in the emergence of an endogenous innovative capacity and of start-ups on the site.

But this was not sufficient to train graduates needed by the developing local labour market for qualified personnel, or to reverse the import of external resources for sustaining local growth. A significant change has been the siting in Sophia-Antipolis of research institutes and doctoral studies from the University of Nice from 1986, which until then was not significantly involved in the experiment. The University of Nice is a young university whose development has largely benefited from the economic development stimulated by Sophia-Antipolis. In turn, the growth of university training for engineers and for doctorates has progressively fitted with the needs of local companies for qualified staff. Specific research capabilities in activities related to computer sciences, telecommunications and electronics have been established and have benefited local firms, mainly high-tech SMEs. This process has been essential to feed the emergence of a local labour market for high-qualified skills and to allow the development of industry-research local linkages, in which students play a significant role.

At least, during the first stage of its development, the Sophia-Antipolis experiment appeared to be a "reverse" technopolis, as regards the usual definition of a science park (a project starting from academics and research institutions and resulting in new industrial activities). However, till the beginning of the 1990s, this project progressively became a more traditional science park model of economic development. In other words, the current stage of the Sophia-Antipolis project now allows for a discussion about the characteristics of its internal working and the identification of specific practices of economic development in the line of a science park model.

To sum-up the characteristics of the second stage of the Sophia-Antipolis experiment, one can figure out the following evolution:

- a continuous increase in the quantity of research institutions and academics;
- a relative decrease of external large companies in locating plants in Sophia-Antipolis;
- a substitutive effect for the increase in local start-ups, specially in activities and services related to information and communication technologies;
- a higher implication of Nice University in the project through the quantitative increase in initial training and doctoral studies located on site;

- a specific "entrepreneurial atmosphere" allowing for peculiar experiments, productive interactions, and mutual benefits for local companies (more especially small and very small companies).

As I have already stressed, the historical structuring of the project has allowed for a local accumulation of diverse resources, i.e. for a very special type of social bio-diversity in the components of the project that makes it a very unique experiment. As a consequence, the main characteristics of the internal working of the Sophia-Antipolis park lie in the originality of this accumulation process and in the diversity of its components.

This is particularly true, as regards the population of companies located on site. The diversity of those companies can be expressed when looking at traditional characteristics (size or sectors). However, what appears as a peculiarity of Sophia-Antipolis is that this variety induces a large diversity in local behavioural patterns. Those companies are participating in the experiment through a huge diversity of practices as regards innovative behaviours. Consequently, there is a need to organize those diverse patterns and to understand the extent to which the experiment behaves as a local system of innovation. At least, one can dissociate three different types of companies strategies, depending on their structural patterns: local R&D plants from large international companies, small and medium-sized companies, and new start-ups or spin-offs from academics or local plants.

What recently appears as a significant shift in the local working of the experiment is the decreasing importance of an external-oriented type of economic development deriving from large companies' strategies and, respectively, an increasing importance of an internal-oriented development due to a huge increase in local start-ups and spin-offs that make the experiment very powerful in its capacity to set up new productive opportunities and markets. This shift is somewhat spectacular because of the recessing context faced at the beginning of the 1990s where ICTs activities were locally confronting important economic trouble. Eight to ten years after, the same activities were at central stake to provide an explanation to this shift in the model of local development. The increase of the e-business economy has largely benefited the site and is part of the explanation for the shift towards a more internal-oriented type of local development. Now, the challenge for the experiment is its ability to capitalize on those local start-ups and spin-offs in order to secure the Sophia-Antipolis viability. Let us now put the local picture into more details.

17.3.1 Local R&D plants from large international companies

I already mentioned that the location of those plants in the Sophia-Antipolis area results from a marketing strategy that has been on an international scale since the end of the 1970s. At that time, many US international companies especially were adapting their products to the national specificities of European countries. As a consequence, they were interested in locating R&D units in Europe in order to solve that difficulty. Sophia-Antipolis was well-suited for such a purpose because of its general facilities (international airport and accommodation facilities, especially). However, those companies have progressively learned from the European market and reduced that need to locate R&D plants in Europe. At least, a significant number of those plants have left the area, and have re-located in other places (Scotland because of the wages differentials, Geneva or Paris because of the proximity to other European decision-centres, Munich because of the proximity to the major industrial customers, and the like...). According to Sophia-Antipolis, no important external R&D unit has located in the project for a long time; on the contrary, those plants have decreased in relative terms, as regards the working of the park. Finally, as they depended on external decision centres, those plants have never had a structural influence on the local economic working: even sub-contracting possibilities have been very weak and what has dominated the working of those plants is a huge mismatch with the rest of the local economy. The transition toward a more technopolis-type of economic development affects this pessimistic picture in two complementary directions. On the one hand, pressure to leave the project from some large companies has motivated a lot of plants to engage in innovative behaviour with local partners in order to justify their location in Sophia-Antipolis. The reason for such behaviour can be explained by the dissatisfaction of employees in leaving the French Riviera. Many of them consider that location as best-suited to their every day life, and they have been very motivated to stay in the area. On the other hand, some of those plants have started original strategies as regards the development of start-ups. A few large companies located in the immediate surrounding of the park (Alcatel, IBM, Texas Instruments) have developed 'private incubators' in order to support intra muros local start-ups and benefit from their capabilities on site. Both evolutions allow us to consider that those initial exogenous components of the Sophia-Antipolis experiment are also participating actively in the transition towards a more technopolis-type of economic development, even if this has to be thought of as a defensive strategy in order to justify the need for locating there.

17.3.2 Small and medium-sized companies

As the Sophia-Antipolis experiment started from scratch, small and medium-sized firms (SMEs) are obviously a critical weakness in this local economy. The reason for that observation lies in the lack of industrial tradition in this geographical part of France. Tourism remains the main economy activity and, apart the perfume industry in the Grasse area, no industrial tradition exists there. Consequently, traditional SMEs are very few and they express the remaining fragility of the Sophia-Antipolis experiment despite its obvious quantitative success. SMEs that exist on site result mainly from a significant success in ICTs applications. Nevertheless, they are not so numerous and are still fragile because their markets are not local. Some have succeeded in expanding because they used the overall South of France as their market space and developed some subsidiaries all around that geographical area. Others succeeded because they engaged in specific partnerships with local large companies. However, the latter are very few and their development can be thought of as very fragile because of that dependency on one dominant customer. In the event of trouble with this dominant partner, they are obliged to leave the area and to establish themselves in more secure areas such as the Parisian metropolis, for instance.

17.3.3 Start-ups or spin-offs from academics or local plants

Apart from the change we already indicated in the local strategy of plants from large international companies, spin-offs and start-ups result from two complementary evolutions in the working of the Sophia-Antipolis experiment.

On the one hand, the breakdown of some large companies located in Sophia-Antipolis has been a major source for new local start-ups. The most famous example remains the local plant from the Digital Equipment Company that reached more than 1,100 jobs on site and that decreased dramatically towards 200 people at the time of the takeover by Compaq. Many of those high-qualified workers refused to leave the South of France and tried to be hired by other local plants or to establish start-ups in the experiment. Those highly qualified workers have contributed to developing local consultancy agencies, software and engineering companies and the like; they also merged with local capabilities stemming from academics and research institutes in order to develop new productive opportunities especially related to the so-called e-business economy.

On the other hand, in the 1990s, a qualitative change appeared very significant in the local scientific infrastructure. Many research institutions developed their own expertise to implement new productive potentialities derived from scientific resources. This took on different forms. If those productive potentialities were embodied in patents and licences sold to companies, they also gave rise to a lot of start-ups that developed in the local area. If this phenomenon is certainly a national one in the sense that this trend is somewhat general in France, it has nevertheless benefited very much the working of the local economy. Many of those start-ups have grown up because of ICTs opportunities, and Sophia-Antipolis has allowed for some success stories that contribute to such a major shift in the experiment from a dominant external-oriented project toward a more endogenous-type of economic development. This phenomenon has become possible because of three essential reasons. The one is the macroeconomic context that became much more favourable for entrepreneurial initiatives. Not only has the infatuation with the so called e-economy benefited Sophia-Antipolis because of the ICTs components accumulated in the experiment, but the general evolution of financial markets and the capability to find cash opportunities to finance new companies was very crucial in the exploding effect of new local spin-offs and start-ups. The second reason is the density of scientific resources progressively reached on site that favours the mixing of complementary resources to establish a start-up. Technical expertise can be coupled with business and finance capabilities and those combinations increase the viability of local start-ups. The third reason is that spin-offs from existing companies have also used research capabilities in order to secure the technical content of their entrepreneurial initiatives.

In other words, the Sophia-Antipolis experiment has started to develop as an actual technopolis-type of local development. One must insist on the peculiar features that underlie this change in local development. Whereas the process of accumulation was driven by the external environment and depended on external decision-making, the evolution toward a more endogenous type of economic development is also focussed on a few selected activities. Obviously, the use of local resources for the emergence of start-ups and spin-offs is crucial, and this contributes to zero in on the type of knowledge in which the area is progressively specializing. Of course, the accumulation of highly qualified people in ICTs sectors largely benefited this structural change of local development (from an exogenous towards a more endogenous type). The local collection of large companies, SMEs and start-ups allows for an experimenting atmosphere, as regards the provision of new ICT applications and services. A localized knowledge in terms of ICTs capabilities is actually becoming one major specific feature of the experiment that is internationally identified and allows for new locations on site. Recently, the Sophia-Antipolis experiment has attracted new locations from companies involved in the telecommunications sectors like Siemens, Lucent, Bay Networks, etc. The latter implement small units in the park because the accumulation of technological knowledge in that area is now visible at European level and obliges those companies to be a part of it. Otherwise, it can be associated in a too highly risky situation that can limit their capabilities and profit potentialities in the future. Thus, Sophia-Antipolis has reached a mass-effect that ranks the site as an inescapable European innovative area for ICTs activities. Consequently, this situation also offers better opportunities for small and very small companies because they are interacting with those large companies, and anticipating on their potential needs and applications.

There is one peculiar aspect of the recent boom of start-ups in Sophia-Antipolis. If they benefit from the so-called e-business economy, they are not solely dedicated to providing new services and uses through the Internet web. A lot of small companies are developing because of their original technological expertise applied in very specific matters, and for which large companies should continue to be interested. In ICTs' sectors, there are local examples of start-ups that are returning to the environment of large companies when they succeed in proving the profitability of new applications or new technological capabilities. For instance, Respublica, created in 1997 and specialized in the organisation and management of chats on the web, was acquired by Libertysurf in 1999 when there were around 60 employees; Echo interactive, specialized in web-search systems, has been acquired by Wanadoo in order to ensure its technological infrastructure; Odisei, a start-up working on telecommunications applications with IP technologies, is now a subsidiary of the US 8X8 company. A much longer list could be made, but the point is that the park is now becoming a laboratory for ICTs' capabilities and applications through the process of acquisition of local start-ups by large companies when they have reached a sufficient level of technological and market credibility. What is now impressive is the ability of the Sophia-Antipolis experiment to produce new technological and market opportunities for those industrial sectors. At least, there are more than 20 local start-ups that are potential candidates to be absorbed by large companies because they are actually performing very well on the market.

Finally, the current start-down phenomenon characterizing the e-business economy that results from the US conjuncture in the end of 2000 is now increasingly sensitive in Europe. However, it is very interesting to note that it actually seems to be not so dramatic in the case of Sophia-Antipolis. The reason lies in the increasing accumulation of localized knowledge in ICTs which is not so much related to the famous "dotcom" revolution. Most local start-ups are implementing technological expertises that require a mixing capability of scientific, technological and market knowledge. This explains why they still succeed in capturing funds from large companies or from financial markets. This also largely contributes to the specific character of the Sophia-Antipolis park at European level and strengthens its attractiveness. To some extent, the experiment, starting from a reverse technopolis is now behaving as an original science park model that succeeds in creating new productive and market opportunities: one can reasonably wonder to what extent the initial myth of Laffitte is currently becoming a reality.

To a large extent, this evolution towards a local science park (or technopolis-type) model of economic development is spontaneous. The local capability to set up localized knowledge and innovative behaviours should be sustained by the local institutional infrastructure and more especially by local policy-making. One must now discuss the actual ability of the local institutional infrastructure to face this obvious qualitative success of the park for some of its sectoral characteristics.

17.4 The requirements to improve localized knowledge: the importance of the local institutional infrastructure.

The mass-effect due to the accumulation of companies in Sophia-Antipolis is not the only source for the quantitative and qualitative success of the experiment. The historical structuring of the area appears as a progressive accumulation of complementary resources that have largely moulded the specific character of local development. Among the latter, one can distinguish three main categories: (1) the influential public policies, (2) specific clubs and associations resulting from the working of the local economy, (3) coaching (private and public) activities including the financial local infrastructure supporting companies' activities.

17.4.1 Influential public policies

For a long time, public policy (local and national) has mainly been oriented towards promoting the area with the aim of capturing external (international) resources. This is particularly obvious during the first stage of the project, but it is still the dominant orientation of policy-making, even during the second phase of the project. In other words, local policy-making has been largely unable to develop a suitable policy to favour the transition towards a more endogenously-driven evolution of the experiment. Apart from promoting the Sophia-Antipolis park, two complementary orientations must be mentioned. One is the local telecommunications network: based on optic fibre technology, it is very suited to the transportation of high levels of EDI within the park. In the 1980s, this was a crucial infrastructure that facilitated experimental applications and contributed to identifying Sophia-Antipolis as one of the main European experimental areas in ICTs technologies. The other influential infrastructure is the CICA (International Centre for Advanced Communication). CICA is an "intelligent building" that concentrates all the infrastructural potentialities currently offered by ICTs technologies. Actually, CICA was initially a very controversial investment because it was very costly and many criticisms arose, because of the huge amount of public money involved (one can estimate this amounted to around 60 millions of FF). Actually, this investment was controversial also because of its high level of functioning costs. Started in the end of the 1980s, this infrastructure was not very suited to the local economic working because of the important crisis in ICT activities that appeared at the beginning of the nineties. To a large extent, a consensual agreement to consider this building as a relative failure could have been obtained at that time. But, interestingly, the positive shock in the economic environment induced by the infatuation for business activities based on Internet applications has considerably changed the assessment of that infrastructure. It progressively appears as a proper location for start-ups and small companies in order to implement new ICTs applications in a favourable technological environment. Moreover, Eurecom, a research and training centre in telecommunications, has located in the immediate surroundings of the CICA and offers the CICA companies a very useful set of capabilities to be activated on site. Finally, the local authorities have sub-contracted the management of this infrastructure to a private company that tried to favour the management and the coaching of start-ups, as regards the disposal of technical and economic facilities.

Thus, CICA appears as a very interesting infrastructure in that it has progressively became one of the central arenas on site for accompanying the transition toward a more technopolis-type of economic development. Even if financial subsidies from local public authorities are still necessary today to ensure its current working, this infrastructure has largely contributed to capturing locally the effects of the general evolution towards the e-business economy. However, the CICA is one of the few examples of an influential role of local policy-making in encouraging the transition toward a technopolis-type of functioning. What prevails, as regards the Sophia-Antipolis experiment, is a huge inability of local policy-makers to favour its economic development. More especially, if promoting the park is correctly assumed, there is an obvious deficit in terms of accompanying its economic working. The transition towards a technopolis-type of development is largely a spontaneous evolution for which no effective incentives from local policy-makers can actually be identified.

17.4.2 Specific clubs and associations resulting from the working of the local economy,

The latter essentially means the emergence of French style of associations and clubs aimed at stimulating local co-ordinated initiatives among the components of the project. Those clubs and associations are various and mainly thematic-oriented.

The two first local associations historically created have been the "Club des dirigeants" and the Persan association. The former informally groups executives from local units of large companies in order to offer a means of discussing and exchanging information about their strategies, problems, and opportunities. In fact, this association had traditionally played a role of lobbying, directed toward public actors, to complain about the insufficient amount of local physical infrastructures. This perfectly matches the first phase of the economic working of the park. Recently, the association has changed its main goals, and is organized in different commissions devoted to specific technological issues and aimed at sharing problems and experience more effectively among its members. As such, they progressively evolve towards an objective of promoting localized knowledge among them. The latter (the Persan association) is a positive experience but only for academics: the Persan association groups all the managers of public research and higher training in order to give a collective image to the research potentials accumulated locally.

These two initial attempts at increasing the internal co-ordination among the Sophia-Antipolis actors have been followed by more specific and thematic-oriented associations. An interesting experiment is the "Association Telecom Valley" which groups the technological users of the telecommunications infrastructure, i.e. public and private, large and small firms. This association aims to give a specific "label" to the park and to strengthen "local experimentations" that favour the identification of Sophia-Antipolis as a specific and remarkable location for the main European actors in the field of telecommunications. The association has been the result of a reactive process to the local crisis and the general process of restructuring the European "high-tech" industry. The local managers of large international firms have to justify their location in Sophia-Antipolis, which has to be defended on the basis of local specific technological advantages. Intending to safeguard the location in Sophia-Antipolis has induced an active process, the large firms beginning to be open with one another and to share technological resources, even with local SMEs. Interestingly, a crisis has been necessary to induce such processes, but again, these private initiatives to favour the emergence of localized knowledge should have to be relayed by a coherent public policy able to strengthen the development of that industrial context. The future of the park lies in the definition of such a new strategy, which appears to be still highly uncertain today.

Various other associations have also been created, mainly oriented towards "high-tech" SMEs and implementation of endogenous processes of creation of resources:

- The "club high-tech" groups mainly the small and medium-sized companies involved in high-tech activities as well as some of the units of large companies with the aim of offering loci of facilitating productive relationships on a local base.
- IMeT, the "Institut Méditerranéen de Téléactivités" tests the contents of ATM services and products. IMeT is led by INRIA; it intends to help the producers (high tech SMEs) of new technologies to "build" their markets, in an activity where the links between producers and users are more difficult to build or imagine than in traditional oligopolistic markets.
- MITSA, directed by Pierre Laffitte, is a specific multimedia club.
- Eurosud 155 is an association gathering all actors involved in the new experimentation on the 155 megabits ATM platform. Telecom Valley and Eurosud 155 play an important role in Sophia-Antipolis for local SMEs. They are effectively led by INRIA, France Telecom and its research institutes; they are used as platforms to promote their new technologies and to make technology transfers easier.

All these collective initiatives aim at exhibiting a clearer focus on specific local knowledge and distinctive capabilities from other "high-tech" agglomerations in the international environment. In fact, it progressively appears that the Sophia-Antipolis park accumulates distinctive capabilities in wireless communications and, more especially, in GSM technologies. Companies like France Telecom, VLSI, Rockwell, Siemens, Lucent, Bay networks are now key-actors in the area to exemplify those distinctive capabilities. Moreover, the latter also benefit from public institutions. For instance, the location of ETSI in the project contributes to those distinctive capabilities because of its influence on the definition of new norms and standards, especially the UMTS (Universal mobile telecommunication system). In a complementary perspective, the location of two professional training centres (Theseus and Eurecom) is also contributing to this specific capability. All those partners are collective ingredients that progressively allow the existence of localized knowledge in those activities. The second area where we found a distinctive capability is computer-aided-design for micro-electronics and its applications. The location in the project of famous companies in these activities like Cadence or Mentor-graphics, as well as an important number of small companies with specific expertise like Euromips System, Studiel, Veriphia, contributes to establishing a complementary field of innovative activities that interact with large local (and regional) companies identified as major customers.

The objective of the club "Energy and Environment" is to play a similar role in another sector which is also significantly represented locally: "Earth activities". However, these activities have never been essential from a quantitative viewpoint. If there is an obvious interest for those companies to group themselves and facilitate the access to common training policies or collective responses to national and international research programmes, the latter is still marginal because of the few quantitative numbers of people involved in those activities. This allows a complementary comment. What appears for the Energy and Earth sciences' activities is that any increase in the local co-ordination of related partners is a good initiative for increasing the opportunity to develop localized knowledge. However, the lack of density of industrial partners is a clear limitation to any attempts. In other words, the notion of mass-effect, even if it is actually difficult to give it a detailed definition, has something to do with the existence of localized knowledge. It appears somewhat as a necessary condition for its emergence. This is explicitly the case for these activities.

To some extent, this situation looks very similar when considering the other dominant industrial and research activities located in the park: those related to life sciences. Here too, even if the context is less drastic than that of energy and earth sciences activities, one can consider that an insufficient mass-effect is still at stake, so that the conditions required to manage a transition from an external towards a more endogenous type of economic development are not satisfied, despite the efforts of their components. In that respect, the "Pôle Santé" aims to play for health and life sciences activities a similar role to the previous associations from the ICTs sectors, but without the same operational effectiveness, due to insufficient pre-conditions for learning. As a consequence, the emergence and the development of internal localized knowledge within these types of activities cannot yet be available, whatever the efforts developed by the actors themselves or by local policy-makers. Here we can consider that the territorial area of Sophia-Antipolis is not of sufficient relevance to allow the implementation of localized knowledge. The latter should require a discussion about the suited territorial definition for localized knowledge in accordance with the required relationships implied by innovative behaviours of these companies. In that perspective, localized knowledge for these activities requires thinking about inter-regional or even national scales of analysis as the relevant territory to consider the coherence of innovative behaviours.

As a whole, all those associations and clubs are nevertheless fundamental elements in the co-ordination of innovative activities of local firms. They play an important role for "high-tech" SMEs to reduce uncertainty and to provide information on technologies as well as related markets. All those collective initiatives aim at exhibiting a clearer focus on specific local knowledge and distinctive capabilities from other high-tech agglomerations in the international environment.

17.4.3 Coaching (private and public) activities including the financial local infrastructure to support companies' activities

The cumulative effect of locating small and very small companies in the Sophia-Antipolis experiment has justified the existence of a coaching infrastructure that contributes to enhancing the effectiveness of the transition toward a technopolis-type of economic development. To some extent, Sophia-Antipolis has progressively concentrated, as any usual French science park, the common infrastructure aimed at favouring science-industry partnerships and technology transfers (Quéré, 1998). Among this infrastructure, one needs to dissociate public from private coaching.

According to public coaching, I already mentioned how local authorities were devoted to the management of the park infrastructure whereas direct public support for the park users has been dramatically weak. Without considering the special case of the CICA building, public coaching essentially includes the urban planning requirements (managed by the SAEM-SACA organisation) and the promoting support of the experiment made by a specific office which is public funded (managed by the "Côte d'Azur Développement" organisation). Both organisations are not concerned with the direct working of the experiment. Some complementary organisations, supported by public funds, contribute to encouraging scientific and technological relationships among the components of the Sophia-Antipolis experiment (for instance, the Fondation Sophia-Antipolis or the local Chamber for Industry and Trade, etc.). However, those components are not very numerous, as regards the scale of the experiment and, consequently, their actions are not significantly structuring the internal economic working. Finally, there is a recent tendency for some scientific public institutions to engage in specific organisations dedicated to technology transfers. But the latter are still largely the result of national-based strategies rather than a spontaneous consequence of the accumulation process started thirty years ago in the park.

According to private coaching, two categories of initiatives are noticeable. On the one hand, large companies have promoted entrepreneurial initiatives by establishing internal infrastructures to develop start-ups from their own human resources. This is the case for companies internal to the experiment, like France Telecom and Lucent, but also for companies located in the immediate surroundings like Alcatel Space, IBM and Texas-Instruments. On the other hand, the success of Sophia-Antipolis has motivated private initiatives to benefit from the localized technological knowledge in two main orientations. The one is the supply of an environment suited to local start-ups. This includes the disposal of offices with usual general services. A few private companies have specialized locally in offering square metres and usual administrative services to local entrepreneurs. The other is venture capital facilities. Offers from a few private companies (local and non local) have recently developed and contributed to increasing the importance of the transition towards a technopolis-type of economic development. Both components structure the local institutional environment and favour the evolution towards a more endogenous production of technological knowledge.

17.5 Conclusive Remarks

In this contribution, I have identified the existence of a two-stage process in the development of the Sophia-Antipolis experiment. The one has been dominated by a very wide accumulation of external resources, be they private companies or public institutions. This first stage results in a very fragile project because of the extremely high dependence of the experiment on the external (international) environment. The internal economic working of Sophia-Antipolis was dominated by external relationships and their related external decision-making. As such, the project can be associated in a "reverse technopolis", as regards the usual definitions. Consequently, the existence of localized knowledge, i.e. the importance of local innovative behaviour from the components of the experiment, was very weak. However, this first stage has also to be thought of as a necessary condition to engage in the second stage of the process which has been much more based on the combination of resources and capabilities accumulated locally. This transition towards a more endogenous-type of economic development appears in the 1990s, and this experiment is now facing a significant change in its internal working. By the behaviour of large companies or the recent exploding growth of local start-ups and spin-offs, structural characteristics of Sophia-Antipolis are changing (Quéré, 1997). The experiment is becoming much more effective in exhibiting innovative behaviour based on the implementation of localized knowledge, i.e. the combination of complementary resources existing on site (Quéré and Ravix, 1997). Finally, a very interesting point lies in the observation that this transition results largely from a spontaneous order in the sense that the ability of public policy-making to sustain this transition of the Sophia-Antipolis experiment has been extremely weak. In other words, local policy-makers are much more concerned with the international attractiveness of the site and with providing it with usual physical infrastructures than with tools and means dedicated to support the existence of localized knowledge and its economic working as a local system of innovation.

References:

Gaffard J.L. and Quéré M. (1996), "The diversity of European regions and the conditions for a sustainable Economic Growth", in Vence Deza X. and Metcalfe J.S. (eds.), *Wealth from Diversity*, Kluwer Academic Publishers, Dordrecht.

Longhi C. and Quéré M. (1997a), "Technopolises and technological development", in Rabkin J. (ed.), *Diffusion of New Technologies in the Post-communist World*, Kluwer Academic Publishers, Dordrecht.

Longhi C. and Quéré M. (1997b), "The Sophia-Antipolis Project or the Uncertain Creation of an Innovative Milieu", in Ratti R, Bramanti A. and Gordon R. (eds.), *The Dynamics of Innovative Regions*, Ashgate, Aldershot.

Quéré M. (1997), "Sophia-Antipolis as a local system of innovation", *Economia & Lavoro,* vol. 31, no. 3-4, p. 259-272.

Quéré M. (1998), (ed.), Les technopoles en Europe, Enjeux et Atouts de la diversité, AFT/DATAR, Paris.

Quéré M. (1999), "Innovation, growth, and co-ordination through institutions: a discussion about 'innovation systems'", in, Fabel O., Farina F. and Punzo L. (eds.), *European Economies in Transition*, Macmillan Press, London.

Quéré M. and Ravix J.L. (1997), "Production de connaissance et institutions innovatrices: le chercheur-entrepreneur", *Revue d'Economie Industrielle*, vol. 79, p. 213-232.

Chapter 18

An ecology based interpretation of district "complexification": the Prato district evolution from 1946 to 1993

Luciana Lazzeretti and Dimitri Storai

18.1 The use of the "organizational population" concept for the analysis of district complexification

The ever growing uncertainty and variability of the environment impose upon economic operators, and not only on them, a constant re-examination of position and constant profile readjustment based upon the oncoming configurations. In a context of this type an ever more complex entity such as the industrial district[1], even if successful, cannot evade facing the changes in act. It must constantly re-examine its position on the market, striving to identify and exploit the opportunities which are created and overcome the difficulties as they arise. A re-definition of the district profile is therefore a particularly elaborate process because, still searching for new configurations, it is necessary to keep solid ties with the tradition from which so many efficient elements of local systems are derived.

Facing this reality, it becomes vital to discover investigation methods that help simplify the interpretation of complex phenomena such as the local industrial district systems. It is to meet these requirements that we have decided to propose the *ecology based* approach for the analysis of the Prato industrial district evolution.

Referring to other contributions (Lazzeretti and Brusoni, 1997) for a whole picture of the different lines of thought following the so-called *organization*

[1] Many recent studies on industrial districts highlight the tendency towards an increasing complexity of districts (Becattini and Menghinello, 1998).

ecology, note that on this occasion we have used only some of the guide concepts of ecology theories (i.e. "organizational form", "organizational variety", and "organizational population") and that methodologically we have carried out only the demographic analysis of vital processes as a preparatory study towards the ecology based analysis "tout court".

From this point of view, even though aware that the process of re-definition of the district profile has a very wide study of a multi-dimensional type, we have limited the examination only to the organizational transformations, particularly studying the modifications of the internal district structure in terms of "birth, performance, decline, disappearance" of some "organizational forms"[2] - intended as different types of firms identified by production specialization[3] - configuring the set of organizations that use the same organizational form and that make use of the same area of resources, such as "organizational populations"[4], or rather as "populations of firms" (Lazzeretti, 1996).

Furthermore, after having reconstructed the evolution of the industrial district through the analysis of the vital flows of the different populations of firms present, we have studied the principal interactions changing the ecological concepts of "symbiosis, competition, prey/predator relationships"[5], departing from the assumption that the district may well be seen as a "community of

[2] The concept of "organizational form" intended as a particular organizational solution is widely used in ecology organization studies (Hannan and Freeman, 1993; p. 52 and p. 94 foll.).

[3] To specify, the different populations have been defined as the whole of the specialized firms in a certain phase of the production process or production processes present in the district. As indicated successively, in the interpretation here proposed, the production process also includes the phases before and after the transformational process in a strict sense - such as the planning and marketing phases - or the collateral activities of the same.

[4] A population may be defined as the "set of organizations characterized by a particular organizational form and dependent upon a common set of material and social resources" (Carroll. 1997; p. 120).

[5] In studies of organizational ecology three types of relationships may be defined that may be established between the populations: competition, symbiosis, and prey/predator. A relationship of competition is established between two populations when the organizations of which they are composed utilize common resources for their activities. In this case an increase of the density of one population cannot have negatives effects on the other. A relationship of symbiosis characterizes the diametrically opposite hypothesis in which the increase of population density stimulates the growth of the other also. Lastly, there may be an asymmetrical relationship in which the growth of one population stimulates the growth of the other, but the increase of density of the second population provokes an erosion of the fundamental resources of the first, reducing its possibilities for development. In this case the relationship is referred to as prey/predator (Hannan and Freeman, 1993).

populations" (Lazzeretti and Storai, 1999)[6]. In fact, inside the district community different types of populations live together and influence each other, such as the manufacturing firms, service firms, the local institutions in their various components - local government, unions, guilds – which, as a whole, confer to the district the characteristics of a socio-economic entity (Becattini, 1991).

Proceeding in this manner we have tried a "de-complexification" of the study of the transformation of the Prato district, returning the focus of analysis to the phenomenon of "organizational variety", still following the ecology perspective[7]. According to this approach we have examined the complexifying process of the district by measuring the degree of increase or decrease of "organizational variety" in terms of absolute (number of firm populations) density and relative (number of organizations belonging to each population of firms[8]) density of the populations present in the district in a determined time period, though limited only to the production system[9].

[6] As is well known, the last level of analysis configured by ecological studies of the populations is the so-called *community level*, the other levels are that of the *population level* - which we have employed in the present paper - and of the *organizational level* (Singh and Lumsden, 1990). Note, for this purpose, a definition of organizational community given by Lomi, which is particularly suited to back our hypothesis of the application of the concept to the district community: "a set of inter-dependent populations, functionally differentiated under an organizational profile" (Lomi, 1995; p. 80).

[7] According to the ecology approach the organizational variety of an environment is determined both by the quantity of organizational forms present inside and by the number of the organizations which adopt each of these forms. Two environments which present the same number of organizational forms may be deeply different if the distribution of the organizations between them has deformed characteristics. An environment which sees that its organizations are distributed almost uniformly between the various organizational forms is profoundly different from an environment that, even though presenting the same organizational forms, sees its organizations concentrate particularly on some of them. The processes which influence the varieties are therefore not only the ones that determine the birth and the disappearance of organizational forms. Variety is modified also by the processes that provoke the increase or the decrease of the number of the organizations which adopt a certain form and those which are at the basis of change of organizational form by the existent organizations (Hannan and Freeman, 1993; p. 50-54).

[8] In organizational ecology studies the number of organizations belonging to each of the populations is specified by using the "population density" concept.

[9] In other terms, all of the organizational populations which lead off from the production system have not been taken into consideration, though they should be considered in a whole socio-economic definition of the industrial district.

18.2 Methodological notes

Following the analytical approach illustrated in the preceding paragraph, we have conducted an empirical research with which it has been possible to follow the evolution of 56 firm populations present in the Prato district[10] in the time between 1946 and 1993.

The analysis was carried out using the data on the registrations and cancellations reported in the "Registro delle Ditte" (Firm Register) kept by the Florence Chamber of Commerce, opportunely integrated with those reported in the "Albo Artigiani" (Artisan Roll) kept by the same institution.

The choice of the source of data fell upon the Firm Register because it permits the reconstruction of a very long historical series, which allows us to follow the principal evolutionary stages of the Prato district production system. Particularly, obtaining the data referring to the years immediately after the Second World War, it has been possible to reconstruct the evolution of the district from its birth during the Fifties. It must also be noted that the Firm Register is more complete than other sources, since it supplies data relating to all the firms operating in the area under examination, irrespective of the activity carried out.

Notwithstanding the merits just highlighted, the data supplied by the archives of the Chamber of Commerce present certain limits which modern literature has not failed to point out (Brandini, Ghilarducci, and Lucchesi, 1990; Rovida, 1990; Fumagalli and Marcora, 1990; Garfoli, 1994b). It was necessary to keep this in mind when gathering and interpreting data. The principal limit pointed out by these authors regards the date of beginning and ending of activity, which not always necessarily coincides with the dates of registration and of cancellation. So as to limit the effect of this phenomenon on the quality of the data, particular attention was placed upon the month and the year of the beginning and the end of activity when the data registered in the archive permitted us to do so. The date of beginning of activity was found for 50% of the firms, while the date of end of activity was available in 94% of the events examined.

For each firm we collected data relating to: the month and year of registration, the legal form, the commune in which it is located, the population of appurtenance, and (eventually) the month and year of cancellation. When available, data relating to the month and year of beginning and (eventually) ending of activity were also gathered.

[10] The Prato industrial district is here identified with the Prato Province.

Altogether 50,00 registrations and about 32,000 cancellations were analysed.

On the basis of the data gathered, an analysis of a demographic type was conducted that allowed the reconstruction, for each population, of the evolution in time of both the vital flows - births and deaths - and of the density. In the final analysis it was possible to follow the principal phases of the life of each population and therefore of the production system of the industrial district as a whole.

In the next paragraph we shall illustrate the principal results of our analysis in relation to the particular aim of our research. To this purpose only some tables relating to the most successful populations will be presented, also making use of aggregations of data relating to more than one population when this is useful to the flow of the discussion.

18.3 The evolution of the production system of the Prato industrial district from 1946 to 1993: the birth of the district and its progressive complexification

The reconstructions of the evolution of the Prato production system agree in placing the birth of the industrial district during the 1950s. Although, in fact, the textile activity is deeply rooted in the most ancient history of the town and forms with it an inextricable unit[11], the organization model adopted up until the second world conflict, and which resumed functioning in the first years after the war, is that which Mori defines a "dual circuit" production system (Mori, 1988; p. 1471). This model is characterized by the simultaneous presence of a nucleus of vertically integrated woollen factories and of a complex system of small and medium sized specialized firms for phases and for functions coordinated in the activities of a typical figure of the Prato industrial world: the "impannatore"[12].

The integrated woollen factories produce mostly standardized textiles, usually destined for poor countries in Africa and Asia or to satisfy the orders of the UNRRA (*United Nations Relief and Rehabilitation Administration*). The second circuit, for the moment, concentrates mainly on the internal market, where it supplies great quantities of fabrics for women's wear. These are able to

[11] Mori speaks of Prato as a "mono-oriented city" (Mori,1988; p. 1470) to indicate that the textile activity pervades the whole life, not only the economy, of all the local community.

[12] The *impannatore* "is a sort of pure entrepreneur who buys the raw materials for textiles; he has the material transformed by specialized local firms through the various phases of manufacturing based upon his own design; and he sells the final product on the open market or in response to orders from buyers" (Dei Ottati, 1994).

affirm themselves thanks to the combination of low cost and "variety of interlacing and designs" (Balestri, 1998; p. 189) that the Prato *impannatori* are able to offer their clients.

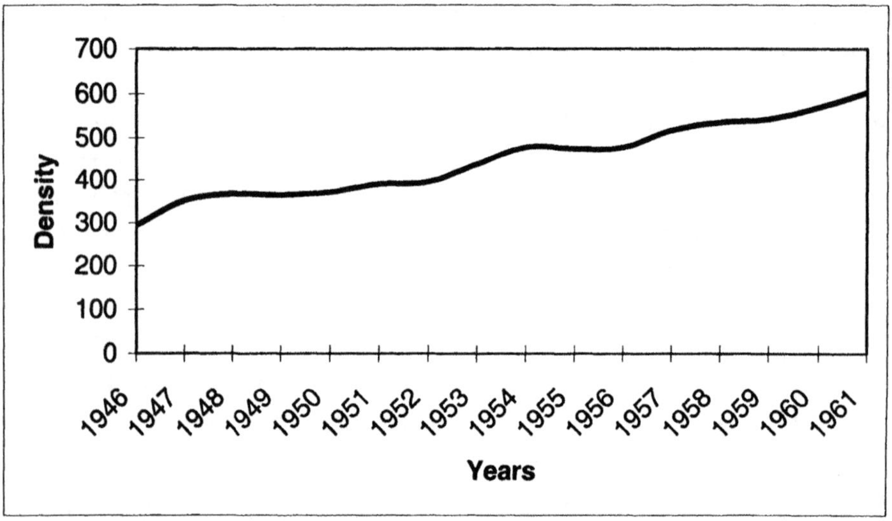

Figure 1 Population density of decentralised woollen factories or *impannatori* (district birth phase)
Source: elaboration of the authors on Florentine CCIAA data.

The turbulence of the post-war period, however, soon caused a deep transformation of the organizational arrangement of the Prato production system: the essential closing of the traditional channels of trade determined a serious crisis of the major firms which reacted by reducing their production structures to the right size and by sending away from the factories entire work units. In substance, the Prato woollen factories end up concentrating on the activities of planning and marketing the product on one side, and on the organization of production on the other, while much of the work previously done inside the firms is "dismantled" and entrusted to external sub-suppliers. In this way the integrated woollen factories disappear from the Prato reality and the dis-integrated production system centred on the duo *impannatori*/subcontractor, typical of the small firm circuit, becomes the organizational model which gives form to the whole city production system. The complex "metamorphosis" (Absalom et al., 1998) of the Prato production system can be said to be concluded around the middle of the 1950s with the birth of the industrial district.

The results of our analysis clearly show the transformation process of the organizational model just illustrated: between 1946 and 1961[13] we notice the steep increase of the density of the *impannatori* population as well as that of the subcontractor firms (figs. 1 and 2).

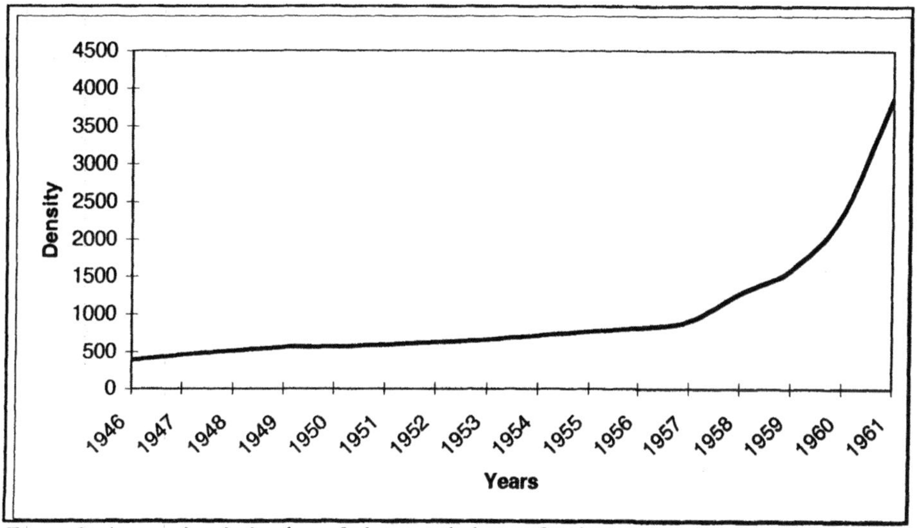

Figure 2 Accumulated density of the populations of subcontractor firms specialized in the manufacture of woven fabrics (phase of district birth)
Source*:* elaboration of the authors on Florentine CCIAA data.

Equally involved in a strong growth process are the populations specialized in the other activity typical of the city of Prato which is represented by the commerce and processing of rags (fig. 3).

In order to define as complete a picture as possible of the district structure in the first years of the 1960s, it is necessary to account for the presence of some populations specialized in activities connected with the production of textiles and with rag processing, among which should first be remembered the firms for the commerce and intermediation of textile raw materials and fabrics that hold

[13] In 1961, the Florentine Chamber of Commerce registered the enrolment of approx. 1500 subcontractor firms. It is not possible to know the date of birth of these firms, but it may be reasonably hypothesized that most part of them were created in the years immediately preceding the dismantling of the large firms. For this reason we suggest the analysis of the metamorphosis of the Prato industrial structure and the birth of the district by examining the evolution of the organizational populations from 1946 to 1961. This time period is wider than that usually taken into consideration by studies on this period that, as we have said, locate the birth of the industrial district around the middle of the Fifties. In our opinion the solution we have adopted seems the one better able to supply less distorted results.

an important role of connecting the production system of the Prato district with its input and output markets (figs. 4 and 5). Finally notice the presence of a good number of textile-mechanics firms which for the moment are represented first of all by the repair shops but which also include firms dedicated to the production of new machinery.

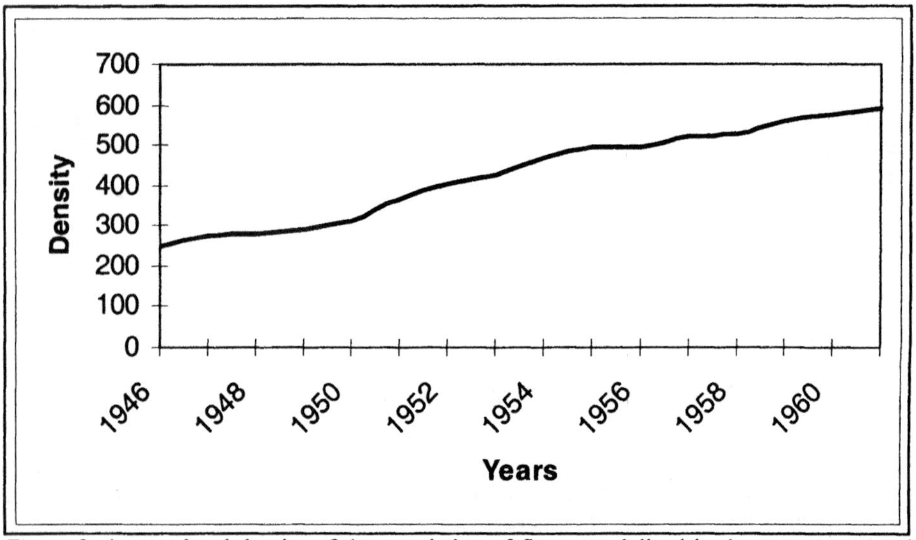

Figure 3: Accumulated density of the population of firms specialized in the commerce, sorting, and processing of rags (phase of district birth)
Source*:* elaboration of the authors on Florentine CCIAA data.

As a whole, therefore, the productive structure of the district a few years from its formation seems relatively simple, with a central body of populations specialized in the production of textiles and in the transformation of rags, with other populations which are occupied by activities closely linked with such production[14].

[14] At the head of the organizational transformations it is also possible to identify deep transformations involving the product and outlet markets. In a few years' time standard productions are substantially abandoned and, still maintaining the ties with the use of regenerated wools, the product typical of the Prato district becomes fabrics for clothing used mainly for the production of winter wear. The introduction of nylon, which joined with the regenerated fibers permits the manufacture of textiles previously obtainable only with the most precious raw materials, opens the way for the process of product range widening which will make itself manifest in the following years. On the market front the change is just as radical, because the industrialized nations become Prato's best customers while the poor states of Asia and Africa disappear, almost, from the list of importers.

With this new organization arrangement, the production system of the Prato district faces the challenge of twenty years characterized by the sustained growth of income per capita in the most industrialized countries and by the demand for textile articles which begins to assume characteristics of fragmentation and variability.

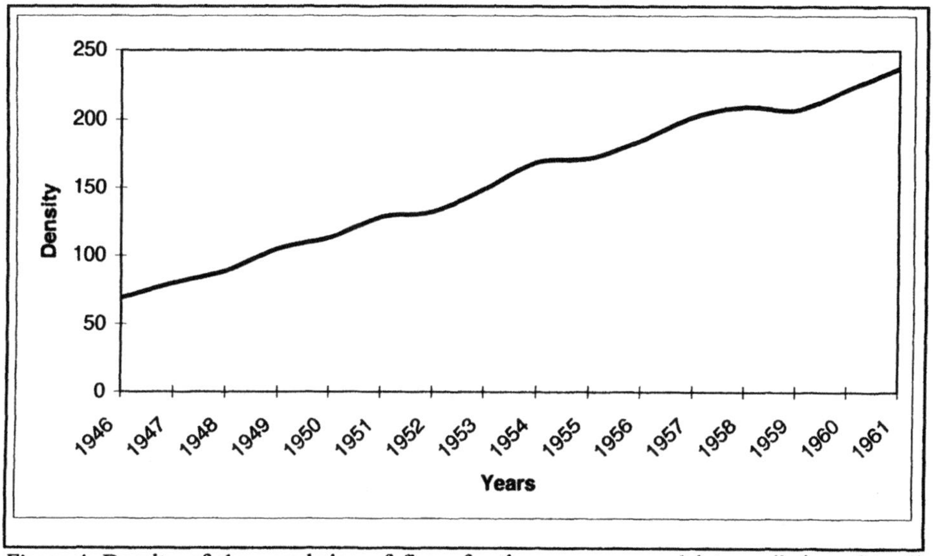

Figure 4 Density of the population of firms for the commerce and intermediation of textiles (phase of district birth)
Source: elaboration of the authors on Florentine CCIAA data.

The district responds well to the changed environmental conditions registering success on both the internal and the international market which is translated into a strong growth of all of its production system.In conformity with the district model the productive growth does not translate into a larger average size of the existing firms, which instead tend to shrink more, but into a larger number of firms present. So the process of the deeper degree of labour division between independent firms typical of the area is confirmed.

A similar reconstruction is confirmed by the results of our analysis that point out a strong increase of density in the existent populations (table 1).

On the whole we may affirm that all of the twenty years that run from the birth of the district to the middle of the 1960s, defined by Becattini as the "canonical development" phase (Becattini, 1998b; p. 465), are characterized by a growth of the prevalently incremental type, where the major phenomenon is the increase of density of the existing populations rather than the birth of new populations.

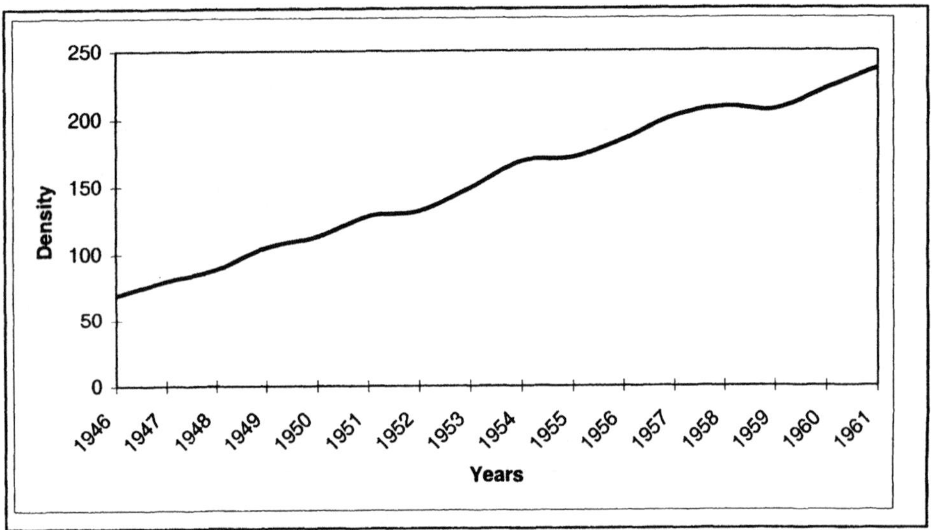

Figure 5: Density of the population of firms for the commerce and intermediation of textile raw materials (phase of district birth)
Source: elaboration of the authors on Florentine CCIAA data.

Truly, upon deeper study of the research results, it is possible to catch the embryos of that process of complexification that will fully show itself in the following years and decades. On one hand, the deepening process of the range of labour division gives place to the birth, in these twenty years, of some populations specialized in the "preparation" of the yarns such as winding, reeling, and coupling, but also of some firms which offer production services of a non-traditional type.

On the other hand, it is possible to catch the first elements of a "physiological diversification" (Becattini and Menghinello, 1998; p. 13) of the district which shows itself with the presence of knitwear factories, of ready-to-wear firms, of some producers of yarns, and, in the last years of the period, some firms specialized in the production of new textile articles such as imitation leather and furs.

Table 1: Average density of the population of firms per evolution phase
Source: elaboration of the authors on Florentine CCIAA data.

Population	**Average density***		
Populations of manufacturing firms	1946-1961	1962-1965	1976-1993
Incomplete woollen factories or *impannatori*	448	696	1082
Carded spinning subcontractors	147	360	560
Worsted spinning subcontractors	5	52	95
Twining	32	165	301
Coupling, unwinding, and skeining	1	4	27
Winding, slubbing and teaseling of yarns	2	97	835
Reeling	1	18	179
Warping	10	68	279
Weaving	772	3443	3384
Dyeing	23	47	93
Finishing	51	124	244
Firms for the gathering, commerce, and sorting of rags	360	483	582
Firms for the commerce and processing of waste fibre	22	38	66
Carbonising, tearing, and fraying	67	131	206
Knitwear factories or knitwear *impannatori*	38	120	440
Sewing firms	14	215	694
Finishing firms for knitwear	29	164	806
Ironing, folding, and packaging firms	1	24	187
Tailoring firms	125	342	289
Firms for ready-to-wear clothing	13	54	298
Subcontractors for clothing	10	55	236
Textile machinery construction firms	4	14	29
Firms for the production and repair of textile machinery	14	22	73
Producers or *impannatori* of imitation furs	0	3	28
Weaving subcontractors of imitation furs	0	2	38
Producers or *impannatori* of imitation leather	0	2	18
Subcontractors for finishing of imitation leather and furs	0	2	12
Producers or *impannatori* of textile for interior decoration	1	4	26
Producers of unwoven textiles	0	1	8
Producers or *impannatori* of carded yarns	4	46	196
Carded spinning subcontractors and own account	19	41	157
Producers or *impannatori* of worsted yarns	1	5	10

Population	Average density*		
Combed spinning subcontractors and own account	1	9	22
Producers or *impannatori* of yarns and textiles	16	43	170
Producers or *impannatori* of textile products and clothing	4	16	145
Producers of textile auxiliary chemicals	8	19	31
Forwarding agencies	202	757	960
Repair workshops for textile machinery	33	69	131
Commerce and intermediation firms for textile raw materials	137	244	506
Commerce and intermediation firms for textile machinery	44	80	118
Commerce and intermediation firms for fabrics	154	235	419
Commerce and intermediation firms for yarns	7	25	92
Commerce and intermediation firms for fabrics and yarns	4	7	39
Commerce and intermediation firms for clothing and knitwear	3	27	174
Commerce and intermediation firms for textile products and clothing	9	35	182
Commerce and intermediation firms for textile chemical products	8	17	19
Firms for planning, designing, and styling	0	1	26
Marketing firms	0	1	19
Management and training consultants	0	0	10
Firms for data processing	0	1	126
Computer and software firms	0	1	39
Real estate firms	8	60	559
Real estate firms with sharing option	0	1	106
Financial institutions	0	1	26
Financial institutions with sharing option	1	9	120

* For each evolution phase the average density of each population was calculated by summing the density at the end of the single years and dividing the end results by the number of years included in each phase. The results have been rounded off to the unit.

However, these are phenomena that, though being potentially considerable under a prospective angle, for the moment assume scanty quantitative importance (fig. 6).

During the middle of the Seventies a radically new phase opened for the district that was characterized by a progressive growth of the organizational variety present within it, due to the birth of new populations or to the affirmation of others that, even though present, had until then carried out a limited role. Speaking of which, it seems proper first of all to note that none of the populations present in the district in the middle of the Seventies disappeared during the following years. The reconstruction phase which characterized the Prato district in the second half of the 1980s hit some populations hard, sometimes reducing their density considerably, but no extinction of any organizational form was registered. The birth and the affirmation of the new populations lead in our opinion, from, two different processes that are at the centre of our analysis.

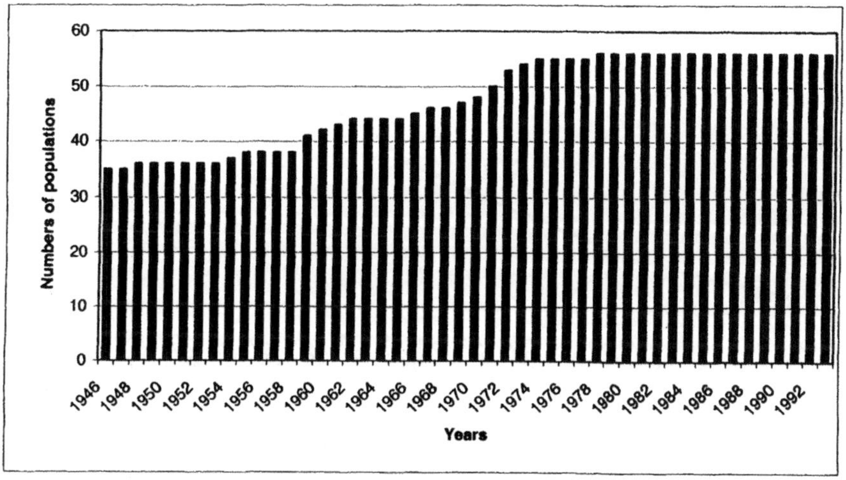

Figure 6: Evolution of the variety of the Prato district: number of populations present inside the district.
Source: elaboration of the authors on Florentine CCIAA data.

First of all there is a process of birth and growth of a number of populations specialized in the supply of production services of a non-traditional type. This process is interpretable, according to much district literature, as a deepening of the degree of labour division among the firms in the district.

Secondly there is a process of physiological diversification of the district productions that has determined the transition from a substantially mono-product district - carded wool textiles - to a district offering a "cluster of variously interrelated products" (Becattini, 1998b; p. 466): textiles, but also yarns, knitted and woven clothing, machinery for the textile industry, and new textile products.

From the second part of the 1960s on, the district offer of services to the firms was enriched with new specializations, thanks to the birth and affirmation of some populations able to offer its clients services of a non-traditional type. On this theme, firm populations have been noted which carry on the following activities: design and styling, advertising and promotion, marketing, management and training consultants, elaboration of data, computer systems and software. Figure 7 shows the course of the accumulated density of all these populations.

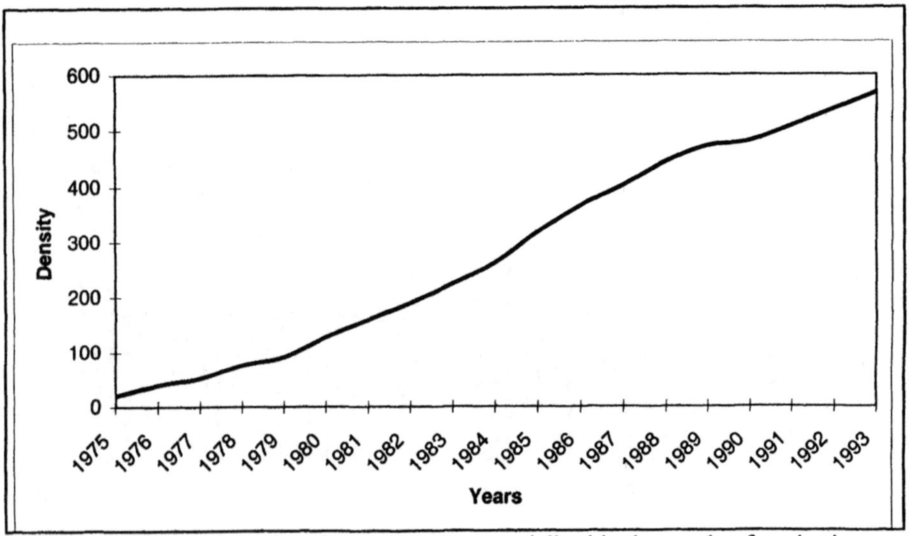

Figure 7: Accumulated density of the populations specialized in the supply of production services of the non-traditional type (complexification phase of the district).
Source*:* elaboration of the authors on Florentine CCIAA data.

Such synthetic data does not undermine the validity of the information, since all the populations are characterized by a virtually analogous course of vital flows with few or small inconsistencies of time: the first firms in fact were born in the first years of the 1970s, but the growth became steep from the second half of the decade on and was maintained as such for all of the 1980s and 1990s. To have an idea of the intensity of the growth of these populations it is sufficient to note one fact: between 1975 and 1993 the total density of all these populations rose from 26 to more than 600 units, with an increment of 2200%.

What goes on in the Prato district does not constitute an anomaly with respect to the trends that emerge from the studies of the whole of the Italian districts. In many researches (Sforzi, 1994 and Dei Ottati, 1996) recently conducted it may be noted that in the years closest to us there has been a strong increase of the number of firms and of the numbers employed in the supply of production services.

This phenomenon is accompanied, in Prato as in other districts, by a decrease of the number of firms and of those employed in the industry[15]. If we accept the idea that the division of labour among the district firms is not limited only to the transformation phases, but that they may be extended to downstream and upstream production processes in a strict sense, and if we abandon the theory circulating with particular insistence during the 1980s according to which the districts had not been able to produce, on their own, the advanced services needed[16], then the phenomenon just described of the growth of services facing the re-dimensioning of the industry must not necessarily be interpreted as a demonstration of "district de-industrialization", intended as the loss of the propulsive role carried out by the manufacturing industry upon the local economy, but may be seen as a demonstration of the indirect and complex ways with which production is carried on in the industrial districts (Sforzi, 1994; p. 213).

The growing complexity of the environment in which the firms operate and the adoption of new technologies require that the production activities in the strict sense be closely supported by ever more qualified and numerous service activities. To satisfy the new needs that are so created, it is not surprising that a growing number of district firms decide to resort to the market instead of creating internal units specialized in this type of activity. Facing this new demand the district has been able to respond adequately with the birth of firms able to guarantee the supply of at least a large part of the services requested. Under this slant, the birth of service firms in the Prato district would be interpretable as a deepening of the labour division among firms that does not involve only the transformation phase in the strict sense, but is extended to the phases upstream and downstream from the production process. It would therefore be "a deepening of the physiological process of progressive specialization and complexification of the Prato district" (Becattini, 1998b; p. 573). The birth of firms specialized in the supply of services of a non-traditional type seems to give value to this type of reconstruction.

[15] According to most recent studies those employed by the industry decrease in the Tuscan districts (Dei Ottati, 1996; p. 98), Emilia-Romagna districts, and Marches districts, while they increase in the districts of the other regions (Sforzi, 1994; p. 214). In Tuscany industrial employment decreases in all of the districts present in the region.

[16] According to the advocates of this theory there would be a division of the localization of the production activities between districts, in which the production in its strict sense would be concentrated, and the urban systems which for their setting in the centre of information flows would be the elected place for the development of the more advanced services.

The second process we intend to account for regards the production diversification that has involved the Prato district from the second half of the 1970s on. In this case also, what has happened in Prato is consistent with the general evolution of the Italian districts. Becattini and Menghinello write (1998; p. 16):

> "The industrial district, in the semi-conscious defence of the conditions of its own social reproduction in an intensely competitive external environment, tends to move its productive barycentre from the products with which they have reached initial success towards productions that place themselves along some branch of the habitual technological path and/or of the proven organizational form and/or of the habitual placement channels on which its products were affirmed previously."

Products of the first type - like fabrics for Prato and knitwear for Carpi - constitute that which the authors call "*direct* made in Italy" while the products of the second type constitute the "*indirect* made in Italy". The results of a research conducted by the authors themselves shows that in the decade between 1985 and 1995 the indirect made in Italy component registered a steeper increase than the direct component, confirming the trend towards "complexification or physiological diversification" of district productions. This supports the idea that the district offer is always more similar to a "cluster of products" interrelated by ties of a technological, organizational, or commercial type.

From its birth the Prato district has tied itself inseparably to the production of woollen goods, almost exclusively of a carded type made from regenerated fibres. Even today the production of carded textiles represents the typical local production. However, in time, not a few transformations have occurred that have deeply changed the district supply. The transformations first of all involved the classic woven fabrics, which have been reinterpreted in various ways through improvement of quality and a wider range of samples present on the market, which today also contain textiles made with fibres not traditional to the Prato district like cotton, linen, silk, and viscose.

However, the accent of our reflection is placed on a second type of transformation which is a true diversification of the district productions. In fact, beside the traditional textile activity is affirmed the production of woven and knitted clothing, machinery for the textile industry, new textile products such as imitation leather and furs, non-woven fabrics, as well as the yarns previously used in the manufacture of knitwear. Figure 8 shows the density trend of purchasing firms distinguished by type of product. From its examination one can clearly see that from the second half of the 1970s these populations have assumed much more importance than in the past.

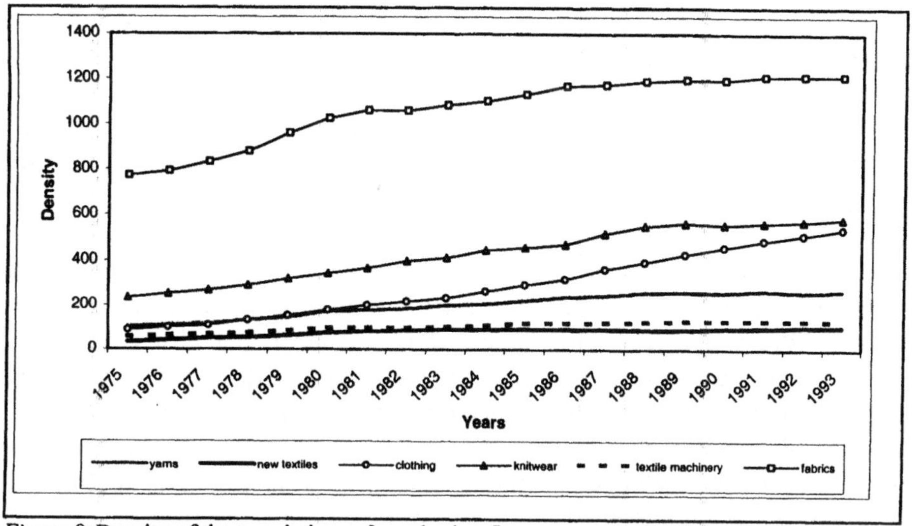

Figure 8: Density of the populations of purchasing firms present in the Prato district distinguished by productive specialization (complexification phase of the district).
Source: elaboration of the authors on Florentine CCIAA data.

It is difficult to say which of the three types of ties - technological, organizational, or market - is the most manifest in the diversification process of the Prato district. From a technological point of view, it is certain that wool working know-how settled in the district represented an important heritage for all of the productions that have taken hold in time. Along this line it is sufficient to remember the case of the imitation furs that, even though presenting a production with a higher level of discontinuity than in the past, has taken advantage of the know-how settled in the district, which has become a world leader in the production of imitation furs using wool yarns.

No less important are the market connections and connections of a commercial type: the most scrupulous attention to the product, the ability to frequently present renewed vastly ranged samples, and the ability to export its products in most the world, represent the abilities localized inside the district, from which the firms specialized in the manufacture of new products have been widely able to benefit.

However, we will focus our attention on the ties of an organizational type. With some exceptions involving the technical non-decomposability of the production cycle, the typical district model, in fact, has given form to the organization of the production of the goods of most recent introduction.

An exemplary phenomenon in this sense is represented by the production of articles of knitwear where an organizational model centred on the *impannatori*/subcontractor concept is affirmed. The great majority of Prato knitwear factories is characterized by a scanty involvement on the production line since the different phases of transformation are entirely or almost entirely entrusted to external sub-suppliers. The knitwear factories, but it would be more correct to speak of knitwear *impannatori,* concentrate (as happens with the woollen factories) on the activities of planning and marketing the products and in the coordination of the production activities. The affirmation of the production of knitwear articles is consequently translated not only into a density increase of the *impannatori* population of this product, but also into a solid development of the subcontractor firms specialized in the different phases of the related production process: sewing, finishing, ironing, and packaging firms.

Similar phenomena may be found concerning the textile-mechanics firms that resort to external workshops for the manufacturing of single mechanical parts and of the ready-to-wear firms that commission to subcontractor firms the cutting, sewing, and ironing of the clothing.

The "districting process of the organization" has also interested the productions that from a technological point of view are the farthest from the local manufacturing traditions and therefore initially must run up against the lack of skills on location necessary for the production in question. An important confirmation in this sense comes from the analysis of the firms specialized in the production of imitation furs.

Figure 9 shows the trend of the two different density curves. The first is related to the population of producers or *impannatori* of imitation furs, while the second represents the accumulated density of the population of firms specialized in the manufacture of imitation furs[17].

As may be clearly gathered from the figure, the first producers of imitation furs were born in Prato in the second half of the 1960s, preceding the birth of the subcontractor firms by a few years.

[17] The manufacturing cycle of imitation furs can be decomposed into two principal phases: the weaving of lengths of fur fabric and the finishing of the fabric. In the district there are firms specialized in both of the single phases of manufacture; however, in order to make the figure more comprehensible, and the reasoning more agile, we have preferred to represent the accumulated density of the two populations.

Therefore it may be hypothesized that, encouraged by conversations with local operators, this type of production has been touched by a "districting process" that witnessed the change from an organizational model of the integrated type to a dis-integrated model of the district type.

The first firms to introduce the production of imitation furs to Prato carried out the whole cycle of production inside their own establishments. Notwithstanding the decomposability of the production process, there are nonetheless no sub-suppliers available to handle some of the phases of production, because they lack the skills necessary for the creation of a subcontractor firm. A similar reconstruction is coherent with that hypothesized by Bellandi and Trigilia (1991), according to which the firms which are farthest from the traditional productions are characterized by a higher degree of vertical integration.

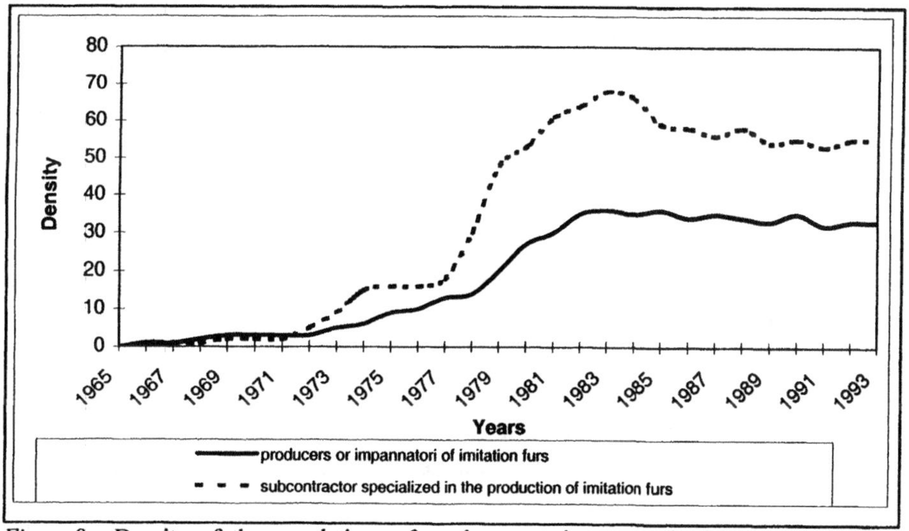

Figure 9: Density of the populations of producers or impannatori of imitation furs and the accumulated density of the populations of subcontractor firms specialized in the production of imitation furs (complexification phase of the district).
Source: elaboration of the authors on Florentine CCIAA data.

During the course of a few years things tend to change and we witness the birth of a certain number of subcontractor firms. The machinery used for the manufacture of imitation fur fabrics are very similar to the circular machines which are used for the manufacture of knitwear fabrics, therefore it was possible to use the knowledge settled in the district for the training of a number of entrepreneurs able to produce the new fabrics.

To facilitate the diffusion of the necessary machinery for these new products the firms often resort to the traditional mechanism of the work discount[18].

The process under question is particularly rapid, attesting the vitality of the Prato district and its capacity to assimilate knowledge coming from the outside in a particularly short time. Already in the last years of the 1960s, that is a short time after the introduction of the first firms for the production of imitation furs, we witness the birth of the first fur weaving upon commission and the first finishers specialized in this type of work[19].

The evolution of these populations seems coherent with the hypothesis according to which the introduction inside the district of a new production collides with the lack of relating know-how; in which case the firm that decides to devote itself to the manufacture of the new products must either undertake the whole cycle of manufacture or resort to sub-suppliers external to the district for the acquisition of semi-processed articles or to have some phases of production done. In time, if the district is vital, the knowledge is diffused and, in the presence of decomposable production processes, we may witness the birth of some subcontractor firms that carry out the phases of processing, previously done inside the final firm or outside of the district. In other words, when the productive knowledge is diffused, the conditions are posed for the expansion of the typical district model for also the productions of more recent introduction.

18.4 The relations of co-evolution between the firm populations of the district

In the analysis of the complexification which has involved the Prato district production system it seems interesting to ask, at this point, if and through which mechanisms the district environment represents a stimulus or a determent to the

[18] The mechanism of the work discount is typical of the Prato reality and consists of an agreement that runs between the owner of the final firm and a subcontractor who is usually an employee. On the basis of this agreement, the entrepreneur transfers to his employee, whose technical skills and moral qualities are a known factor, the property of a machine in his possession, receiving in exchange the pledge of the now ex-employee to work as a sub-supplier until the complete payment of his debt. The result of this process is the transformation of the entrepreneur from employer to commissioner, while an employee is transformed into a sub-supplier.

[19] A similar phenomenon interested the production of imitation leather that witnessed the birth of firms specialized in the finishing treatment of these particular products. The production of non-woven fabrics instead occurs inside the integrated productive establishments, since the production process is not technically decomposable.

birth of new organizational populations. In other words, we may ask if the probability of birth and growth of a new population inside a production system like that of the Prato district is equal, greater or smaller than that of a population which is situated in a less self-contained environment. And yet, if we hypothesize that the district environment influences the probabilities of birth and of growth of a new population, we may ask if the latter latches on with the existing populations a relationship of mutual support or that it is instead hampered and deterred in its growth.

On this line it is possible first to recall the considerations advanced in relation to the birth of the populations specialized in the supply of services to production. The conditions for an ever more pronounced division of labour among firms are created with the growth of the district production system which in this way tends to adopt a deeper specialization. The introduction of new technologies that determine the appearance of new needs and the growing complexity of some activities represents a very strong stimulus for the firms to externalise some phases of the production process, entrusting their execution to subcontracting firms specialized in this type of activity.

From this point of view the presence of a system of populations tied to each other by a "filière" relationship stimulates the birth of new populations that are specialized in the execution of a specific activity connected with the same production "filière" - here intended in a broad sense and inclusive of the activities upstream and downstream from the transformation process in the strict sense. Therefore, in this case a relationship of symbiosis is created between the new population and the already existing populations.

Also when referring to the physiological diversification process we may advance some considerations concerning the type of relationships which tie the populations specialized in the productions typical of the district and those which instead produce different kinds of goods. We have already underlined how the diversifying path followed by the Prato district reflects the theoretic scheme proposed by Becattini and Menghinello since the new productions present ties of a technological, organizational, and commercial type with the traditional manufacturing of textiles. Therefore a verisimilar hypothesis may be that the existence of these connections with the past experience well-embedded in the firms and in the inhabitants of the district has constituted a stimulus for the birth and growth of new populations specialized in the production of goods more or less far from the local production tradition. In this case also there seems to exist a relationship of symbiosis.

Along this line we may first of all note how the district is able to stimulate the birth of populations specialized in *instrumental* goods with respect to the typical local production. We have already said that through the deepening of the degree of the division of labour between firms we witness the birth of populations that supply products or services necessary to the principal activity. Then with the passing of time these new populations may abandon the role of being strictly instrumental to the typical local production to assume its own features, acquire markets external to the district, and therefore affirm itself as a population that is at least partially independent. This process is favoured by the same district mechanisms since the final firms request that the firms specialized in instrumental activities improve products and services continuously to always higher levels, to compete on markets outside the local one.

An illuminating example to this point is represented, for the case of Prato, by the population of firms that produce machinery for the textile industry. Born as an activity mainly for the repair and improvement of machinery bought outside the district to meet the requirements of local production, the Prato textile mechanics industry assumes, from the middle of the 1970s on, features of increasing independence with respect to the district production system by widening its range of action far outside the local boundaries. A similar phenomenon seems to be happening in more recent times to some populations specialized in the supply of services to the production. Born to satisfy the requirements of local firms, they have widened their activities also to clients outside the district.

A partially different phenomenon regards the birth of populations specialized in the manufacture of products, which in the beginning are made in the district as goods *intermediate* to the typical local production and then later become established as an at least partially independent activity. Also in this case it is not difficult to hypothesize a relationship of symbiosis.

In the case of Prato an example of this is represented by the affirmation of the production of yarns. This type of production was born in Prato as an activity strictly instrumental to the manufacture of fabrics. However, in time the accumulation of skills that were not only technical and technological, but also commercial - particularly tied to the ability to readily interpret and follow market evolution - made it possible to win new markets of its own, outside the district, reaching leadership in the production of yarns for knitwear.

Stimuli to the birth of new populations may nevertheless spring from interdependencies of a more general character connected with the presence of productions organized according to the typical district model. Just to make a few examples, it is sufficient to remember a particular approach to the market adopted by the local firms, the productive culture embedded in the inhabitants of the locality, the social climate that characterizes the function of these local small firm systems, the skills diffused among the workmen, as well as the presence of mechanics and intermediaries that are capable of adapting their skills to the requirements of the new productions if these present a more or less high degree of continuity with the traditional productions. Knitwear and new textile products represent for Prato two symbolic cases of productions that, even though lacking a tradition embedded in the district, have been able to take advantage of the already existing skills, not only in production.

If so far we have concentrated our attention upon the mechanisms which facilitate the birth of new firm populations, it appears however opportune to underline that from the simultaneous presence inside the local system of populations specialized in the production of interrelated goods there spring forth synergies that represent a reciprocal stimulus to growth. Let us once again remember the relationships, which are made between populations of final product firms and the populations of mechanical firms that make goods instrumental to this type of production. From the interaction between these populations spring innovations of products and processes which permit both to gather benefits. The manufacturing operators can obtain more efficient machinery or are able to speed up production. The mechanical firms can utilize the different techniques improved by themselves or by other district firms to offer machinery better than that offered by their competitors on the market. With reference to the Prato district this is certainly true for the production of woven fabrics, but not less important for the other productions. On this topic remember the importance of the action of the Prato local textile mechanics firms in their improvement of the machinery for the manufacture of new textile products to make possible the use of the fibres and production techniques typical of the district.

18.5 Conclusions

The evolution of the Prato district appears marked by a trend towards a growing complexification characterized by an increase of the organizational variety leading from two fundamental processes:

- a deepening of the degree of labour division among the firms which leads to the birth of new types of economic activities previously carried on autonomously inside the different firms;
- a process of physiological diversification that determines the appearance of firms specialized in productions not typical to the district.

According to the analytical perspective we have used, it has been possible to point out the existence of a progressive complexification in the Prato district from the second half of the 1970s on, in terms of the increase in degree of organizational variety. This is represented on one hand by the birth of new populations of firms, and on the other hand by the strong increase of the density of those already present. Furthermore, none of the emerging results lead us to foresee a trend reversal, that is lead us to foresee (in perspective) a reduction of the variety of district organizations and therefore a reduction of the number of firm populations. A reduction of the relative density may be eventually hypothesized, that is of the size of the number of firm populations. In fact, in the course of the fifty years examined none of the types of organizational populations became extinct. The district's competitive environment therefore turns out to be highly "self-conservational". In a historical period such as today's, characterized by strong processes of industrial restructuring which often penalizes the small and medium size firms in a significant way, the district formula holds the roll of "socio-economic shock absorbers" attenuating the process of selection of the species. This is the exact opposite of what happens in the firms not in districts as, for example, commercial firms.

The analytical approach that we have used should be extended to other industrial districts. This would permit us, besides the verification of the validity of the research method, to evaluate if the increase of organizational variety found in the Prato district is a phenomenon present in other districts, and eventually to proceed to a comparison between the degree of complexity noted in the different districts investigated.

Finally, a comment on the system of relationships which is established between the firm populations. The most interesting analysis perspective which presents itself for the future is naturally represented by the relationships of co-evolution of symbiosis, since the physiological diversification of the district does not seem to lead to the extinction of any of the species of firms. It is our intention to follow up this aspect by verifying the relationships crossed between several types of populations, considered both in a binary and systemic manner, utilizing birth models of the *density dependence* type belonging to the organizational ecology studies.

References

Absalom R., Becattini G., Dei Ottati G. and Giovannini P. (1998), Il Bruco e la farfalla. Ragionamenti su di un decennio di vita pratese (1943-1953), in Becattini, G. (ed.) 1998a, *Prato, storia di una città*, vol. 4, Le Monnier, Florence.

Balestri A. (1997), La deindustrializzazione del distretto pratese, in, Varaldo R. and Ferrucci L. (eds.), *Il distretto industriale tra logiche di impresa e logiche di sistema*, Franco Angeli, Milan.

Becattini G. (1991), Il distretto industriale marshalliano come concetto socioeconomico, in Becattini G., Pyke F. and Segenberger W., (eds.) 1991, "Distretti industriali e cooperazione fra imprese in Italia", supplement to *Studi & informazioni* no. 3, Banca Toscana, Florence.

Becattini G. (ed.) 1998a, *Prato, storia di una città*, vol. 4, Le Monnier, Florence.

Becattini G. (1998b), Prato nel mondo che cambia (1954-1993), in Becattini, Becattini, G. (ed.) 1998a, *Prato, storia di una città*, vol. 4, Le Monnier, Florence.

Becattini G. and Menghiniello S. (1998), "Contributo e ruolo del made in Italy "distrettuale" nelle esportazioni nazionali di manufatti", *Sviluppo locale*, no. 9, vol. V, p. 5-41.

Becattini G., Pyke F., Segenberger W., (eds.) 1991, "Distretti industriali e cooperazione fra imprese in Italia", supplement to *Studi & informazioni* no. 3, Banca Toscana, Florence.

Bellandi M. and Russo M. (1994), (eds.), *Distretti industriali e cambiamento economico locale*, Rosenberg & Sellier, Turin.

Bellandi M. and Trigilia C. (1991), "Come cambia un distretto industriale: strategie di riaggiustamento e tecnologie informatiche nell'industria tessile pratese", *Economia e politica industriale*, no. 70, p. 121-152.

Brandani G., Ghilarducci O. and Lucchesi P. (1990), *Metodologia e struttura del progetto*, in Mariti P. (eds.), *La formazione di nuove imprese*, Franco Angeli, Milan.

Carroll G. R. (1997), "Long-term evolutionary change in organizational populations: theory, models and empirical findings in industrial demography", *Industrial and Corporate Change*, Vol. 6, no. 1, p. 120-143.

CCIAA – Camera di Commercio Industria Artigianato e Agricoltura di Firenze, *Bollettini economici della provincia di Firenze*, August 1946 – June 1964.

CCIAA – Camera di Commercio Industria Artigianato e Agricoltura di Firenze, *Movimento verificatosi nel Registro Imprese e Albo Imprese Artigiane*, July 1964 - December 1993.

Crestanello P. (1993), *Le trasformazioni in 10 distretti industriali durante gli anni '80*, in, Varaldo R. and Ferrucci L. (1997), *Il distretto industriale tra logiche di impresa e logiche di sistema*, Franco Angeli, Milan.

Dei Ottati G. (1995), *Tra mercato e comunità: aspetti concettuali e ricerche empiriche sul distretto industriale*, Franco Angeli, Milan.

Dei Ottati G. (1996), "La recente evoluzione economica dei distretti industriali toscani", *Sviluppo locale*, vol. II-III, p. 92-124.

Fumagalli A. and Marcora L. (1990), *Le nuove imprese in Provincia di Milano*, in Mussati G., 1990.

Garofoli G. (1994a), (ed.), *Formazione di nuove imprese: un'analisi comparata a livello internazionale*, Franco Angeli, Milan.

Garofoli G. (1994b), *Formazione di nuove imprese in Italia: analisi dei differenziali territoriali*, in, Garofoli G. (ed.), *Formazione di nuove imprese: un'analisi comparata a livello internazionale*, Franco Angeli, Milan.

Hannan M. T. and Freeman J. (1993), *Ecologia Organizzativa*, Etas Libri, Milan.
Lazzeretti L. (1995), Dal gruppo strategico alla popolazione di imprese. Per uno spostamento dell'unità di analisi nei settori emergenti: il caso dell'ingegneria ambientale, Working paper University of Florence ,republished in *Economia e diritto del terziario* no. 4/1996.
Lazzeretti L. and Brusoni G. (1997), Dinamica e competizione localizzata della popolazione delle strutture alberghiere fiorentine *(1964-1994)*, Quaderni del dipartimento di statistica "G. Parenti", Florence.
Lazzeretti L. and Storai D. (1999), *Il distretto come comunità di popolazioni organizzative. Il caso Prato*, Forthcoming.
Lomi A. (1995), "The Population and Community Ecology of Organizational Founding: Italian Co-operative Banks, 1936-1989, *European Sociological Review*, no.1, vol. 11.
Mariti P. (1990), (ed.), *La formazione di nuove imprese*, Franco Angeli, Milan.
Mori G. (1988a), (ed.), *Prato storia di una città,* vol. 3, Le Monnier, Florence.
Mori G. (1988b), Il tempo dell'industria (1815-1943), in, Mori, G. (1988a), (ed.), *Prato storia di una città,* vol. 3, Le Monnier, Florence.
Mussati G. (1990), (ed.), *Alle origini dell'imprenditorialità*, Etas Libri, Milan.
Rovida F. (1990), *La differenziazione territoriale della nascita di nuove imprese in Italia: risultati per il periodo 1985-1987,* in Mussati, G. (1990), (ed.), *Alle origini dell'imprenditorialità*, Etas Libri, Milan.
Sforzi F. (1994), La terziarizzazione dei distretti industriali in Toscana, in Bellandi M. and Russo M, (eds.), *Distretti industriali e cambiamento economico locale*, Rosenberg & Sellier, Turin.
Singh J. and Lumsden C. (1990), Theory and research in organizational ecology, *Annual Review of Sociology,* vol. 16, p. 161-195.
Varaldo R. and Ferrucci L. (1997), (eds.), *Il distretto industriale tra logiche di impresa e logiche di sistema*, Franco Angeli, Milan.
Zagnoli P. (1993), *Percorsi di diversificazione dei distretti industriali: il caso Prato*, Giappichelli, Turin.

Chapter 19

New forms of knowledge creation and diffusion in the industrial district of Matera-Altamura-Santeramo

Vito Albino and Gianni Schiuma

19.1 Introduction

In the Italian economic scenario the industrial district (ID) represents a successful industrial development model. Arisen as a local productive reality, nowadays it is planning its development processes in accordance with internationalisation approaches in a more and more complex competitive context. There are various reasons for this complexity, such as the globalisation of markets, the continuous social-cultural changes of society, the events of political-financial instability, the development of technology. In this context, the competitiveness of the ID seems to present some shortcomings in efficiency as well as, in some cases, in effectiveness, due to structural and/or contingent causes (Albino and Garavelli, 1996; Ferrucci, 1996; Gottardi, 1997).

Some restrictions on ID competitiveness are related to the same features that have characterised its development and success. In particular, the nature of its innovation processes, based on incremental and discontinuous activities, no longer appears to be sufficient to support the acquisition of competitive advantages in the market. The present competitive context requires the adoption of strategic approaches based on a proactive nature rather than those based on a reactive and/or adaptive nature.

The IDs need to develop abilities of continuous innovation as well as to combine the incremental innovation processes with those of a breakthrough nature. They need to develop the ability to scan and adopt innovations developed outside the local area that can be implemented in the district production and management processes, with the result of improving the firms' competitiveness. Furthermore they need to increase the added value of their products by continuously improving the quantity of information and scientific and technological content embedded in the products (Anastasia and Corò, 1996).

Thus IDs have to develop new competitive strategies. For this reason the role of knowledge is of fundamental importance within the IDs. As a widespread resource in the local area and adopted by firms, it has traditionally represented a key strategic factor for the success of the ID (Albino *et al.*, 1998b; Becattini and Rullani, 1993; Bellandi and Russo, 1994; Vaccà, 1995a). In fact, although it is possible to identify different reasons for the development of IDs, e.g. the development of agglomeration economies in a local area resulting in a specific social-cultural context (Becattini, 1987; Camagni, 1989), the development of a flexible specialisation and the growth of an intensive networking activity between small and medium firms (Piore and Sabel, 1984), the development of a widespread innovation capability (Bellandi, 1989), and the capability to operate in international markets (Albino *et al.*, 1998c; Gottardi, 1997), at the basis of all these factors there is always the use and the exploitation of specific individual and collective knowledge.

Thus knowledge represents, at the same time, a viewpoint and a strategic resource for analysing and interpreting the competitiveness of the ID. Consequently the analysis of knowledge management processes within the local area of IDs has great relevance. In particular it is important to investigate the forms of creation and diffusion of the district knowledge, pointing out the nature that knowledge management processes are assuming at present. In fact as a result of the evolution of IDs, from a structural configuration of integrated system-area to one characterised by the development of some actors with a co-ordination function, the forms of knowledge creation and diffusion appear modified.

An example of this phenomenon is provided by the industrial district of Matera-Altamura-Santeramo. It represents one of the successful Italian examples of development in a local production system. After a first stage of formation, involving the start up of small and medium enterprises, this district has been characterised by an extraordinary growth. The most important trait of this process is the success of some large firms with a leadership position within the local area. These firms, aiming to improve their business performance, have continuously developed and internalised commercial and productive competencies (Albino *et al.*, 1998a; Schiuma, 2000). They have adopted strategies that are steering the development of the ID, changing its specific structural and cognitive characteristics. In particular the leader firms are operating in such a way as to control the supply chain, modifying the relationships with suppliers and developing hierarchically governed networks of firms. Thus the leader firms, to improve their performance continuously, are restructuring the inter-firm relationships as well as the borders of the ID. Furthermore, in order to be more innovative, these firms are changing the specific cognitive features of the local area by adopting new knowledge management approaches based on fresh forms of knowledge creation and diffusion. What is emerging is a gradual process of identification of the ID with the leader firms.

This research contribution aims to propose a possible interpretation, in accordance with a cognitive viewpoint, of the evolution process of IDs looking in particular at the experience of the ID of Matera-Altamura-Santeramo, which is the world leader in the leather sofa market. Two structural ID configurations are taken into account to analyse the forms of knowledge creation within IDs. The first analyses the ID as an integrated local system characterised by small and medium firms, while the second looks at those IDs in which it is possible to recognise leader firms operating as coordination centres.

Adopting a cognitive approach, we are going to argue that the evolution of IDs, with the development of some firms with a leadership position within the local area, involves a deep change of the forms of knowledge creation. These new forms of knowledge creation can result in the shifting of IDs towards new configurations that differ greatly from the traditional ID model.

19.2 The industrial district as a cognitive system

In order to analyse the characteristics of IDs, different interpretation models have been proposed. In particular, the operations of IDs have been analysed, adopting the flexible specialisation model (Piore and Sabel, 1984). According to this model the division of labour into productive stages and the specialisation of a large number of different firms in the execution of each productive stage, have provided the district model with production elasticity and flexibility[1]. This has allowed the district production model to be able to respond quickly and effectively to the changing of the market demand.

In order to analyse the inter-firm relationships of IDs, the interpretative metaphor of the network has been proposed (Butera, 1991; Camagni 1989;

[1] Production elasticity and flexibility have been traditionally adopted to describe the adaptive capability of an ID to the changes of market demand. In particular, the elasticity represents the capability of the district firms to change their production volume keeping nearly unchanged the production unit cost (Gandolfi, 1988), while the flexibility represents the capability of an ID to respond to the market demand by new and different products (Sabel, 1989).

Lorenzoni, 1987; Vaccà, 1986). This has allowed an analysis of both the characteristics of the relationships between the district firms, and the nature of the negotiation processes driving the inter-organisational allocation of production resources.

With the aim of pointing out the social-cultural nature of IDs, i.e. the deep integration between the production system and the social components of a local area, the conceptual frameworks of "district atmosphere" (Bagnasco, 1996; Becattini, 1987; 1989; Bellandi, 1982; Brusco, 1989; Marshall, 1920) and of "milieu innovateur" (Camagni, 1989; 1991; Cooke and Morgan, 1996; Maillat *et al.*, 1995) have been adopted.

The above conceptual models have been used by the district literature to interpret the success of IDs. It is interesting to point out that the analysis of this literature shows how knowledge represents a common interpretation variable to explain the specific features of IDs. In other words knowledge is considered as a fundamental resource. Often it is identified with the human capital or the intellectual capital of district firms (Vaccà, 1995a; 1995b; Vaccà *et al.*, 1997). Thus it is possible to state that knowledge represents one of the most important factors for the competitive success of IDs (Rullani, 1994).

The analysis of the start up of IDs shows that, at the basis of the development of IDs, it is always possible to recognise the growth and diffusion, at a local level, of practical and technical know-how on which virtuous circles of entrepreneurial diffusion have been triggered (Anastasia and Corò, 1996). In particular, the availability, in a bordered geographic space, of a historic heritage of social-cultural values and of technical-operative knowledge, mainly of craftsman nature, was of fundamental importance. This has laid the basis for the development of specific specialised production competencies as well as of a district entrepreneurial atmosphere.

In order to stress the cognitive nature of the ID, Becattini and Rullani (1993) have proposed an effective interpretation metaphor outlining the district model as a "cognitive laboratory", i.e. as a social-productive system in which productive, organisational and social experiences are generated and accumulated over time.

This interpretation metaphor allows an analysis of the production processes which take place within an ID as "complete production processes" (Becattini and Rullani, 1993): the production within an ID involves, besides the transformation of input factors into outputs in accordance with specific technical procedures, also the "reproduction of the production organism - the ID -". This is the same as stating that an ID can be compared to a cognitive system, i.e. an organism in which information is processed and technical knowledge, culture, and social values are generated.

The interpretation of the ID as a cognitive system provides the knowledge creation processes with a relevant role, recognising them as the main "engine" of the success of a local area. In order to analyse the knowledge processes of an ID it is important to distinguish two cognitive spheres: one of an "internal" nature and the other of an "external" nature. The first corresponds to the cognitive heritage of the district area, the second is the overall knowledge identifiable in the external environment of the ID in the form both of "public knowledge", e.g. the scientific-technological knowledge, and of "private knowledge", e.g. knowledge linked to the know-how of the firms operating outside the local area of a district.

Consistently with the above distinction of the district cognitive spheres it is possible to identify two cognitive dynamics of IDs. The first, labelled "internal cognitive dynamic", is linked to the local capabilities of a district to generate and diffuse knowledge, sometimes also by acquiring information from the external environment and processing it within the district. The second, labelled "external cognitive dynamic", is related to the capabilities of a district to acquire knowledge from the external environment or to generate knowledge through relationships established with cognitive systems external to the local area. For example, this can be achieved by setting up collaborative relationships between district firms and other firms operating outside the border of the local area[2].

Knowledge creation processes are strictly related to the characteristics of the actors operating within the local area. By adopting the network interpretation model, it is possible to analyse the district cognitive dynamics, taking into account two aspects: the actors involved in the knowledge processes and the relationships between the actors (Lipparini and Lorenzoni, 1996; Lomi, 1997).

[2] The distinction between a mere information acquisition process and a knowledge absorption process, which characterises the two alternative forms of cognitive dynamics (i.e. internal vs. external), is based on the conceptual difference between information and knowledge (Davenport and Prusak, 1998). Information can be defined as a set of data, i.e. as a set of recordings of a specific event and/or variable, combined in accordance with specific aggregation rules. Knowledge is the result of an interpretation process by a cognitive system of a set of information. Thus, information can be considered the structural component of knowledge since knowledge is an interpreted information. Therefore it is possible to analyse the internal cognitive dynamics of an ID as interpretation processes applied to a set of information acquired from the external environment to the local area. Instead the external cognitive dynamics of IDs equal knowledge absorption processes. They are performed mainly by knowledge transfer processes.

As regards the actors it seems possible to identify three main classes: the district firms, i.e. all those economic units with production functions, the district social structure, built up by individuals and families located within the local area, and, finally, the institutions operating in the district, such as, for example the local public administrations. Each actor takes part in developing and supporting the district knowledge processes.

In the following the focus will be mainly on the activities and cognitive features of the district firms, since they are considered the main actors in developing the district cognitive dynamics. The social structure and the institutions will be considered as the same actor, labelled "district social system".

19.3 Knowledge creation within integrated and local systems of firms

The cognitive dynamics within those IDs which present the structural configuration of an integrated-local system[3] appear mainly as the result of two processes: the activities of the district firms and the development of an inter-personal network centred on the firms, but extended outside the organisational borders with a direct involvement of the district social system.

The role of district firms in the development of knowledge creation processes within the local area is particularly important. These processes can be interpreted by two main classes of learning mechanisms: the "individual learning" processes related to each firm[4] and the "relational learning" processes associated with cognitive relationships of district firms.

[3] In this study, by ID characterised by the configuration of integrated and local system we mean the traditional ID as it has been described by the district literature. Thus, the configuration of an integrated and local system represents the set of small and medium firms characterised by high production specialisation, by common geographic localisation, and by deep integration with the local social-cultural structure.

[4] In accordance with the cognitive approach a district firm is considered as a "cognitive system", i.e. as an organism whose distinctive nature is represented by its ability to generate and accumulate knowledge. A firm is interpreted as an entity able to learn, transfer, codify and stock knowledge. This knowledge can be embedded either into specific hardware, e.g. technologies, technical equipment and any other tangible support, or into personal capabilities, or, again, into organisation routines. Thus the overall components of a firm, the tangible and intangible assets, are interpreted as a "repository" and/or "vector" of knowledge (Krogh and Vicari, 1992; Vicari, 1991).

The class of "individual learning" processes involves all those knowledge creation processes of the individuals working in the district firms and directly involved in the production processes[5]. These processes are performed mainly by two cognitive dynamics: those of a "trial and error" nature, mainly based on learning-by-doing and learning-by-using mechanisms (Rosenberg, 1982), and those of a social nature. The latter are mainly based on learning-by-socialising processes and take the form of face-to-face relationships between individuals operating within the organisational context. In face-to-face relationships, individuals share their technical-operative experiences as well as their "views of the world". The individual learning mechanisms are based on the high specialisation level of district firms. The specialisation allows the development of specific district firms' core competence, involving the growth of learning mechanisms based on learning-by-specialisation processes.

The development of individual learning processes of the district firms is strongly affected by the role of entrepreneurs, with which the firms are identified. These have a twofold role. On the one hand, they start up the learning processes of the firm[6], and on the other hand, they steer the evolution process of the firm involving a cognitive growth of path-dependent nature by their decision-making processes usually based on specific technical know-how (Levinthal, 1996; Nelson and Winter, 1982).

Instead, relational learning is strictly related to the characteristics of the relationships between district firms within the local area. It is possible to distinguish three main typologies of relationship: a) inter-firm relationships; b) relationships with the social system; c) relationships with the outside district environment.

The first class of relationships (inter-firm relationships) involves all inter-firm networking processes developed within IDs. These are due to the decomposition of the production process in operative sub-processes in which clusters of firms are specialised. The specialisation of district firms in specific stages of the production process has a fundamental role in the development of inter-firm relational learning processes (Sabel, 1989). In fact the specialisation

[5] This is coherent with the main results of the studies on the characteristics of the learning organisation. According to the literature, an organisation learns by its individuals (Huber, 1991; Garvin, 1993; Kim, 1993).

[6] Analysing the alternative forms of learning organisation, Huber (1991) defines this learning process as "congenital learning". It represents the knowledge generation process involved during the process of start up of the firm. This process is of a fundamental importance for the district firms since both the building up and the running of the firms are strictly related to the technical and operative knowledge of the entrepreneur.

involves the integration of different competencies, which fosters an intensive networking activity of the firms aiming to coordinate their operations.

The networking processes of the district firms involve an exchange of operation experiences that leads to the generation of new knowledge by the development of cross fertilisation processes (Nevis *et al.*, 1995; Sanchez and Mahoney, 1996; Tsoukas, 1996). Furthermore the cognitive growth of the district firms is linked to the content of the inter-firm relationships, involving numerous technical, economical, commercial, social and personal aspects (Håkansson, 1987).

The networking processes within the local area of the district enable the firms to develop relational abilities. These abilities allow firms to reduce their coordination costs thanks to a sharing of a set of tacit behavioural and interpretational rules, which reduces the uncertainty and the ambiguity of the informative-cognitive exchanges related to the inter-firm relationships (Albino *et al.*, 1999). This ability of district firms is the result of a knowledge generation process, based on relational learning mechanisms, that is labelled "learning by inter-firm relationship" (Fagerberg, 1995; Inkpen, 1996; Lipparini 1998; Lipparini and Lorenzoni, 1996; Nishiguchi, 1994; Powell *et al.*, 1996).

In order to reveal the integration between the district production system and the local social system, different conceptual frameworks have been proposed in the literature. However, two main interpretation metaphors can be identified, namely the "district atmosphere" and the "milieu innovateur" (Becattini, 1989; Bellandi, 1989; Brusco, 1989; Camagni, 1989; 1991). Both allow an analysis of the relationships linking the district firms and the social system of the local area. Moreover they allow an explanation of the origin of the external economies that have supported the development of IDs (Dei Ottati, 1995). The network of the district relationships involves the formation of a "relational cognitive area" that allows learning-by-localising or collective learning processes (Camagni, 1991; Dogson, 1993; Maskell, 1996). The district firms operate in an economic environment characterised by multiple social relationships, which appear mainly of an informal nature. They are based on the spread of inter-personal relationships within the local area. These relationships allow knowledge diffusion supporting both collective learning processes and the formation of an innovative context (Ferrucci, 1996; Maillat *et al.*, 1995). The collective learning processes involve knowledge creation processes based on learning-by-imitation or on local benchmarking processes, even if these present specific characteristics. At a local level, the learning processes based on the imitation of the products as well as of the behaviour of successful district firms, represent one of the most important internal boosters for extensive growth of IDs. In fact emulation processes involve the growth of an entrepreneurial structure characterised at the same time by cooperation and competition.

For the integrated and local system, the relationships between the district firms and the outside environment present mainly a commercial nature. In particular, adopting the agents' categories proposed by Lombardi (1994) in order to analyse the informative features of an ID[7], it is possible to state that for traditional IDs the relationships with the external environment present mainly the form of information acquisition from the market. These relationships are managed by a set of firms, which translate the market information into production orders for the district firms. Information flows appear mainly based on an adaptive approach. Thus the district firms, interacting with the outside environment, have the ability both to identify signals of the market demands and to respond to the market requests, but they generally lack the ability to generate new market outlets opportunities in a systematic way. Consequently, from a cognitive point of view, the district firms show a learning ability based on learning by external adapting processes.

The above learning mechanisms appear strongly related to the context in which they take place. Thus the knowledge they generate presents a "context" nature, i.e. it is strictly linked to the social-productive environment of generation[8] (Antonelli, 1999). Furthermore, since the learning mechanisms are mainly based on personal experience processes, knowledge presents a tacit nature (Polanyi, 1966). It is embodied in the individual action and in the inter-personal relational dynamics, and it can be transferred only by socialisation processes based on face-to-face interaction.

The contextual nature of the district knowledge affects the knowledge transfer and diffusion processes taking place within IDs. It allows communication processes to be performed within a local area with a high level of effectiveness and efficiency. This is true for communication processes involving both codified knowledge, e.g. technical standards, and tacit knowledge, e.g. specific individual operation practices. In fact the structural and functional characteristics of a district ensure that the actors operating in the local

[7] Analysing an ID as an informative system, Lombardi (1994) proposes five categories of agents: the "end firms" which represent an interface between the market and the local system, the "suppliers" which represent the production units specialised in a specific production stage, the "workers", i.e. all individuals working for the district firms and characterised by great commitment to their work, and the "institutions".

[8] The term "contextual" expresses the belonging of knowledge to a specific context. The context can coincide either with a social-economical system localised in a specific geographic area, like an ID, or with a community of individuals.

area share a cognitive background. This involves a sharing among the district actors of cognitive maps and communication codes which allow the efficiency of the information processes to be increased thanks to a reduction in the ambiguity related to the interpretation processes of the exchanged information (Albino *et al.*, 1999; Cohen and Levinthal, 1990; Daft and Weick, 1984). However it is important to point out that the tacit nature of the district knowledge imposes some limits on the wideness of knowledge transfer processes. In fact, since knowledge is not expressed in codes, knowledge transfer can be affected by ambiguity (Daft and Lengel, 1986; Kogut and Zander, 1995). Also, requiring socialisation processes to be transferred, it constrains the number of actors involved, and at the same time requires a medium-long period of interaction.

It is possible to classify the knowledge generated by individual learning processes into two main categories: the personal-embodied knowledge or individual skills, and the organisation routines. The first is the result of an individual knowledge accumulation process of an empirical know-how developed by carrying out specific specialised operation practices. The second involves all that knowledge of an inter-personal nature developed by individual face-to-face relationships within an organisation (Nelson and Winter, 1982). The organisation routines can present the form of non-formalised organisational rules and procedures, of organisational social-economical values, and of trust as well as of relationships of esteem between individuals.

The relational learning processes lead to the generation of inter-organisational knowledge, in the form of inter-firm routines, as well as to the growth of local social-cultural values (Bagnasco, 1996). The inter-organisational routines correspond to the overall knowledge developed by district firms in their networking processes. They present mainly a tacit nature, and are grounded in the firms' abilities to establish flexible relationships and to solve any emergent relationship conflict quickly and effectively. The inter-organisational routines can be interpreted as the result of inter-firm socialisation processes, which take the form of direct relationships between both entrepreneurs and workers operating in different firms integrated along the district supply chains.

On the contrary, the local social-cultural values involve a wide range of aspects related to the psychological and sociological features of an ID. They appear the result of an extension of inter-personal networks, characterising district firms, outside organisational borders with an involvement of the district social system. The local social-cultural values present the form of personal evaluating frameworks shared among the individuals living in a district area. They involve the sharing of beliefs, of cultural and political background, of behavioural codes, of life styles, and of "work ethics". The last, in particular, has been recognised by many authors as a fundamental viewpoint for interpreting the success of IDs (Anastasia and Corò, 1996; Becattini, 1992; Brusco *et al.*, 1995). In fact, the high productivity of district firms, as well as the extensive start up of small firms within the district area, can be linked to the interpretation of the work, not only as a means of subsistence, but mainly as a social means of identification[9].

Knowledge generated within an ID appears strictly and inseparably linked to the firms and to the social structure of the district. Thus each ID is characterised by a specific cognitive heritage, in other words it presents an idiosyncratic nature. This feature points out that each ID is unique. Although it is possible to identify a set of similar factors at the basis of the growth, each ID presents specific and distinctive features, which appear mainly based on the district cognitive dynamics. In accordance with an autopoiesis theory (Krogh and Vicari, 1992; Vicari, 1991), it is possible to assert that an ID as a cognitive laboratory presents a self-referential nature, i.e. it is a cognitive system with its specific and distinctive features (Corò, 1997).

Due to its mainly tacit nature, the knowledge characterising the integrated system-area can be exchanged exclusively by socialisation processes (Nonaka, 1991; 1994). This limits the transferability of knowledge. While codified knowledge can be represented and communicated by information codes, tacit knowledge can be communicated only by sharing working or life experiences. This matter represents an important viewpoint for understanding the knowledge diffusion processes within IDs.

In order to analyse these processes, two approaches can be adopted: one looking at the district firm and the other referring to the local system-area. By analysing the knowledge diffusion processes in a district firm, two main socialisation mechanisms can be identified: the apprenticeship in order to transfer worker technical-operative knowledge, and the inter-personal networks to share organisational and social-cultural values. Referring to the knowledge diffusion processes within a district, three main socialisation mechanisms can be considered: the mobility of workers between the district firms, which allows the diffusion of operation best-practices among the district firms, the extended inter-

[9] The value of the work for the individuals operating in district firms involves the strong commitment of workers to the production processes, which contributes to support entrepreneurial atmosphere. Brusco (1994) states that the ID has a great advantage, that many large firms are not able to capture: the ability to involve and motivate all workers in the operations.

personal networks, which allow an informal exchange of information between individuals operating in different firms, and the number of unstable inter-firm relationships, which push different firms to interact, sharing productive problems and operative experiences.

The above analysis of the knowledge creation and diffusion processes within an integrated system-area allows us to stress some important cognitive properties for traditional IDs. First, an integrated system-area is mainly a closed cognitive system, characterised essentially by internal cognitive dynamics. In fact, the external learning processes are limited to the acquisition of information about the market demand. The internal cognitive dynamics are based on the formation of virtuous circles of tacit and contextual knowledge creation and diffusion, fed by the activities of multiple specialised and integrated firms as well as by the intensive networking processes between the individuals operating in the firms. Thus, the continuous regeneration process of an integrated system-area appears mainly based on the creation and diffusion of tacit knowledge.

Second, the internal nature of the cognitive dynamics of an integrated system-area generates endogenous innovation capabilities. These have represented one of the most important resources for the competitive advantage of traditional IDs. Due to their distributed and diffused nature within an ID, the innovation capabilities have been labelled in the literature as "diffuse innovation capability (DIC)" (Bellandi, 1989). They have been recognised, at the level of the local system of firms, as a substitute for the traditional R&D function of large firms (Gottardi, 1996). However, due to the characteristics of the knowledge creation processes at the basis of the DIC, the innovations generated within IDs present an incremental and non-systemic nature. This can contribute to understanding why the IDs with the configuration of an integrated system-area have shown high performance in mainly static competitive contexts. In such contexts, adaptive learning processes, characterised by a behaviour of "stimulus-response" (McGill *et al.*, 1992) and based on a DIC (i.e. on an incremental improvement process of the products, services, technologies and marketing activities) are enough to acquire competitive advantages.

In recent years the competitive context has changed with an erosion of the competitive advantage of traditional IDs. Diverse factors have caused this change, such as the increased dynamics of the market with a demand which appears more and more differentiated, unstable and characterised by quick saturation (Becattini, 1999). Furthermore, many markets in which IDs operate are characterised by structural economical changes, and the competitive pressure of firms operating in developing countries is increasing. These firms can rely on lower labour costs, as well as on an increase in their technical-operative know-how with a consequent improvement in product and process quality. An example of this phenomenon is provided by the industrial district of Matera-Altamura-Santeramo, labelled "Distretto Murgiano del Mobile Imbottito". This system of firms is located in a geographic area between the provinces of Matera and Bari, and it is the world leader in the production of leather sofas. At the end of 1999, in the local area 293 firms, involved in the different stages of leather sofa production, were identified[10]. After an extraordinary economic and productive growth, the district is now dealing with an ongoing increase in the competitive pressure of firms located in the Far-East (China). The start up of these firms has been possible thanks to the initiative of some district entrepreneurs, who have decentralised some production units. These entrepreneurs, looking for new market opportunities and aiming to reach cost competitiveness, have transferred production technologies and operation know-how, performing apprenticeship processes of the local workforce. This results in risky technical-operative knowledge creation processes in an area outside the district. In fact, although leather sofas produced in the Far-East are still of a lower quality compared with the district products, they are competitive because of their low price. Moreover these products present margins for a qualitative improvement due to the local availability of high quality raw material and of the support provided by some specialised workers, who are moving to the new productive areas, motivated by economic incentives.

As this case shows, in this new context the integrated system-areas present some shortcomings which need to be faced by organisational-productive and inter-organisational processes reengineering aimed at increasing the added value embedded into the products. In particular, IDs need to focus on "non-price" competitive factors, e.g. quality and services. Furthermore they need to develop pro-active strategies in order to anticipate and/or generate new market demand (Anastasia and Corò, 1996; Gottardi, 1997; Lombardi, 1994). Then, IDs need to improve their innovation abilities. The traditional DIC has to be accompanied by new management approaches to innovation processes. These have to be less fortuitous and incremental, and much more systemic and continuous. We believe that the recent structural changes in successful Italian IDs can be interpreted as an effort to improve their innovation capabilities. This appears as one of the drivers of the evolution characterising the ID of Matera-Altamura-Santeramo. In

[10] All data and insights about the features of the ID of Matera-Altamura-Santeramo, provided in this chapter, are the results of empirical researches performed by authors in the last five years, performed both by collecting public statistics and by direct questionnaires and interviews.

this district some leader firms are gradually changing the district cognitive dynamics in order to improve their business performance.

19.4 Coordination in industrial districts

The evolution of IDs has been analysed by several authors in the last decade (see, for instance, Albino *et al.*, 1998a; Nuti, 1992; Gandolfi, 1988; Unioncamere, 1995; Varaldo, 1994).

In the most dynamic IDs new structures of coordination are arising. The emerging trend seems to show a change from distributed coordination, i.e. based on an "intelligent and autonomous behaviour" of the district firms, to a governed coordination, i.e. based on the structural integration of the district firms performed by some actors operating as coordination centres. These actors coordinate and manage the district firms' activities, by restructuring the inter-firm networks, and changing the district cognitive dynamics.

The features of the coordination actors can be different in the diverse district areas. However, it is possible to identify two main categories: *a)* actors with a "meta-management" function; *b)* firms with a leader position within the district area.

The rising of some actors with a function of meta-management (Pilotti, 1998) within integrated system-areas mainly corresponds with a strengthening of the role of the institutions and/or of the district firms' entrepreneur association. These actors offer services to support the innovation processes of the small and medium district firms, and coordinate district firms, setting and managing the inter-firms and inter-industries coordination processes both within the local area and between the district and the outside environment. Furthermore these actors perform fundamental commercial brokerage by monitoring the markets and developing marketing activities of the district products. All the activities carried out by the actors with a meta-management function support internal and external district cognitive dynamics. In fact, the actors operate as an interface between the internal and the external cognitive context of an ID. They support both the strengthening of the commercial competencies of an ID and the acquisition of scientific-technological and organisational-managerial knowledge providing innovation-oriented services.

The alternative development process of actors with a role of coordination is represented by the action of some firms which acquire a leader position in the ID (Albino and Garavelli, 1996; Lorenzoni and Baden-Fuller, 1995; Varaldo, 1994). This process can be the result of the growth of some local firms and of the localisation in the district of large and external firms which control strategic production and commercial knowledge.

This process involves the restructuring of the inter-firm networks (Albino *et al.*, 1998a), as well as new approaches to manage the district cognitive dynamics. In fact, leader firms develop higher capabilities of interaction with both external and internal district contexts. In the district they operate integrating the different production units and modifying inter-firm relationships, forcing district firms to improve performance. Furthermore, leader firms adopt new management approaches to the innovation processes which can increase their dynamic efficiency (Mariotti, 1989).

Table 1: Number of firms operating in the ID of Matera-Altamura-Santeramo grouped by provinces.

	1984	1985	1986	1987	1988	1989	1990	1991
Matera	10	13	16	19	24	29	34	39
Bari	24	29	34	44	52	67	81	90
Total	**34**	**42**	**50**	**63**	**76**	**96**	**115**	**129**
	1992	1993	1994	1995	1996	1997	1998	1999
Matera	40	43	44	46	51	57	54	55
Bari	104	117	144	158	182	211	231	238
Total	**144**	**160**	**188**	**204**	**233**	**268**	**285**	**293**

The ID of Matera-Altamura-Santeramo represents a case example characterised by the growth of some leader firms. In the 1950s, the origin of this district is identified with the presence of some small firms producing sofas and by a set of small carpenters' workshops, knitwear and leather craftsman-like laboratories. These production activities have built up the original favourable conditions for the development of the ID. In Table 1 the number of firms of the ID of Matera-Altamura-Santeramo, grouped by geographic area, is presented.

However the role of a local entrepreneur - Pasquale Natuzzi - has been of fundamental importance for the growth of the district. In fact, this dynamic entrepreneur has drawn the evolution of the district. Through a spin-off from his firm many firms have been generated and many entrepreneurs, imitating him, set up their own firms. Natuzzi's development started in 1977, and in the last two decades it has moved from an income of 29 billion Italian lira (in 1980) to an income of 1436 billion Italian lira (in the year 2000), with an average annual growth rate of 36%.

At present the ID of Matera-Altamura-Santeramo is characterised by three main leader firms: Natuzzi, Nicoletti and Calia, which centralise both commercial and productive competencies. However, Natuzzi, as depicted in Figure 1, represents the major firm in the local area, performing 60% of the total income of the district.

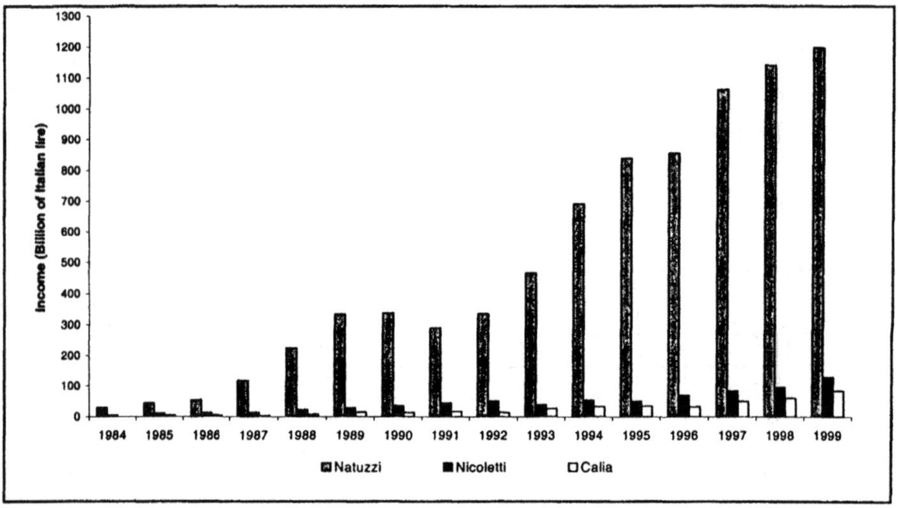

Figure 1: Growth of the district leader firms' income

In the following, referring to this case example, the main approaches adopted by leader firms in managing their own, as well as the district's knowledge to improve their performance are analysed. Also the impact of these approaches on the district cognitive dynamics is investigated.

19.5 The role of leader firms for knowledge creation and diffusion

Leader firms tend to modify the ID organisation towards a more hierarchical structure. This can be the result of two different development processes: *i*) "internal" growth of the leader firm, carried out by direct investments and/or by merging or acquisition processes; *ii*) "external" growth of the leader firm that, focusing on its core competencies, increases the number of partnership links by alliances and agreements with local and external firms owning strategic complementary competencies (Albino *et al.*, 1998a; Lipparini, 1996).

The ID of Matera-Altamura-Santeramo represents a case example of the growth of some leader firms. The development of the district can be mainly identified with the formation of those firms, which at present detain a leadership position within the local area.

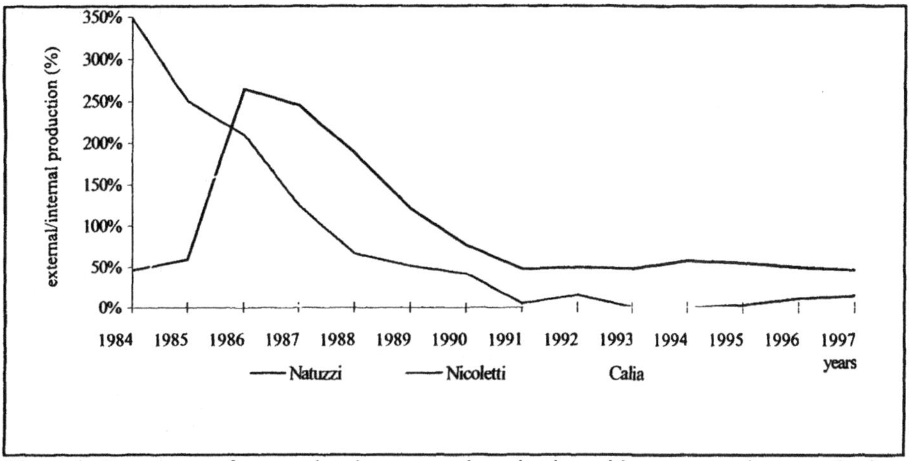

Figure 2 Percentage of costs related to external production with respect to the costs for internal production.

In recent years, following a development process of the district characterised by an extensive formation of small and medium firms, the gradual restructuring process of firm networks has been occurring. In particular, to improve their performance, leader firms are restructuring their inter-firm relationships. This development process appears mainly based on internal growth (Albino *et al.*, 1998a) as is shown by Figure 2. In fact, for instance, Natuzzi is internalising production phases and is acquiring district firms. By direct investments Natuzzi has increased its dimension and created an industrial group, which at the present is the world leader in leather sofa production. In this way this leader firm is able to control all the production stages of the sofa productive cycle.

Integrating the overall production process, Natuzzi directly controls purchasing, leather tanning, polyurethane transformation, assembly, logistics, production development and marketing activities. This has allowed new approaches to innovation, for example by improving the quality of the most strategic raw materials: leather and polyurethane (Albino *et al.*, 1998a).

Also two other leader firms (Nicoletti and Calia), even if with different characteristics, seem to follow the same growth strategy as Natuzzi. At the present, Nicoletti has stable relationships with a set of firms, which are autonomous; however, in the near future, this company foresees structural reorganisation with the setting up of a hierarchically integrated industrial group. Calia is adopting a development strategy based mainly on internal growth performed by direct investment.

Furthermore the leader firms are creating intensive relationships with the external district environment through the definition of partnership relationships with other specialised firms owning either complementary competencies or know-how, which can be combined with the district knowledge. Leader firms are operating as the interface with public and private R&D centres.

Due to their dimension and position within the ID, the leader firms perform the critical role of governing internal and external knowledge creation processes (Albino *et al.*, 1998c; 1999). In fact, being an interface between the district and the external environment, they run both the commercial interactions and the acquisition of knowledge developed outside the district area. Within the local area, the leader firms, operating as coordination centres, manage and affect knowledge creation and diffusion of technical-operative knowledge.

Analysing the evolution of this ID, it is possible to recognise a progressive identification of the district firms' cognitive dynamics with those of the leader firms, which appear to focus on managing district knowledge creation and diffusion processes. This is performed by the modification of their individual and relational learning mechanisms. In particular, to increase quality and productivity, leader firms improve district technical-managerial knowledge, also by adopting scientific-technological results developed outside the district.

Focusing on individual learning processes, the most interesting aspect is the implementation of the codification processes of the operative knowledge, considered highly strategic. The codification processes affect the internal cognitive dynamics of district firms. From learning mechanisms based mainly on the generation and diffusion within the organisational context of tacit knowledge, the firms move to knowledge creation processes characterised by a continuous "dialogue" and "interaction" between tacit and codified knowledge. This interaction can be modelled using the knowledge-creating company framework[11] proposed by Nonaka and Takeuchi (1995). In fact, leader firms are

[11] This is an interpretation model of the learning organisation dynamics. It interprets the knowledge within an organisation as the result of a spiral process, characterised by a continuous combination between "tacit" elements and "explicit/codified" ones. The framework is based on four stages: "socialisation", which is the process of transferring tacit knowledge within an organisation by face-to-face relationships between individuals; "externalisation", in which the tacit knowledge is codified by a process based on three conceptual tools, which are the metaphor, the analogy and the codification model; "combination", which is the process by which different codified knowledge is combined or integrated to generate a new codified

engaged in planning, designing and implementing knowledge codification projects. These involve organisational knowledge as well as some operative best-practices grounded in the individual abilities. For example, within the ID of Matera-Altamura-Santeramo some codification projects have been developed to codify the technical-operative best practices owned by individual experts involved in the developing of prototypes, in leather cutting and in the upholstering of sofas.

The aims of these codification projects are different, but they can be fundamentally traced back to the leader firms' need to make the technical-operative knowledge more accessible, exploitable, transferable and combinable with other codified knowledge developed outside the district area; this knowledge is at the basis of their productive and innovative abilities.

The implementation of codification processes allows the leader firms to increase their innovative capabilities, supporting the increase in the effectiveness of the knowledge creating and sharing processes within the organisation. It also allows them to improve the capabilities to absorb the scientific-technological and organisational-managerial knowledge available in the external district environment (Cohen and Levinthal, 1990).

Also, productive best-practices (e.g. leather cutting or sofa upholstering), developed by individuals during long experiential learning processes and embodied in their activities in the form of tacit knowledge can be codified. This allows the knowledge to be easily transferable to a large number of individuals, increasing the organisational productivity[12], as well as the innovation potentiality of the organisation. In fact, since it is expressed by a specific set of information codes (e.g. natural language descriptions, graphs, flow-charts, models, etc.), codified knowledge represents an effective base of discussion and analysis for individuals with either the same specialisation or a different know-how background. This increases the chances of generating new knowledge. Furthermore, by making the knowledge more widespread, it is possible to increase the number of its possible interpretations with a consequent increase in the generative richness of the organisation.

---

knowledge; and finally "internalisation", which is the process of transforming codified knowledge into individual tacit knowledge (Nonaka and Takeuchi, 1995).

[12] A best-practice can be interpreted, looking at a specific technical-operative task, as the best way to carry out the activity related to that task. Of course a best-practice can involve different aspects of the performance such as, for example, the speed and/or quality. For a specific aspect of the performance the identification and codification of the related best-practice, owned by some expert individuals, allow that best-practice to be made easily transferable to a higher number of individuals, increasing the firm performance.

The implementation of codification projects allowed the leader firms of the ID of Matera-Altamura-Santeramo to support reengineering and industrialisation processes. For example, the codification of the know-how at the basis of leather cutting has allowed the development of new technology for the automatic cutting of leather. In this way the firms have improved their productivity. In fact, in the past an expert was not able to cut leather for more than four sofas daily, whereas today he is able to cut leather for about forty sofas. Moreover, the codification of the organisational knowledge allows firms to adopt the knowledge and technologies developed outside the district area.

The codification processes of the district technical-operative knowledge appear as a strategic lever, used by the leader firms to move from an innovation process only based on the DCI (essentially characterised by socialisation processes and linked to the diffusion and increase of tacit knowledge) to more systematic and planned innovation processes.

The adoption of knowledge codification processes involves new individual learning processes in the leader firms. These are the mechanisms of learning by R&D and of learning by combining. The first corresponds to a knowledge creation process based on the traditional research and development activities characterising large firms. In fact, the growth of leader firms allows resources on their internal department of R&D to be invested, with the aim of researching and testing new technologies based essentially on codified knowledge. For example, Natuzzi has recently set up a research centre for testing raw materials and sofa components as well as for looking for new technology aimed at increasing production performance. This centre is connected with other local, national and international research centres and provides the results of its research to all firms in Natuzzi's industrial group. The second learning mechanism consists of the knowledge creation process based on the combination and/or absorption of the codified knowledge developed within the organisation (Cohen and Levinthal, 1990; Kogut and Zander, 1992).

The relational learning processes of the leader firm are modified too. In fact, these are strictly related to the renewal of the inter-firm relationships managed by leader firms. In particular, looking at the relationships with the external district environment the leader firms can adopt different strategies. These can be aimed at improving interaction with the markets, developing commercial competencies which allow the implementation of a pro-active strategy, based on product marketing, market research, and improvement in the network of deliverers and dealers. For example, as far as its dealers are concerned, Natuzzi has developed a network of Italian franchising dealers (Divani&Divani) in order to improve the interface with the market. It offers more services to the sale of products, such as interior design services, and recently furniture accessories. Other strategies of interaction with the external environment aim at acquiring scientific-technological and organisational-managerial knowledge available in the form of patents, trademarks, and technologies. Interaction can be based on partnership relationships with other firms and/or with research and development centres owning strategic know-how for the product and process development. Recognising the external district environment as a rich cognitive context, leader firms are engaged in establishing relationships to acquire new knowledge heritages adding value to district products. The relationships of the leader firms with the external district environment define relational learning processes, which can be labelled learning-by-interacting processes because of their pro-active nature and high intensity (Lipparini and Lorenzoni, 1996).

Within the ID leader firms tend to establish hierarchical relationships with the district firms (Albino *et al.*, 1998a). These are based on a higher exchange of codified knowledge. In fact, codified knowledge allows leader firms to improve their coordination abilities by the clearer definition of the production tasks as well as by the greater control of district firms' operations. This is possible thanks to an increase both in the information exchanged, and in the speed of the knowledge transfer. This also involves a change in the internal cognitive dynamics of the district firms operating with the leader firms. In fact, to increase the interaction effectiveness of the organisational-operative processes, leader firms force their suppliers to implement codification processes of their specific knowledge. However, these codification processes appear not to be completely structured, and fundamentally based on prescriptions provided by leader firms. In particular, they require a quality certification from their suppliers and the adoption of specific production and information-communication technologies. The exchange and sharing of codified knowledge in the inter-firm relationships between leader firms and district firms reduce the role of "learning by inter-firm relationship" mechanisms. Relationships tend to be based on structured co-ordination mechanisms regulated by formalised norms and procedures.

The exchange of codified knowledge in inter-firm relationships involves a relational learning process, based on the abilities of the district firms to create knowledge within each organisational context, to codify it and subsequently to combine it in a synergetic way, creating new knowledge[13].

[13] It is important to point out that the way in which innovation is generated is strictly related to its nature. Referring to the cognitive forms of interaction between the firms involved in the innovation process, it is possible to distinguish two main typologies: the autonomous innovations and those of a systematic nature (Chesbrough and Teece, 1996). The first category

In inter-firm relationships, this learning mechanism can be labelled "learning-by-combining" to point out how the inter-organisational creation of knowledge is mainly based on the integration of codified knowledge (Grant, 1996; Kogut and Zander, 1992).

Knowledge codification processes modify knowledge diffusion processes within an ID. In fact, the codification process makes knowledge easily imitable. Thus, to protect their know-how from competitors, and in particular from those operating in the same district area, leader firms try to control knowledge diffusion processes. This is particularly true in the ID of Matera-Altamura-Santeramo where leader firms protect their know-how.

Reducing the number of firms related to a leader firm, the number of possible knowledge transfer channels is reduced. Furthermore, the leader firm requires dedicated relationships from its suppliers (Albino *et al.*, 1998c). Also, particular forms of contract are adopted with employees. For example, a leader firm in the ID of Matera-Altamura-Santeramo has introduced a particular contract which prevents its employees from leaving the firm to move to another firm operating in the sofa industry, without at least one year's notice during which the employee cannot work any more in the firm even if he/she continues to be paid.

Based on the above analysis, it seems possible to claim that, for those IDs characterised by the development of some leader firms, disintegrating forces, at least from a cognitive point of view, can arise. In fact, the lack of virtuous knowledge diffusion can involve a "suffocation" of the ID, which can take the form of a niche withdrawal in the district's small and medium enterprises which are not related to the leader firms. Also, technological investments made by leader firms create barriers to knowledge imitation with a consequent reduction in the starting up of new firms within the district area.

The growth of leader firms within an ID can involve the modification of the specific characteristics of the integrated system-area as well as the reduction of the importance of the link between the production system and the social-cultural aspect of the local context. However, it is important to point out that external economies continue to represent a fundamental reason for the leader firms to remain localised within the district area. These economies are mainly the result

---

involves those innovations, which can be autonomously developed by a firm, and subsequently integrated with the innovations developed by other firms. In this case the development of an innovation can be based on tacit knowledge, but the end result has to be codified. In particular, the knowledge of interface, i.e. knowledge which enables different firms to communicate, has to be codified. The systemic innovations require an integration of different complementary competencies. In this case, to overcome the problems related to the ambiguity and uncertainty, firms have to operate together mainly on the basis of tacit knowledge.

of the integration, within the limited geographic area, of the district production system and of the local socio-cultural context. In accordance with this idea all those functions related to the local area, such as the infrastructures, the local culture, the language, the collective identity, and the entrepreneurial culture spread in the form of initiatives of innovative nature will become more and more important (Becattini, 1999).

Within the IDs the development of the industry producing technologies to support the production carried out in the district will continue to be particularly important for the creation of external economies based on a subdivision of the "innovative work" (Arora and Gambardella, 1994).

The role of the economical externalities tied to the formation of social links (Becattini, 1999) is also shown by the investments carried out by leader firms in the local area. These seem aimed at supporting those socio-cultural and professional development processes which allow the common base of knowledge and values supporting the identity of the district social structure to be maintained. Then, all those initiatives aimed at creating, for example, museums about the industry technologies, the development of cultural events, the creation and support of technical schools, and the establishment of collaborative relationships with the local institutions, can be interpreted as efforts to increase such a social identity.

19.6 Conclusion

To face new competitive contexts, IDs are modifying their traditional structural and cognitive features. The most important aspect of the development process of IDs is the growth within the district area of actors with a coordination function. In fact, the growth of leader firms seems to be the most important change in ID evolution. Looking for performance improvement, leader firms are restructuring their inter-firm networks and are modifying knowledge creation and diffusion processes.

A framework to analyse these cognitive dynamics is provided and applied to the ID of Matera-Altamura-Santeramo. The analysis of its cognitive dynamics shows that the most important action of the leader firms appears to be the codification projects of the technical-operative knowledge. These result in the improvement of the effectiveness of both district knowledge creation and diffusion processes, and the acquisition of scientific-technological knowledge from the external district environment. In this way leader firms are able to manage the district innovation processes better, improving their own and the district's performance.

The investments in the acquisition of technologies and the formation of hierarchical networks of suppliers centred on the leader firms determine barriers for the spreading of knowledge within the area. Furthermore, the leader firms, particularly when they are engaged in the codification of their strategic know-how, adopt protection mechanisms of the organisational knowledge. In particular, this is carried out by creating dedicated suppliers networks that can involve the disintegration of the district.

Analysing the evolution processes of IDs some questions arise about the future of this economical-productive model. In particular, it is important to understand whether the activities of leader firms, changing the traditional cognitive-relational balances of the district areas, can mark either the end of the district model or its restructuring towards new and more effective organisation.

We believe that a fundamental role will be played by external economies, which will motivate the leader firms to remain localised and to support the growth of the local area. In fact, localisation will continue to be a resource of competitive advantage if it is able to exploit the positive effects of the economical externalities due to the agglomeration of a productive, institutional and social-cultural structure in the area. Furthermore the district's capability to capture external innovation to be implemented in the local area will be particularly important. District firms need to monitor the evolution of technological innovations in order to identify those technologies that can be adopted, improving product and process performance. In this way IDs will be able to exploit internal as well as external knowledge.

References

Albino V. and Garavelli A.C. (1996), "Alcuni aspetti della crescita di un'impresa guida in un distretto industriale del Mezzogiorno", *L'Industria*, no. 4, p. 731-747.

Albino V., Carbonara N. and Schiuma G. (1998°), "Le strategie di sviluppo dei network locali", *Sviluppo & Organizzazione*, no. 170, p. 33-46.

Albino V., Garavelli A.C. and Schiuma G. (1998b), "Relazioni inter-organizzative e gestione della conoscenza nei distretti industriali", *Convegno sulla Subfornitura e Competitività dei Sistemi di Impresa*, 11 December, Caserta.

Albino V., Garavelli A.C. and Schiuma G. (1998c), "Internazionalizzazione dei distretti industriali e codifica della conoscenza: il ruolo dell'impresa leader", in, Albino V. (ed.) *Competitività delle imprese e sviluppo locale*, Quaderni AiIG, Vol. IV, Potenza.

Albino V., Garavelli A.C. and Schiuma G. (1999), "Knowledge Transfer and Inter-Firm Relationship: The role of the Leader Firm", *Technovation*, no. 19, p. 53-63.

Anastasia B. and Corò G. (1996), *Evoluzione di un'economia regionale – Il Nordest dopo il successo*, Nuova dimensione ediciclo, Venezia.

Antonelli C. (1999), *The Microdynamics of Technological Change*, Routledge, London.

Arora A. and Gambardella A. (1994), "The changing technology of technological change: general and abstract knowledge and division of innovative labour", *Research Policy*, vol. 23, no. 5, p. 523-532.

Becattini G. and Rullani E. (1993), "Sistema locale e mercato globale", *Economia e politica industriale*, no. 80, p. 25-48.

Becattini G. (1987), (ed.), *Mercato e forze locali: il distretto industriale*, Il Mulino, Bologna.

Becattini G. (1989), "Riflessioni sul distretto industriale marshalliano come concetto socio-economico", *Stato e mercato*, no. 25, p. 111-128.

Becattini G. (1992), "Concorrenza e cooperazione: la formula italiana", *Il Ponte*, no. 4, p. 69-80.

Becattini G. (1999), "La fioritura della piccola impresa e il ritorno dei distretti industriali", *Economia e politica industriale*, no. 103, p. 5-16.

Bellandi M. and Russo M. (1994), (ed.), *Distretti industriali e cambiamento economico locale*, Rosenberg & Sellier, Turin.

Bellandi M. (1982), "Il distretto industriale in Alfred Marshall", *L'industria*, no. 3, p. 355-375.

Bellandi M. (1989), "Capacità innovativa diffusa e sistemi locali di imprese", in, Becattini G. (ed.), *Modelli locali di sviluppo*, Il Mulino, Bologna.

Brusco S. (1989), "Piccole imprese e distretti industriali", in Brusco S. (ed.), *Piccole imprese e distretti industriali. Una raccolta di saggi*, Rosenberg & Sellier, Turin.

Butera F. (1991), "Gestire la crescita di una impresa inconoscibile: la piccola e media impresa come pacchetto di organizzazioni coesistenti e come nodo di reti organizzative", *L'Industria*, no. 2, p. 319-335.

Camagni R. (1989), "Cambiamento tecnologico, milieu locale e reti di imprese: verso una teoria dinamica dello spazio economico", *Economia e politica industriale*, no. 64, p. 209-236.

Camagni R. (1991), *Innovation Networks. Spatial Perspectives*, Belhaven Press, London.

CERTeT (1995), *Il distretto industriale brianzolo del legno-arredo*, Collana OPES – Unione delle Camere di Commercio della Lombardia, Quaderno 6.

Chesbrough H.W. and Teece D. J. (1996), "When is virtual virtuous?", *Harvard Business Review*, vol. 74, no. 1, p. 65-73.

Cohen W.M. and Levinthal D.A. (1990), "Absorptive capacity: A new perspective on learning and innovation", *Administrative Science Quarterly*, March, vol. 35, p. 128-152.

Cooke P. and Morgan K. (1994), "The Creative Milieu: A Regional Perspective on Innovation", in Dodgson M. and Rothwell R. (ed.), *The handbook of industrial innovation*, Edward Elgar, Brookfield (US).

Corò G. (1997), "Competenze contestuali e regolazione economica", in, Belfanti C.M. and Maccabelli T. (ed.), *Un paradigma per i distretti industriali*, Grafo Edizioni, Brescia.

Daft R.L. and Lengel R.H. (1986), "Organizational information requirements, media richness and structural design", *Management Science*, vol. 32, no. 5, p. 554-571.

Daft R.L. and Weick K.E. (1984), "Toward a model of organizations as interpretation systems", *Academy of Management Review*, no. 9, p. 284-295.

Dagnino G.B. and Albanese M.A. (1997), "Efficienza e produzione di valore nelle reti e nei sistemi di imprese: un'interpretazione in termini di cicli economico-cognitivi", *Economia e politica industriale*, no. 96, p. 55-94.

Davenport T.H. and Prusak L. (1998), *Working Knowledge*, Harvard Business School Press, Boston, Massachusetts.

Dei Ottati G. (1995), *Tra mercato e comunità: aspetti concettuali e ricerche empiriche sul distretto industriale*, Franco Angeli, Milan.

Dogson M. (1993), "Learning, Trust and Technological Collaboration", *Human Relations*, no. 46, p. 77-95.

Fagerberg J. (1995), "User-producer interaction, learning and comparative advantage", *Cambridge Journal of Economics*, vol. 19, no. 1, p. 243-256.

Ferrucci L. (1996), *I processi evolutivi nei sistemi di piccole imprese*, Guerini Scientifica, Milan.

Gandolfi V. (1988), *Aree sistema: internazionalizzazione e reti telematiche*, Franco Angeli, Milan.

Gandolfi V. (1990), "Relazionalità e cooperazione nelle aree-sistema", *Economia e politica industriale*, no. 65, p. 95-117.

Garvin D.A. (1993), "Building a learning organization", *Harvard Business Review*, Vol. 71, no. 4, p. 78-91.

Gottardi G. (1996), "Technology strategies and innovation without R&D", *Journal of Industry Studies*, vol. 3, no. 2, p. 119-134.

Gottardi G. (1997), "I distretti industriali", in U. Bertelè (ed.), *Cambiamento e innovazione: strategie e politiche per le imprese e per le aree sistema*, Quaderni AiIG, Studi e Ricerche, vol. 3.

Grant R.M. (1996), "Prospering in dynamically-competitive environments: organisational capability as knowledge integration", *Organisation Science*, vol. 7, no. 3, p. 375-387.

Håkasson H. (1987), *Industrial Technological Development – A Network Approach*, Croom Helm.

Huber G. (1991), "Organizational Learning: The Contributing Processes and the Literatures", *Organization Science*, vol. 2, no. 1, p. 88-115.

Inkpen A.C. (1996), "Creating Knowledge Through Collaboration", *California Management Review*, vol. 39, no. 1, p. 123-140.

Kim H., (1993), "The Link between Individual and Organizational Learning", *Sloan Management Review*, fall, 37-50.

Kogut B. and Zander U. (1992), "Knowledge of the firm, combinative capabilities and replication of technology", *Organisation Science*, vol. 3, no. 3, p. 383-97.

Kogut B. and Zander U. (1995), "Knowledge and the speed of the transfer and imitation of organizational capabilities: An empirical test", *Organization Science*, vol. 6, no. 1, p. 76-92.

Krogh G. and Vicari S. (1992), "L'approccio autopoietico all'apprendimento strategico sperimentale", *Economia e politica industriale*, no. 74/76, p. 67-87.

Levinthal D. (1996), "Learning and Schumpeterian dynamics", in Dosi G. and Malerba F. (ed.), *Organisation and Strategy in the Evolution of the Enterprise*, McMillan Press.

Lipparini A. and Lorenzoni G. (1996), "Le organizzazioni ad alta intensità relazionale. Riflessioni sui processi di learning by interacting nelle aree ad alta concentrazione di imprese", *L'Industria*, no. 4, p. 817-839.

Lipparini A. (1996), "Architetture relazionali inter-impresa: Promuovere l'innovazione attraverso le reti d'impresa", *Sviluppo & Organizzazione*, no. 153, p. 33-51.

Lipparini A. (1998), "L'apprendimento relazionale", *Sviluppo & Organizzazione*, no. 166, p. 75-89.

Lombardi M. (1994), "L'evoluzione del distretto industriale come sistema informativo: alcuni spunti di riflessione", *L'Industria*, no. 3, p. 523-535.

Lorenzoni G. and Baden-Fuller C. (1995), "Creating a strategic center to manage a web of partners", *California Management Review*, vol. 37, no. 3, p. 146-163.

Lorenzoni G. (1987), "Costellazione di imprese e processi di sviluppo", *Sviluppo & Organizzazione*, no. 102, p. 59-72.

Maillat D., Lecoq B., Nemeti F. and Pfister M. (1995), "Technology district and innovation: the case of the Swiss Jura Arc", *Regional Studies*, vol. 29, no. 3, p. 251-263.

Mariotti S. (1989), "Efficienza dinamica e sistemi di imprese", *Economia e politica industriale*, no. 64, p. 91-124.

Marshall A. (1920), *Principles of Economics*, Macmillan, London.

Maskell P. (1996), "Learning in the village economy of Denmark. The role of instituitions and policy in sustaining competitiveness", *DRUID working paper*, no. 96-6.

McGill M.E., Slocum J.W. and Lei D. (1992), "Management Practices in Learning Organizations", *Organizational Dynamics*, vol. 21, p. 5-17.

Nelson R.R. and Winter S.G. (1982), *An Evolutionary Theory of Economic Change*, Bellknap, Cambridge (Mass.).

Nevis E.C., DiBella A.J. and Gould J.M. (1995), "Understanding Organisations as Learning Systems", *Sloan Management Review*, vol. 32, no.2, p. 75-85.

Nishiguchi T. (1994), *Strategic industrial sourcing. The Japanese Adavantage*, Oxford University Press, New York.

Nonaka I. and Takeuchi H. (1995), *The Knowledge-creating Company: How Japanese Companies Create the Dynamics of Innovation*, Oxford University Press, Oxford.

Nonaka I. (1991), "The Knowledge-Creating Company", *Harvard Business Review*, Vol. 69, no. 6, p. 96-104.

Nonaka I. (1994), "A Dynamic Theory of Organizational Knowledge Creation", *Organization Science*, vol. 5, no. 1, p. 14-37.

Nuti F. (1992), *I distretti dell'industria manifatturiera in Italia*, Franco Angeli, Milan.

Pilotti L. (1998), "I distretti innovativi del Nordest", *Sviluppo & Organizzazione*, no. 167, p. 15-32.

Piore M.J. and Sabel C. (1984), *The Second Industrial Divide. Possibilities for Prosperity*, Basic Books, New York.

Polanyi M., (1966), *The Tacit Dimension*, Garden City (N.Y.), Doubleday.

Powell W.W., Koput K.W. and Smith-Doerr L. (1996), "Interorganisational Collaboration and the Locus of Innovation: Networks of Learning in Biotechnology", *Administrative Science Quarterly*, no. 41, p. 116-145.

Rosenberg N. (1982), *Inside the Black Box: Technology and Economics*, Cambridge University Press.

Rullani E. (1994), "Il valore della conoscenza", *Economia e politica industriale*, no. 82, 47-73.

Sabel C. F. (1989), "Flexible Specialisation and the Re-emerge of Regional Economies", in, Hirst P. and Zaitlin J. (ed.), *Reversing Industrial Decline? Industrial Structure and Policy in Britain and her Competitors*, Berg, Oxford.

Sanchez R. and Mahoney J.T. (1996), "Modularity, flexibility, and knowledge management in product and organisation design", *Strategic Management Journal*, vol. 17, p. 63-76.

Schiuma G. (2000), "Dinamiche Cognitive dei Distretti Industriali in Evoluzione", *Economia e Politica Industriale*, no. 106, p. 105-138.

Schiuma G. (1998), "Knowledge codification in Industrial Districts", *Proceeding of the Workshop on SMEs and Districts: Hybrid governance forms, knowledge creation and technology transfer*, 5-7 November, Castellanza.

Silvestrelli S. (1984), "L'impianto", in, Rispoli M. (ed.), *L'impresa industriale*, Il Mulino, Bologna.

Tsoukas H. (1996), "The Firm as a Distributed Knowledge System: A Constructionist Approach", *Strategic Management Journal*, Special Issue, vol. 17, p. 11-25.

Unioncamere (1995), Imprese e istituzioni nei distretti industriali che cambiano, Franco Angeli, Milan.

Vaccà S. (1986), "L'economia delle relazioni tra imprese: dall'espansione dimensionale allo sviluppo per reti esterne", Economia e politica industriale, no. 51, p. 3-41.

Vaccà S. (1995a), "Impresa locale distrettuale e transnazionale", Economia e politica industriale, no. 86, p. 53-68.

Vaccà S. (1995b), L'impresa transnazionale tra passato e futuro, Angeli, Milan.

Vaccà S., Torrisi S. and Zanfei A. (1997), "Cambiamento tecnologico, dinamica industriale e impresa transnazionale", *Economia e politica industriale*, no. 94, p. 239-254.

Varaldo R. (1994), "La natura e la dinamica dell'impresa distrettuale", in, Dioguardi G. (ed.), Sistemi di imprese. Le nuove configurazioni dell'impresa e dei mercati, Etaslibri, Milan.

Vicari S. (1991), L'impresa vivente – Itinerario in una diversa concezione, Etaslibri, Milan.

Chapter 20

The chair manufacturing district of Manzano: evolutionary processes and the role of the institutions

Roberto Grandinetti

20.1 Introduction

Over time the chair district of Manzano has taken on all the distinguishing features of a typical Italian industrial district: a widespread closely-knit production network, large-scale labour division among small and medium-sized firms that control various stages of the value chain, a strong link between production and the local community, a plentiful supply of skilled labour, a wealth of technical knowledge and competence, a wide range of products within the district's sector of operation and export-driven production (Becattini, 1987).

Like other districts in Italy, in the past the chair district of Manzano also operated as a closed local network whose only point of contact with the outside was at the extremities of the production chain: raw materials supply, capital goods acquisition and the sale of finished goods in the final market. In general, up until the 1980s the level of involvement of the district in the markets was low and, therefore, local firms were barely able to monitor and control them.

However, this closure to the external environment did not damage the competitive advantage of industrial districts. There was considerable development and the chair district was able to increase its share in the world chair market. Indeed, the district managed to generate most of the resources required for development independently.

In the early 1980s, after a particularly strong economic crisis, pressure for change started to grow (Tamisari, 2000), and as global competition increased the pressure became stronger. Today the chair district is dynamic. Many district firms have adopted very different strategies from those of the closed district network. They have become leaders in the local context. A similar trend was observed in the nearby furniture district known as Livenza, situated on the border between the provinces of Pordenone and Treviso, as well as in other districts in N.E. Italy (Corò and Grandinetti, 1999).

This study will focus on the evolutionary processes that have affected the district in its recent history and that are still operating today. As well as looking at statistics, we have also based our findings on the following recent empirical works:

1. a broad study conducted in 1998 in a large number of firms, commissioned by The Chamber of Commerce of Udine (Grandinetti, 1999)[1];
2. a survey commissioned by the Area Science Park of Trieste and carried out in 2000, which specifically aimed to produce an in-depth analysis of production chain relations by observing a limited number of firms (De Toni, Grandinetti, 2001);
3. several case studies of district firms and institutions included in final dissertations produced by Economics students at the University of Udine.

20.2 Firm population

Regional law no. 27/1999 established that the chair district covered an area comprising 11 municipalities in the Province of Udine: Aiello del Friuli, Buttrio, Chiopris-Viscone, Corno di Rosazzo, Manzano, Moimacco, Pavia di Udine, Premariacco, S. Giovanni al Natisone. S. Vito al Torre and Trivignano Udinese. In 1996 the intermediate industrial census recorded 1,069 local production units in the area and a slightly lower number of firms. These units either produced finished goods or operated in the intermediate stages of the production chain, employing a total of 10,976 people. Other production units not included in the group above supply firms with materials (adhesives, paints and small metal parts) and capital goods and services. If we compare the 1996 census with the 1991 census and the firms registered at the Chamber of Commerce in 2001, we can see that the number of district firms has remained stable (Table 1)[2].

[1] The firms in the universe, which were solely corporations, were divided into 2 groups according to turnover size (above and below 10 billion Italian lira based on 1996 financial results). The first group represented the "smaller" firms in the district and a sample of 113 firms was constructed. The second group represented the "larger" firms and the 31 firms out of 39 to which it was possible to submit a questionnaire were all surveyed.

[2] It would not be appropriate to take the comparison further since the data sources (ISTAT and Infocamere) are very different and the information regarding the number of employees from the latter source is not reliable.

From the 1998 study mentioned above we can extract some interesting information regarding the demography of district development. If we divide the firms according to their age we will find that smaller firms are younger than larger ones. This is to be expected, given the usual correlation between size and stage in the life-cycle of a firm. Most firms with a turnover of under 10 billion lira (almost the entire population of the district) were founded in the 1980s, whereas almost the same number of firms were founded in the 1990s as in the 1970s. The youngest production units operate inside the production chain. In fact, if we compare firms according to production type, we can see that 62.8% of firms founded in the 1980s or later are involved in processing, 67.8% make components, 72.7% manufacture finished products for other firms but only 35.7% produce finished goods for the final market.

Table 1: Local production units and employees in the chair district
Source: ISTAT (1991 and 1996), Infocamere (2001).

		Wood		Furniture		Total
	L.P.U	Empl.	L.P.U	Empl.	L.P.U	Empl.
1991	319	2,134	703	8,054	1,022	10,188
1996	261	2,023	808	8,953	1,069	10,976
2001	319	2,226	970	8,458	1,289	10,684

The high number of young firms and the fact that the population of district firms has remained virtually unchanged since the expansion of the 1970s are evidence of a high demographic turnover in the chair district. Indeed, a high firm birth/mortality rate is a typical feature of Italian industrial districts. In other words, thanks to employees choosing or being forced to set up their own businesses after firms have closed down, districts are capable of generating new entrepreneurship. The low average age of firms operating in the intermediate stages of the production cycle, and especially of those that manufacture finished products for other firms, is evidence that enterprise creation mechanisms are working since these kinds of businesses are usually started by former employees.

Another noteworthy feature is the large number of family businesses. The vast majority of firms are small firms in which there is more than one entrepreneur and in 78.8% of these cases the entrepreneurs are related. In larger firms the figure is only slightly lower. In the socio-economic system of the district there is a strong drive to set up one's own business and it is often former employees that do so, together with relatives and acquaintances. This enterprise creation impulse is based on the shared values that structure the collective identity of the district (Rullani, 1995), on widespread social legitimisation and on extensive social circuits that encourage business culture[3].

20.3 The division of labour among district firms: the district network

One of the main features of the district is a high degree of vertical division of labour among specialised firms in the various stages of production. Only a quarter of the small firms (24.8%) mainly manufacture finished goods for the final market. In fact most of these firms operate inside the production chain: 38.1% do contract work for other firms and 27.4% produce partly-finished goods or components. The remaining 9.7% produce finished goods for other firms[4]. In the latter case firm business could be said to be almost inside the district production. In fact, the turnover of these firms is mainly absorbed by the market inside the district.

The figures regarding larger firms are the exact opposite of those regarding smaller firms: 80.7% mainly manufacture finished goods for the final market and 16.1% produce partly-finished goods or components, whereas the other two forms of output are almost non-existent. The fact that larger firms are oriented towards the final market is to be expected. However, a number of subcontractors have also become quite large.

It is useful to compare these figures with the situation at the start of the 1980s. We can see that the high degree of vertical labour division that characterises the district today has been constant over the past 20 years. In fact, a survey carried out in 1981 by the "Ente per lo Sviluppo dell'Artigianato del Friuli-Venezia Giulia" (a local body that supports the development of the crafts) on a similar sample of firms to that of the universe in question showed that: 29.0% of firms manufactured finished chairs or other furnishings (armchairs, tables, beds, magazine racks and clothes hangers); 23.3% produced partly-finished goods or components (including partly-finished chairs) and 47.7% carried out specialist work. Bearing in mind that this survey also included

[3] We should, however, also mention the negative effects of family ownership. Due to the high number of entrepreneurs and family-run firms, ownership structures are small and share capital is insufficient. This can be detrimental to firm evolution if the family firm is unable to grow or face up to the strategic complexity of the markets because it is not prepared to take on new business partners and change the firm power structure.

[4] Bearing in mind that the survey only studied district firms that were partnerships or joint-stock companies, if we were to add individual concerns, the weight of firms operating in the intermediate stages of the production chain would be even greater.

individual concerns and these firms operate prevalently in intermediate production, we can reasonably estimate that the relationship between the production of finished goods and the production of intermediate goods in the district has remained largely unchanged. There was, however, an adjustment in subcontract work (from processing to component manufacture) after certain traditional processes such as straw and reed covering started to decline.

The greatest change, however, involved finished-goods manufacturers. A survey carried out in the late 1970s identified the following firm types in the chair district (Grandinetti, 1979):

1. small firms of craftsmen producing semi-finished goods (cut or squared timber) or carrying out specialist processing (turning, bending, smoothing, carving, painting and upholstering) on contract for local chair manufacturers, on which they are almost totally dependent;
2. slightly larger firms of craftsmen that purchase semi-finished goods, complete the work (assembly and finishing) and sell the finished product;
3. small industrial firms (between 20 and 100 employees) that are the mainstay of the district and complete almost the entire production process alone, except for specific processing that is contracted out to craftsmen and home-workers;
4. a few medium-sized firms that, though capable of carrying out all the stages of the production cycle, in their growth period tend to decentralise some processing, which then returns to the firm in times of difficulty.

The main buyers of chairs manufactured in the Manzano district were furniture businesses located in the region, in other Italian industrial districts and abroad, above all Germany.

Today the district has changed. As we have seen, the number of firms producing finished goods for the final market or other manufacturers is still the same. However, on average they pass work on more systematically to other firms inside the district that are specialised in various stages of the production. Moreover, the number of firms that operate independently in the market has overtaken those that remain in the shadow of the furniture producers. Today, the latter mainly work for other district manufacturers.

The second empirical study we have referred to in this paper will help us to analyse production chain relations.

One of the most important features is the great variety of entrepreneurial formulae that exists in district subcontracting. Alongside numerous firms working in specific stages of the production chain, there is an even greater number of firms that carry out a wide range of activities. The technological characteristics of the production cycle not only permit its breakdown into separate activities and therefore the sharing of work among firms, they also permit activities to be combined flexibly in various ways. Consequently, the district firms can specialise in a particular kind of work if the economies of specialisation and the technological economies of scale in a particular production stage are significant. Alternatively, they can combine different activities when the costs of production, logistics and co-ordination, deriving from separate production and a high degree of division of labour, are greater than those incurred by a single firm doing the same work.

In the latter case district firms should be able to take full advantage of the modular nature of the production cycle. In fact this does not always happen. However, there are many examples of firms that have succeeded in setting up entrepreneurial formulae that have been able to exploit the potential advantages of combining activities.

Another feature regards relations with buyer firms. It may be advantageous for the principal to reduce the number of interfaces it has with the network of specialist subcontractors. It will therefore attempt to select a limited number of permanent suppliers, which in turn will either carry out various activities themselves or co-ordinate the work of other firms by referring to the vast supply of specialist district firms available.

Despite the vastly heterogeneous nature of the make-up of district firms, it is possible to outline recurring firm types and entrepreneurial formulae.

The first type regards firms that carry out a single activity in the production chain. The most highly specialised work is the following: upholstering, smoothing and painting. Other kinds of work are also specialised, especially bending. Less frequently, there are suppliers that are specialised in semi-finished goods and components.

The second type, homogeneous integrated supply, includes firms that carry out various fundamental production chain activities i.e. by using machinery to produce semi-finished goods (planing, profiling, turning etc.). The choice of this kind of subcontractor can be strategic rather than technological. Their competitive advantage may be based on their ability to offer their customers a specific product quality, short delivery times, the skills required to meet a specific order; in short, a more complete, higher quality service. A variation on this model regards the supplier that offers a complete service to the principal, but that actually contracts out a large part of the work to other firms, which it co-ordinates.

The third type, heterogeneous integrated supply, seems to be the product of a specific path (not necessarily evolutionary). This is the case of suppliers that take on additional specialised activities so as to meet demand coming from inside the district or, on the contrary, in order to survive when there is a crisis in the market.

The last type regards firms that sell finished goods to other manufacturers who include them in their product range. These suppliers in turn make use of other producers. The large variety of products in the ranges of many manufacturers in the chair district is often the result of the purchase of finished products. In other cases, the wide-scale purchase of finished products is due to the fact that the firm is more oriented towards sales than production.

On the basis of this categorisation we can attempt to describe the main network relations.

In the district system a first-level group of suppliers can be easily identified. These are mainly firms belonging to the homogeneous integrated supply category. Though working as subcontractors, these firms often contract out work to other subcontractors in order to supply the principal with a complete service: parts, components, assembled parts and unfinished chairs. For example a firm producing semi-finished parts (first-level supply) may carry out machine processing inside the firm, but contract out the smoothing process (second-level processing supply) to another firm so that it can provide the principal with a part that is ready to assemble. Below second-level supply we can sometimes find other kinds of subcontractors that carry out highly labour-intensive work, such as sanders who prepare the parts for painters.

At a slightly more advanced level than first-level subcontractors we can find suppliers of finished products. At times they may occupy a position that is higher than the others in the hierarchy of supply levels or, alternatively, they may constitute an independent model as a result of the choices of the principal and the suppliers that make the finished product for them.

Therefore, inside the district there are supply networks that are structured on different levels in a similar way to the models that have recently developed in other contexts in the wake of the Japanese example.

However, in reality the district supply network is more complex, varied and extensive than a multi-level hierarchy. In addition to working for first-level suppliers, often second-level subcontractors also work directly for the principal, e.g. they smooth semi-finished parts that the principal has given them. In some cases the processing subcontractors may themselves contract out work to other firms. Some component manufacturers may be specialists in a particular kind of processing and not contract out any work at all.

Finally, it is important to verify if any of the supply types have been deliberately chosen by firms as a path towards competitive evolution in respect to subordinate supply.

It must be said that "subordinate" suppliers have very little independence in their relations with the principal and there is a very low level of complexity to govern: the latter firm fixes the order and the former firm carries it out. Today, this is the simplest of supply configurations, whereas in the 1970s it was the predominant formula, not only in the chair district of Manzano, but also throughout Italy. Currently, there is a composite universe in which a modern supply type has taken over from the very different subordinate supply type. Today the supplier can face the competitive environment and operate in the market independently.

A number of empirical studies have traced the many evolutionary paths taken by supply firms. In Italy the variety of paths described by scholars can follow three main axes:

1. the gradual diversification of customers and market outlets;
2. the shift in core competences from production to outlet planning, supply chain management and business to business marketing;
3. the movement away from traditional operation methods based on tacit knowledge and informal relations and processes towards a model that allows for a higher level of codified knowledge, relations and processes, such as when a firm takes a path to quality certification.

Following the above considerations it is easier for us to examine the evolutionary potential associated with supply formulae adopted in the chair district.

If we exclude the third supply type because of its excessively composite nature, in the other three formulae there are cases in which the subordinate supply configuration has been discarded. However, there are differences between the evolutionary paths if we consider them according to the three axes described above.

If highly specialised firms and suppliers producing finished goods extend their customer portfolio outside the district, this may be a strategic option for their evolution. On the other hand, homogeneous integrated supply tends to be more strongly bound to the district market. Firms adopting extreme specialisation may be rewarded if they also have specific competences allowing them to hold a sustainable position in a global niche.

The development of competences in planning, organisation and interactive marketing can lead to competitive evolution in the case of homogeneous integrated suppliers and to a greater extent for suppliers of finished goods. Competitive advantage is based on the ability to manage a network of second-level suppliers and complex interactions with business customers.

The same could also be true as regards the codification of knowledge, relations and processes, due to the nature of the cognitive base (which is usually codified) supporting competences that describe the previous evolutionary axis. In the case of specialist suppliers, the development of tacit knowledge through learning by doing processes and socialisation mechanisms maintains much of its importance. If we also consider the codification of knowledge deriving from the in-company development of technology and machinery, which is quite common in the various stages of the production chain, the situation appears to be more complex. Moreover, inside the district as well as an intra-firm dimension there is also a district dimension that adds to the classic dichotomy between tacit/explicit knowledge and the mechanisms that convert one form into the other (Rullani, 1994; Belussi and Gottardi, 2000).

In conclusion, an analysis of vertical labour division inside the district and the entrepreneurial formulae adopted by suppliers clearly shows a high level of complexity in the network of district relations. This systemic complexity may be considered one of the fundamental factors that permit the district to generate value and produce sustainable competitive advantage. We must underline that it is very difficult to replicate this factor outside the district. Currently, the reproduction of systemic quality in the district, together with the dynamism of leader-firms that are beginning to take on the structure of global corporations (which will be described in the next section), have helped the chair district to face the challenges of a global economy.

20.4 The district's level of internationalisation

A further distinguishing feature of the chair district in Friuli is the high level of exports. Chamber of Commerce figures show that in 87.1% of firms with a turnover of more than 10 billion Lira more than half of the total sales come from foreign markets. Moreover, the 50% figure is reached by all the firms producing mainly finished goods for the final market, and a cluster of firms manufacture exclusively for the export market.

Naturally the statistics for smaller firms are rather different, since they are mainly involved in subcontracting. Overall the percentage of small firms operating abroad is 40.7%, but if we examine the figures according to output type we can see the following results:

- 92.9% of manufacturers of finished goods have some form of contact with the international market, a figure that is similar to that of large firms. However, the percentage of firms with a strong export trend is lower;

- on the contrary, in 93.0% of firms that carry out processing work under contract, sales are achieved on the domestic market, i.e. inside the district itself;
- most manufacturers of semi-finished goods and components (64.5%) do not export. However, the number of suppliers that also supply foreign customers with intermediate goods has become quite significant. In this way, they have gained some independence from the chair district market or even other Italian furniture districts. The export of semi-finished goods and components is not a novelty in the district, but over the years the suppliers' level of internationalisation has grown and is continuing to grow.

In addition to export sales, a further indication of a firm's level of internationalisation is related to the number of different countries it sells to. In general, larger firms operate in the markets of a wide range of countries; more than five in 75.9% of companies. Smaller exporters limit the number of their markets so as to reduce the amount of complexity they have to manage. However, on a systemic level, district products reach a highly widespread and diversified range of outlet markets. Germany is by far the largest outlet for district exports, followed by Austria, France, the United Kingdom and Holland for the smaller firms, and France, Austria, the USA, the United Kingdom, Switzerland and Japan in the case of the larger ones. It is important to note that the USA and Japan, whose cultures are very different from that of the district, have become consolidated export destinations, especially for larger firms due to their more structured approach to internationalisation.

The changes described above show that the chair district in Friuli has made great progress in selling its products. In many Italian districts links with the final market have traditionally resulted in a structural dependence on external customers and intermediaries. The case of the Manzano chair district was emblematic, since the chronic weakness that threatened to lead the area to decline in the second half of the 1980s was due to a general lack of market orientation and a dependence on external trade (Mattioni, 1992). By the 1990s, however, the situation was very different.

Indeed, the level of district internationalisation through exports has grown and is continuing to grow. This process has also involved some intermediate stages of the district value chain.

In addition, the outlets cover a much wider geographic area: firms either specialised in specific locations or extended their customer-base to other countries.

Furthermore, there has been an increase in the number of firms that approach a market independently, rather than simply include their product in the product range of other furniture manufacturers. If during the initial stage this re-orientation has only meant firm dependence has shifted from furniture manufacturers to sales intermediaries, it has, however, set down a base on which entry strategies and greater firm involvement in distribution can be developed. As a result, market information can be controlled.

20.5 The competitive positioning of district firms

To continue our analysis of the changes taking place in the chair manufacturing system, we should discuss the competitive positioning of district firms in relation to their competitors. In order to determine the main strengths and weaknesses of local firms in our empirical study, we have taken a wide range of factors into consideration. The firms carried out the evaluation themselves using a scale of 1 to 5.

An important result is that the score given to production costs is lower than the average of the scores of all the other competitive factors. This is true both for firms with a turnover of more than 10 billion lira (3.20 against 3.60) and for those with a turnover under that figure (3.29 against 3.51).

We can compare this result expressing weakness on the costs side with scores for other factors of competition that are higher than the general average both for large and small firms: flexible production (3.96 and 3.93), intrinsic product quality (3.99 and 3.81), a large, diversified product range (3.98 and 3.84) and finally customer service as a factor for district firm excellence (4.26 and 4.00).

These data may lead to a long-term analysis showing a fundamental change in the district competitive advantage structure. In the past, the competitive advantage structure supporting the development of the chair district was based on two factors:

1. firms were able to keep their production costs low;
2. firms were able to react to changes in the quality and quantity of demand.

These factors were not very different from those that supported the development of industry in Friuli in general (Grandinetti, 1998). However, it must be said that in the district there were strong economies thanks to the division of labour among firms and to the "collective" accumulation of useful knowledge.

Today the structure of district competitive advantage is no longer supported by a differentiation in production costs. This is due both to exogenous factors (the fact that countries with lower labour costs that have recently industrialised or re-industrialised are operating on international markets) and endogenous factors (changes in the local environment that limit an excessively flexible use of the workforce). On the other hand, flexible production has become an advantage factor bringing a complex web of variables into play: human resource flexibility, production organisation, acquired processing and product knowledge and the technology available. The novelty is that the integration of flexible production has resulted from the strengthening of product and service factors: intrinsic quality, the supply of a wide range of products, customer service (reliability, delivery times, flexible relations). It is in this area that the district has made most progress. On the one hand, progress has resulted from the repositioning of many local firms in the middle and middle to high segments of the market in terms of price and quality, whereas in the past firms focused on the lower level segments. On the other hand, suppliers have tended to follow the evolutionary trajectories of leader-firms: even firms that do not produce goods for the final market gave themselves high scores as regards the four factors of competition mentioned above.

Chair design has also seen many changes in the district. Many firms, large and small, producing finished products for the final market have realised that this factor is an opportunity for them to differentiate, as well as being an important source of competitive advantage.

It must be stated, however, that the district has not yet progressed from the concept of product quality to the concept of total quality. Currently, few companies have applied for and obtained ISO 9000 quality certification.

The factors for competitive advantage in which firm scores are much lower than the general average are marketing communication and the control of distribution channels, even if the difference is much smaller in the case of large firms. Therefore, the critical point for district firms is related to marketing factors. In effect, the development of a stronger market orientation is the greatest challenge that the chair district must face in the current competitive scenario. However, up until now only a minority of district firms has started to meet this challenge, even among those that sell mainly to the final markets.

One last aspect that should be mentioned regards the diffusion of communication and information technology in the district. In this area the chair district exemplifies the concept of "network without network technology". Indeed, today district firms still use informal procedures and traditional methods of communication to manage internal relations and those with the supply chain and with the market. Here, too, there is a significant difference between large and small firms and, moreover, the situation is evolving rapidly[5].

20.6 Evolutionary trends in firm approach to the markets

In order to study marketing policy more carefully, we will focus on manufacturers of finished goods for the final market or other manufacturing firms.

We have mentioned both the limitations that still affect local firms in this field and, on the contrary, the progress made in the district as regards the sale of goods, which is reflected in the steady increase in the quantity of exports and the number of countries that goods are sold in.

From our analysis of firm behaviour, both as regards the traditional components of the marketing mix (product, place, promotion and price) and the positioning of products in the various demand segments, a general weakness is confirmed, as well as the presence of some evolutionary processes.

First of all, the district product range has expanded. This is partly due to the fact that firms produce different types of similar or complementary goods. Another reason is that some firms have specialised in specific production types as an alternative to chair production, which is still, however, the main product manufactured both as regards small firms (76.7%) and large ones (88.5%). Nonetheless, local firms are also manufacturing other products: stools, benches, tables, upholstered goods and other furnishings. These goods are either included in the product ranges of large firms or are produced by small specialist firms.

Our analysis has also shown that the number of models and alternatives has increased and that the product range, both for single firms and for the district in general, covers more segments of the market. An exception to this trend regards market segmentation according to quality and price. Following changes in the competitive environment, a drastic repositioning process has taken place and district goods have gradually moved towards the middle and middle to high segments of the market. The shift was not only due to a change in customer

[5] This picture does not only apply to the chair district. In general, Italian industrial districts still operate as "networks without technology". There are intense relations among firms that are immersed in a specific geographical context with barely any support from advanced network technology (Micelli and Di Maria, 2000). This situation results from lacks in the supply and demand of technology. On the one hand, the management model in local systems of small firms systematically under-estimates communication costs and, on the other, the existing learning and work culture finds it difficult to create a "structural link" with network technology.

preferences, but also to the fact that since the 1980s the lower segment of the final market had been penetrated by goods from developing countries, resulting in the erosion of profit margins on district products.

Finally, we should also mention that distribution channels have evolved. Today firms are more involved and therefore there is more control over market information. In addition, there has been marketing channel diversification both as regards customer type and the people or organisations that carry out sales. Although direct sales through company salesmen or even the owner himself are still the norm, more advanced methods have been introduced, especially in foreign markets. While in the case of small firms the second most common way is through sales agents representing a number of different firms, for large firms the use of this kind of agents is followed by direct investment in sales offices abroad. If we consider that in the past only companies with the size of multinationals would have been able to be control foreign markets by direct investment, this trend is highly significant. In effect, the title of "small multinationals" used in literature to describe some district leader firms does not appear entirely inappropriate.

Therefore, district firms have shown that they are capable of dealing with higher levels of complexity in comparison to the very simple production-oriented approach that prevailed in the past. Evidence shows that the behaviour of larger firms is more advanced. Smaller firms still tend to use the more traditional approach, which we can summarise as passive selling or passive exporting in the case of sales abroad. However, we have observed that, in some smaller districts, firms' behaviour is oriented towards greater market involvement and control. Nonetheless, to face up to an acceleration in market globalisation it is imperative for the district to improve marketing substantially. It is no surprise that both large and small district firms perceive themselves to be at a disadvantage, when compared to their competitors as concerns the factors of distribution channel control and marketing communication.

20.7 Change phenomenology: leader firms, niche firms, global suppliers and the institutions

From our study so far the chair district in Friuli can be described as an evolving production system. There are many signals that change is taking place.

The leader-firms have chosen different entrepreneurial formulae and evolutionary paths. For example, "Gruppo Lovato" has grown in foreign markets independently and today, by means of direct investment in supply, assembly and sale to the final markets, it has taken on a typically transnational structure. While maintaining a strong presence in the domestic market, "Calligaris" has increased its export volumes substantially by diversifying its product range, controlling distribution channels and improving customer communication. "ID Export" has set up a co-ordinating and marketing service for a group of small finished goods manufacturers for whom, otherwise, it would have been very difficult to control market outlets independently. Despite the wide variety of formulae and paths chosen, it is still possible to identify some common firm strategies linked to the international projection of their value chain.

Although using different approaches to the market, all leader firms have diversified their markets as regards geographical location and strengthened/structured strategic and operational marketing competence.

Another common trait concerns relations with suppliers of key services such as technology innovation, information technology, market research, product design, and management consultancy. In these fields leader-firms usually deal with service companies outside the district.

The final path that the firm value chain has taken, which has had an impact on the system of firm relations inside the district, is to delocalise supply work at the initial stages of the production chain to Eastern European countries. The competitive advantage of this choice derives from low labour costs and proximity to raw materials sources.

However, the evolutionary potential of the district is not limited to leader firms, but also involves several small firms that operate on the final market. Some of them have repositioned their production towards high-level market segments thanks to quality materials, product innovation and advanced design. Others have adopted highly sustainable niche strategies based on products designed for specific kinds of use. Others still have specialised in specific country markets. In addition to traditional sales and promotion consortiums sponsored by state institutions, alternative forms of horizontal co-operation among small manufacturers have been set up. The best example is the case of "Grup Sedia". This company was founded by three small manufacturers so that they could promote sales of their own products together.

As we have mentioned above, district supply firms are not static either. Several suppliers have used the knowledge and competences they have accumulated in a specific stage of the production chain to search for customers outside the district and, in turn, become international firms. Learning through relations with leader firms has had an important role in this evolutionary process.

Lastly, we should examine the behaviour of the institutions inside the district. In the 1990s some of these institutions had an active role in supporting the local production system. Municipal Councils opened up a debate on what industrial policies should be adopted to remove chronic structural deficiencies and promote the district's competitive evolution. In 1999 their role was fundamental in convincing the Regional Assembly to pass a vital law for the development of industrial districts.

The IPSIA of San Giovanni al Natisone, a school specialised in technical subjects, has renovated itself and brought its training up to date so as to meet the needs of the local production system.

A local district bank, "Banca di Credito Cooperativo di Manzano", has recently set up a company with leader firms, "Rete Dis", to support the development of I.C.T. and network services in the district (Miorin, 2000).

"Promosedia" is a company based in Manzano whose share capital is made up of the Chamber of Commerce in Udine and numerous chair manufacturers. Its role is to organise the annual chair exhibition (Salone Internazionale della Sedia), but it also carries out collective marketing.

Finally, the "CATAS" in San Giovanni al Natisone supports district firms in product quality and its certification[6]. The "CATAS" acts as a cognitive interface between local context and global economy. It is part of a group of service organisations, which have also been observed in other industrial districts (Becattini and Rullani, 1993), supplying specialist value chain services that district firms do not have the means to invest in. The organisations provide the district with the knowledge and competences they have accumulated, even through international links. To a certain extent, they create specific channels so that they can interact with the outside environment "on behalf of" district firms.

In Italian industrial districts, the role of local institutions has been important for two reasons (Dei Ottati, 1995): they regulate competition and they provide fundamental public goods (infrastructures and collective services). However, in the latter case, the role of the institutions is very different in each local context. For example, in the foothills of the Alps, from Lumezzane in Valgobbia to the chair district of Manzano, firms have complained about the inadequacy of the local road system and the connections with the main motorways.

In the chair district, apart from the access problem, we have seen that several institutions have given the local production system effective support, and this is important. Indeed, the removal of the evolutionary inertia that still exists in

[6] In 2000, 18,000 tests were carried out for 1,200 industrial clients, 835 of which were regular clients, inside and outside the district.

Manzano and other Italian industrial districts also depends on the ability of local institutions to plan the competitive evolution of the district they operate in (Albertini and Pilotti, 1996; Pilotti, 1998; Rullani, 1998).

References

Albertini S. and Pilotti L. (1996), *Reti di reti. Apprendimento, comunicazione e cooperazione nel Nordest*, Cedam, Padua.

Becattini G. (1987), (ed.), *Mercato e forze locali: il distretto industriale*, Il Mulino, Bologna.

Becattini G. and Rullani E. (1993), "Sistema locale e mercato globale", *Economia e Politica Industriale*, no. 80, p. 25-48.

Belussi F. and Gottardi G. (2000), (ed.), *Evolutionary Patterns of Local Industrial Systems. Towards a Cognitive Approach to the Industrial District*, Ashgate, Aldershot.

Corò G. and Grandinetti R. (1999), "Evolutionary patterns of Italian industrial districts", *Human Systems Management*, vol. 18, no. 2, p. 117-129.

De Toni A. and Grandinetti R. (2001), (eds.), *Conoscenze, relazioni e tecnologie di rete nelle filiere distrettuali. Il caso del distretto della sedia*, Franco Angeli, Milan.

Grandinetti R. (1998), "L'evoluzione del sistema industriale dell'area friulana: vantaggi competitivi delle imprese e dell'ambiente locale", *Economia e Società Regionale-Oltre il Ponte*, no. 2, p. 5-27.

Grandinetti R. (1999), (ed.), *Il seggiolaio e l'economia globale. La transizione evolutiva del distretto friulano della sedia attraverso i risultati di una indagine sul campo*, Cedam, Padua.

Grandinetti P. and Grandinetti R. (1979), *Il caso Friuli. Arretratezza o sviluppo?*, Il Campo, Udine.

Mattioni F. (1992), "L'area friulana della sedia: un distretto a rischio", *Economia e Società Regionale-Oltre il Ponte*, no. 40, p. 153-187.

Micelli S. and Di Maria E. (2000), (ed.), Distretti industriali e tecnologie di rete: progettare la convergenza, Franco Angeli, Milan.

Miorin T. (2000), "Strategie per una nuova logistica di distretto", *Economia e Società Regionale-Oltre il Ponte*, no. 1, p. 17-41.

Pilotti L. (1998), "I distretti innovativi del Nordest", *Sviluppo & Organizzazione*, no. 167, p. 15-32.

Rullani E. (1994), "Il valore della conoscenza", *Economia e Politica Industriale*, no. 82, p. 47-73.

Rullani E. (1995), "Distretti industriali ed economia globale", *Economia e Società Regionale-Oltre il Ponte*, no. 50, p. 5-61.

Rullani E. (1998), "Riforma delle istituzioni e sviluppo locale", *Sviluppo Locale*, no. 8, p. 5-46.

Tamisari M. (2000), "The evolution of the chair-manufacturing industry in Friuli-Venezia Giulia", in, Belussi F. and Gottardi G. (eds.), *Evolutionary Patterns of Local Industrial Systems. Towards a Cognitive Approach to the Industrial District*, Aldershot, Ashgate

Chapter 21

The role of academic spin-offs in connecting local knowledge

Guido Capaldo, Luca Iandoli, Mario Raffa, and Giuseppe Zollo

21.1 Introduction: the role of academic spin-offs for local development in less developed areas

Geographical areas with a consistent delay in development are characterized by a set of extremely weak links among the places where technical and scientific knowledge are produced, i.e. Universities, Research Centres, innovative large companies, etc., and the places where this knowledge might be utilized, i.e. local SMEs, local Public Administration, etc. (Corti, 1997).

The traditional interventions of regional policy, which brought huge public investments towards weak areas in order to increase local production and employment, produced very poor results. The reasons for this failure lie in the logic of this kind of intervention characterized by a strong contrast with the principles of any model of endogenous development, according to which the underdevelopment of the Southern Italian Regions is mainly due to their inability to organize efficient relationships between the production sector and other key elements in the environment (Cappellin, 1996).

This aspect is strongly pointed out in literature on small innovative firms: the development of small enterprises depends on the strength of the relationships which they are able to establish with the local available resources (OECD, 1982; Oakey, 1984; Rothwell, 1984; Raffa and Zollo, 1994). In particular, Less Developed Areas (LDAs) generally show a low degree of co-operative relationships between local Universities and small firms, despite the presence of endogenous resources and the availability of consistent public funds. Also in this case, the problem seems to regard more the connecting system than the presence of local qualified actors.

The good quality of scientific and technological activities carried out by the local research institutions is a necessary but not sufficient condition to strengthen the competitive advantages of local SMEs. The development of effective networking relationships between local sources of scientific knowledge, such as universities and research centres, and SMEs is necessary. These relationships facilitate the construction of the *technological environment* (Staudenmaier, 1985) that sustains the innovative capability of small firms to renew their own skills, once the initial creative impulse of the entrepreneur/founder has been exhausted (Meyer and Roberts, 1986; Raffa and Zollo, 1994).

In LDAs the technological environment of small firms is poor, but there are actors, such as universities, research centres, large companies, which could become potential sources of technological resources. In those cases it is possible to build up an artificial technological environment through networking (Bellini *et al.*, 1996).

In this paper the authors propose to analyse the role that Academic Spin-offs (ASOs) could play in connecting local knowledge and favouring knowledge diffusion in innovative small firms. The paper is structured in the following way. In Section 21.2 we propose a networking model to connect local knowledge; in Section 21.3 some factors influencing the birth and the development of ASOs are analysed; in Section 21.4 the results obtained from the analysis of two case-studies concerning the identification and evaluation of competencies for knowledge transfer and ASOs creation are presented; Section 21.5 contains the discussion of the results obtained.

21.2 A networking model to strengthen the technological environment of SMEs

A model describing the role of networking in creating and strengthening the technological environment of SMEs is shown in Figure 1. The model, developed starting from previous research of some of the authors (Capaldo *et al.*, 1997), is based on two main assumptions:

1. In LDAs, the demand of services for innovation and of competitive technological solutions is mostly tacit. Thus, according to the model described in Figure 1, only few firms are positioned in the left cell (b.1) and are concerned with demand of competitive technological solutions.
2. The supplier of scientific knowledge can be divided into two categories:

i. *Committed to knowledge advance* (cell a.1 in Figure 1). They include mainly university departments and public research.
ii. *Interested in technological innovation* (cell a.2 in Figure 1). They usually include private research centres, consulting firms, technological services firms.

Based upon the above-mentioned assumptions, relationships between small firms and research centres need to be facilitated by a *co-ordination facility* acting as *facilitator of the development processes* of local small firms. Such a facility, as illustrated in Figure 1, should hold constant relationships both with "manufacturers" of scientific findings (suppliers of scientific knowledge committed to knowledge advances) and with advanced service enterprises, such as organization/finance consultancy companies, advanced design companies, human resources recruiting and training companies (suppliers of scientific knowledge committed interested in technological innovation).

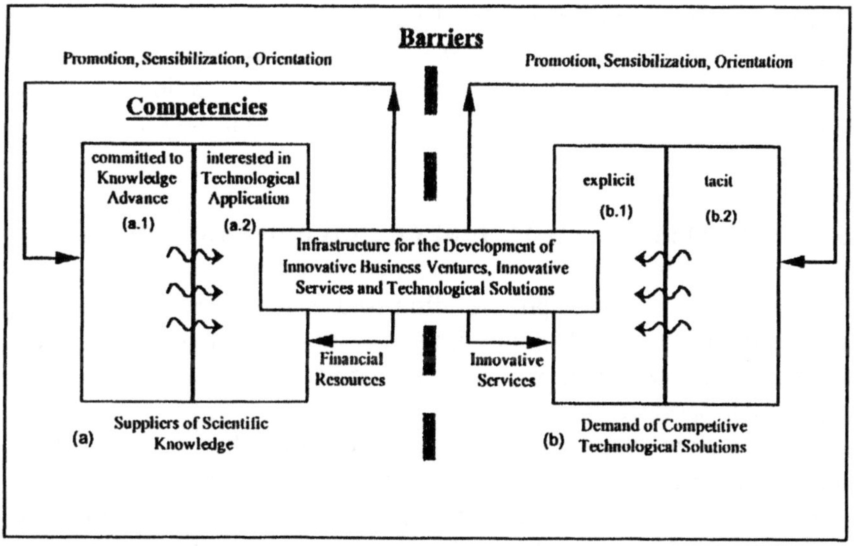

Figure 1: A model for the management of research centres - SMEs relationships

Activities performed by this co-ordination facility are clearly aimed at facilitating the intersection of demand for and supply of technological knowledge. In particular, in LDAs the co-ordination facility should focus its efforts to devise and implement actions aimed at increasing firms' awareness of the advantages deriving from innovation. Other activities should be concerned with education and training, technological consulting and support to the creation of new firms. In particular, the co-ordination facility ought also to promote the creation of high-technology firms by means of the support of university teachers or researchers interested in ASOs creation (Raffa and Zollo, 1997).

Two further elements appear in the model in Figure 1:

- *Barriers*. These elements represent the obstacles that can hinder technology transfer as they are perceived by academic researchers involved in a new start-up. The analysis of barriers to technological transfer as perceived by companies is beyond the scope of this paper.
- *Competencies*. These elements are related to universities' capabilities of ensuring effective actions for technology transfer. To this aim, and restricting our attention to the role of academic spin-offs, it is essential to analyse which competencies should possess or actually possess academic people involved in the start-up of new companies born from the results of their research activity. Competencies are both individual and organizational, that is linked to an academic structure such as a department or a faculty.

In order to investigate the role played by competencies and barriers, we present in Section 21.4 the results of the analysis of two case-studies concerning two departments of an Italian University. In the next section the theoretical background and the methodological approach followed during the field research is presented.

21.3 Knowledge transformation in the networking model

In the previous paragraph we discussed the necessity of creating an Infrastructure that should help firms to make explicit the demand of technological services, and we also highlighted the role of universities in delivering such services.

Anyway, we are aware that any organizational solution aimed at promoting innovation and change should be able to remove cultural barriers, which are very often among the most relevant obstacles to the economic exploitation of scientific results. Consequently, in this paragraph we analyse the relationship between University and Industry as a cognitive issue, following a recent tradition that focuses on the interpretation of the world of experience, that is on categories, semantic networks and heuristics that a community uses in framing events and making evaluations and decisions (Weick, 1979; Barley, 1983; Daft and Weick, 1984; Smirmich and Stubbart, 1985; Porac, Howard and Baden-Fuller, 1989).

According to Porac, Howard and Baden-Fuller (1989; p. 398),

> "the interpretative approach rests upon four longstanding assumptions. First, activities and structures of organizations are assumed to be determined in part by the micro-momentary actions of their members. Second, such actions are assumed to be based upon an information-processing sequence in which individuals attend to cues in the environment, interpret the meaning of such cues, and then externalize these interpretations via concrete activities. Third, it is assumed that the 'meaning' is problematic, and that individuals must construct actively an interpretation by linking received cues with well-learned and/or developing cognitive structures. Finally, individuals are assumed to possess a reflective capability such that they are able to verbalize at least the content of their interpretations if not the processes through which such interpretations were generated".

As a whole, those assumptions portray human activity as an ongoing cycle of subjective interpretations and objective projections of behaviours, which eventually generate interpretations shared by several people, which are consolidated in activities and social structures, which in turn reinforce dominant interpretations and behaviours. An organization is one of the most important social structures that stabilize interpretations shared by organizational members.

According to the interpretative approach, any organization operating within a stable environment selects and stores in a collective memory a set of interpretations, which actually shape the behaviour of organizational members. By applying this approach to the analysis of academic spin-offs, some relevant research questions are: what are the categories, semantic networks, and heuristics that academic staff use to define the goals of their activities? Which role is played by the economic exploitation of scientific results? What are the principal cognitive barriers to the formation of effective relationships between University and Industry?

But these are only a subset of possible questions. The interpretative approach suggests also that the reinforcement of the prevalent view of the world is continuously threatened by the turbulence of the world, which forces individuals to change their interpretations.

Consequently, the set of interpretations selected by the organization is far from a coherent system of meanings. Thus, by assuming that the University is operating in a very complex and turbulent environment, we expect to find partial and contradictory interpretations concerning the definition of research objectives, or, in the worst case, alternative viewpoints challenging the dominant vision. Other research questions become relevant: how many different systems of interpretations regarding University-Industry cooperation are present in the

University? Who are the organizational members supporting alternative viewpoints? What is the context shaping those alternative viewpoints?

The meaning system by which professors frame the business start-up is centred around three themes: What do professors think their job is? What do professors think benefits from technology transfer are? What do professors think about their involvement in business spin-offs?

By applying this approach to the analysis of ASOs, our research questions are:

According to academic people perception, what are the main *barriers* to the formation of effective relationships between University and Industry?

According to academic people's perception, what are the most relevant *competencies* which play a major role in influencing universities' capability to give life to commercial exploitation of the results of the research activity?

21.4 Barriers and competencies for knowledge transfer

21.4.1 Barriers

In their study, BenDaniel and Szafara (1998) present the result of a research aimed at identifying what aspects of the culture of technical Professors and the structure of research universities help or hinder the transfer of technology to start-up ventures. In their study they compare some departments of two of the most important American Universities: MIT and Cornell Universities. Through a survey based on an ethnographic study, they proposed the following classification for factors linked to organizational culture supporting or hindering technology transfer from Academy to new start-ups:

- *Attitudes*: an investigation into meaning system which seems to be present in the academic occupational culture;
- *Structure and Resources*: a description of the formal and informal rules of the research university and the organizational support for professorial entrepreneurship;
- *Motivation and De-motivations*: a discussion concerning the personal pulls on professors toward or away from entrepreneurship.

Their findings showed that there were many factors, both objective and subjective, that can help distinguish professors who have started business from those who have not: age, professional activities, social networks, business resources linked to location, but also perception of actual evaluation of profession activities, perception of business start-up policies, of resources allocation policies.

Based on these results, we identified the following typologies of barriers that can potentially hinder technological transfer to new start-up:

A) *Cultural barriers*: The meaning system by which professors frame the business start-up is centred around the way they conceive their job, essentially intended as the production and dissemination of knowledge. Consequently, most professors see technology transfer as a continuation of the main commitment to spread knowledge and control their discoveries. Thus they focus on some aspects of technology transfer that are not fully compatible with running a business. Usually a professor is satisfied when he/she is able to construct the best explanation of an event, while an entrepreneur is satisfied when he/she achieves market and productive success; professors care for the dissemination of their ideas, while an entrepreneur's concern is the retention of relevant knowledge; finally, the time horizon of a professor is the long period, while that of an entrepreneur is the short one.

B) *Organizational barriers* concern organizational aspects linked to the lack of structural and systematic support of activities needed for technology transfer such as patents requests, incubation of new start-ups, procedural support in business creation.

C) *Normative barriers* concern policies aimed at regulating activities of professors involved in business activities, intellectual properties, control, and conflict of interest.

D) *Financial barriers* are related to the difficulties of finding and managing funding sources, capitals.

21.4.2 Competencies

According to our approach, competency is the capability of an individual to activate his/her own organizational and environmental resources in order to achieve superior performances in a given job situation (Capaldo and Zollo, 1999). By adopting this view, competency is a multi-dimensional concept that incorporates all factors identified by BenDaniel and Szafara (1998): attitudes, resources, and motivation. Thus, according to our perspective, identifying competencies implies achieving a more synthetic and complete view of the relationships among those factors and the ways they are interpreted and put into practice by individuals.

Following a constructionist approach (Berger and Luckmann, 1966), the analysis of competencies of people working in an organization cannot leave aside the way in which they give meaning to their job and that those meanings are context dependent and constructed in the course of actions within given job situations.

For these reasons, in order to identify which competencies are needed to ensure an effective transfer of technology from universities to new start-up ventures, we analysed two concrete cases concerning two departments of an Italian University producing quite different outcomes with respect to the effectiveness of technology transfer.

By adopting the methodology followed by BenDaniel and Szafara (1998), in-depth interviews were conducted with researchers and professors in order to identify which competencies characterize successful academic entrepreneurs or, more in general, Academy's capability of supporting the commercial exploitation of the results of research. The results and the characteristics of the study are presented in the next section.

21.5 Two case-studies

For our study we selected two departments belonging to the same Italian University: Physical Sciences (dep. A) and Materials Engineering (dep. B). The criteria we adopted for the selection of the departments were the typology of research activity (mainly theoretical in A, mainly applied in B), the location (both operate in a less developed area of the country), the presence of some interesting success stories and of a stronger tradition of relationships with the industrial and productive world in dep. B. The size of the selected departments is the same. When reading results of our analysis, one should keep into mind that, in general, in Italy there is a lack of favourable context for academic start-ups, if compared with levels achieved by other countries such as the United States, the U.K. and France.

About 50 researchers and professors were interviewed in order to obtain information concerning the perception of barriers and the identification and evaluation of competencies. Documentation concerning reports of the main research projects conducted in both departments was also collected and analysed.

Interviews were organized in two rounds. In the first round, qualitative and unstructured interviews were aimed at identifying and describing a set of competencies that the interviewees considered as necessary for effective technology transfer practices. A comparison among the competencies identified in each department was effected in order to determine distinguishing factors.

Table 1: A comparison among the analysed departments

	Dep. A	Dep. B
Age	51.17 years	42.31 years
Sex	21 M / 3 F	19 M / 7 F
Qualification	15 MS Degree/ 8 Ph.D.	14 MS degree/ 11 Ph.D./ 1 MBA
Academic position	5 Researchers/ 9 Associated Prof./ 9 Tenured Prof. / 1 technician.	7 Researchers/ 4 Associated Professors/ 5 Tenured Prof./ 10 technicians.
Teaching years	15.59	13.55
Previous work experience in companies	23 No / 1 Yes	23 No / 3 Yes
Job-based collaboration with companies (ex. Consulting)	17 Never/ 6 Sometime/ 1 often	6 Never / 11 sometime / 9 often
Business and management skills	23 No / 1 Yes	25 No / 1 Yes
Patents	1 Person = 2 patents	8 Persons = 18 patents
Research Typology (0 completely theoretical, 10 completely applied)	5.21/10	7.14/10
Typology of research funds	Mainly public (80%)	50% public and 50% private
Start-up ventures	2	4

Once competencies were identified and described, in the second round of interviews, individuals were asked to assess themselves and their structure with respect to the proposed competencies. A survey concerning the evaluation of barriers was also distributed in this phase. The main differences among A and B concerning some demographic factors are shown in Table 1.

At the end of the first round interview analysis, the following competencies were identified:

- *Team-working*: the activity of technology transfer seems to require a very collaborative work environment both in order to cope with the difficulties usually arising from running start-up projects and to aggregate and integrate the right mix of necessary skills.
- *Favouring collegial support.* The collaborative environment is not only the fruit of personal dispositions, but it is also the result of the creation of a friendly environment that encourages knowledge sharing, that contributes to

create an atmosphere in which entrepreneurial success is perceived as effective to professional evaluation and career advancement, and in which people are inclined to align themselves to an overall strategy of the department or of their research group.

- *Internal networking* is referred to the capability of collaborating on common research projects with colleagues, to aggregate and involve people on common research projects, to integrate and look for synergies with peers working in different but complementary research areas.
- *External networking* is the capability of setting up and managing long term relationships with the industrial and productive context, as well as with the main public institutions together with the capability of fostering these relationships thanks to a high degree of attention to the needs of industrial partners.
- *Commitment to technology transfer* is related to the capability of undertaking concrete initiatives, making extraordinary efforts to develop a feeling that one can put his/her knowledge and ideas into practice, to the desire to see technology in use thanks to the results of research activities.
- *Research oriented to technology transfer* is concerned with the capabilities of identifying sources of information concerning market needs and technology monitoring.
- *Experience and skills* concern excellence in technical knowledge as well as project management skills and previous experience in managing innovation projects aimed at transferring technology to companies.
- *Achievement* is a transversal competence concerning personal characteristics such as propensity to continuous improvement, awareness of the impact and the importance of one's efforts and actions, strong will to complete and finish the task in the best possible way.

As can be observed, the competencies identified after the first phase of the field analysis seem to be related more to a group or to an organization than to a single individual.

In the second round of interviews people were asked to assess themselves and the structure in which they work with respect to the list of competencies. Individual responses were collected and aggregated through a multi-attribute decision-making model developed by Zollo *et al.* (1999) in order to obtain both aggregated opinions and a measure of consensus of the aggregated judgements. A comparison of the results among dept. A and dept. B is shown in Figure 2.

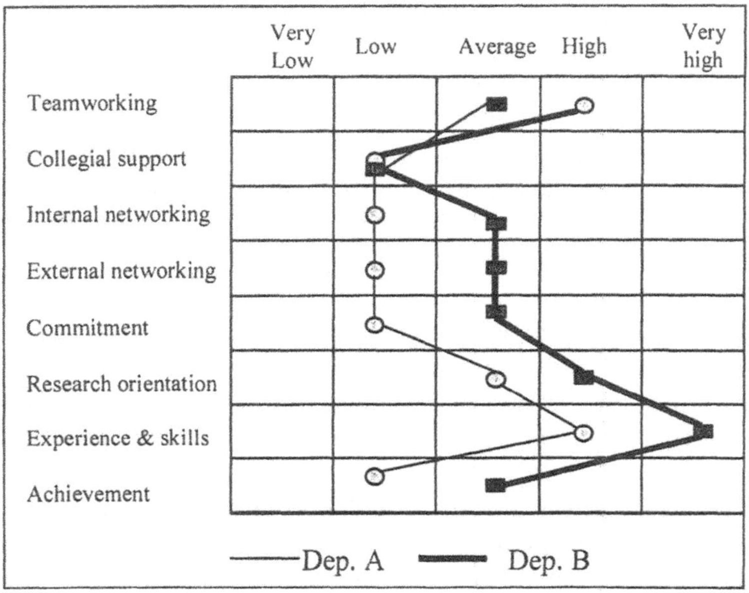

Figure 2: A comparison between competencies of the two departments

As Figure 2 shows, dep. B, the more effective in technology transfer, is also the one with a better competencies profile.

The same assessment procedure was effected with respect to barriers. The results showed that both departments perceive the barriers as quite high, especially the cultural and organizational ones, but in dep. B, the one with a higher level of competencies, barriers are perceived at a relatively lower level than A.

21.6 Results discussion and implications

The results presented in the previous section seem to suggest that, though the effectiveness of technology transfer depends on a complex mix of interrelated factors (Figure 1), the role played by some variables internal to the academic world and shaped by its cultural, organizational and human background is also crucial. Of course it is out of the scope of this study to determine the relative importance of internal and external factors. Nevertheless, by limiting our attention solely to internal factors such as competencies and the perceived barriers, and starting from the results of our field analysis and of other studies carried out in other contexts (BenDaniel and Szafara, 1998; Grimaldi, 2001), it is possible to draw some considerations.

First of all, it should be observed that among the elicited competencies there is a lack of competencies strictly related to entrepreneurship, especially in so far as concerns the propensity to risk assumption. Cultural differences between the entrepreneur and the professor seem to play a major role. The former is strongly business-committed, the latter is more oriented to scientific commitment (knowledge dissemination, intellectual curiosity).

If we look at the most successful spin-offs (Grimaldi, 2001), we find the surprising results that they are grounded on technological and scientific excellence. Indeed, we face a paradox: how is it possible to conciliate both a strong scientific and business commitment? How is it possible to prevent a business orientation from reducing the scientific commitment of research groups?

Any possible answer to this paradox excludes the trivial solution of integrating the scientific commitment of research groups with the business one. An effective solution should connect the two points of view of professors and entrepreneurs, leaving their identity unchanged.

Some suggestions about the characteristics of the organization able to connect different requirements of research and business come from the garbage can model (Cohen, March and Olsen, 1972), which tries to define the working model of incoherent organizations, the so-called "organized anarchies". Those organizations are characterized by three general properties. The first is problematic preferences. The organization operates on the basis of a variety of inconsistent and ill-defined preferences; it discovers preferences through action. The second property is unclear technology; it operates on the basis of simple trial-and error procedures, the residue of learning from the accidents of past experience, and pragmatic invention of necessity. The third property is fluid participation. Participants vary in the amount of time and effort they devote to different domains. Most of the cases of cooperative relationships between University and Industry have the properties of the organized anarchies of the garbage can model.

Those properties characterize the network organizations with weak links, where organizational units are primarily concerned with their specific objectives and do not recognize the necessity of a strong coordination with other organizational units. This is the case of professors and entrepreneurs. Usually, professors and researchers do not recognize that papers contain possible solutions to technological gaps of transformation processes, so they can sell them to entrepreneurs. On the other side, entrepreneurs do not recognize that their competitive weakness is due to technological gaps that they can fill by buying competencies and knowledge from professors.

The conclusion is that Universities that want to promote ASOs cannot operate efficiently by simply pushing professors and researchers to go to the market. It is necessary to create a new organizational structure whose mission is to run entrepreneurial projects, which develop technological solutions based on research advances in order to fill technological gaps resulting from the analysis of transformation processes and market needs of clusters of private companies and public organizations. Three requisites are important:

i. The organizational unit that is in charge of promoting entrepreneurial projects is not a University department, but a business oriented company strongly linked with both University and the economic environment;
ii. The entrepreneur running the technological project should not be a professor or a researcher, but an individual from outside the University. He/she should work in close relationship with technology originators and with venture investors;
iii. As it is necessary to identify the demand for the technological project, the project should include a market champion, which tests the intermediate and the final results of the project. Strong relationships with local industry are highly recommended.

21.7 Conclusion

According to requisites pointed out at the end of the previous paragraph and to the results obtained from the two case-studies, successful ASOs require a complex mix of competencies that it is very difficult to find in academic people. For this reason, ASOs should be considered as a business venture promoted by three critical figures: an *entrepreneur* with an industrial background, a *technological originator* with an academic background, a *market champion* involved in current business. Of course the technology originator and market champion could be both individuals and groups. Within the business venture the technological originator is required to transform scientific knowledge into technological solutions, the market champion is committed to transform his/her knowledge of the market into identification of technological gaps, while the entrepreneur should ensure the link between the technological needs and the technological solutions.

The central actor of the technology transfer is the independent entrepreneur, who is engaged in the management of the three properties of the "garbage can model" characterizing the relationships between professors and entrepreneurs. First, the entrepreneur has to face the "inconsistent and ill-defined preferences" of the technological business venture, and most of his management efforts will be devoted to discovering those preferences through actions. Second, he/she has to manage the "unclear transformation process" of scientific knowledge into operational technology by means of trial-and-error activities. Finally, the entrepreneur has to manage the "fluid participation" of the technology originator and of the market champion in the new business venture.

Franklin, Wright and Lockett (2001) define this entrepreneur as a "surrogate entrepreneur", and describe him as an individual "from outside the university assuming the role of entrepreneur with the technology originator maintaining their positioning in the university. In principle, surrogate entrepreneurs may also assume the role of, or may work in close partnership with venture investors" (p. 128).

They suggest three different options for universities, which want to run the surrogate entrepreneur program. They can confer the rights of a technology on an external, independent entrepreneur, or they can pay the surrogate entrepreneur a basic salary commensurate to what he/she earned in the industry, or they can utilize the services of a firm specialized in this type of activity. The first two options are difficult to run in LDAs, where entrepreneurial and technological culture is not widespread. Thus the only possibility is to utilize a service firm, which also operates as an incubator of entrepreneurial competencies. It seems to us that the best option is that the service firm should be partially owned by the university, in order to facilitate the exchange between University and economic environment. This service firm should operate with the mission of networking research and market through business ventures, and developing consulting services to overcome the cognitive, behavioural and institutional barriers that restrict the development of effective relationships between University and Industry.

Notes

Although the paper is the fruit of the collaboration of all the authors, in this version, Section 21.2 and 21.3 are by G. Capaldo, Section 21.4 and 21.5 are by G. Zollo, Section 21.6 is by L. Iandoli, and Section 21.7 is by M. Raffa. The remaining parts are the fruit of collaborative work.

References

Bellini E., Capaldo G., Coppola A. and Zollo G. (1996), "Problemi di acquisizione e sviluppo delle competenze delle piccole e medie imprese operanti nel Mezzogiorno", *L'Industria*, no. 4, p. 749-766.

BenDaniel D.J. and Szafara K. (1998), "What aspects of the Culture of technical Professors and the structure of Research Universities help or hinder the transfer of technology", Report of Center for entrepreneurial leadership, Kauffman Foundation, Cornell Universities.

Berger L. and Luckmann T. (1966), *The social construction of the reality*, Harmondsworth, Penguin.

Capaldo G., Corti E. and Greco O. (1997), "A Coordinated Network of Different Actors to offer Services for innovation to Develop Local SMEs inside Areas with a delay of Development", ERSA - European Science Association - 37th European Congress, Rome, Italy, August 26-29.

Capaldo G. and Zollo G. (1999), "La valutazione delle risorse umane basata sulle competenze: aspetti teorici ed evidenze empiriche", *Studi Organizzativi*, no. 3, p. 61-94.

Cappellin R. (1996), "Una politica per il Mezzogiorno orientata al mercato e la proposta di un Progetto dalla Calabria", in, Cersosimo D. (ed.), *Una Politica Economica per la Calabria,* Rubbettino Editore, Messina.

Cohen M.D., March J.G. and Olsen J.P. (1972), A Garbage Can Model of Organizational Choice, *Administrative Science Quarterly*, vol. 17, p. 1-25.

Corti E. (1997), "Il Parco Tecnologico come strumento per lo sviluppo delle economie locali", 1997 TII Annual Conference, Naples, 28-29 April.

Daft R.L. and Weick K.W. (1984), "Toward a model of organizations as interpretative systems", *Academy of Management Review,* vol. 9, p. 284-295.

Franklin S., Wright D. and Lockett A. (2001), "Academic and Surrogate Entrepreneurs in University Spin-out Companies", *Journal of Technology Transfer*, Special Issue in Organizational Issues in University-Industry Technology Transfer, p. 127-141.

Grimaldi R. (2001), University Spin-out Companies", mimeo.

Meyer M.H. and Roberts E.B., (1986), "New Product Strategy in Small Technology-Based Firms: a Pilot Study", *Management Science,* vol. 32, no. 7, p. 806-21.

Oakey, R. (1984), *High Technology Small Firms*, Frances Pinter, London.

OECD, (1982), *Innovation in Small and Medium Firms*, Paris.

Porac J.F., Howard T. and Baden-Fuller C. (1989), "Competitive Groups as Cognitive Communities: the Case of Scottish Knitwear Manufacturers", *Journal of Management Studies*, vol. 26, no. 4, p. 397-416.

Raffa M. and Zollo G. (1994), "The Role of Professionals in Small Italian Software Firms", *Journal Systems Software,* no. 26, p. 19-30.

Rothwell R. (1984), "The Role of Small Firms in the Emergence of new Technologies", *Omega*, vol. 12, no. 1, p. 19-24.

Smirmich L. and Stubbart C. (1985) "Strategic management in an enacted world", *Academy of Management Review*, vol. 10, p. 724-36.

Staudenmaier J.M. (1985), *Technology's Storytellers: Reweaving the Human Fabric*, The MIT Press, Cambridge (MA).

Stubbart C.I. (1989), "Managerial Cognition: a missing link in strategic management research", *Journal of Management Studies,* vol. 26, no. 4, p. 325-347.

Weick K.E. (1979), *The Social Psychology of Organizing*, Addison-Wesley, Reading, Mass.
Zollo G., Iandoli L. and Cannavacciuolo A. (1999), "The performance Requirements Analysis with Fuzzy Logic", *Fuzzy Economic Review*, vol. 4, no. 1, p. 35-73.

Index

1. A. Phillips, A.P. Phillips and T.R. Phillips:
   Biz Jets. Technology and Market Structure in the Corporate Jet Aircraft Industry. 1994 — ISBN 0-7923-2660-1
2. M.P. Feldman:
   The Geography of Innovation. 1994 — ISBN 0-7923-2698-9
3. C. Antonelli:
   The Economics of Localized Technological Change and Industrial Dynamics. 1995 — ISBN 0-7923-2910-4
4. G. Becher and S. Kuhlmann (eds.):
   Evaluation of Technology Policy Programmes in Germany. 1995 — ISBN 0-7923-3115-X
5. B. Carlsson (ed.): *Technological Systems and Economic Performance: The Case of Factory Automation.* 1995 — ISBN 0-7923-3512-0
6. G.E. Flueckiger: *Control, Information, and Technological Change.* 1995 — ISBN 0-7923-3667-4
7. M. Teubal, D. Foray, M. Justman and E. Zuscovitch (eds.):
   Technological Infrastructure Policy. An International Perspective. 1996 — ISBN 0-7923-3835-9
8. G. Eliasson:
   Firm Objectives, Controls and Organization. The Use of Information and the Transfer of Knowledge within the Firm. 1996 — ISBN 0-7923-3870-7
9. X. Vence-Deza and J.S. Metcalfe (eds.):
   Wealth from Diversity. Innovation, Structural Change and Finance for Regional Development in Europe. 1996 — ISBN 0-7923-4115-5
10. B. Carlsson (ed.):
    Technological Systems and Industrial Dynamics. 1997 — ISBN 0-7923-9940-4
11. N.S. Vonortas:
    Cooperation in Research and Development. 1997 — ISBN 0-7923-8042-8
12. P. Braunerhjelm and K. Ekholm (eds.):
    The Geography of Multinational Firms. 1998 — ISBN 0-7923-8133-5
13. A. Varga:
    University Research and Regional Innovation: A Spatial Econometric Analysis of Academic Technology Transfers. 1998 — ISBN 0-7923-8248-X
14. J. de la Mothe and G. Paquet (eds.):
    Local and Regional Systems of Innovation, 1998 — ISBN 0-7923-8287-0
15. D. Gerbarg (ed.):
    The Economics, Technology and Content of Digital T V, 1999 — ISBN 0-7923-8325-7
16. C. Edquist, L. Hommen and L. Tsipouri
    Public Technology Procurement and Innovation, 1999 — ISBN 0-7923-8685-X
17. J. de la Mothe and G. Paquet (eds.):
    Information, Innovation and Impacts, 1999 — ISBN 0-7923-8692-2

18. J. S. Metcalfe and I. Miles (eds.):
Innovation Systems in the Service Economy: Measurement and Case Study Analysis, 2000 ISBN 0-7923-7730-3
19. R. Svensson:
Success Strategies and Knowledge Transfer in Cross-Border Consulting Operations, 2000 ISBN 0-7923-7776-1
20. P. Braunerhjelm:
Knowledge Capital and the "New Economy": Firm Size, Performance and Network Production, 2000 ISBN 0-7923-7801-6
21. J. de la Mothe and J. Niosi (eds.):
The Economic and Social Dynamics of Biotechnology, 2000 ISBN 0-7923-7922-5
22. B. Guilhon, (ed.):
Technology and Markets for Knowledge: Knowledge Creation, Diffusion and Exchange within a Growing Economy, 2000 ISBN 0-7923-7202-6
23. M. Feldman and A. Link (eds.):
Innovation Policy in the Knowledge-Based Economy, 2001 ISBN 0-7923-7296-4
24. J. de la Mothe and D. Foray (eds.):
Knowledge Management in the Innovation Process ISBN 0-7923-7464-9
2001
25. M. Feldman and N. Massard (eds.):
Institutions and Systems in the Geography of Innovation, ISBN: 0-7923-7614-5
2002
26. B. Carlsson (ed.):
Tehnological Systems in the Bio Industries: An International Study, 2002 ISBN: 0-7923-7633-1
27. D. Audretsch, R. Thurik, I. Verheul, and S. Wennekers (eds.): *Entrepreneurship: Determinants and Policy in a European-US Comparison*, 2002 ISBN: 0-7923-7685-4
28. J. de la Mothe and A. Link (eds.):
Networks, Alliances and Partnerships in the Innovation Process ISBN: 1-4020-7172-8
29. F. Belussi, G. Gottardi, and E. Rullani (eds.):
The Technological Evolution of Industrial Districts ISBN: 1-4020-7555-3

KLUWER ACADEMIC PUBLISHERS – BOSTON / DORDRECHT / LONDON

Batch number: 10755248

Printed by Printforce, the Netherlands